# Quantitative Genetics

Shizhong Xu

# Quantitative Genetics

 Springer

Shizhong Xu
Department of Botany and Plant Sciences
University of California, Riverside
Riverside, CA, USA

ISBN 978-3-030-83942-0    ISBN 978-3-030-83940-6  (eBook)
https://doi.org/10.1007/978-3-030-83940-6

© Springer Nature Switzerland AG 2022
This work is subject to copyright. All rights are reserved by the Publisher, whether the whole or part of the material is concerned, specifically the rights of translation, reprinting, reuse of illustrations, recitation, broadcasting, reproduction on microfilms or in any other physical way, and transmission or information storage and retrieval, electronic adaptation, computer software, or by similar or dissimilar methodology now known or hereafter developed.
The use of general descriptive names, registered names, trademarks, service marks, etc. in this publication does not imply, even in the absence of a specific statement, that such names are exempt from the relevant protective laws and regulations and therefore free for general use.
The publisher, the authors and the editors are safe to assume that the advice and information in this book are believed to be true and accurate at the date of publication. Neither the publisher nor the authors or the editors give a warranty, expressed or implied, with respect to the material contained herein or for any errors or omissions that may have been made. The publisher remains neutral with regard to jurisdictional claims in published maps and institutional affiliations.

This Springer imprint is published by the registered company Springer Nature Switzerland AG
The registered company address is: Gewerbestrasse 11, 6330 Cham, Switzerland

# Preface

Polygenic traits are those controlled by multiple loci. Most economically important traits in plants and animals are polygenic in nature. Many disease traits are also polygenic although some disease phenotypes may appear to be categorical. Quantitative genetics is an area of genetics that studies the genetic basis of polygenic traits. Because the variation of a polygenic trait, by definition, is collectively controlled by multiple loci, tools commonly used in classical Mendelian genetics are not sufficient for analysis of polygenic traits. Advanced statistical tools play an important role in quantitative genetics. Although the term "expectation" in statistics is easy to understand, terminology such as "variance" and "covariance" are the foci of quantitative genetics. Therefore, the properties of variance and covariance are fundamentally important in understanding all equations presented in the textbook. A whole chapter (Chap. 4) is dedicated to statistical review, especially variance (covariance) and its properties.

I have been teaching Quantitative Genetics (BPSC 148) as an undergraduate course at the University of California, Riverside, since 1995. This textbook is a collection of lecture notes for this course. Material and methods are mainly extracted from Falconer and Mackay (1996) and Lynch and Walsh (1998). Some topics are adopted from the 1987 handouts of Quantitative Genetics (AGR/ANSC 611) by Professor Wyman E. Nyquist at Purdue University. The Quantitative Genetics course at UCR is offered for one quarter (10 weeks) annually. The Lynch and Walsh (1998) book is too much as a textbook for this one-quarter course. The Falconer and Mackay (1996) book is about the right size but derivations of important equations are often not provided. I selectively adopted topics from both books and extended these topics in detail by providing derivations for some fundamentally important equations. More importantly, the book provides sample data and SAS code for analyses of these data. The statistical analysis system (SAS) is a suite of analytics software. It is the largest private software company in the world with the best technical support system. The SAS products are more reliable than products from all other software companies. This is why I chose SAS as the analytic tool for data analysis.

Unlike the introductions of other quantitative genetics textbooks, the introduction (Chap. 1) of this book presents two examples about breeding values and their applications to plant and animal breeding. In addition, the chapter describes the relationship between quantitative genetics and statistics. Major statistical methods that are commonly used in quantitative genetics are mentioned briefly in the introduction chapter. Chapters 2–4 review Mendelian genetics, population genetics, and statistics. The actual quantitative genetics starts at Chap. 5 by defining various kinds of genetic effects followed by definitions of genetic variance and its variance components in Chap. 6. Genotype by environment (G×E) interaction is discussed in Chap. 7 where environmental variation is classified into systematic error and random error. Much of the content of G×E interaction is adopted from Professor Nyquist's 1987 lecture notes for Quantitative Genetics. Chapter 8 introduces the concept of major genes and describes methods of major gene detection and segregation analysis, a topic not often seen in quantitative genetics textbooks. Chaps. 9–11 cover theory and methods for estimation of heritability, primarily adopted from Falconer and Mackay (1996). Methods of kinship matrix calculation are added to Chap. 11 where the Inbreed Procedure in SAS is used to calculate kinship matrices for arbitrarily complicated pedigrees. Linear mixed models (LMM) are introduced in Chap. 12 for estimation of variance components from pedigree data. PROC MIXED in SAS is used to implement the mixed model analysis. Multiple traits and genetic correlation between multiple traits are discussed in Chap. 13. Theory and methods of artificial selection are covered by Chaps. 14–16, followed by methods of multiple trait selection by Chap. 17. The last three chapters cover quantitative trait locus (QTL) mapping, genome-wide association studies (GWAS), and genomic selection (GS). The three topics belong to molecular quantitative genetics and molecular breeding and are often ignored in classical quantitative genetics.

This book can be used as a textbook for quantitative genetics courses for both undergraduate and graduate teaching. For undergraduate teaching, instructors can skip the derivation steps and directly present the final forms of equations. Some contents can also be ignored for undergraduate teaching, e.g., calculation of the standard error of an estimated heritability from family data. For graduate teaching, derivations are important part of the teaching and can improve the depth of understanding of the course material, and thus should not be skipped. The book can also be used as a reference book for experimental quantitative geneticists, plant and animal breeders, theoretical quantitative geneticists, population geneticists, and even biologists who are interested in quantitative methodology and statisticians who are looking for examples of statistical methods that are often applied to biology. The book may be particularly useful as a reference book for junior faculty as instructors of quantitative genetics, even if it is not used as the textbook for students.

I sincerely thank my graduate advisor for my MS degree, Professor Zhilian Sheng, who introduced me to this wonderful area of genetics (quantitative genetics). I learned his pooled analysis of sib correlation from multiple farms for estimating heritability and genetic correlation when I was a graduate student at Northeast Agricultural University under his supervision. Professor Sheng called his method "the within-unit sib analysis." After I started my PhD program in the USA, I learned the method of restricted maximum likelihood method (REML) for estimation of variance components and immediately realized that the pooled analysis of sib correlation is identical to the REML method. The REML method was formally published in 1971 by Patterson and Thompson while Professor Sheng already used the idea of REML for estimating genetic parameters in the mid-1960s. Professor Sheng formally published his "within-unit method for sib correlation analysis" in 1980 when REML was not well known to the majority of statisticians. My whole quantitative genetics career benefited from the 3-year training under Professor Sheng's supervision. I continued my training in quantitative genetics under the supervision of Professor William M. Muir at Purdue University. Not only did Professor Muir teach me all the analytical skills of quantitative genetics, but he also taught me the SAS programming, which allowed me to successfully compete for a student statistical consulting position at Purdue to support my family during the difficult years as a PhD student. I am extremely grateful to Professor Muir for his continued supervision during my whole quantitative genetics career. The third person who was important to my quantitative genetics research and teaching career was Professor Wyman Nyquist at Purdue University. He taught me Quantitative Genetics (AGR/ANSC 611) and Statistical Genetics (AGR/ANSC 615). He was the greatest educator I have ever met and his influence on my academic career was lifelong. My written and teaching styles are primarily inherited from his. Without the above three influential mentors, I might have taken a different career path other than quantitative genetics. Without a career in quantitative genetics, I would not have the opportunity to teach quantitative genetics and to write this textbook.

Two of my PhD dissertation committee members also deserve special appreciation: Professor Thomas Kuczek in Statistics and Professor Truman Martin in Animal Sciences at Purdue University. When I learned the Bulmer effect (selection on the reduction of genetic variance), I was trying to extend the method to see how selection affect the genetic covariance between two traits that are related to the target trait of selection. I consulted with Professor Thomas Kuczek, who had an office next to Bill Muir's laboratory in Lilly Hall. Although he did not tell me the answer immediately, he pointed me to a direction where we can express both traits as linear functions of the target trait of selection. Following his direction, I successfully derived the formula the next day. I could not wait to tell Professor Thomas Kuczek about the answer and how grateful I was to him for the valuable suggestion. A few days later, I had a chance to meet with Professor Truman Martin. I told him about my derivation on the change of the genetic covariance between two traits related to the target trait. He did not say a word but opened a draw of his file cabinets, pulled a paper published by Alan Robertson in 1977 and told me that the formula was already presented a decade ago. I was sad in the beginning that I was not the first person to derive the formula, but so amazed by Professor Truman Martin's quick response and clear memory about the formula. My lessons learned from this event are (1) always searching literature thoroughly for a project before trying to investigate the project by yourself and (2) consulting with your advisors, colleagues, classmates, even roommates for a project before actually starting the project.

All UCR students taking this course (BPSC 148) for the last 25 years or so have contributed to the book in one way or another. Errors in the lecture notes were periodically identified by students and then corrected during the quarters when the course was offered. Special appreciation goes to all my former students who contributed to my research in quantitative genetics; some of their work was eventually integrated into the textbook. I particularly thank Meiyue Wang, a recently graduated student from my lab, for her help during the book writing process by drawing all the figures, communicating with the Springer Publisher for copyrights issues and making sure the book production process was smooth. Appreciation also goes to two more graduate students in the UCR Quantitative Genetics Program, Fangjie Xie from Xu's lab and Han Qu from Jia's lab, who offered help in the book preparation process. Dr. Xiangyu Wang, a visiting scientist from The People's Republic of China, audited the class (BPSC 148) in the 2020 Winter quarter and identified a few errors in the textbook. The errors have been fixed and she deserves special appreciation.

Before the manuscript was submitted to the Springer Publisher, I distributed each chapter to one current or former member of the big Xu & Jia laboratory for proof reading. They all put significant effort to correct any errors occurring in an earlier version of the book manuscript. These contributors include John Chater, Ruidong Li, Meiyue Wang, Shibo Wang, Qiong Jia, Tiantian Zhu, Fuping Zhao, Xuesong Wang, Xiangyu Wang, Julong Wei, Han Qu, Chen Lin, Saka Sigdel, Arthur Jia, Ryan Traband, Yanru Cui, Lei Yu, Huijiang Gao, Fangjie Xie, and Yang Xu. They all deserve special appreciation.

Finally, I thank my wife (Yuhuan Wang), daughters (Mumu Xu and Nicole Xu), and grandson (Max Borson) for their support and patience. My wife originally thought that I could make a lot extra money by writing a book and only realized that the return was not worth for the significant effort I put forth.

Riverside, CA, USA                                                                          Shizhong Xu

# Contents

**1 Introduction to Quantitative Genetics** ........................................... 1
   1.1    Breeding Value and Its Application ................................. 1
         1.1.1    Genetic Gain via Artificial Selection (Example 1) .......... 1
         1.1.2    Predicting Human Height (Example 2) ..................... 1
         1.1.3    Breeding Value .......................................... 3
   1.2    Complicated Behavior Traits ...................................... 4
   1.3    Qualitative and Quantitative Traits ................................ 5
   1.4    The Relationship Between Statistics and Genetics ................. 6
   1.5    Fundamental Statistical Methods in Quantitative Genetics ........ 8
   1.6    Statistical Software Packages ..................................... 9
   1.7    Classical Quantitative Genetics and Modern Quantitative Genetics ...... 10
   References ............................................................. 11

**2 Review of Mendelian Genetics** ...................................... 13
   2.1    Mendel's Experiments ........................................... 13
   2.2    Mendel's Laws of Inheritance .................................... 14
   2.3    Vocabulary ..................................................... 19
   2.4    Departure from Mendelian Ratio ................................. 19
   2.5    Two Loci Without Environmental Effects ........................ 20
   2.6    Multiple Loci with or Without Environmental Effects ............. 21
   2.7    Environmental Errors Can Lead to a Normal Distribution ........ 22
   References ............................................................. 24

**3 Basic Concept of Population Genetics** ............................... 25
   3.1    Gene Frequencies and Genotype Frequencies .................... 25
   3.2    Hardy-Weinberg Equilibrium .................................... 25
         3.2.1    Proof of the H-W Law .................................... 26
         3.2.2    Applications of the H-W Law .............................. 27
         3.2.3    Test of Hardy-Weinberg Equilibrium ...................... 28
   3.3    Genetic Drift ................................................... 30
   3.4    Wright's $F$ Statistics ........................................... 33
   3.5    Estimation of $F$ Statistics ...................................... 34
   References ............................................................. 37

**4 Review of Elementary Statistics** .................................... 39
   4.1    Expectation ..................................................... 39
         4.1.1    Definition ............................................... 39
         4.1.2    Properties of Expectation ................................. 39
         4.1.3    Estimating the Mean ..................................... 40
   4.2    Variance ....................................................... 40
         4.2.1    Definition ............................................... 40
         4.2.2    Properties of Variance .................................... 41

|       | 4.2.3  | Normal Distribution | 41 |
|       | 4.2.4  | Estimating Variance from a Sample | 41 |
|       | 4.2.5  | An Application of the Variance Property | 42 |
| 4.3   | Covariance | | 43 |
|       | 4.3.1  | Definition | 43 |
|       | 4.3.2  | Properties of Covariance | 43 |
|       | 4.3.3  | Estimating Covariance from a Sample | 44 |
|       | 4.3.4  | Conditional Expectation and Conditional Variance | 45 |
| 4.4   | Sample Estimates of Variance and Covariance | | 47 |
| 4.5   | Linear Model | | 48 |
|       | 4.5.1  | Regression | 48 |
|       | 4.5.2  | Correlation | 48 |
|       | 4.5.3  | Estimation of Regression and Correlation Coefficients | 49 |
| 4.6   | Matrix Algebra | | 51 |
|       | 4.6.1  | Definitions | 51 |
|       | 4.6.2  | Matrix Addition and Subtraction | 52 |
|       | 4.6.3  | Matrix Multiplication | 52 |
|       | 4.6.4  | Matrix Transpose | 55 |
|       | 4.6.5  | Matrix Inverse | 55 |
|       | 4.6.6  | Generalized Inverse | 56 |
|       | 4.6.7  | Determinant of a Matrix | 57 |
|       | 4.6.8  | Trace of a Matrix | 57 |
|       | 4.6.9  | Orthogonal Matrices | 57 |
|       | 4.6.10 | Eigenvalues and Eigenvectors | 58 |
| 4.7   | Linear Combination, Quadratic Form, and Covariance Matrix | | 59 |

**5    Genetic Effects of Quantitative Traits** ............................ **63**
| 5.1   | Phenotype, Genotype, and Environmental Error | 63 |
| 5.2   | The First Genetic Model | 64 |
| 5.3   | Population Mean | 66 |
| 5.4   | Average Effect of Gene or Average Effect of Allele | 67 |
| 5.5   | Average Effect of Gene Substitution | 67 |
| 5.6   | Alternative Definition of Average Effect of Gene Substitution | 67 |
| 5.7   | Breeding Value | 68 |
| 5.8   | Dominance Deviation | 68 |
| 5.9   | Epistatic Effects Involving Two Loci | 69 |
| 5.10  | An Example of Epistatic Effects | 72 |
| 5.11  | Population Mean of Multiple Loci | 75 |
|       | References | 76 |

**6    Genetic Variances of Quantitative Traits** ......................... **77**
| 6.1   | Total Genetic Variance | 77 |
| 6.2   | Additive and Dominance Variances | 78 |
| 6.3   | Epistatic Variances (General Definition) | 79 |
| 6.4   | Epistatic Variance Between Two Loci | 79 |
| 6.5   | Average Effect of Gene Substitution and Regression Coefficient | 81 |
|       | References | 83 |

**7    Environmental Effects and Environmental Errors** ................ **85**
| 7.1   | Environmental Effects | 85 |
| 7.2   | Environmental Errors | 85 |
| 7.3   | Repeatability | 86 |
|       | 7.3.1  | Estimation of Repeatability | 86 |
|       | 7.3.2  | Proof of the Intra-Class Correlation Coefficient | 89 |

7.3.3 Estimation of Repeatability with Variable Numbers
of Repeats.................................... 90
7.3.4 An Example for Estimating Repeatability................. 91
7.3.5 Application of Repeatability........................ 94
7.4 Genotype by Environment (G × E) Interaction................... 95
7.4.1 Definition of G × E Interaction..................... 95
7.4.2 Theoretical Evaluation of G × E Interaction............. 97
7.4.3 Significance Test of G × E Interaction................. 98
7.4.4 Partitioning of Phenotypic Variance................... 100
7.4.5 Tukey's One Degree of Freedom G × E Interaction Test..... 102
References........................................... 106

8 Major Gene Detection and Segregation Analysis................ 107
8.1 Two Sample t-Test or F-Test.......................... 107
8.2 F-Test for Multiple Samples (ANOVA)..................... 110
8.3 Regression Analysis................................ 113
8.3.1 Two Genotypes............................... 113
8.3.2 Three Genotypes.............................. 116
8.4 Major Gene Detection Involving Epistatic Effects............... 118
8.4.1 Test of Epistasis.............................. 118
8.4.2 Epistatic Variance Components and Significance Test for
each Type of Effects............................ 119
8.5 Segregation Analysis............................... 122
8.5.1 Qualitative Traits............................. 122
8.5.2 Quantitative Traits............................ 128
References........................................... 133

9 Resemblance between Relatives.......................... 135
9.1 Genetic Covariance Between Offspring and One Parent............. 136
9.1.1 Short Derivation.............................. 136
9.1.2 Long Derivation.............................. 137
9.2 Genetic Covariance between Offspring and Mid-Parent............. 139
9.3 Genetic Covariance between Half-Sibs..................... 139
9.4 Genetic Covariance between Full-Sibs..................... 140
9.4.1 Short Way of Derivation......................... 140
9.4.2 Long Way of Derivation......................... 141
9.5 Genetic Covariance between Monozygotic Twins (Identical Twins)..... 142
9.6 Summary....................................... 142
9.7 Environmental Covariance............................ 143
9.8 Phenotypic Resemblance............................. 143
9.9 Derivation of within Family Segregation Variance............... 143
References........................................... 146

10 Estimation of Heritability............................. 147
10.1 $F_2$ Derived from a Cross of Two Inbred Parents............... 148
10.2 Multiple Inbred Lines or Multiple Hybrids................... 148
10.2.1 With Replications............................. 149
10.2.2 Without Replications........................... 150
10.3 Parent-Offspring Regression........................... 152
10.3.1 Single Parent Vs. Single Offspring................... 152
10.3.2 Middle Parent Vs. Single Offspring................... 153
10.3.3 Single Parent Vs. Mean Offspring................... 154

          10.3.4    Middle Parent Vs. Mean Offspring . . . . . . . . . . . . . . . . . . . . . 155

          10.3.5    Estimate Heritability Using Parent-Offspring Correlation . . . . . . 155

    10.4    Sib Analysis . . . . . . . . . . . . . . . . . . . . . . . . . . . . . . . . . . . . . . . . . 158

          10.4.1    Full-Sib Analysis . . . . . . . . . . . . . . . . . . . . . . . . . . . . . . . . . 158

          10.4.2    Half-Sib Analysis . . . . . . . . . . . . . . . . . . . . . . . . . . . . . . . . . 160

          10.4.3    Nested or Hierarchical Mating Design . . . . . . . . . . . . . . . . . . . 161

    10.5    Standard Error of an Estimated Heritability . . . . . . . . . . . . . . . . . . . 164

          10.5.1    Regression Method (Parent Vs. Progeny Regression) . . . . . . . . 164

          10.5.2    Analysis of Variances (Sib Analysis) . . . . . . . . . . . . . . . . . . . 166

    10.6    Examples . . . . . . . . . . . . . . . . . . . . . . . . . . . . . . . . . . . . . . . . . . . . 167

          10.6.1    Regression Analysis . . . . . . . . . . . . . . . . . . . . . . . . . . . . . . . . 167

          10.6.2    Analysis of Variances . . . . . . . . . . . . . . . . . . . . . . . . . . . . . . 168

          10.6.3    A Nested Mating Design . . . . . . . . . . . . . . . . . . . . . . . . . . . . 171

    References . . . . . . . . . . . . . . . . . . . . . . . . . . . . . . . . . . . . . . . . . . . . . . . . 175

**11 Identity-by-Descent and Coancestry Coefficient** . . . . . . . . . . . . . . . . . . . . 177

    11.1    Allelic Variance . . . . . . . . . . . . . . . . . . . . . . . . . . . . . . . . . . . . . . . 177

    11.2    Genetic Covariance Between Relatives . . . . . . . . . . . . . . . . . . . . . . 178

    11.3    Genetic Covariance of an Individual with Itself . . . . . . . . . . . . . . . . 178

    11.4    Terminology . . . . . . . . . . . . . . . . . . . . . . . . . . . . . . . . . . . . . . . . . 179

    11.5    Computing Coancestry Coefficients . . . . . . . . . . . . . . . . . . . . . . . . . 181

          11.5.1    Path Analysis . . . . . . . . . . . . . . . . . . . . . . . . . . . . . . . . . . . . 182

          11.5.2    Tabular Method . . . . . . . . . . . . . . . . . . . . . . . . . . . . . . . . . . . 185

    11.6    R Package to Calculate a Coancestry Matrix . . . . . . . . . . . . . . . . . . 189

          11.6.1    Path Analysis . . . . . . . . . . . . . . . . . . . . . . . . . . . . . . . . . . . . 189

          11.6.2    Tabular Method . . . . . . . . . . . . . . . . . . . . . . . . . . . . . . . . . . . 190

    11.7    SAS Program for Calculating a Coancestry Matrix . . . . . . . . . . . . . . 191

    References . . . . . . . . . . . . . . . . . . . . . . . . . . . . . . . . . . . . . . . . . . . . . . . . 193

**12 Mixed Model Analysis of Genetic Variances** . . . . . . . . . . . . . . . . . . . . . . . 195

    12.1    Mixed Model . . . . . . . . . . . . . . . . . . . . . . . . . . . . . . . . . . . . . . . . . 195

    12.2    Maximum Likelihood (ML) Estimation of Parameters . . . . . . . . . . . . 198

    12.3    Restricted Maximum Likelihood (REML) Estimation of Parameters . . . . . 200

    12.4    Likelihood Ratio Test . . . . . . . . . . . . . . . . . . . . . . . . . . . . . . . . . . . 201

    12.5    Examples . . . . . . . . . . . . . . . . . . . . . . . . . . . . . . . . . . . . . . . . . . . . 202

          12.5.1    Example 1 . . . . . . . . . . . . . . . . . . . . . . . . . . . . . . . . . . . . . . . 202

          12.5.2    Example 2 . . . . . . . . . . . . . . . . . . . . . . . . . . . . . . . . . . . . . . . 206

    12.6    Monte Carlo Simulation . . . . . . . . . . . . . . . . . . . . . . . . . . . . . . . . . 207

    References . . . . . . . . . . . . . . . . . . . . . . . . . . . . . . . . . . . . . . . . . . . . . . . . 213

**13 Multiple Traits and Genetic Correlation** . . . . . . . . . . . . . . . . . . . . . . . . . . . 215

    13.1    Definition of Genetic Correlation . . . . . . . . . . . . . . . . . . . . . . . . . . . 215

    13.2    Causes of Genetic Correlation . . . . . . . . . . . . . . . . . . . . . . . . . . . . . 216

    13.3    Cross Covariance between Relatives . . . . . . . . . . . . . . . . . . . . . . . . . 217

    13.4    Estimation of Genetic Correlation . . . . . . . . . . . . . . . . . . . . . . . . . . 217

          13.4.1    Estimate Genetic Correlation Using Parent-Offspring

                     Correlation (Path Analysis) . . . . . . . . . . . . . . . . . . . . . . . . . . 217

          13.4.2    Estimating Genetic Correlation from Sib Data . . . . . . . . . . . . . 221

          13.4.3    Estimating Genetic Correlation Using a Nested Mating

                     Design . . . . . . . . . . . . . . . . . . . . . . . . . . . . . . . . . . . . . . . . . . 227

    References . . . . . . . . . . . . . . . . . . . . . . . . . . . . . . . . . . . . . . . . . . . . . . . . 232

**14 Concept and Theory of Selection** . . . . . . . . . . . . . . . . . . . . . . . . . . . . . . . . . 233

    14.1    Evolutionary Forces . . . . . . . . . . . . . . . . . . . . . . . . . . . . . . . . . . . . 233

    14.2    Change in Gene Frequency and Genotype Frequencies . . . . . . . . . . . . 233

14.3   Artificial Selection................................................ 234
      14.3.1   Directional Selection.................................. 234
      14.3.2   Stabilizing Selection.................................. 240
      14.3.3   Disruptive Selection.................................. 242
14.4   Natural Selection.................................................. 243
      14.4.1   Directional Selection.................................. 243
      14.4.2   Stabilizing Selection.................................. 244
      14.4.3   Disruptive Selection.................................. 245
14.5   Change of Genetic Variance After Selection..................... 245
      14.5.1   Single Trait.......................................... 245
      14.5.2   Multiple Traits....................................... 246
      14.5.3   The $\kappa$ Values in Other Types of Selection.............. 248
      14.5.4   Derivation of Change in Covariance.................... 250
References............................................................ 253

**15   Methods of Artificial Selection**..................................... 255
15.1   Objective of Selection............................................ 255
15.2   Criteria of Selection.............................................. 255
15.3   Methods of Selection............................................. 255
      15.3.1   Individual Selection................................... 255
      15.3.2   Family Selection...................................... 256
      15.3.3   Sib Selection......................................... 256
      15.3.4   Progeny Testing...................................... 256
      15.3.5   Within-Family Selection.............................. 257
      15.3.6   Pedigree Selection................................... 257
      15.3.7   Combined Selection.................................. 257
15.4   Evaluation of a Selection Method................................ 257
15.5   Examples........................................................ 258
      15.5.1   Family Selection...................................... 258
      15.5.2   Within-Family Selection.............................. 260
      15.5.3   Sib Selection......................................... 261
      15.5.4   Pedigree Selection................................... 262
      15.5.5   Combined Selection.................................. 263
References............................................................ 264

**16   Selection Index and the Best Linear Unbiased Prediction**......... 265
16.1   Selection Index.................................................. 265
      16.1.1   Derivation of the Index Weights...................... 265
      16.1.2   Evaluation of Index Selection........................ 267
      16.1.3   Comparison of Index Selection with a Simple Combined
                Selection........................................... 267
      16.1.4   Index Selection Combining Candidate Phenotype
                with the Family Mean................................ 269
16.2   Best Linear Unbiased Prediction (BLUP)......................... 271
      16.2.1   Relationship Between Selection Index and BLUP........... 271
      16.2.2   Theory of BLUP..................................... 272
16.3   Examples and SAS Programs.................................... 275
      16.3.1   Example 1........................................... 275
      16.3.2   Example 2........................................... 279
References............................................................ 282

**17   Methods of Multiple Trait Selection**................................ 283
17.1   Common Methods of Multiple Trait Selection..................... 283
      17.1.1   Tandem Selection.................................... 283
      17.1.2   Independent Culling Level Selection.................... 283

17.1.3 Index Selection . . . . . . . . . . . . . . . . . . . . . . . . . . . . . . . 283
17.1.4 Multistage Index Selection . . . . . . . . . . . . . . . . . . . . . . . 284
17.1.5 Other Types of Selection . . . . . . . . . . . . . . . . . . . . . . . . 284
17.2 Index Selection . . . . . . . . . . . . . . . . . . . . . . . . . . . . . . . . . . . . . 285
17.2.1 Variance and Covariance Matrices . . . . . . . . . . . . . . . . . . 285
17.2.2 Selection Index (Smith-Hazel Index) . . . . . . . . . . . . . . . . 287
17.2.3 Response to Index Selection . . . . . . . . . . . . . . . . . . . . . . 288
17.2.4 Derivation of the Smith-Hazel Selection Index . . . . . . . . . 289
17.2.5 An Example of Index Selection . . . . . . . . . . . . . . . . . . . . 291
17.3 Restricted Selection Index . . . . . . . . . . . . . . . . . . . . . . . . . . . . . 292
17.4 Desired Gain Selection Index . . . . . . . . . . . . . . . . . . . . . . . . . . 295
17.5 Multistage Index Selection . . . . . . . . . . . . . . . . . . . . . . . . . . . . . 297
17.5.1 Concept of Multistage Index Selection . . . . . . . . . . . . . . . 297
17.5.2 Cunningham's Weights of Multistage Selection Indices . . . . . 298
17.5.3 Xu-Muir's Weights of Multistage Selection Indices . . . . . . . . 299
17.5.4 An Example for Multistage Index Selection . . . . . . . . . . . . 300
17.6 Selection Gradients . . . . . . . . . . . . . . . . . . . . . . . . . . . . . . . . . . 305
References . . . . . . . . . . . . . . . . . . . . . . . . . . . . . . . . . . . . . . . . . . . . . 305

18 Mapping Quantitative Trait Loci . . . . . . . . . . . . . . . . . . . . . . . . . . 307
18.1 Linkage Disequilibrium . . . . . . . . . . . . . . . . . . . . . . . . . . . . . . . . 307
18.2 Interval Mapping . . . . . . . . . . . . . . . . . . . . . . . . . . . . . . . . . . . . 310
18.2.1 Least Squares Method (the Haley-Knott Method) . . . . . . . . 313
18.2.2 Iteratively Reweighted Least Squares (IRWLS) . . . . . . . . . 315
18.2.3 Maximum Likelihood Method . . . . . . . . . . . . . . . . . . . . . 317
18.3 Composite Interval Mapping . . . . . . . . . . . . . . . . . . . . . . . . . . . . 319
18.4 Control of Polygenic Background . . . . . . . . . . . . . . . . . . . . . . . . . 321
18.5 Ridge Regression . . . . . . . . . . . . . . . . . . . . . . . . . . . . . . . . . . . . 323
18.6 An Example (a Mouse Data) . . . . . . . . . . . . . . . . . . . . . . . . . . . . 325
18.6.1 Technical Detail . . . . . . . . . . . . . . . . . . . . . . . . . . . . . . 325
18.6.2 Results of Different Methods . . . . . . . . . . . . . . . . . . . . . . 327
18.6.3 Remarks on the Interval Mapping Procedures . . . . . . . . . . . 330
18.7 Bonferroni Correction of Threshold for Multiple Tests . . . . . . . . . 332
18.8 Permutation Test . . . . . . . . . . . . . . . . . . . . . . . . . . . . . . . . . . . . . 333
18.9 Quantification of QTL Size . . . . . . . . . . . . . . . . . . . . . . . . . . . . . 334
18.10 The Beavis Effect . . . . . . . . . . . . . . . . . . . . . . . . . . . . . . . . . . . . 339
References . . . . . . . . . . . . . . . . . . . . . . . . . . . . . . . . . . . . . . . . . . . . . 344

19 Genome-Wide Association Studies . . . . . . . . . . . . . . . . . . . . . . . . . 347
19.1 Introduction . . . . . . . . . . . . . . . . . . . . . . . . . . . . . . . . . . . . . . . . 347
19.2 Simple Regression Analysis . . . . . . . . . . . . . . . . . . . . . . . . . . . . . 347
19.3 Mixed Model Methodology Incorporating Pedigree Information . . . . 349
19.4 Mixed Linear Model (MLM) Using Marker-Inferred Kinship . . . . . . 349
19.5 Efficient Mixed Model Association (EMMA) . . . . . . . . . . . . . . . . . 350
19.6 Decontaminated Efficient Mixed Model Association (DEMMA) . . . . 353
19.7 Manhattan Plot and Q-Q Plot . . . . . . . . . . . . . . . . . . . . . . . . . . . 355
19.8 Population Structure . . . . . . . . . . . . . . . . . . . . . . . . . . . . . . . . . . . 358
19.9 Genome-Wide Association Study in Rice—A Case Study . . . . . . . . . 360
19.10 Efficient Mixed Model Association Studies Expedited (EMMAX) . . . . 363
References . . . . . . . . . . . . . . . . . . . . . . . . . . . . . . . . . . . . . . . . . . . . . 365

20 Genomic Selection . . . . . . . . . . . . . . . . . . . . . . . . . . . . . . . . . . . . . 367
20.1 Genomic Best Linear Unbiased Prediction . . . . . . . . . . . . . . . . . . . 367
20.1.1 Ridge Regression . . . . . . . . . . . . . . . . . . . . . . . . . . . . . . 367

20.1.2 Best Linear Unbiased Prediction of Random Effects . . . . . . . . . 370
20.1.3 Predicting Genomic Values of Future Individuals . . . . . . . . . . . 371
20.1.4 Estimating Variance Parameters . . . . . . . . . . . . . . . . . . . . . . 372
20.1.5 Eigenvalue Decomposition for Fast Computing . . . . . . . . . . . . 374
20.1.6 Kinship Matrix . . . . . . . . . . . . . . . . . . . . . . . . . . . . . . . . . . 375
20.2 Reproducing Kernel Hilbert Spaces (RKHS) Regression . . . . . . . . . . . 386
20.2.1 RKHS Prediction . . . . . . . . . . . . . . . . . . . . . . . . . . . . . . . . 386
20.2.2 Estimation of Variance Components . . . . . . . . . . . . . . . . . . . . 388
20.2.3 Prediction of Future Individuals . . . . . . . . . . . . . . . . . . . . . . 389
20.2.4 Bayesian RKHS Regression . . . . . . . . . . . . . . . . . . . . . . . . . 390
20.2.5 Kernel Selection . . . . . . . . . . . . . . . . . . . . . . . . . . . . . . . . 392
20.3 Predictability . . . . . . . . . . . . . . . . . . . . . . . . . . . . . . . . . . . . . . . 394
20.3.1 Data Centering . . . . . . . . . . . . . . . . . . . . . . . . . . . . . . . . . 394
20.3.2 Maximum Likelihood Estimate of Parameters in Random
Models . . . . . . . . . . . . . . . . . . . . . . . . . . . . . . . . . . . . . . . 395
20.3.3 Predicted Residual Error Sum of Squares . . . . . . . . . . . . . . . . 395
20.3.4 Cross Validation . . . . . . . . . . . . . . . . . . . . . . . . . . . . . . . . 397
20.3.5 The HAT Method . . . . . . . . . . . . . . . . . . . . . . . . . . . . . . . . 397
20.3.6 Generalized Cross Validation (GCV) . . . . . . . . . . . . . . . . . . . 400
20.4 An Example of Hybrid Prediction . . . . . . . . . . . . . . . . . . . . . . . . . . 400
20.4.1 The Hybrid Data . . . . . . . . . . . . . . . . . . . . . . . . . . . . . . . . 400
20.4.2 Proc Mixed . . . . . . . . . . . . . . . . . . . . . . . . . . . . . . . . . . . . 401
References . . . . . . . . . . . . . . . . . . . . . . . . . . . . . . . . . . . . . . . . . . . . . . . 406

Index . . . . . . . . . . . . . . . . . . . . . . . . . . . . . . . . . . . . . . . . . . . . . . . . . . . 407

# About the Authors

**Shizhong Xu** is a Professor of Quantitative Genetics at the University of California, Riverside. He earned a BS degree in Animal Sciences at Shenyang Agricultural University (China), an MS degree in Quantitative Genetics at Northeast Agricultural University (China), and a PhD in Quantitative Genetics at Purdue University. He teaches two courses at UCR: Statistics for Life Sciences (STAT231B) and Quantitative Genetics (BPSC148). His research covers theoretical quantitative genetics, biostatistics, and statistical genomics. He specializes in linear mixed models, generalized linear mixed models, and Bayesian analysis. He published about 170 technical journal articles and two textbooks (including this one).

# Introduction to Quantitative Genetics

Before we formally introduce the book, let us define an important terminology in quantitative genetics, the "breeding value." We will first see two examples of how to apply quantitative genetics to solve real-life problems. The first example is about improvement of plant production through artificial selection. In the second example, we will show how to predict the height of a child from the heights of parents in humans.

## 1.1 Breeding Value and Its Application

### 1.1.1 Genetic Gain via Artificial Selection (Example 1)

Heritability is the most important genetic parameter in quantitative genetics. It is defined as the proportion of the phenotypic variance contributed by the genetic variance for a quantitative trait. The concept of variance will be formally introduced in a later chapter. You have to understand variance before understanding heritability. However, you can interpret variance as a sort of difference. Then heritability can be interpreted as the proportion of the difference between the mean of a group of selected parents and the mean of the whole population of the parental generation that can be passed to the next generation. In other words, the ratio of the difference between the mean of the children of the selected parents and the mean of the whole population of the younger generation to the difference between the mean of the selected parents and the mean of the entire parental population.

In the first example, we will demonstrate how to use heritability to predict genetic gain of yield in maize (*Zea mays*). Assume that the mean yield of maize is 6939 kg/ha and the heritability is 0.67 (Ali et al. 2011). Assume that we select a group of plants with very high yield for breeding and the mean yield of these selected plants is 7500 kg/ha. We then plant these selected seeds and allow them to pollinate randomly to produce seeds for the next generation. We want to predict the average yield of the next generation after one generation of selection. The formula for such a prediction is

$$\text{Predicted yield} = 0.67 \times (7500 - 6939) + 6939 = 7314.87$$

The difference between the mean of the selected parents and the mean of the whole population is $7500 - 6939 = 561$ kg/ha. This difference is called the selection differential. Only 0.67 of the selection differential will be passed to the next generation as the genetic gain (selection response). Therefore, we expect to see a gain of $0.67 \times 561 = 375.87$ kg/ha. The actual predicted yield for the new generation is the gain plus the mean yield of the original population, $375.87 + 6939.00 = 7314.87$ kg/ha.

### 1.1.2 Predicting Human Height (Example 2)

Once we understand the concept of heritability, we can use heritability to predict the height of Yao Ming's daughter (Yao Qinlei) from the heights of Yao Ming and his wife Ye Li.

© Springer Nature Switzerland AG 2022
S. Xu, *Quantitative Genetics*, https://doi.org/10.1007/978-3-030-83940-6_1

**Fig. 1.1** Yao Ming (Former NBA Player with the Houston Rockets). Image obtained from Yao Ming Wikipedia (Keith Allison, https://en. wikipedia.org/wiki/Yao_Ming)

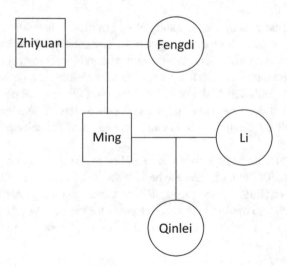

**Fig. 1.2** Pedigree of Yao's family

Figure 1.1 is a photo of Yao Ming. Figure 1.2 shows the pedigree of the Yao's family (a pedigree is a graphical representation of the relationships among members of a family). A tabular presentation of Yao's family is given in Table 1.1, which also contains heights of all members of Yao's family. The average Chinese man is 1.72 m (5ft8in) and the average Chinese woman is 1.60 m (5ft3in). Let us first predict Yao Ming's height from the heights of his parents. Given the heritability of human height being approximately 0.8 for both men and women (Fisher 1918; Yang et al. 2010), the predicted height of Yao Ming is

$$\widehat{H}_{\text{YaoMing}} = \frac{1}{2}\left[0.8 \times (2.01 - 1.72) + 0.8 \times (1.91 - 1.60)\right] + 1.72 = 1.96 \ \text{m}$$

which is way below the observed height of 2.26 m. We now predict the height of Yao Ming's daughter, Yao Qinlei, from the heights of Yao Ming and his wife Ye Li. The predicted height of Yao Qinlei is

**Table 1.1** Tabular presentation of Yao's pedigree and the heights of Yao's family[a]

| Member | Father | Mother | Observed height | Predicted height |
|---|---|---|---|---|
| Yao Zhiyuan | – | – | 2.01 m (6ft7in) | – |
| Fang Fengdi | – | – | 1.91 m (6ft3in) | – |
| Yao Ming | Yao Zhiyuan | Fang Fengdi | 2.26 m (7ft6in) | 1.96 m |
| Ye Li | – | – | 1.93 m (6ft4in) | – |
| Yao Qinlei | Yao Ming | Ye Li | – | 1.95 m |

[a]Data from Yao Ming Wikipedia (https://en.wikipedia.org/wiki/Yao_Ming)

$$\widehat{H}_{\text{YaoQinlei}} = \frac{1}{2}[0.8 \times (2.26 - 1.72) + 0.8 \times (1.93 - 1.60)] + 1.60 = 1.95 \text{ m}$$

Yao Qinlei (born on May 21, 2010) is now just 11 years old in the year of this book's publication. How close this prediction to the actual height is will not be known until about 10 years later when she reaches the adulthood. If Qinlei were a boy, the predicted height would be

$$\widehat{H}_{\text{BoyYaoQinlei}} = \frac{1}{2}[0.8 \times (2.26 - 1.72) + 0.8 \times (1.93 - 1.60)] + 1.72 = 2.07 \text{ m}$$

The 0.8 heritability value refers to the human population on average. Therefore, the predicted heights of Yao Ming and his daughter may not be close to their actual heights. But if we have millions of such predictions, on average, the squared correlation between the observed heights and the actual heights will be approximately 0.8.

### 1.1.3 Breeding Value

In animal and plant breeding, we often need to estimate the "breeding value" of an individual for a quantitative trait of interest. We first define the breeding value as the heritable part of the difference between the individual's phenotypic value and the mean phenotypic value of the population. For example, the mature body weight of bulls (beef cattle, *Bos taurus* L.) is a heritable trait. Assume that the average mature body weight of bulls is 900 lb. and the average mature body weight of many progeny (sons) of this bull is 1000 lb. What is the breeding value of mature body weight for this bull? Assume that each son is born from a cow randomly sampled from the population. *The breeding value of the bull is twice the difference between the average body weight of the bull's sons and the average body weight of all bulls in the population.* In this example, the breeding value of the bull is $2 \times (1000 - 900) = 200$ lb. We now explain why the breeding value is calculated this way. Let the breeding value of the bull be $A_{\text{Bull}} = h^2 (P_{\text{Bull}} - \overline{P})$, let the average breeding value of the cows mated with the bull be $A_{\text{Cow}} = h^2 (\overline{P}_{\text{Cow}} - \overline{P})$. The predicted mature body weight of all sons of the bull is

$$\widehat{W}_{\text{Son}} = \frac{1}{2}\left[h^2 \times \left(P_{\text{Bull}} - \overline{P}\right) + h^2 \times \left(\overline{P}_{\text{Cow}} - \overline{P}\right)\right] + \overline{P} = \frac{1}{2}(A_{\text{Bull}} + A_{\text{Cow}}) + \overline{P}$$

If the cows are randomly selected from the population, $\overline{P}_{\text{Cow}} - \overline{P} \approx 0$ and thus $A_{\text{Cow}} \approx 0$. As a result,

$$\widehat{W}_{\text{Son}} = \frac{1}{2}A_{\text{Bull}} + \overline{P}$$

Given the average body weight of the sons being 1000 lb. and the average body weight of the population being 900 lb., we have

$$1000 = \frac{1}{2}A_{\text{Bull}} + 900$$

Rearranging the above equation leads to

$$\widehat{A}_{\text{Bull}} = 2 \times (1000 - 900) = 200\,\text{lb}$$

Therefore, the breeding value of this bull is 200 lb.

## 1.2    Complicated Behavior Traits

What all living organisms do throughout their lives is partly determined by their genes. No matter whether the organisms are viruses, vegetables, bacteria, bats, house flies, or human beings, their structures, physiologies, and lifestyles are ordained by the proteins produced by their genetic material. The genes do not act independently, but do act in response to triggers in their internal and external environments. However, these environmental stimuli can only alter events to a limited degree and may simply act to initiate or to stop any activities, or to change their directions (Kearsey and Pooni 1996).

Consider birds like the emperor penguins (Fig. 1.3), *Aptenodytes forsteri*, for example. To avoid predators, they decide to live on the Antarctic ice and in the frigid surrounding waters. They employ physiological adaptations and cooperative behaviors in order to deal with the incredibly harsh environment, where wind chills can reach −76 °F (−60 °C). The genes instruct them to spend the long winter on the open ice and breed during this harsh season. They instruct them to find and mate with individuals of the opposite sex of their own species. The genes also tell them that each individual can only mate with one other individual of the opposite sex, not two or more. Female penguins lay a single egg (occasionally two eggs) and then promptly leave the egg(s) to their male partners. They then take an extended hunting trip that lasts for some two months and travel for some 50 miles just to reach the open sea, where they feed on fish, squid, and krill. The genes tell the male penguins to keep the newly laid eggs warm and hatch them. The male penguins are instructed not to sit on the eggs, as many other birds do,

**Fig. 1.3**  Emperor penguins (Ian Duffy, https://en.wikipedia.org/wiki/Emperor_penguin)

but stand and protect their eggs from the elements by balancing them on their feet and covering them with their brood pouches (feathered skin). During this two-month bout of babysitting, the males eat nothing and are at the mercy of the Antarctic elements. The genes instruct the male penguins to huddle together to escape wind and conserve warmth. Individuals take turns moving to the group's protected and relatively toasty interior. Once a penguin has warmed a bit, the genes tell it to move to the perimeter of the group so that others can enjoy protection from the icy elements. The genes tell the female penguins to return to the breeding site just at the time when the eggs begin to hatch. When the female penguins return to the breeding site, the male penguins transfer very carefully the newly hatched chicks to their female partners (not to other females). Mothers care for their young chicks and protect them with the warmth of their own brood pouches. Outside of this warm cocoon, a chick could die in just a few minutes. The female penguins bring a belly full of food that they regurgitate for the newly hatched chicks. Meanwhile, after their duty is done, male penguins go to the sea in search of food for themselves. Occasionally, a female penguin lays two eggs and the male partner has to transfer two chicks (if both hatched) to the mother. However, a mother only has enough food to feed one chick. The mother must choose one out of the two. The genes tell the mother to run away from the two chicks and choose the one who runs fast enough to follow the mother. The mother knows that the faster chick will have a better chance to survive than the slower one, a hard but smart decision. In December, Antarctic summer, the pack ice begins to break up and open water appears near the breeding site, just as young emperor penguins are ready to swim and fish on their own. All these behaviors and activities of the emperor penguins are achieved by proteins determined by genes; they have no other instruction manual to guide them through this unimaginably complex set of behaviors, none of which make any sense to the penguins themselves.

Not surprisingly, it requires many thousands of genes to produce and manage such a complex organism as an emperor penguin, as indeed it does to produce a vegetable or a human being. It would be impossible to study all or even a small proportion of these genes at once, and so geneticists normally concentrate on just one characteristic or trait at a time. In order to study how these genes act, geneticists have to look at genetic differences; variation, in other words. It would be impossible to study how a penguin's genes instruct it to survive in the Antarctic harsh environment without having access to penguins which have genes resulting in a different behavior. Typically, the geneticist's approach to studying how a particular characteristic controlled is to look for, or induce by mutation, genetic variants which behave abnormally or at least behave differently; variation is the essential prerequisite of genetic analysis. Enormous advances have been made in our knowledge of genetics the last and current centuries and the pace is accelerating all the time. It is a subject which has been fortunate in attracting the interests of highly intelligent people in all branches of science, including not only biology but also chemistry, physics, and mathematics, and this has resulted in genetics holding a central role in late twentieth century biology (Kearsey and Pooni 1996).

## 1.3 Qualitative and Quantitative Traits

Qualitative traits are categorical in nature and their phenotypic differences are clearly visible. Qualitative traits are described by discrete distributions. Typical examples of qualitative traits include disease status (resistant or susceptible), the seven Mendelian traits in common peas, horned *versus* polled cattle or tall *versus* dwarf rice. Many qualitative traits are single gene-controlled traits, e.g., flower color (purple and white) in common pea of the Mendelian hybridization experiments. These single gene-controlled qualitative traits do not belong to the traits targeted by quantitative genetics, which deal primarily with quantitative traits with a continuous distribution. Mendel defined quantitative traits as traits that differ in a "more or less" nature (Mendel 1866; Mendel and Bateson 1901). Quantitative traits are "often" controlled by the segregation of multiple genes. More importantly, their phenotypic differences are affected by environmental factors. By "often" we mean that there are exceptions where a single gene-controlled trait may vary in a continuous manner if the trait is strongly affected by environments. All continuously distributed traits are target traits studied in quantitative genetics.

Some qualitative traits are controlled by a single gene or at most a few genes, but many qualitative traits are controlled by multiple genes and multiple environmental factors. Although they are not called quantitative traits, their polygenic natures (controlled by multiple genes) do not allow them to be studied using classical Mendelian genetic tools. Sewall Wright (1934) defined such traits as threshold traits. He examined the variation of the number of digits on the hind feet of guinea pigs (*Cavia porcellus*). The typical number of digits of a hind foot is three, but guinea pigs with four digits on a hind foot appeared very often in his guinea pig population. He hypothesized that there was an underlying continuously distributed "quantitative trait," now called the liability (Falconer and Mackay 1996; Roff 1996, 2008), and a hypothetical threshold on the scale of the liability that controls the number of digits. The liability is often less than the threshold and thus most individuals are three-toed guinea pigs. However, if the liability is greater than the threshold, the guinea pig will develop four digits. The hypothetical liability is a quantitative trait and the binary phenotype just reflects partial information of the underlying trait. The same argument can be used to describe many discrete traits, e.g., twinning in lambs. Regardless of whether the phenotypes are continuous or discrete, these traits are all called polygenic traits and are the targets aimed for by quantitative geneticists. Typical statistical tools for threshold trait analyses are generalized linear models (GLM) and generalized linear mixed models (GLMM).

## 1.4    The Relationship Between Statistics and Genetics

Recognizing the fact that quantitative traits are controlled by the segregation of multiple genes, the Mendelian method that analyzes one gene at a time is not the way to study the inheritance of quantitative traits. Instead, all genes must be studied simultaneously, which requires more advanced statistical tools. An area called biometrical genetics then took place. Biometrical genetics is the science dealing with the inheritance of metrical characters (Mather 1949; Owen 1949; Mather and Jinks 1982; Evans et al. 2002). Theoretically, if all other loci are fixed, the segregation pattern of a single locus and its effect on a quantitative trait are still guided under Mendel's laws of inheritance. However, the collective effect of all genes completely masks the discrete nature of the Mendelian loci. Therefore, statistical tools must be adopted to study the inheritance of quantitative traits. For example, the phenotypic value of an individual is controlled by both the genetic effect and the environmental effect. The genetic variance and the environmental variance are confounded, and their combination contributes to the phenotypic variance. Separating the genetic variance from the environmental variance is one of the important tasks faced by quantitative geneticists. Statistics now plays a key role in this separation. Relatives share a partial amount of their genetic material, depending on the degree of relatedness. For example, parents and progeny share half of their genetic material. Therefore, the covariance between a parent and a progeny equals half of the genetic variance. Twice the parent-progeny covariance equals the genetic variance. Therefore, by evaluating the covariance between relatives, one can estimate the genetic variance and thus separate the genetic variance from the environmental variance.

Biometrical genetics, presently more often called quantitative genetics, is a branch of genetics. However, based on my own experience of teaching and research in quantitative genetics, it is more accurate to say that quantitative genetics is a branch of statistics. It does not require substantial knowledge of genetics to study quantitative genetics, but requires a decent and diverse statistical background to study quantitative genetics. Understanding Mendel's laws of inheritance and the concept of linkage disequilibrium between loci is perhaps sufficient to study quantitative genetics. However, studying quantitative genetics requires much more statistical knowledge beyond the means and the variances. Basic knowledge of statistics required by quantitative genetics include, but are not limited to: distributions, statistical tests, expectations, variances, covariance, correlation, regression, analysis of variances, linear models, linear mixed models, generalized linear models, generalized linear mixed models, matrix algebra, matrix calculus, moment estimation, maximum likelihood estimation, Bayesian statistics, structural equation modeling, best linear unbiased estimation (BLUE), best linear unbiased prediction (BLUP), and many more. To be a good quantitative geneticist, you must be a decent statistician first, not the other way around.

Population genetics is very close to quantitative genetics. In fact, it is the foundation of quantitative genetics. For example, Hardy-Weinberg equilibrium is the basic principle on which the total genetic variance is partitioned into the additive variance and the dominance variance. Linkage disequilibrium is the basis for quantitative trait locus (QTL) mapping. Identity-by-descent (IBD) that describes the relationship between two alleles is the basic element for us to calculate the kinship matrix of genetically related individuals and use the numerator relationship matrix (two times the kinship matrix) to estimate the genetic variance for a quantitative trait. The three founders of population genetics (R. A. Fisher, J. B. S. Haldane, and Sewall Wright) were all famous mathematicians and statisticians. They were also the founders of quantitative genetics. Many statistical theories and methods were developed by statisticians with problems originated from genetics. For example, Fisher (1918), the father of modern statistics, developed the technology of analysis of variances (ANOVA) by partitioning the variance ($\sigma^2$) of human stature into the between-family variance ($\sigma_B^2$) and the within-family variance ($\sigma_W^2$), and described the standard deviation of human stature by $\sigma = \sqrt{\sigma_B^2 + \sigma_W^2}$.

The concept of regression was proposed by Francis Galton in studying the inheritance of human stature and, in general, the law of ancestral heredity. Galton compared the height of children to that of their parents. He found that the adult children are closer to the average height than their parents are. He called the phenomenon regression towards the mean (Galton 1886; Senn 2011). A much clearer description of the phenomenon is that "deviation of the average offspring stature of tall parents from the population mean is less than the deviation of the average stature of their parents from the mean." The ratio of the deviation of the offspring (from the mean) to the deviation of the parents (from the mean) later is called the realized heritability (Falconer and Mackay 1996; Lynch and Walsh 1998). The realized heritability is simply obtained by a rearrangement of the breeders' equation where the deviation of the parents is called the selection differential and the deviation of the progeny is called the response to selection.

Similar to regression, Galton also coined another important concept "correlation" although it was Karl Pearson who mathematically formulated the correlation coefficient in evaluation of the law of ancestral heredity. The original formulation of the correlation coefficient is called Pearson's product-moment correlation coefficient or Pearson correlation in short. Pearson compared the correlation of human stature between the progeny and the mid-parent, between the progeny to the mid-grandparent, and so on. He found that the individual ancestral correlations decrease in geometrical progressions (Pearson 1903). Regression and correlation analyses are fundamentally important tools in statistical analyses, but they were originated from studying inheritance of human characters.

The theory and method of linear mixed models were developed by C. R. Henderson in the 1950s for estimating genetic parameters and predicting breeding values in unbalanced populations of large animals, e.g., beef and dairy cattle (Henderson 1950, 1953). Henderson started his career as an animal breeder, but his contribution in statistics made him one of the greatest statisticians in the world. The linear mixed model he developed has revolutionized both statistics and genetics. Prior to the mixed model era, regression analysis was mainly used for data analysis in models consisting of fixed effects only. The analysis of variance (ANOVA) was primarily applied to data analysis in models consisting of only random effects. His mixed model can handle both the fixed and the random effects simultaneously in the same model. With the availability of the MIXED procedure in SAS (SAS Institute Inc. 2009), the mixed model is really the "one model fits all" model. The regression analysis, the ANOVA analysis, the general linear model analysis are all special cases of the linear mixed model.

The expectation and maximization (EM) algorithm developed by Dempster et al. (1977) was illustrated with an example in genetic linkage study, a special case of the general maximum likelihood estimation of multinomial distribution (Rao 1965). The EM algorithm is a general algorithm for parameter estimation with missing values. The algorithm was initially triggered by a problem in linkage analysis where the recombination fraction between two loci (A and B) is to be estimated from an $F_2$ population. The class of double heterozygotes consists of two different phases, the coupling phase (AB/ab) and the repulsion phase (Ab/aB). The coupling phase involves no recombination from both gametes $(1 - r)^2$ while the repulsion phase requires two recombination events $(r^2)$, one from each gamete, where $r$ is the recombination fraction between the two loci. When a double heterozygote is observed, say $n_C + n_R = n_D$, where $n_D$ is the number of double heterozygotes, $n_C$ is the number of double heterozygotes with the coupling phase and $n_R$ is the number of double heterozygotes with the repulsion phase, we do not know $n_C$ and $n_R$. However, if $r$ were known, $n_R$ can be predicted by

$$\widehat{n}_R = \frac{r^2}{r^2 + (1 - r)^2} n_D$$

and $\widehat{n}_C = n_D - \widehat{n}_R$. Treating $\widehat{n}_R$ and $\widehat{n}_C$ as if they were observed values, the recombination fraction $r$ would be estimated with an explicit and mathematically attractive formula. The EM algorithm requires several cycles of iterations to provide a numerical solution for the estimated recombination fraction.

The mixture model, not to be confused with the mixed model, was developed by Karl Pearson (1894) from a problem originated from evolution, speciation to be specific. The asymmetry in the histogram for the forehead to body length ratios in a female shore crab population was being considered. Walker Weldon, a zoologist, provided the specimens of the crabs to Pearson for analysis. Weldon speculated that the asymmetry in the histogram of the ratios could signal evolutionary divergence. Pearson's approach was to fit a univariate mixture of two normal distributions by choosing five parameters of the mixture $(\mu_1, \sigma_1^2, \mu_2, \sigma_2^2, \pi)$ such that the empirical moments matched that of the model. Although the mixture model analysis was not a common tool in classical quantitative genetics, it is crucial in the segregation analysis of major genes (Heiba et al. 1994) and the maximum likelihood method of interval mapping for quantitative trait loci (Lander and Botstein 1989; Jansen 1992). The discrete and discontinuous distribution of Mendelian factors can be described by the mixture model in statistics (Wang and Xu 2019). Ironically, the mixture distribution Pearson developed can explain the Mendelian segregation variance, but Pearson stood in the other side of the debate between the Mendelian school and the biometrical school (Piegorsch 1990), where the former supported discrete inheritance and the latter was in favor of continuous inheritance.

## 1.5    Fundamental Statistical Methods in Quantitative Genetics

Knowing how to download software packages and analyze data are not enough to make you a good quantitative geneticist. It requires knowledge from as simple as elementary statistics such as expectation and variance to as comprehensive as generalized linear mixed models for pedigree analysis of discrete polygenic traits. The basic statistical skills are fundamentally important in understanding how quantitative genetic formulas were originally derived. Concepts and operations of basic statistics include the expectations and the variances of discrete and continuous distributions, the properties of expectations, variances, and covariance between two variables.

Regression analyses, analyses of variances, and linear models are the most commonly used statistical tools for quantitative genetics studies. In classical quantitative genetics, however, they were often expressed in the forms of complicated summations, requiring multiple layers of loops and thus making them error prone. When they are expressed in matrix notation, not only are they clean and beautiful but also error proof when being coded in the form of computer programs. It is hard to imagine how to learn mixed models and the best linear unbiased prediction (BLUP) without any knowledge of matrix algebra. Therefore, it is very important to understand matrix algebra and to be able to use matrix manipulation skills to perform derivations and to write scripts of computer programs.

Significance tests have become more and more important in modern quantitative genetics, where detection of major genes, mapping quantitative trait loci and genome-wide association studies have been added to the contents of quantitative genetics. In classical quantitative genetics, other than segregation analysis and tests for genotype by environment interactions, significance tests are rarely conducted for an estimated repeatability, heritability, and genetic correlation. In modern quantitative genetics, significance tests are essential in QTL mapping and in genome-wide association studies. Because the sample sizes are often relatively large, e.g., often greater than 200, the $t$-test and F-test have been replaced by a more robust Wald test, which has a Chi-square distribution under the null model.

Linear models and linear mixed models are always the main tools of data analysis in quantitative genetics. Parent-progeny regression analysis (linear models), sib analysis (analysis of variances, represented by general linear models), and pedigree analysis (linear mixed models) are all essential tools for estimating heritability and genetic correlation of quantitative traits. Emphases have been shifted from understanding the technical details of analytical tools to calling professional software packages for data analysis; from estimating genetic parameters with intermediate results (summary data) to analyzing the original data; from analysis of real-life data to analysis of simulated data.

Regarding to estimation of genetic parameters from the analyses of variance tables, traditional methods of estimating variance components are various moment methods, i.e., equating the observed mean squares to the expected mean squares and solving for variance components embodied in the expected mean squares. For balanced data in the common designs of experiments in plant and animal breeding, the expected mean squares are simple linear functions of the variance components. Our job was to solve for the variance components and then convert the variance components into genetic variances and environmental variances. The current mixed model software packages, e.g., the mixed procedure in SAS and the lmer() function in the lme4 package of R, already deliver the final estimated variance components, requiring no actions from the users' end. The mixed procedure in SAS has an option to produce ANOVA tables, which contain the degrees of freedom, the sums of squares and the mean squares. When the maximum likelihood method or the restricted maximum likelihood method is used, no degrees of freedom are produced because these methods do not need the degrees of freedom. Without the intermediate steps of variance component estimation, there is little we need to do other than to understand how to run PROC MIXED in SAS and lmer() in R. The time saved allows students to learn more in depth statistical tools, e.g., Bayesian statistics and the Markov chain Monte Carlo (MCMC) algorithm.

The best linear unbiased prediction (BLUP) is an essential tool to estimate the breeding values of candidate individuals for selection. Since the breeding values are random effects, Henderson called the estimated breeding values "predictions" as opposed to the estimated fixed effects "estimations," while the estimated fixed effects are called the best linear unbiased estimations (BLUE). The BLUP model is an advanced and general technology to predict breeding values of candidate individuals of a random population containing animals with arbitrarily complicated relatedness. The BLUP and selection index are the same thing, simply being presented differently. There are two kinds of linear selection indices: (1) selection index for multiple traits and (2) selection index for a single trait collected from multiple relatives. Let us compare selection index of the second kind with the BLUP. The index for individual $j$ is

$$I_j = b_1 y_{j1} + b_2 y_{j2} + \cdots + b_m y_{jm}$$

where $b_k$ is the index weight for the $k$th relative of individual $j$, $y_{jk}$ is the deviation of the trait value of the $k$th relative of individual $j$ from the grand mean and $m$ is the number of relatives included in the index. The phenotypic deviation, $y_{jk}$, is called the centered phenotypic value, to be exact, because it is the phenotypic value adjusted by all fixed effects, not just by the grand mean, if other fixed effects are present. In matrix notation, $I_j = b^T Y_j$, where $b$ is an $m \times 1$ vector of index weights and $Y_j$ is an $m \times 1$ vector of the trait values from all $m$ relatives of individual $j$. The selection index for individual $i$ is $I_i = b^T Y_i$ where $Y_i = \{y_{i1}, y_{i2}, \cdots, y_{im}\}$. The index weights are the same for individuals $j$ and $i$, because $Y_j$ and $Y_i$ represent the same set of relatives. For example, if $Y_j = \{y_{j1}, y_{j2}\}$ are the phenotypic values of the sire and a progeny of individual $j$, $Y_i = \{y_{i1}, y_{i2}\}$ also represent the phenotypic values of the sire and a progeny of individual $i$. The BLUP of an individual's breeding value, however, is a linear combination of the phenotypic values of ALL individuals in the population. For a population consisting of only individuals $i$ and $j$ as an example, the BLUP of the two individuals may be expressed by

$$\begin{bmatrix} I_i \\ I_j \end{bmatrix} = \begin{bmatrix} b^T & 0 \\ 0 & b^T \end{bmatrix} \begin{bmatrix} Y_i \\ Y_j \end{bmatrix}$$

Therefore,

$$I_i = \begin{bmatrix} b^T & 0 \end{bmatrix} \begin{bmatrix} Y_i \\ Y_j \end{bmatrix} = b_i^T Y$$

and

$$I_j = \begin{bmatrix} 0 & b^T \end{bmatrix} \begin{bmatrix} Y_i \\ Y_j \end{bmatrix} = b_j^T Y$$

where $b_i^T = \begin{bmatrix} b^T & 0 \end{bmatrix}$, $b_j^T = \begin{bmatrix} 0 & b^T \end{bmatrix}$ and $Y = \begin{bmatrix} Y_i^T & Y_j^T \end{bmatrix}^T$. We now conclude that in the context of selection index, the index weights are the same, but the phenotypic values are different for different individuals. In the context of BLUP, the index weights are different, but the phenotypic values are the same for different individuals. Therefore, the BLUP and the selection index are the same both conceptually and technically.

## 1.6    Statistical Software Packages

Statistical software packages are essential tools for modern quantitative genetics. Many statistical packages are available for quantitative trait analysis, typical examples include SPSS (Statistical Package for the Social Sciences), MATLAB (the Mathworks), R (R Foundation for Statistical Computing), SAS (Statistical Analysis System), Microsoft Excel, and so on. SPSS is the most widely used statistics software package within human behavior research. SPSS offers the ability to easily compile descriptive statistics, parametric and non-parametric analyses, as well as graphical depictions of results through the graphical user interface (GUI). It also includes the option to create scripts to automate analysis, or to carry out more advanced statistical processing. Another software is BMDP (Bio-Medical Data Package), which is no longer available after 2017. SAS is a statistical analysis platform that offers options to use either the GUI, or to create scripts for more advanced analyses. It is a premium solution that is widely used in business, healthcare, and human behavior research alike. It is possible to carry out advanced analyses and produce publication-worthy graphs and charts although the coding can be a difficult adjustment for those not used to this approach. SAS is the most comprehensive statistical package in the world with the best technical support. The quality of SAS products is the highest among all statistical software packages. The reason I listed the three statistical packages here is that these were the three statistical software packages available at Purdue University in the 1980s when I was a Ph.D. student serving as a student statistical consultant in the Department of Statistics. I chose SAS as my primary software package and helped Purdue students and faculty analyze their data. I only stopped using SAS during the one-year postdoc at Rutgers University where I wrote FORTRAN programs instead.

R is a free statistical software package that is widely used across both social sciences and natural sciences. The plain R and R Studio are available for a great range of applications, which can simplify various aspects of data processing. While R is a very powerful software, it also has a steep learning curve, requiring a certain degree of coding. It does, however, come with an active community engaged in building and improving R and the associated plugins, which ensures that help is never too far away. The R platform allows users to submit their own packages to share with other users. Therefore, one can virtually find everything from R, for problems as small as a single function to problems as large as generalized linear mixed models. MATLAB is an analytical platform and programming language that is widely used by engineers and scientists. As with R, the learning path is steep, and you will be required to create your own code at some point. While not a cutting-edge solution for statistical analysis, MS Excel offers a wide variety of tools for data visualization and simple statistics. It is simple to generate summary metrics and customizable graphics and figures, making it a useful tool for many who want to see the basics of their data. As many individuals and companies both own and know how to use Excel, it also makes it an accessible option for those looking to get started with statistics.

In this course and within this textbook, you will have the opportunity to learn and use SAS for data analysis. Depending on whether your universities provide free SAS licenses for students or not, R may be introduced for the course if free licenses for SAS are not available. The MS Excel is always available for all students, as long as their computers are installed with the MS Windows system. The MS Excel program has many built in functions, and they are very handy to use. For example, you can calculate the p-value for your F-test from the F-distribution function. We have been using the MS Excel to manually calculate the coancestry matrix for a very complicated pedigree using the tabular method (Chap. 11). The tabular method is extremely flexible to allow users to incorporate inbred and correlated founders of any complicated pedigree. The resultant coancestry matrix from the tabular method can be validated with the kinship matrix produced from PROC INBREED in SAS. The GLIMMIX procedure in SAS is a general procedure to analyze all linear mixed models for both quantitative traits and categorical traits. It handles any generalized linear mixed models (GLMM) and is really a "one-model-fits-all" model. Procedures like REG, ANOVA, GLM, MIXED, GENMOD, LOGISTIC, and CATMOD are all subsets of the generalized linear mixed model procedure. It was converted from an SAS MACRO that repeatedly calls PROC MIXED to analyze a linearly transformed intermediate variable from an observed categorical data point and its likelihood. The NLMIXED procedure has a lot of overlap with GLIMMIX, but non-linear models are not as common as linear models, thus making GLIMMIX the real universal model.

## 1.7    Classical Quantitative Genetics and Modern Quantitative Genetics

Quantitative genetics can be categorized into classical quantitative genetics and modern quantitative genetics. Prior to the genome era, behaviors of individual genes for quantitative traits could not be directly observed. They had to be studied collectively via statistical inference. Estimating genetic variance and partitioning the total genetic variance into different variance components are the cores of classical quantitative genetics. The genetic variance and their components must be estimated via the covariance between relatives, either between parents and progeny or between siblings. Three major genetic parameters are involved in classical quantitative genetics: (1) repeatability, (2) heritability, and (3) genetic correlation. Selection theory and methods in classical quantitative genetics consist of mass selection, pedigree selection, family selection, progeny testing, and selection via BLUP (a generalized index selection).

As the advancement of molecular technology in late 1980s of the last century continued, various molecular markers were available and their linkage maps were constructed with MAPMAKER (Lander et al. 1987), a software package for construction of linkage maps in experimental and natural populations. With a complete linkage map, we are able to capture information more directly about the genes of quantitative traits. A gene for a quantitative trait segregates guided by Mendel's laws of inheritance, but its segregation cannot be observed. A marker, on the other hand, does not affect the quantitative trait, but its segregation can be observed. If a marker overlaps with the gene, we can directly observe the segregation of the gene via the marker. If the marker does not overlap with the gene but sits nearby the gene, partial information of the segregation of the gene can be captured. Co-segregation or partial co-segregation between a gene and a marker is called linkage disequilibrium (LD), which is the foundation of QTL mapping. QTL mapping overcomes the bottleneck of quantitative genetics. We are now in the genome era and the classical quantitative genetics has evolved into the modern quantitative genetics (also called molecular quantitative genetics). Detected QTL can facilitate selection, which is called marker-assisted selection (Lande and Thompson 1990). QTL mapping and marker-assisted selection are based on the assumption that quantitative traits are controlled by a few major genes plus numerous modifying genes, the so-called oligogenic model. If a quantitative trait is controlled by the infinitesimal model (Fisher 1918), QTL mapping and marker-assisted selection cannot be applied to such traits. The marker-assisted selection scheme is then replaced by the so-called genomic selection (Meuwissen et al. 2001),

**Table 1.2** Additional genetic parameters and methods of selection for modern quantitative genetics compared with classical quantitative genetics

|  | Modern quantitative genetics | Classical quantitative genetics |
|---|---|---|
| Genetic parameter | Repeatability | Repeatability |
|  | Heritability | Heritability |
|  | Genetic correlation | Genetic correlation |
|  | Number of QTL |  |
|  | Locations of QTL |  |
|  | QTL heritability (QTL size) |  |
| Selection method | Mass selection | Mass selection |
|  | Group selection | Group selection |
|  | Index selection (BLUP) | Index selection (BLUP) |
|  | Marker-assisted selection |  |
|  | Genomic selection |  |

where the QTL mapping step is skipped and markers of the whole genome are used to predict the genomic values of candidate individuals for selection. Table 1.2 summarizes the differences between the classical quantitative genetics and the modern quantitative genetics.

The last three chapters (Chaps. 18, 19, and 20) of the textbook cover the principles of modern quantitative genetics. Chapter 18 introduces the concept of QTL mapping and the statistical methods commonly used in QTL mapping. Chapter 19 covers topics related to genome-wide association studies and primarily focuses on the linear mixed model approach that incorporates a marker inferred kinship matrix to capture the random polygenic effects. Chapter 20 describes the concept and methods of genomic selection using markers of the whole genome via the best linear unbiased prediction (BLUP) mechanism and the reproducing kernel Hilbert spaces (RKHS) regression.

# References

Ali F, Mareeya M, Hidayatur R, Muhammad N, Sabina S, Yan J. Heritability estimates for yield and related traits based on testcross progeny performance of resistant maize inbred lines. J Food Agric Environ. 2011;9:438–43.

Dempster AP, Laird MN, Rubin DB. Maximum likelihood from incomplete data via the EM algorithm. J Royal Statist Soc Ser B (Methodological). 1977;39:1–38.

Evans DM, Gillespie NA, Martin NG. Biometrical genetics. Biol Psychol. 2002;61:33–51.

Falconer DS, Mackay TFC. Introduction to quantitative genetics. Harlow, Essex, UK: Addison Wesley Longman; 1996.

Fisher RA. The correlation between relatives on the supposition of Mendelian inheritance. Trans Royal Soc Edinburgh. 1918;52:399–433.

Galton F. Regression towards mediocrity in hereditary stature. J Anthropol Inst G B Irel. 1886;15:246–63.

Heiba IM, Elston RC, Klein BEK, Klein R. Sibling correlations and segregation analysis of age-related maculopathy: the beaver dam eye study. Genet Epidemiol. 1994;11:51–67.

Henderson CR. Estimation of genetic parameters (Abstract). Ann Math Stat. 1950;21:309–10.

Henderson CR. Estimation of variance and covariance components. Biometrics. 1953;9:226–52.

Jansen RC. A general mixture model for mapping quantitative trait loci by using molecular markers. Theor Appl Genet. 1992;85:252–60.

Kearsey MG, Pooni HS. The genetical analysis of quantitative traits. London: Chapman & Hall; 1996.

Lande R, Thompson R. Efficiency of marker-assisted selection in the improvement of quantitative traits. Genetics. 1990;124:743–56.

Lander ES, Botstein D. Mapping Mendelian factors underlying quantitative traits using RFLP linkage maps. Genetics. 1989;121:185–99.

Lander ES, Green P, Abrahamson J, Barlow A, Daly MJ, Lincoln SE, Newburg L. MAPMAKER: an interactive computer package for constructing primary genetic linkage maps of experimental and natural populations. Genomics. 1987;1:174–81.

Lynch M, Walsh B. Genetics and analysis of quantitative traits. Sunderland: Sinauer Associates, Inc.; 1998.

Mather K. Biometrical genetics: the study of continuous variation. London: Methuen; 1949.

Mather K, Jinks JL. Biometrical genetics: the study of continuous variation. London: Chapman & Hall; 1982.

Mendel JG. Versuche über Pflanzen-Hybriden Verhandlungen des Naturforschenden Vereines in Brünn, , bd. IV für das jahr 1865. Abhandlungen. 1866;4, 3:–47.

Mendel JG, Bateson W. Experiments in plant hybridization (Translated by William Bateson in 1901). J R Hortic Soc. 1901;16:1–32.

Meuwissen THE, Hayes BJ, Goddard ME. Prediction of total genetic value using genome-wide dense marker maps. Genetics. 2001;157 (4):1819–29.

Owen ARG. Biometrical genetics. Nature. 1949;164:420–1.

Pearson K. III. Contributions to the mathematical theory of evolution. Philos Trans R Soc Lond A. 1894;185:71–110.

Pearson K. The law of ancestral heredity*. Biometrika. 1903;2:211–28.

Piegorsch WW. Fisher's contributions to genetics and heredity, with special emphasis on the Gregor Mendel controversy. Biometrics. 1990;46:915–24.

Rao CR. Linear Statistical Inference and Its Applications. New York: Wiley; 1965.

Roff DA. The evolution of threshold traits in animals. Q Rev Biol. 1996;71:3–35.

Roff DA. Dimorphisms and threshold traits. Nature Education. 2008;1:211.

SAS Institute Inc. SAS/STAT: Users' Guide, Version 9.3. Cary, NC: SAS Institute Inc.; 2009.

Senn S. Francis Galton and regression to the mean. Significance. 2011;8:124–6.

Wang M, Xu S. Statistics of Mendelian segregation: a mixture model. J Anim Breed Genet. 2019; https://doi.org/10.1111/jbg.12394.

Wright S. An analysis of variability in number of digits in an inbred strain of Guinea Pigs. Genetics. 1934;19:506–36.

Yang J, Benyamin B, McEvoy BP, Gordon S, Henders AK, Nyholt DR, Madden PA, Heath AC, Martin NG, Montgomery GW, et al. Common SNPs explain a large proportion of the heritability for human height. Nat Genet. 2010;42:565–9.

# Review of Mendelian Genetics

## 2.1 Mendel's Experiments

Quantitative genetics deals with quantitative traits that are often controlled by multiple genes. However, the behavior of each gene still follows Mendel's laws of inheritance. Therefore, it is essential to review Mendel's laws before going further to discuss the principles of quantitative genetics. Gregor Mendel (1822–1884) was the founder of Genetics (see Fig. 2.1). He published the laws of inheritance in 1866 (Mendel 1866; Mendel and Bateson 1901), known as Mendel's first and second laws. His ideas had been published in 1866 but largely went unrecognized until 1900 when his findings were rediscovered independently by three botanists in three different countries (Hugo DeVries, Carl Correns, and Erich von Tschermak) in a short span of 2 months within the same year, one generation after Mendel published his paper. They helped expand awareness of Mendel's laws of inheritance in the scientific world. The three Europeans, unknown to each other, were working on different plant hybrids when they each worked out the laws of inheritance. When they reviewed the literature before publishing their own results, they were startled to find Mendel's old paper spelling out those laws in detail. Each man announced Mendel's discoveries and his own work as confirmation of them. During the 35 years between Mendel's publication and the rediscovery of his laws, Mendel's paper was only cited three times. However, Mendel's laws have become the foundation of modern genetics, not just for plants, but for animals, humans, and all living organisms.

For thousands of years, farmers and herders have been selectively breeding their plants and animals to produce more useful hybrids. It was somewhat of a hit or miss process since the actual mechanisms governing inheritance were unknown. Knowledge of these genetic mechanisms finally came as a result of careful laboratory breeding experiments carried out over the last century and a half. Through the selective cross-breeding of common pea plants (*Pisum sativum*) over many generations, Mendel discovered that certain traits show up in offspring without any blending of parental characteristics. For instance, the pea flowers are either purple or white—intermediate colors do not exist in the offspring of cross-pollinated pea plants. Mendel selected seven traits that are easily recognized and apparently only occur in one of two forms. These traits are called Mendel's seven traits in common pea: (1) seed shape (round or wrinkled), (2) seed color (yellow or green), (3) flower color (purple or white), (4) pod shape (inflated or constricted), (5) pod color (yellow or green), (6) flower position (axil or terminal), and (7) stem length (long or short). Figure 2.2 shows Mendel's seven traits of common pea.

Mendel observed that these traits do not show up in offspring plants with intermediate forms. This observation was critically important because the leading theory in biology at the time was that inherited traits blend from generation to generation. Most of the leading scientists in the nineteenth century accepted this "blending theory." Charles Darwin proposed another equally wrong theory known as "pangenesis"—the whole of parental organisms participate to heredity. This held that hereditary "particles" in our bodies are affected by the things we do during our lifetime. These modified particles were thought to migrate via blood to the reproductive cells and subsequently passed to the next generation. This was essentially a variation of Lamarck's incorrect idea of the "inheritance of acquired characteristics."

© Springer Nature Switzerland AG 2022
S. Xu, *Quantitative Genetics*, https://doi.org/10.1007/978-3-030-83940-6_2

**Fig. 2.1** Gregor Mendel (1822–1884) (Wikipedia contributor, https://en.wikipedia.org/wiki/Gregor_Mendel)

## 2.2   Mendel's Laws of Inheritance

Now let us go back to Mendel's experiments. Mendel picked common garden pea plants for the focus of his research because they can be grown easily in large numbers and their reproduction can be manipulated by human. Pea plants have both male and female reproductive organs. As a result, they can either self-pollinate themselves or cross-pollinate with another plant. In his experiments, Mendel was able to selectively cross-pollinate purebred plants with particular traits and observe the outcome over many generations. This was the basis for his conclusions about the nature of genetic inheritance. Mendel performed seven hybrid experiments, one for each trait. His findings from the seven experiments are summarized in Table 2.1.

In Table 2.1, $F_1$ stands for *Filial* 1, the first filial generation seeds resulting from a cross of distinctly different parental types. $F_2$ stands for *Filial* 2, the second filial generation of the cross. An $F_2$ is resulted from self- or cross-pollination of an $F_1$. Each parent used to initiate the crosses was "pure bred" or "bred true," i.e., it consistently produced the same phenotype for many generations (forever). Mendel made the following observations: (1) All $F_1$ plants from the same cross showed the phenotype of only one parent, not both. (2) The phenotype of the other parent reappeared in the $F_2$ plants. (3) The phenotype ratios were close to 3:1 in all seven experiments. This 3:1 ratio was very peculiar and Mendel then proposed a theory to explain it.

Before we introduce Mendel's theory, let us use a statistical method to test whether the 3:1 ratio is true or not. The test is called the Chi-square test. Mendel's original study did not include such a test. Chi-square test was invented by Pearson (Pearson 1900) 35 years after Mendel's publication of his laws. Consider the seed shape trait, for example, the round seed to wrinkled seed ratio is 2.96:1, not exactly 3:1. Under the 3:1 ratio, the expected numbers of round and wrinkled seeds are $E_R = 0.75 \times (5474 + 1850) = 5493$ and $E_W = 0.25 \times (5474 + 1850) = 1831$, respectively. The Chi-square test statistic is

$$\chi^2 = \frac{(O_R - E_R)^2}{E_R} + \frac{(O_W - E_W)^2}{E_W} = \frac{(5474 - 5493)^2}{5493} + \frac{(1850 - 1831)^2}{1831} = 0.2629 \tag{1}$$

where $O_R = 5474$ and $O_W = 1850$ are the observed numbers of the round and wrinkled seeds. The calculated Chi-square test statistic is 0.2629, which is less than 3.84, the critical value of Chi-square distribution with one degree of freedom $(1 - \Pr(\chi_1^2 \leq 3.84) = 0.05)$. The upper tail probability is $\Pr(\chi_1^2 > 0.2629) = 0.6081$, which is larger than 0.05. The

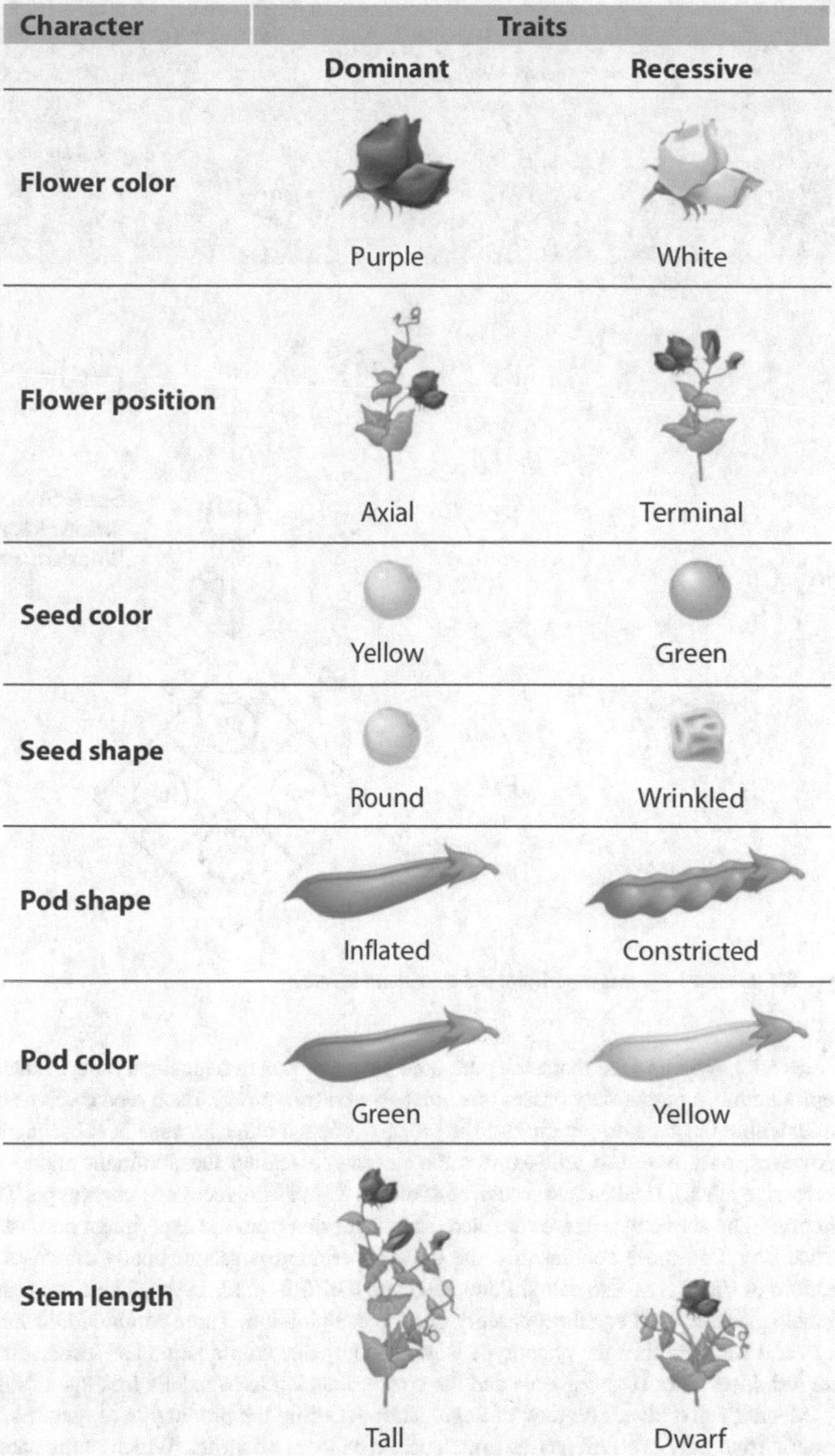

**Fig. 2.2** Mendel's seven traits of common pea. (Figure adopted from Reece, Jane B., Taylor, Martha R., Simon, Eric J. and Dickey, Jean L. 2011. Campbell Biology: Concepts & Connections. Seventh Edition, Benjamin Cummings, San Francisco, USA)

| Character | Traits | |
|---|---|---|
| | Dominant | Recessive |
| Flower color | Purple | White |
| Flower position | Axial | Terminal |
| Seed color | Yellow | Green |
| Seed shape | Round | Wrinkled |
| Pod shape | Inflated | Constricted |
| Pod color | Green | Yellow |
| Stem length | Tall | Dwarf |

conclusion is that this ratio (2.96:1) is not significantly different from the theoretical 3:1 ratio. The Chi-square test statistics of the remaining six experiments are all less than 3.84 and thus all the observed ratios are claimed to be 3:1.

You can write an R program to calculate the Chi-square tests for all the seven traits. Alternatively, you may use the excel spreadsheet to perform the Chi-square test manually using the formula given in Eq. (1).

**Table 2.1** Mendel's seven cross experiments in common pea (Pisum sativum) led to the 3:1 ratio

| Experiment | Parental cross | F₁ Phenotype | F₂ Phenotype | F₂ ratio | Chi-square |
|---|---|---|---|---|---|
| 1. Seed shape | Round × Wrinkled | Round | 5474R:1850 W | 2.96:1 | 0.2629 |
| 2. Seed color | Yellow × Green | Yellow | 6022Y: 2001G | 3.01:1 | 0.0150 |
| 3. Flower color | Purple × White | Purple | 705P:224 W | 3.15:1 | 0.3907 |
| 4. Pod shape | Inflated × Constricted | Inflated | 882I:299C | 2.95:1 | 0.0635 |
| 5. Pod color | Green × Yellow | Green | 428G:152Y | 2.82:1 | 0.4506 |
| 6. Flower position | Axial × Terminal | Axial | 651A:207 T | 3.14:1 | 0.3497 |
| 7. Stem length | Long × Short | Long | 787 L:277S | 2.84:1 | 0.6065 |

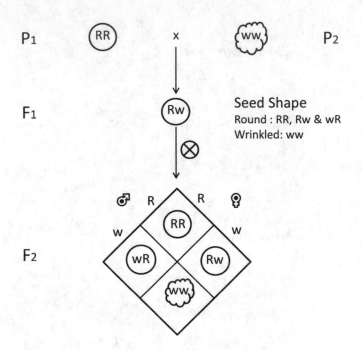

**Fig. 2.3** Mendel's F₂ cross experiment and the Punnett square

Mendel hypothesized that each pure bred parent carries two identical units or factors (they are called alleles now). During reproduction, a parent only passes one unit to a seed (offspring). Each seed receives two units, one from each parent. The two units within the seed do not change the property of each other because in subsequent generations the two units can separate. However, only one unit will express the phenotype (called the dominant allele) and the other allele will be suppressed (recessive allele). It takes two recessive alleles to express the recessive phenotype. This is clearly in contrast to the "blending theory." The above hypotheses are also the conclusions from the experiment because they allowed Mendel to explain the 3:1 ratio. Based on these conclusions, the cross experiment is schematically drawn as shown in Fig. 2.3. The diagram at the bottom of Fig. 2.3 is also called Punnett square (Griffiths et al. 1996). There are four possible combinations of the male and female gametes with equal probability of each combination. Three combinations have seeds with at least one R factor (RR, Rw, and wR) and thus the phenotype is round. Only one combination has seeds with both w factors (ww) and thus the seeds are wrinkled. These experiments and the conclusions led to Mendel's first law of inheritance.

Mendel's first law: **The** Law of Segregation—During the production of gametes, the pair of units (factors) of each parent separate and only one unit passes from a parent on to an offspring. Which of the parent's units inherited is merely a matter of chance as shown in Fig. 2.4).

To further validate the principle of segregation, Mendel continued to observe the next generation after F₂, i.e., the F₃ generation (Mendel called it the second generation from the hybrid). Mendel found that the plants with recessive character from the F₂ (ww) did not further vary in F₃; they remained constant in their offspring. However, among those who possess the dominant character in F₂ (RR or Rw or wR), two-thirds yielded offspring which display the dominant and recessive characters in the proportion of three to one (Rw and wR) while only one-third remained with the dominant character (RR). In other

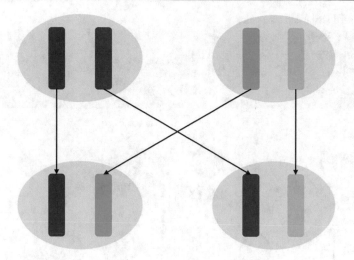

**Fig. 2.4** Segregation of alleles in the production of sex cells

**Table 2.2** Results of the next-generation Mendel's seven cross experiments in common pea (*Pisum sativum*) showed the 2:1 ratio

| Experiment | Dominant $F_2$ | Segregation | Constant | $F_2$ ratio | Chi-square |
| --- | --- | --- | --- | --- | --- |
| 1. Seed shape | 565 | 372 | 193 | 1.93:1 | 0.1734 |
| 2. Seed color | 519 | 353 | 166 | 2.13:1 | 0.4248 |
| 3. Flower color | 100 | 64 | 36 | 1.78:1 | 0.3200 |
| 4. Pod shape | 100 | 71 | 29 | 2.45:1 | 0.8450 |
| 5. Pod color | 100 | 60 | 40 | 1.50:1 | 2.0000 |
| 6. Flower position | 100 | 67 | 33 | 2.03:1 | 0.0050 |
| 7. Stem length | 100 | 72 | 28 | 2.57:1 | 1.2800 |

words, the heterozygote dominance seeds (Rw and wR) and the "true bred" dominant seeds (RR) should have a ratio of 2:1. Mendel's results of the $F_3$ analysis from the $F_2$ seeds with the dominance character showed the 2:1 ratio for all the seven traits. The results for the 2:1 ratio analysis are given in Table 2.2.

Mendel could have used a test cross to confirm or validate his first law. The test cross is a cross between the $F_1$ hybrid with the recessive parent. If the law is true, he should be able to observe a 1:1 ratio for a trait in the progeny. Take the seed shape (round vs. wrinkled), for example, the $F_1$ hybrid has a genotype of Rw and the recessive parent has a genotype of ww. The progeny should have two types, round (Rw) and wrinkled (ww), with an equal probability. This test cross is much easier than the $F_3$ experiments conducted by Mendel. Gregor Mendel was a genius, but he was still a human being. He did not take the easier test cross experiment but chose a complicated $F_3$ experiment. The test cross was postulated by geneticists half century later after Mendel's laws were rediscovered. From Table 2.2, we see that the Chi-square test statistics are all smaller than 3.84, the critical value at 0.05 Type I error. Therefore, the 2:1 ratio is true. For experiments 3–7, the sample size was 100, which is small and we expect to see relatively large Chi-square test statistics. The largest departure from the 2:1 ratio appeared in experiment 5 with an observed ratio of 1.5:1. The Chi-square test statistic is

$$\chi^2 = \frac{(60 - 66.67)^2}{66.67} + \frac{(40 - 33.33)^2}{33.33} = 2.00$$

Had Mendel performed the Chi-square test, he would not have to repeat this experiment to obtain a result closer to the 2:1 ratio. He could have easily convinced the reviewers that the large departure was due to the small sample size. Instead, he repeated the experiment for the pod color trait and observed 65 segregations vs. 35 constants with a ratio of 1.86:1. He was very satisfied with this ratio. You can perform your own Chi-square test to see whether the observed 65:35 ratio from the repeated experiment is indeed not significantly different from the 2:1 ratio. You can also calculate the *p*-value of this test.

**Table 2.3** Joint observations of seed shape and seed color

| Outcome | Shape | Color | Observed | Expected | Ratio |
|---|---|---|---|---|---|
| 1 | Round | Yellow | 315 | 312.75 | 9 |
| 2 | Wrinkled | Yellow | 101 | 104.25 | 3 |
| 3 | Round | Green | 108 | 104.25 | 3 |
| 4 | Wrinkled | Green | 32 | 34.75 | 1 |

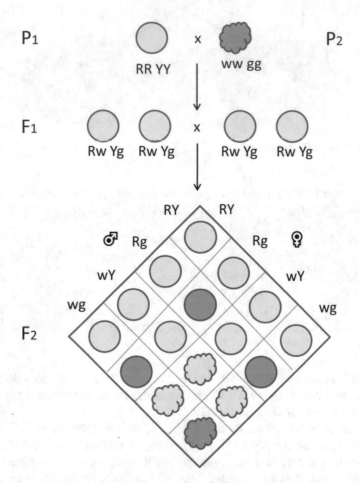

**Fig. 2.5** $F_2$ mating design for two traits (seed shape and seed color). Each of the four types of gametes has an equal probability of 1/4 and each of the 16 two-trait combinations has a probability of $(1/4)^2 = 1/16$

Mendel did not stop here with the 3:1 ratio for a single character in his $F_2$ cross. He observed a pair of characters, say seed color and seed shape, simultaneously in the $F_2$ population. This led to Mendel's second law of inheritance: the principle of independent assortment. When he observed two traits simultaneously, he found that there are four outcomes with ratios close to 9:3:3:1. Take the seed shape and seed color, for example. Among a total of 556 seeds yielded by 15 plants, he found the following distribution of the seeds (Table 2.3): The 9:3:3:1 ratio is explained by independent assortment of the two traits. Schematically, the two-trait $F_2$ mating design is shown in Fig. 2.5. Let us perform a Chi-square test to see whether the data to follow the 9:3:3:1 ratio. The test statistic is 0.47, smaller than 7.8147 (the critical value of $\chi_3^2$ distribution). The $p$-value is $\Pr(\chi_3^2 > 0.47) = 0.9254$, larger than 0.05. Therefore, the data do follow the 9:3:3:1 ratio. If we ignore the color and just focus on the shape, the Round to Wrinkled ratio remains 12:4 = 3:1. Likewise, the yellow to green color ratio remains 12:4 = 3:1. The two traits segregate independently. This led to Mendel's second law.

Mendel's second law: **The** Law of Independent Assortment—different pairs of units are passed to offspring independently of each other. The result is that new combinations of units present in neither parent are possible. For example, a pea plant's inheritance of the ability to produce round seeds instead of wrinkled ones does not make it more likely inherit the ability to produce yellow seeds in contrast to green ones. In other words, seed shape and seed color are completely independent. The principle of independent assortment explains why the human inheritance of a particular eye color does not increase or decrease the likelihood of having six fingers on each hand. Today, we know this is due to the fact that the genes for independently assorted traits are located on different chromosomes.

The following paragraph lists some common terminologies related to Mendelian genetics and quantitative genetics. They are fundamentally important and thus are given here before they are formally introduced in the main text.

## 2.3  Vocabulary

*Gene*

*Gene*: A piece of DNA involved in producing a polypeptide chain. Sometimes, it is used as a synonym for allele. The term "gene" is an ambiguous term and it has been used in so many different contexts that it is not very useful for our purposes.

*Allele*:

One of several alternative forms of a gene occupying a given position of a chromosome. Mendel called it a unit or a factor. With this convention, a diploid individual may be said to have two alleles at a particular locus, one from the mother and one from the father.

*Locus*:

The position on a chromosome at which an allele resides. An allele is just a piece of DNA at that place. A locus is a template for an allele. The word "loci" is a plural form of locus, e.g., one locus and two loci.

*Genotype*:

The genetic constitution of an organism. The genotype of an individual at a locus is the type of combination of the alleles. If $A_1$ and $A_2$ are two alleles in a population, there will be three genotypes in the population: two homozygous genotypes, $A_1A_1$ and $A_2A_2$, and one heterozygous genotype, $A_1A_2$.

*Phenotype*:

The appearance of an organism or a trait determined by the genotype. Later on, we will learn that the phenotype may also be affected by environment. It is a result of the interaction between the genotype and the environment.

*Qualitative character*:

Numerical measurement is not possible. There is no natural order of ranking. For example, the seed coat color, yellow and green, of Mendel's experiments.

*Quantitative character*:

Trait value differs by degree rather than by kind. Mendel defined such a trait as one that differs in "more or less" nature. A quantitative trait is measurable or countable. There are two kinds of quantitative characters: (a) continuous, all values in some range are possible, e.g., body weight; (b) discrete, some values are not observable, e.g., litter size (number of progenies) can only be counted as integer, e.g., 1, 2, …, $n$. It is impossible to have a value like 1.5. The number of categories in class (b) must be large in order to qualify the trait as a quantitative trait.

*Quantitative genetics*:

Also called Biometrical Genetics, it is the study of the genetic mechanism of quantitative characters or quantitative traits.

*Population genetics*:

The study of the composition of natural or artificial populations (e.g., gene and genotype frequencies) for qualitative characters, the forces affecting this composition, and implications with respect to evolution and speciation.

## 2.4  Departure from Mendelian Ratio

What we have learned from the Mendelian experiments is that the genotype exclusively determines the phenotype. For example, the RR, Rw, and wR genotypes show the round seed phenotype and the ww genotype shows the wrinkled seed phenotype. There is no environmental effect. Whether you plant the seed in the field or in a green house, the genotype ww will never produce the round phenotype. We now show that if the trait is affected by environment, we may see a ratio deviating from 3:1.

**Table 2.4** Mendelian inheritance plus environmental effects (conditional probability table)

| Genotype | Black | White | Marginal |
|---|---|---|---|
| $AA$ | $\alpha$ | $1 - \alpha$ | 1/4 |
| $Aa$ | $\alpha$ | $1 - \alpha$ | 2/4 |
| $aa$ | $1 - \beta$ | $\beta$ | 1/4 |
| Marginal | $\frac{3}{4}\alpha + \frac{1}{4}(1 - \beta)$ | $\frac{3}{4}(1 - \alpha) + \frac{1}{4}\beta$ | |

**Table 2.5** Mendelian inheritance plus environmental effects (joint probability table)

| Genotype | Black | White | Marginal |
|---|---|---|---|
| $AA$ | $\frac{1}{4}\alpha$ | $\frac{1}{4}(1 - \alpha)$ | $\frac{1}{4}$ |
| $Aa$ | $\frac{2}{4}\alpha$ | $\frac{2}{4}(1 - \alpha)$ | $\frac{2}{4}$ |
| $aa$ | $\frac{1}{4}(1 - \beta)$ | $\frac{1}{4}\beta$ | $\frac{1}{4}$ |
| Marginal | $\frac{3}{4}\alpha + \frac{1}{4}(1 - \beta)$ | $\frac{3}{4}(1 - \alpha) + \frac{1}{4}\beta$ | |

The black to white ratio is

$$\left[\frac{3}{4}\alpha + \frac{1}{4}(1 - \beta)\right] : \left[\frac{3}{4}(1 - \alpha) + \frac{1}{4}\beta\right]$$

which can be any ratio, depending on the values of $\alpha$ and $\beta$. The 3:1 ratio occurs only if $\alpha = \beta = 1$. Suppose that the "black" phenotype is the reference phenotype, $\alpha$ is called the "penetrance" of genotype $AA$ or $Aa$ and $1 - \beta$ is the "penetrance" of genotype $aa$. So, penetrance is defined as the probability that a genotype shows the reference phenotype. For example, if $\alpha = 0.9$ and $\beta = 0.85$, then the $AA$ and $Aa$ genotypes will show the black phenotype 90% of the time and the white phenotype 10% of the time. The $aa$ genotype will show a black phenotype 15% of the time and the white phenotype 85% of the time. The proportion of black individuals is $0.75 \times 0.90 + 0.25 \times (1 - 0.85) = 0.7125$ and the proportion of the white individuals is $0.75 \times (1 - 0.90) + 0.25 \times 0.85 = 0.2875$, which yields a 2.4783 : 1 ratio. The penetrance of $AA$ and $Aa$ are all equal to $\alpha$, representing dominance of $A$ over $a$. In reality, they can be different, a situation called co-dominance.

Table 2.4 is a $3 \times 2$ table (excluding the last row and the last column), where 3 is the number of genotypes and 2 is the number of phenotypes. In general, we may have $p$ genotypes and $q$ phenotypes. This $3 \times 2$ table is called the conditional probability table. The last column gives the marginal probabilities of the three genotypes. First, we need to construct a $3 \times 2$ joint probability table whose elements equal the conditional probabilities multiplied by the corresponding marginal probabilities of the genotypes (Table 2.5). The marginal probability of each phenotype equals the sum of all elements of the corresponding column of the joint probability table.

How do we tell whether a table is a conditional probability table or a joint probability table? The rules are: (1) If the sum of all elements of each row equals one, then the table is a conditional probability table; (2) If the sum of all elements of the entire table equals one, then the table is a joint probability table. The last row and last column represent the marginal probabilities and should not be confused with the joint probabilities.

## 2.5  Two Loci Without Environmental Effects

We now evaluate a trait controlled by two independent loci, A and B, where $A$ and $a$ are the two alleles of locus A and $B$ and $b$ are the two alleles of locus B. The mating design and the $F_2$ genotypes are given in Fig. 2.6. If we define a trait (genotype) value as the number of capital letters in the genotype. The distribution of the trait value is shown below:

| Trait value | 0 | 1 | 2 | 3 | 4 |
|---|---|---|---|---|---|
| Frequency | 1/16 | 4/16 | 6/16 | 4/16 | 1/16 |

The distribution is graphically presented in Fig. 2.7, which is roughly a "bell-shaped" distribution.

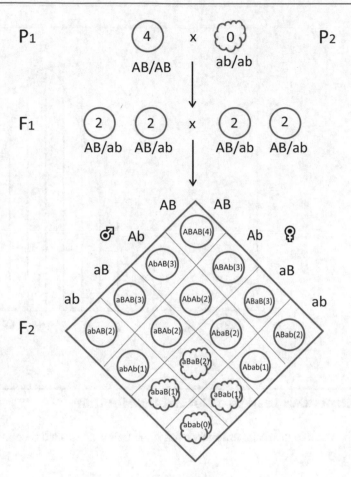

**Fig. 2.6** Mating design and distribution of the $F_2$ genotypes. Genotypic values (number of capital letters) are shown in parentheses within the circles

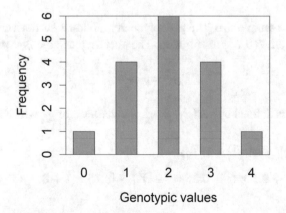

**Fig. 2.7** The distribution of the trait value for two independent loci

## 2.6   Multiple Loci with or Without Environmental Effects

For more than two loci, the distribution is close to a continuous distribution (Fig. 2.8). The observed continuous distribution of a quantitative trait is caused by (1) multiple loci, (2) environmental effect, or (3) both. The second factor is very important because even if a trait is controlled by a single locus, it can still show a normal distribution provided that the environmental effect is large compared to the genetic variance.

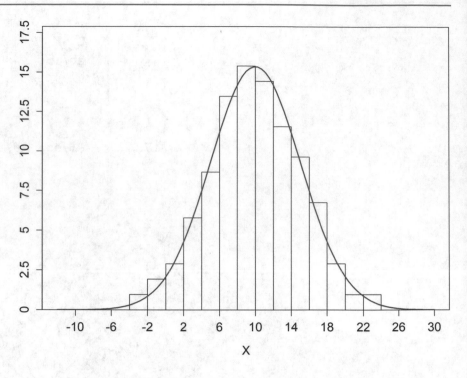

**Fig. 2.8** Distribution of a trait controlled by multiple loci

## 2.7   Environmental Errors Can Lead to a Normal Distribution

Environmental errors can cause a trait to distribute normally even if the trait is controlled by only a single locus. Let us define the genotypic value for a plant by

$$
g_j = \begin{cases} +1 \text{ for } AA \text{ with probability } 0.25 \\ \ \ \ 0 \text{ for } Aa \text{ with probability } 0.50 \\ -1 \text{ for } aa \text{ with probability } 0.25 \end{cases}
$$

where $j$ indexes the plant and $n$ is the sample size. The phenotype of individual $j$ is determined by the genotypic value plus an environmental error $e_j$ that follows an $N(0, \sigma_e^2)$ distribution. The phenotypic value is expressed by

$$
y_j = g_j + e_j
$$

Theoretically, the genetic variance is defined by $\mathrm{var}(g_j) = \sigma_g^2$, the variance of $g_j$, which has a form like

$$
\begin{aligned}
\sigma_g^2 &= \mathrm{E}[g - \mathrm{E}(g)]^2 = \mathrm{E}(g^2) - \mathrm{E}^2(g) \\
&= \left[ 0.25 \times 1^2 + 0.5 \times 0^2 + 0.25 \times (-1)^2 \right] - \left[ 0.25 \times 1 + 0.5 \times 0 + 0.25 \times (-1) \right]^2 \\
&= 0.5
\end{aligned}
$$

The variance of the trait value is denoted by $\mathrm{var}(y_j) = \sigma_y^2$, which is

$$
\begin{aligned}
\mathrm{var}(y_j) &= \mathrm{var}(g_j) + \mathrm{var}(e_j) \\
\sigma_y^2 &= \sigma_g^2 + \sigma_e^2
\end{aligned}
$$

The proportion of the trait contributed by the genetic effect (called heritability) is defined by

**Fig. 2.9** Trait value distributions under three different proportions of the genetic variances contributed to the phenotypic variance. The data were simulated from 10,000 individuals

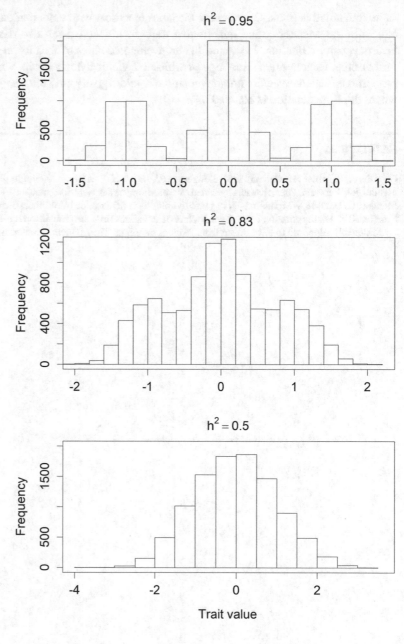

$$h^2 = \frac{\sigma_g^2}{\sigma_y^2} = \frac{\sigma_g^2}{\sigma_g^2 + \sigma_e^2} = \frac{0.5}{0.5 + \sigma_e^2} = \frac{1}{1 + 2\sigma_e^2}$$

When $\sigma_e^2$ takes values of 0.025, 0.1, and 0.5, the corresponding $h^2$ will be 0.95, 0.83, and 0.50. The phenotypic distributions under the three scenarios are depicted in Fig. 2.9 (drawn from simulated data with $n = 10000$). We can see that when $h^2 = 0.95$, the environmental proportion $e^2 = 1 - h^2 = 1 - 0.95 = 0.05$ is very small and we can clearly distinguish the three genotypes based on the distribution (the top panel of Fig. 2.9). As we reduce the genetic proportion to $h^2 = 0.83$, the three genotype classes are blurred, but we can still tell them apart (the panel in the middle). When $h^2 = 0.50$, the phenotypic distribution becomes completely normal and we can no longer tell the three genotypes apart (the panel at the bottom).

Before we exit from the review on Mendelian Genetics, it is interesting to mention that Mendel purposely selected traits that had demonstrated a clearly discrete distribution. He said "Some of the characters noted do not permit of a sharp and certain separation, since the difference is of a "more or less" nature (Mcndcl 1866; Mendel and Bateson 1901), which is often difficult to define. Such characters could not be utilized for the separate experiments; these could only be applied to characters which stand out clearly and definitely in the plants." The traits that differ in a "more or less" nature defined by Mendel are now called quantitative traits. Quantitative Genetics is to deal with traits that cannot be studied using the Mendelian approaches. Not only did Mendel define quantitative traits in his study, but also described a phenomenon in population genetics— heterozygosity reduction. He stated "If an average equality of fertility in all plants in all generations be assumed, and if, furthermore, each hybrid forms seed of which 1/2 yields hybrids again, while the 1/2 is constant to both characters in equal proportions." He derived the homozygosity to heterozygosity ratio after $n$ generations of continuous selfing to be $(2^n - 1) : 1$, where the $F_2$ generation is counted as $n = 1$.

## References

Griffiths AJF, Miller JH, Gebhart WM, Lewontin RC, Suzuki DT. An introduction to genetic analysis. New York: W.H. Freeman & Co.; 1996.

Mendel JG. Versuche über pflanzen-hybriden. Verhandlungen des Naturforschenden Vereines in Brünn. 1866;4:3–47.

Mendel JG, Bateson W. Experiments in plant hybridization (Translated by William Bateson in 1901). J R Hortic Soc. 1901;16:1–32.

Pearson K. X. On the criterion that a given system of deviations from the probable in the case of a correlated system of variables is such that it can be reasonably supposed to have arisen from random sampling. The London, Edinburgh, Dublin Philos Mag J Sci. 1900;50:157–75.

# Basic Concept of Population Genetics

**3**

Population genetics is another branch of genetics. It mainly deals with the distribution of genes and genotypes in a population. It also deals with changes of genetic composition of populations under common evolutionary forces such as natural selection and genetic drift. The primary founders of population genetics were Sewall Wright (1889–1988), J. B. S. Haldane (1892–1964), and Ronald Fisher (1890–1962), who also laid the foundations for quantitative genetics. Population genetics is traditionally a highly mathematical discipline. Modern population genetics, however, encompasses theoretical, lab, and field work. Some material of this chapter was adopted from Falconer and Mackay (Falconer and Mackay 1996).

## 3.1 Gene Frequencies and Genotype Frequencies

Consider an autosomal locus of a diploid organism with two alleles, $A_1$ and $A_2$. There are three possible genotypes, $A_1A_1$, $A_1A_2$, and $A_2A_2$. Let $N$ be the total number of individuals sampled from the population among which $N_{11}$ are of genotype $A_1A_1$, $N_{12}$ of $A_1A_2$ and $N_{22}$ of $A_2A_2$. The sample size is $N = N_{11} + N_{12} + N_{22}$. The genotype frequencies are defined and estimated in a way given in Table 3.1. For example, the frequency of genotype $A_1A_1$ is defined as the proportion of individuals carrying this genotype in the entire population.

The gene frequencies (also called allele frequencies) are defined and estimated from Table 3.2. The number 2 appearing in the last column of the table is due to the fact that each homozygote carries two copies of the same alleles in diploid organisms. Note that $N_1 = 2N_{11} + N_{12}$ and $N_2 = 2N_{22} + N_{12}$. Therefore, the total number of alleles in the population is

$$N_1 + N_2 = 2(N_{11} + N_{12} + N_{22}) = 2N$$

which is twice the sample size. The true genotype and gene frequencies are never known (they are the theoretical values), but they can be estimated from the sample. For example, the estimated frequency for genotype $A_1A_1$ is denoted by $\widehat{P}$ and the estimated allele frequency for $A_1$ is denoted by $\widehat{p}$. A parameter with a hat means an estimate of the parameter.

## 3.2 Hardy-Weinberg Equilibrium

In a large random mating population with no selection, no mutation, and no migration, the gene frequencies and genotype frequencies are constant from generation to generation (the Hardy-Weinberg law). The assumption of large random mating population means that there is no genetic drift (another evolutionary force), and there is no assortative mating (mating choice). Genetic drift is caused by small population size and the gene and genotype frequencies will fluctuate across generations for a finite sample size. Under HW equilibrium, there is a simple relationship between gene frequencies and genotype frequencies. The relationship follows the "square law." Hardy-Weinberg equilibrium can be reached at one generation after random mating. The square law can be mathematically expressed as

© Springer Nature Switzerland AG 2022
S. Xu, *Quantitative Genetics*, https://doi.org/10.1007/978-3-030-83940-6_3

**Table 3.1**  Genotypes, the counts, and the frequencies in a biallelic population

| Genotype | Count | Frequency | Estimate |
|----------|-------|-----------|----------|
| $A_1A_1$ | $N_{11}$ | $P$ | $\widehat{P} = N_{11}/N$ |
| $A_1A_2$ | $N_{12}$ | $H$ | $\widehat{H} = N_{12}/N$ |
| $A_2A_2$ | $N_{22}$ | $Q$ | $\widehat{Q} = N_{22}/N$ |

**Table 3.2**  Alleles, counts and frequencies in a biallelic population

| Gene | Count | Frequency | Estimate |
|------|-------|-----------|----------|
| $A_1$ | $N_1$ | $p$ | $\widehat{p} = (2N_{11} + N_{12})/(2N) = \widehat{P} + \widehat{H}/2$ |
| $A_2$ | $N_2$ | $q$ | $\widehat{q} = (2N_{22} + N_{12})/(2N) = \widehat{Q} + \widehat{H}/2$ |

**Table 3.3**  From gene frequencies to genotype frequencies

| | | Female gamete | |
|---|---|---|---|
| Male gamete | | $A_1(p)$ | $A_2(q)$ |
| $A_1(p)$ | | $A_1A_1(p^2)$ | $A_1A_2(pq)$ |
| $A_2(q)$ | | $A_2A_1(qp)$ | $A_2A_2(q^2)$ |

$$(p+q)^2 = p^2 + 2pq + q^2$$

Logically, it is interpreted by

$$(A_1 + A_2)^2 = A_1A_1 + 2A_1A_2 + A_2A_2$$

If $p$ is the allele frequency for $A_1$ and $q = 1 - p$ is the frequency of $A_2$, under the square law, $p^2$ is the frequency of genotype $A_1A_1$, $2pq$ is the frequency of genotype $A_1A_2$ and $q^2$ is the frequency of genotype $A_2A_2$.

### 3.2.1  Proof of the H-W Law

1. From gene frequencies to genotype frequencies

   If the allelic frequencies are $p$ and $q$ for the two alleles, under random mating, the genotype frequencies for the three genotypes are $p^2$, $2pq$, and $q^2$, respectively. This can be shown in Table 3.3.
   The square law is formed by the product of male and female gamete,

$$(p+q)(p+q) = (p+q)^2 = p^2 + 2pq + q^2$$

   Random mating means that the female and male gametes are united randomly (independently). Therefore, the $A_1A_1$ genotype frequency is the product of frequency of $A_1$ from the male and the frequency of $A_1$ from the female. The heterozygote has two forms, $A_1A_2$ and $A_2A_1$. Since we cannot tell the difference, we simply lump them together as $A_1A_2$. This explains why we have $2pq$ as the frequency of the heterozygote.

2. From genotype frequencies to gene frequencies
   Recall that gene frequencies are determined by genotype frequencies by the following relationship, regardless what the population status is,

$$p = P + H/2$$
$$q = Q + H/2$$

**Table 3.4** Different frequencies for male and female populations

| Male gamete | Female gamete $A_1(p_f)$ | $A_2(q_f)$ |
|---|---|---|
| $A_1(p_m)$ | $A_1A_1(p_mp_f)$ | $A_1A_2(p_mq_f)$ |
| $A_2(q_m)$ | $A_2A_1(q_mp_f)$ | $A_2A_2(q_mq_f)$ |

If the population is in HW equilibrium, we know that $P = p^2$, $H = 2pq$, and $Q = q^2$. We will show that the gene frequencies in the next generation remain $p$ and $q$ for the two alleles.

$$\text{Frequency for } A_1 : P + H/2 = p^2 + 2pq/2 = p^2 + pq = p(p+q) = p$$
$$\text{Frequency for } A_2 : Q + H/2 = q^2 + 2pq/2 = q^2 + pq = q(p+q) = q$$

Note that $p + q = 1$ always holds.

3. If a population is not in HW equilibrium, one generation of random mating suffices to make the population reach HW equilibrium.

Suppose that the male and female populations have different gene frequencies, H-W equilibrium can be reached in one generation of random mating (Table 3.4).

The male and female gene frequencies are $p_m$ and $p_f$, respectively, for allele $A_1$, and $q_m$ and $q_f$, respectively, for allele $A_2$. Note that $p_m + q_m = p_f + q_f = 1$. The frequency of $A_1$ in the offspring is $p = p_mp_f + \frac{1}{2}(p_mq_f + q_mp_f) = \frac{1}{2}(p_m + p_f)$. The frequency of $A_2$ in the offspring is $q = q_mq_f + \frac{1}{2}(p_mq_f + q_mp_f) = \frac{1}{2}(q_m + q_f)$. When the offspring population is randomly mated, the gene frequencies will remain at $p$ and $q$ because the frequencies of the males and females will be the same thereafter.

### 3.2.2 Applications of the H-W Law

1. Estimating frequency of the recessive allele

One application of the H-W law is to estimate frequency of a recessive allele. Recall that given the frequencies of the three genotypes, $P$, $H$, and $Q$, we can estimate the recessive allele frequency by $\widehat{q} = \widehat{Q} + \widehat{H}/2$. If a particular trait, e.g., a disease, is controlled by two recessive allele ($A_2A_2$), we can estimate $Q$ by counting the proportion of individuals with the recessive trait. However, the normal individuals (with the dominant phenotype) consist of both $A_1A_1$ and $A_1A_2$. We cannot tell the frequencies of the dominant homozygote and the heterozygote; rather, we only observe the normal individuals with frequency of $P + H$. Without specifically knowing $H$, we cannot use $\widehat{q} = \widehat{Q} + \widehat{H}/2$ to estimate $q$. However, if the population is in H-W equilibrium, we know the relationship $Q = q^2$. Since we already knew $\widehat{Q} = N_{22}/N$, the estimated $q$ is simply obtained by $\widehat{q} = \sqrt{\widehat{Q}}$. Do not use this formula to estimate gene frequencies unless the population is in HW equilibrium! This problem can be used to demonstrate a famous algorithm in solving parameters with a maximum likelihood method, called the expectation and maximization algorithm, i.e., the EM algorithm (Dempster et al. 1977). We observed $\widehat{Q}$ but not $H$ because it is confounded with $P$. Given the total sample size $N$, if we know the allele frequency, say $q^{(0)}$ and $p^{(0)} = 1 - q^{(0)}$, we can calculate the expected value of $H$ under the HW law, which is $\widehat{H} = 2p^{(0)}q^{(0)}$. Therefore, the allele frequency can be updated with

$$q^{(1)} = \widehat{Q} + \widehat{H}/2 = \widehat{Q} + p^{(0)}q^{(0)}$$

In general, let $q^{(t)}$ be the frequency at the $t$th iteration, the frequency at iteration $t + 1$ is

**Table 3.5** Iteration process of the EM algorithm

| Iteration (t) | Estimate ($q^{(t)}$) | Error ($\varepsilon^{(t)}$) |
|---|---|---|
| 0 | 0.500000 | 1.00E+03 |
| 1 | 0.478395 | 2.16E−02 |
| 2 | 0.477928 | 4.67E−04 |
| 3 | 0.477908 | 2.04E−05 |
| 4 | 0.477907 | 9.00E−07 |
| 5 | 0.477907 | 3.98E−08 |
| 6 | 0.477907 | 1.76E−09 |

$$q^{(t+1)} = \widehat{Q} + p^{(t)}q^{(t)}$$

When the iteration process converges, i.e., $\varepsilon = |q^{(T+1)} - q^{(T)}|$ is sufficiently small, the maximum likelihood estimate of $q$ takes $q^{(T)}$, where $T$ is the number of iterations. The process may take just a few iterations to converge. Note that an explicit solution is already known, $\widehat{q} = \sqrt{\widehat{Q}}$, and there is no need to use the EM algorithm. But we may use this problem to teach the EM algorithm. For example, if a population has a proportion of $\widehat{Q} = 0.2283951$ individuals with a recessive disease, the estimated frequency of the recessive allele is

$$\widehat{q} = \sqrt{\widehat{Q}} = \sqrt{0.2283951} = 0.477907$$

If we use the EM algorithm and take $q^{(0)} = 0.5$ as the initial value, the iteration equation is

$$q^{(t+1)} = \widehat{Q} + p^{(t)}q^{(t)} = 0.2283951 + p^{(t)}q^{(t)}$$

Table 3.5 shows the iteration history and the process took six iterations to converge to a criterion of $\varepsilon < 10^{-8}$.

2. Calculate the frequency of carriers

   Carriers are individuals who look normal but carry one recessive allele, i.e., $Aa$. We cannot distinguish $Aa$ from $AA$ because of dominance. The frequency of carriers is defined as the frequency of heterozygote among NORMAL individuals. It is not the heterozygote frequency $H$. Whenever we mention the word carrier, we are actually focusing on the normal individuals only, excluding the affected individuals. Verbally, the frequency of carriers is defined as the conditional frequency of heterozygote among the normal individuals. It is a conditional frequency. Mathematically, we define the frequency of normal individuals by $P + H$ and the frequency of heterozygotes by $H$. Therefore, the frequency of carriers is

$$f = \frac{H}{P+H} = \frac{2pq}{p^2 + 2pq} = \frac{2(1-q)q}{(1-q)^2 + 2(1-q)q} = \frac{2q}{1+q}$$

Again, a carrier is a heterozygote, but the frequency of carriers is not the frequency of heterozygotes.

### 3.2.3 Test of Hardy-Weinberg Equilibrium

1. Chi-square test for goodness of fit

   The following example shows the MN blood group frequencies with two alleles (M and N) and three genotypes (MM, MN, and NN). This is a Chi-square goodness of fit test with three categories. The important difference between this Chi-square test and the usual Chi-square test with given probabilities is that, here, we have to estimate the HW predicted genotype frequencies.

First, we need to estimate the allele frequencies and then use the estimated allele frequencies to calculate the expected genotype frequencies using the HW law. Let $p$ be the frequency of allele M and $q = 1 - p$ be the frequency of allele N. The estimated allele frequencies are (using the rule defined early)

$$\widehat{p} = \frac{2 \times 233 + 385}{2 \times 747} = \frac{233}{747} + \frac{385}{2 \times 747} = 0.5696$$
$$\widehat{q} = \frac{2 \times 129 + 385}{2 \times 747} = \frac{129}{747} + \frac{385}{2 \times 747} = 0.4304$$

The expected number for MM is $\widehat{p}^2 \times 747 = 242.36$. The expected number for MN is $2\widehat{p}\widehat{q} \times 747 = 366.26$ and the expected number for NN is $\widehat{q}^2 \times 747 = 138.38$. The Chi-square test statistic for the goodness of fit with three genotypes is

$$\chi^2 = \sum \frac{(O - E)^2}{E} = \frac{(233 - 242.36)^2}{242.36} + \frac{(385 - 366.26)^2}{366.26} + \frac{(129 - 138.38)^2}{138.38} = 1.96$$

which is smaller than $\chi^2_{.95,1} = 3.84$ (the 95% of the Chi-square one distribution). The test is not significant and thus the population is in Hardy-Weinberg equilibrium. Note that the symbol O stands for the observed number and E stands for the expected number. The expected numbers are given in the second row of Table 3.6. One would think that the degree of freedom should be $3 - 1 = 2$, but it is not. The actual degree of freedom is one because another degree of freedom has been lost by estimating the allele frequency using the data.

2. The 2×2 contingency table association test

Let us construct the following $2 \times 2$ contingency table for the observed genotype counts (Table 3.7). The columns and rows represent the male and female gametes, respectively. The association test is to test whether the male and female gametes are independent or not. Independence of the female and male gametes means random mating and the population is in HW equilibrium. If the hypothesis is rejected, the population is not in HW equilibrium. The allele frequencies are

$$\widehat{p} = \frac{233 + \frac{1}{2} \times 385}{747} = 0.5696 \quad \text{and} \quad \widehat{q} = \frac{129 + \frac{1}{2} \times 385}{747} = 0.4304$$

The number of heterozygotes is 385, which is split evenly for the two forms (MN and NM). Therefore, the observed cell count for MN is $0.5 \times 385 = 192.5$ and the count for NM is also 192.5. The expected counts are calculated as follows:

The expected number for MM is $\widehat{p}^2 \times 747 = 242.36$
The expected number for MN is $\widehat{p}\widehat{q} \times 747 = 183.13$
The expected number for NM is $\widehat{p}\widehat{q} \times 747 = 183.13$
The expected number for MM is $\widehat{q}^2 \times 747 = 138.38$

The Chi-square test statistic for association is

Table 3.6 The MN blood group example for testing Hardy-Weinberg equilibrium

| | Genotype | | | | Gene frequency | |
|---|---|---|---|---|---|---|
| | MM | MN | NN | Total | $M$ | $N$ |
| Observed | 233 | 385 | 129 | 747 | 0.5696 | 0.4304 |
| Expected | 242.36 | 366.26 | 138.38 | 747 | | |

Table 3.7 The $2 \times 2$ contingency table for the genotype frequencies

| ♀\♂ | $M$ | $N$ | Marginal |
|---|---|---|---|
| M | MM 233 (242.36) | MN ½ × 385(183.13) | $\widehat{p} = 0.5696$ |
| N | NM ½ × 385(183.13) | NN 129 (138.38) | $\widehat{q} = 0.4304$ |
| Marginal | $\widehat{p} = 0.569$ | $\widehat{q} = 0.430$ | 747 |

$$\chi^2 = \frac{(233 - 242.36)^2}{242.36} + 2\frac{(0.5 \times 385 - 0.5 \times 366.26)^2}{0.5 \times 366.26} + \frac{(129 - 138.38)^2}{138.38}$$
$$= \frac{(233 - 242.36)^2}{242.36} + \frac{(385 - 366.26)^2}{366.26} + \frac{(129 - 138.38)^2}{138.38} = 1.96$$

which is smaller than $\chi^2_{0.95,1} = 3.84$ and thus we cannot reject the null hypothesis that the population is in HW equilibrium. The $p$-value corresponding to $\chi^2 = 1.96$ is $p = 1 - \Pr(\chi^2_1 < 1.96) = 0.1615$, greater than 0.05. Note that the degree of freedom for a $2 \times 2$ contingency table is $(2 - 1)(2 - 1) = 1$. This is a more intuitive way to explain the one degree of freedom test for HW equilibrium.

3. Yates' continuity correction for $2 \times 2$ contingency table test

   Using the Chi-squared distribution to interpret Pearson's Chi-square statistic requires one to assume that the discrete probability of observed binomial frequencies in the table can be approximated by the continuous Chi-square distribution. This assumption is not quite correct and it is subject to some error. To reduce the error in the approximation, Frank Yates, an English statistician, suggested a correction for continuity that adjusts the formula for *Pearson's Chi-square test* by subtracting 0.5 from the difference between each observed value and its expected value in a $2 \times 2$ contingency table. This reduces the Chi-square value and thus increases its p-value.

$$\chi^2_{\text{Yates}} = \sum \frac{(|O - E| - 0.5)^2}{E}$$
$$= \frac{(|233 - 242.36| - 0.5)^2}{242.36} + 2 \times \frac{(|0.5 \times 385 - 0.5 \times 366.26| - 0.5)^2}{0.5 \times 366.26} + \frac{(|129 - 138.38| - 0.5)^2}{138.38}$$
$$= 1.7530$$

which is smaller than $\chi^2_{.95,1} = 3.84$. The corresponding $p$-value is

$$p = 1 - \Pr(\chi^2_1 < \chi^2_{\text{Yates}}) = 0.1855$$

larger than 0.1615 (the $p$-value without the Yate's correction).

## 3.3  Genetic Drift

Genetic drift describes a phenomenon that the frequency of an allele in a population changes due to random sampling of organisms. The correct mathematical model and the term "genetic drift" was first coined by a founder of population genetics, Sewall Wright in 1929 (Wright 1929). Genetic drift is caused by genetic sampling error in a finite population. For example, assume that a population has a frequency of $p = 0.5$ for allele $A_1$ and $q = 0.5$ for allele $A_2$. If the population size is infinitely large, we expect the allele frequency to remain 0.5 for each allele in the next population (guided by the HW equilibrium law). However, if we randomly sample $N = 5$ individual as parents to generate the next generation, the allele frequency of the next generation will be determined by the genotypes of the five individuals selected. If the five individuals randomly mate to generate an infinite number of progenies for the next generation, the allele frequency of $A_1$ will be determined by the number of $A_1$ alleles carried by the five parents sampled. For example, if the genotypes of the five individuals are $A_1A_1$, $A_1A_2$, $A_1A_2$, $A_2A_2$, and $A_2A_2$, the number of $A_1$ alleles is four and thus $p = 4/10 = 0.4$, which is not 0.5. The change of $p$ from 0.5 to 0.4 is due to genetic drift. For the $N = 5$ parents sampled, the probability of having $k$ alleles of type $A_1$ is defined by the Wright-Fisher model (Hartl and Clark 2007)

$$\frac{(2N)!}{k!(2N - k)!} p^k q^{2N-k}$$

In the above example, $k = 4$ and $N = 5$, and thus the probability is

$$\frac{(2 \times 5)!}{4!(2 \times 5 - 4)!} \left(\frac{1}{2}\right)^4 \left(\frac{1}{2}\right)^{2 \times 5 - 4} = 0.2050781$$

To maintain the 0.5 frequency for the next generation, there must be $k = 5$ $A_1$ alleles for the five parents, which has a probability of

$$\frac{(2 \times 5)!}{5!(2 \times 5 - 5)!} \left(\frac{1}{2}\right)^5 \left(\frac{1}{2}\right)^{2 \times 5 - 5} = 0.4921875$$

So, for a sample size of this small, the probability that the allele frequency remains unchanged is less than 50%.

Genetic drift can cause fixation or loss of an allele, even if there is no selection. For example, the five individuals sampled may all be of type $A_1A_1$, a situation where $A_1$ is fixed ($A_2$ is lost). The probability of $A_1$ being fixed is

$$\frac{(2 \times 5)!}{10!(2 \times 5 - 10)!} \left(\frac{1}{2}\right)^{10} \left(\frac{1}{2}\right)^{2 \times 5 - 10} = 0.0009765625$$

Although the probability looks very small, taking into consideration multiple generations, the fixation probability will eventually be 0.5, which is the initial gene frequency of allele $A_1$. Figure 3.1 Shows the results of a simulation experiment (Hartl and Clark 2007). The initial frequency of $A_1$ was 0.5. Ten replicated simulations were conducted for three different population sizes ($N = 20, 200, 2000$) over 50 generations. Fixation only occurred for the small population of $N = 20$. The amount of drift can be measured by the variance of allele frequencies over the repeats.

At any given point (generation), the amount of random drift can be measured by the variance of allele frequencies over the subpopulations (lines). Let $q_k$ be the allele frequency for $A_2$ from the $k$th lines. The variance is

$$\sigma_q^2 = \frac{1}{m} \sum_{k=1}^{m} (q_i - q)^2$$

where $m$ is the number of subpopulations and $q = 1 - p$ is the initial frequency of allele $A_2$. If $q$ is not known but replaced by the mean over the $m$ subpopulations, the variance is defined as

$$\sigma_q^2 = \frac{1}{m - 1} \sum_{k=1}^{m} (q_i - \overline{q})^2$$

where $\overline{q}$ is the mean allele frequency calculated from the whole population.

The consequence of genetic drift is the reduction in heterozygosity. In basic statistics, the variance of a variable is defined as the expectation of the squared variable minus the square of the expectation of the variable. In this case, the variable is denoted by $q$ (allele frequency of a subpopulation, not the initial frequency here) and the variance of this variable is expressed by

$$\sigma_q^2 = \mathrm{E}\left(q^2\right) - \mathrm{E}^2(q) \tag{3.1}$$

If there is no population subdivision, we expect that $\sigma_q^2 = 0$, i.e., both $\mathrm{E}(q^2)$ and $\mathrm{E}^2(q) = [\mathrm{E}(q)]^2$ are the frequency of homozygosity for genotype $A_2A_2$. With population subdivision, however, $\mathrm{E}(q^2)$ remains the frequency of homozygosity but $\mathrm{E}^2(q)$ is not. Rearranging equation (3.1) leads to

$$\mathrm{E}\left(q^2\right) = \mathrm{E}^2(q) + \sigma_q^2$$

If we do not know that there is population subdivision, we would expect that the homozygosity of the whole population be $\mathrm{E}^2(q)$. Since $\mathrm{E}(q^2)$ is larger than $\mathrm{E}^2(q)$, we say that there is a gain of homozygosity or loss of heterozygosity. This phenomenon was first observed by Swedish geneticist Sten Wahlund in 1928 (Wahlund 1928) and thus it is called the

**Fig. 3.1** Ten simulations (populations) of random genetic drift of allele $A_1$ in a biallelic system with an initial frequency distribution $p = 0.5$ measured over the course of 50 generations, repeated in three reproductively synchronous populations of different sizes. In these simulations, alleles drift to loss or fixation (frequency of 0.0 or 1.0) only occurred in the smallest population (Marginalia, https://en.wikipedia.org/wiki/Genetic_drift)

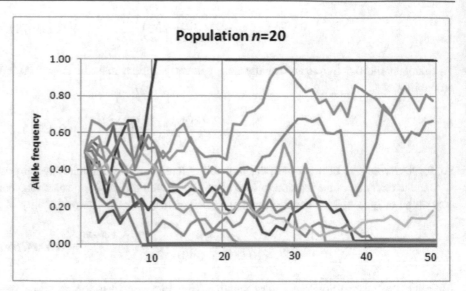

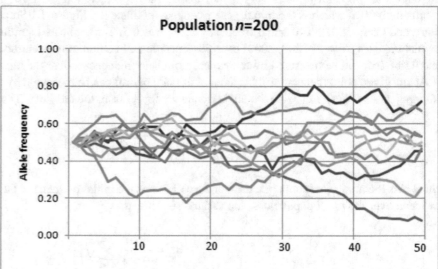

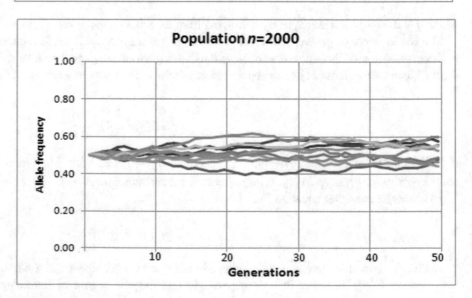

**Table 3.8** Genotype frequencies of a subdivided population, a hypothetic pooled population and a population with inbreeding

| Genotype | Subdivided | Pooled | Inbreeding |
|----------|-----------|--------|-----------|
| $A_1A_1$ | $\bar{p}^2 + \sigma_q^2$ | $\bar{p}^2$ | $\bar{p}^2 + \overline{pq}F$ |
| $A_1A_2$ | $2\overline{pq} - 2\sigma_q^2$ | $2\overline{pq}$ | $2\overline{pq} - 2\overline{pq}F$ |
| $A_2A_2$ | $\bar{q}^2 + \sigma_q^2$ | $\bar{q}^2$ | $\bar{q}^2 + \overline{pq}F$ |

Wahlund effect (Hartl and Clark 2007). To simplify the notation, we let $\bar{q} = \mathrm{E}(q)$. Table 3.8 shows the difference in genotype frequencies between the subdivided populations and the hypothetic pooled population (Hartl and Clark 2007).

Genotype frequencies of the pooled population are hypothetical because it only holds if the whole population is in Hardy-Weinberg equilibrium. In the actual subdivided population, the genotypic frequencies deviate from the Hardy-Weinberg equilibrium.

Genetic drift in a subdivided population has the same behavior as inbreeding. The last column of Table 3.8 shows the genotypic frequencies in a population with an average inbreeding coefficient of $F$ (Weir 1996). We can see that $\sigma_q^2 = \overline{pq}F$. Although inbreeding and genetic drift are conceptually different, they lead to the same consequence to the target population. Because of this, we often use $F$ to indicate the degree of population divergence,

$$F = \frac{\sigma_q^2}{\overline{pq}} \tag{3.2}$$

which is called fixation index or $F_{ST}$ (Weir 1996). Inbreeding is caused by mating between relatives. As a result, the two alleles carried by a progeny may be a copy of the same allele in the history. Such two alleles are said to be identity by descent (IBD). The probability of IBD of the two alleles in the progeny is called the inbreeding coefficient. Details about IBD and inbreeding coefficient will be discussed in a later chapter.

## 3.4 Wright's F Statistics

The $F_{ST}$ introduced above is one of the $F$ statistics developed by Wrights (Wright 1951) to describe evolutionary behaviors of structured populations. Wright provided three $F$ statistics, which are $F_{IT}$, $F_{ST}$ and $F_{IS}$, where $F_{IT}$ is called the correlation coefficient of the two alleles within individuals, $F_{ST}$ is the correlation coefficient between two alleles from different individuals within a population and $F_{IS}$ is the correlation of the two alleles within individuals within a population. It is hard to understand the definitions before we use a hierarchical ANOVA model to describe them (Weir and Cockerham 1984).

Malécot (1948) defined the correlation coefficient between two alleles as the probability that the two alleles are "identical by descent" (IBD). Such a definition is also called the fixation index. Under this definition, $F_{IT}$ is the probability that the two alleles from the same individuals are IBD, with the "base population" defined as the ancestors of all individuals in all populations. $F_{IS}$ is the probability that the two alleles from the same individuals are IBD, with the "base population" defined as the population when isolation just started. $F_{ST}$ is the probability that a random allele from one population is IBD to a random allele from another population.

When the $F$ statistics are defined as fixation indices, $1 - F$ becomes a heterozygosity reduction. The relationship of the three $F$ statistics in terms of heterozygosity reduction is depicted in Fig. 3.2. Let $H_0$ be the heterozygosity (proportion of heterozygotes) in the beginning (at time $t_0$). The heterozygosity at the point where the population split into two populations (at time $t_1$) is $H_1 = (1 - F_{ST})H_0$. The heterozygosity in the end (at time $t_2$) is $H_2 = (1 - F_{IS})H_1 = (1 - F_{IS})(1 - F_{ST})H_0$. Since we also know that $H_2 = (1 - F_{IT})H_0$. Therefore,

$$(1 - F_{IT}) = (1 - F_{IS})(1 - F_{ST})$$

This is the very famous equation in population genetics and evolution. These $F$-statistics can be estimated from molecular data using Cockerham (1969) hierarchical analysis of variances (ANOVA). Cockerham used the $(f, F, \theta)$ notation, where $f = F_{IS}$, $F = F_{IT}$ and $\theta = F_{ST}$.

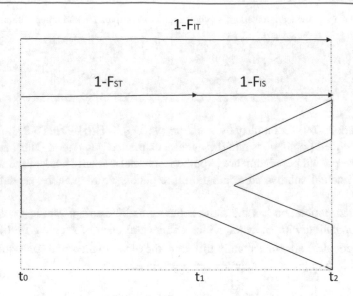

**Fig. 3.2** Relationship of Wright's $F$ statistics: $1 - F_{ST}$ represents heterozygosity reduction from $t_0$ to $t_1$, $1 - F_{IS}$ represents heterozygosity reduction from $t_1$ to $t_2$, $1 - F_{IT}$ represents heterozygosity reduction from $t_0$ to $t_2$. Therefore, $(1 - F_{IT}) = (1 - F_{IS})(1 - F_{ST})$

## 3.5 Estimation of *F* Statistics

Equation (3.2) is the definition formula for population differentiation (fixation index). Although the formula itself can be used to estimate $F_{ST}$, a far more better estimate of $F_{ST}$ is the one developed by (Cockerham 1969; Weir and Cockerham 1984), who used a nested design of analysis of variances. Let $i = 1, \ldots, m$ indexes population, $j = 1, \ldots, n_i$ indexes individual within the $i$th population and $k = 1, 2$ indexes an allele within the same individual. Let

$$y_{ijk} = \begin{cases} 1 & \text{for} & A_1 \\ 0 & \text{for} & A_2 \end{cases}$$

be the allelic indicator for the $k$th allele of the $j$th individual within population $i$. Cockerham (1969) used the following linear model to describe this allelic indicator variable,

$$y_{ijk} = \mu + \alpha_i + \beta_{(i)j} + \gamma_{(ij)k}$$

where $\mu$ is the population mean (frequency of $A_1$ in the whole population), $\alpha_i$ is the mean of $y$ in the $i$th population expressed as a deviation from the mean of the whole population, $\beta_{(i)j}$ is the mean of $y$ for the $j$th individual within the $i$th population expressed as a deviation from the mean of this population and $\gamma_{(ij)k}$ is the residual (the allelic indicator expressed as deviation from all previous terms). Let us assume that all terms in the model except $\mu$ are random so that $\text{var}(\alpha_i) = \sigma_\alpha^2$, $\text{var}\left(\beta_{(i)j}\right) = \sigma_\beta^2$ and $\text{var}\left(\gamma_{(ij)k}\right) = \sigma_\gamma^2$. Therefore, the total variance of $y$ is

$$\text{var}\left(y_{ijk}\right) = \sigma_y^2 = \sigma_\alpha^2 + \sigma_\beta^2 + \sigma_\gamma^2$$

Let us look at the definition of the $F$ statistics in terms of correlation coefficients. $F_{IT}$ is the correlation coefficient between the two alleles with an individual. Statistically, this correlation is defined as

$$r_{y_{ijk}y_{ijk'}} = \frac{\text{cov}\left(y_{ijk}, y_{ijk'}\right)}{\sqrt{\text{var}\left(y_{ijk}\right)\text{var}\left(y_{ijk'}\right)}}$$

where $k' \neq k$ and both $y$'s have the same subscript $j$ (meaning the same individual). The denominator is simply $\sigma_y^2 = \sigma_\alpha^2 + \sigma_\beta^2 + \sigma_\gamma^2$. The numerator is

$$\begin{aligned}
\text{cov}\left(y_{ijk}, y_{ijk'}\right) &= \text{cov}\left(\mu + \alpha_i + \beta_{(i)j} + \gamma_{(ij)k}, \mu + \alpha_i + \beta_{(i)j} + \gamma_{(ij)k'}\right) \\
&= \text{cov}(\alpha_i, \alpha_i) + \text{cov}\left(\beta_{(i)j}, \beta_{(i)j}\right) \\
&= \text{var}(\alpha_i) + \text{var}\left(\beta_{(i)j}\right) \\
&= \sigma_\alpha^2 + \sigma_\beta^2
\end{aligned}$$

Therefore,

$$F_{IT} = r_{y_{ijk}y_{ijk'}} = \frac{\sigma_\alpha^2 + \sigma_\beta^2}{\sigma_\alpha^2 + \sigma_\beta^2 + \sigma_\gamma^2}$$

The $F_{ST}$ parameter is defined as the correlation coefficient between two alleles from two different individuals within the same population. This correlation is

$$r_{y_{ijk}y_{ij'k'}} = \frac{\text{cov}\left(y_{ijk}, y_{ij'k'}\right)}{\sqrt{\text{var}\left(y_{ijk}\right)\text{var}\left(y_{ij'k'}\right)}}$$

The variances in the denominator are all the same, i.e., $\sigma_y^2 = \sigma_\alpha^2 + \sigma_\beta^2 + \sigma_\gamma^2$. The covariance in the numerator is

$$\begin{aligned}
\text{cov}\left(y_{ijk}, y_{ij'k'}\right) &= \text{cov}\left(\mu + \alpha_i + \beta_{(i)j} + \gamma_{(ij)k}, \mu + \alpha_i + \beta_{(i)j'} + \gamma_{(ij)k'}\right) \\
&= \text{cov}(\alpha_i, \alpha_i) \\
&= \sigma_\alpha^2
\end{aligned}$$

Therefore,

$$F_{ST} = r_{y_{ijk}y_{ij'k'}} = \frac{\sigma_\alpha^2}{\sigma_\alpha^2 + \sigma_\beta^2 + \sigma_\gamma^2}$$

The $F_{IS}$ parameter is defined as the correlation coefficient between two alleles from the same individual within the same population. This means that you only consider one population and the parameter is the average of the parameter across all populations. This time we revise the model by focusing on one population only and thus drop subscript $i$,

$$y_{jk} = \mu + \beta_j + \gamma_{(j)k}$$

The total variance here is the variance among all individuals with the same population, which is

$$\text{var}\left(y_{jk}\right) = \text{var}\left(\beta_j\right) + \text{var}\left(\gamma_{(j)k}\right) = \sigma_\beta^2 + \sigma_\gamma^2$$

The correlation coefficient between the two alleles from the same individual is

$$r_{y_{jk}y_{jk'}} = \frac{\text{cov}\left(y_{jk}, y_{jk'}\right)}{\sqrt{\text{var}\left(y_{jk}\right)\text{var}\left(y_{jk'}\right)}}$$

The variances in the denominator are all the same, i.e., $\sigma_\beta^2 + \sigma_\gamma^2$. The covariance in the numerator is

$$\text{cov}\left(y_{jk}, y_{jk'}\right) = \text{cov}\left(\mu + \beta_j + \gamma_{(j)k}, \mu + \beta_j + \gamma_{(j)k'}\right) = \text{cov}\left(\beta_j, \beta_j\right) = \sigma_\beta^2$$

Therefore,

$$F_{\text{IS}} = r_{y_{jk}y_{jk'}} = \frac{\sigma_\beta^2}{\sigma_\beta^2 + \sigma_\gamma^2}$$

We can rearrange the following equation

$$(1 - F_{\text{IT}}) = (1 - F_{\text{IS}})(1 - F_{\text{ST}})$$

into

$$F_{\text{IS}} = \frac{F_{\text{IT}} - F_{\text{ST}}}{1 - F_{\text{ST}}}$$

and substitute the $F$ statistics in the right hand side by the variance ratios. This manipulation will help you verify the equation, as shown below.

$$F_{\text{IS}} = \frac{F_{\text{IT}} - F_{\text{ST}}}{1 - F_{\text{ST}}} = \frac{\sigma_\beta^2}{\sigma_\beta^2 + \sigma_\gamma^2}$$

Given the linear model described above, we can sample allelic data from individuals of the populations. Again, each data point is an allelic state, a binary variable taking either 0 or 1. In SNP data, there are only two alleles (multiple alleles are very rare) and the "reference allele" is coded 1. Which allele is the reference allele is entirely arbitrary, depending on the investigator's preference. The ANOVA table is shown in Table 3.9. In this ANOVA table, when the number of individuals within a population is different across different populations, the data are called unbalanced. In fact, in population differentiation analysis, population is always unbalanced. The "average" number of individuals of a population is calculated differently from the usual definition of average. It is calculated using

$$k_0 = \frac{1}{m-1}\left(\sum_{i=1}^m n_i - \frac{\sum_{i=1}^m n_i^2}{\sum_{i=1}^m n_i}\right)$$

The three variance components are then estimated using

Table 3.9  The ANOVA table for the hierarchical model of a structured whole population

| Source of variation | df | SS | MS | $E$ (MS) |
|---|---|---|---|---|
| Between populations | $df_\alpha = m - 1$ | $SS_\alpha$ | $MS_\alpha = SS_\alpha/df_\alpha$ | $\sigma_\gamma^2 + 2\sigma_\beta^2 + 2k_0\sigma_\alpha^2$ |
| Between individuals within populations | $df_\beta = \sum_{i=1}^p (n_i - 1)$ | $SS_\beta$ | $MS_\beta = SS_\beta/df_\beta$ | $\sigma_\gamma^2 + 2\sigma_\beta^2$ |
| Between alleles within individuals | $df_\gamma = \sum_{i=1}^p n_i$ | $SS_\gamma$ | $MS_\gamma = SS_\gamma/df_\gamma$ | $\sigma_\gamma^2$ |

$$\widehat{\sigma}_\gamma^2 = MS_\gamma$$

$$\widehat{\sigma}_\beta^2 = \frac{1}{2}\left(MS_\beta - MS_\gamma\right)$$

$$\widehat{\sigma}_\alpha^2 = \frac{1}{2k_0}\left(MS_\alpha - MS_\beta\right)$$

The three estimated variance components are used to infer the $F$ statistics. In reality, you do not need to know the formulas to perform the ANOVA. The SAS package contains several procedures for analysis of variances under the hierarchical model. There is even a procedure called PROC ALLELE particularly designed for estimating the $F$ statistics.

## References

Cockerham CC. Variance of gene frequencies. Evolution. 1969;23:72–84.

Dempster AP, Laird MN, Rubin DB. Maximum likelihood from incomplete data via the EM algorithm. J Royal Stat Soc Ser B (Methodological). 1977;39:1–38.

Falconer DS, Mackay TFC. Introduction to quantitative genetics. Harlow, Essex: Addison Wesley Longman; 1996.

Hartl DL, Clark AG. Principles of population genetics. Sunderland, MA: Sinauer Associates; 2007.

Malécot G. Les mathématiques de l'hérédité. Paris: Masson; 1948.

Wahlund S. Zusammensetzung von Population und Korrelationserscheinung vom Standpunkt der Vererbungslehre aus betrachtet. Hereditas. 1928;11:65–106.

Weir BS. Genetic data analysis II—Methods for discrete population genetic data. Synderland, MA: Sinauer Associates, Inc. Publishers; 1996.

Weir BS, Cockerham CC. Estimating F-statistics for the analysis of population structure. Evolution. 1984;38:1358–70.

Wright S. The evolution of dominance. Am Nat. 1929;63:556–61.

Wright S. The genetic structure of populations. Ann Eugen. 1951;15:323–54.

# Review of Elementary Statistics

**4**

Quantitative genetics heavily involves statistics, especially variance and covariance. Therefore, a solid background in statistics is essential to learning quantitative genetics. In this chapter, we are going to review some basic concepts of statistics. Some chapters require specialized statistical theories and computational algorithms, which will be reviewed case by case prior to the lectures of those chapters.

## 4.1 Expectation

### 4.1.1 Definition

Expectation is defined as the central position of a distribution. If a random sample is obtained from the distribution, we can calculate the mean or the average of the sample. The sample mean is usually different from the expectation unless the sample size is infinitely large. However, expectation and mean are often used interchangeably. Assume that $x$ is a random variable with a distribution (probability density) denoted by $f(x)$. If the distribution is discrete (variable with a finite number of values), the expectation is defined by

$$E(x) = \sum_{k=1}^{m} x_k f(x_k)$$

where $\sum_{k=1}^{m} f(x_k) = 1$. If $f(x)$ is continuous, then the expectation is

$$E(x) = \int_{-\infty}^{\infty} x f(x) dx$$

where $\int_{-\infty}^{\infty} f(x) dx = 1$. The summation notation has a special property that is $\sum_{j=1}^{n} c = cn$ for a constant $c$. For example, if $n = 5$ and $c = 2$, then $\sum_{j=1}^{5} 2 = 2 + 2 + 2 + 2 + 2 = 2 \times 5 = 10$.

### 4.1.2 Properties of Expectation

1. If $c$ is a constant, then $E(c) = c$.
2. If $y = a + bx$, then $E(y) = a + bE(x)$, where $a$ and $b$ are constants.
3. If $x$ and $y$ are two variables and each has its own distribution, then

© Springer Nature Switzerland AG 2022
S. Xu, *Quantitative Genetics*, https://doi.org/10.1007/978-3-030-83940-6_4

$$E(x + y) = E(x) + E(y)$$

4. If $x$ and $y$ are independent, then

$$E(xy) = E(x)E(y)$$

5. If $x_1, \ldots, x_n$ are $n$ random variables each having its own expectation, then

$$E\left(\sum_{j=1}^{n} x_j\right) = \sum_{j=1}^{n} E(x_j)$$

6. If $x_1, \ldots, x_n$ are $n$ independent random variables and each has its own expectation, then

$$E\left(\prod_{j=1}^{n} x_j\right) = \prod_{j=1}^{n} E(x_j)$$

### 4.1.3  Estimating the Mean

If $x_1, \ldots, x_n$ are $n$ observations (a sample with size $n$) from a variable $x$ whose expectation is $E(x) = \mu$, then the estimated value of the expectation (also called the mean or the average) is

$$\bar{x} = \widehat{\mu} = \frac{1}{n} \sum_{j=1}^{n} x_j$$

Note that $\widehat{\mu}$ is called an estimate and $\mu$ is called a parameter. A parameter is a constant, not a variable, but it is unknown. An estimate of a parameter, however, is a variable in the sense that it varies from sample to sample. According to the properties of the expectation, we immediately know that

$$E(\bar{x}) = \frac{1}{n} \sum_{j=1}^{n} E(x_j) = \frac{1}{n} \sum_{j=1}^{n} \mu = \frac{n\mu}{n} = \mu$$

The expectation of a sample mean equals the expectation of the original variable.

## 4.2    Variance

### 4.2.1  Definition

Variance reflects the variation of a random variable around the central position. If $x$ is a discrete variable with $m$ possible categories, the variance is defined by

$$\text{var}(x) = \sum_{k=1}^{m} (x_k - \mu)^2 f(x_k) = \sum_{k=1}^{m} x_k^2 f(x_k) - \sum_{k=1}^{m} \mu^2 f(x_k) = E(x^2) - E^2(x)$$

where $\mu = E(x)$ is the expectation of $x$ and $\mu^2 = E^2(x) = [E(x)]^2$ is just an alternative expression for the squared expectation. We have learned $E(x)$ but not $E(x^2)$, which is the expectation of $x^2$ and is defined by

$$E(x^2) = \sum_{k=1}^{m} x_k^2 f(x_k)$$

For a continuous variable, the variance is defined by

$$var(x) = \int_{-\infty}^{\infty} (x - \mu)^2 f(x) dx = \int_{-\infty}^{\infty} x^2 f(x) dx - \int_{-\infty}^{\infty} \mu^2 f(x) dx = E(x^2) - E^2(x)$$

where $E(x^2) = \int_{-\infty}^{\infty} x^2 f(x) dx$ is the expectation of squared $x$.

## 4.2.2 Properties of Variance

1. If $c$ is a constant, then $var(c) = 0$.
2. If $y = a + bx$, then $var(y) = b^2 var(x)$, where $a$ and $b$ are constants.
3. If $x$ and $y$ are independent, then $var(x + y) = var(x) + var(y)$.
4. If $x_1, \ldots, x_n$ are independent and each has its own variance, then

$$var\left(\sum_{j=1}^{n} x_j\right) = \sum_{j=1}^{n} var(x_j)$$

## 4.2.3 Normal Distribution

Normal distribution is a very important distribution and will be used repeatedly in this course. This distribution is determined by the expectation and the variance (two parameters). Since we have learned the expectation and the variance, it is time to introduce the normal distribution. Any variable that has a probability density of the following form is said to be normally distributed,

$$f(x) = N(x|\mu, \sigma^2) = \frac{1}{\sqrt{2\pi}\sigma} \exp\left[-\frac{1}{2\sigma^2}(x - \mu)^2\right] \qquad (4.1)$$

where $\mu$ and $\sigma^2$ are the expectation and the variance of the variable, respectively. Normal distribution is often denoted by $x \sim N(\mu, \sigma^2)$. The first expression in Eq. (4.1) is a short notation for normal distribution. A standardized normal distribution is a normal distribution with expectation 0 and variance 1 and is denoted by $x \sim N(0, 1)$, whose density is

$$f(x) = N(x|0, 1) = \frac{1}{\sqrt{2\pi}} \exp\left(-\frac{1}{2}x^2\right)$$

Normal distribution is also called Gaussian distribution or bell-shaped distribution.

## 4.2.4 Estimating Variance from a Sample

Let $x_1, \ldots, x_n$ be $n$ randomly sampled observations from variable $x$ with a particular distribution. If the expectation of variable $x$ is known to be $\mu$, the estimated variance (also called the sample variance) is

$$s^2 = \widehat{\sigma}^2 = \frac{1}{n} \sum_{j=1}^{n} (x_j - \mu)^2 \qquad (4.2)$$

If the expectation is not known and it is estimated from the sample, then

$$s^2 = \widehat{\sigma}^2 = \frac{1}{n-1} \sum_{j=1}^{n} (x_j - \widehat{\mu})^2 \qquad (4.3)$$

where

$$\widehat{\mu} = \frac{1}{n} \sum_{j=1}^{n} x_j$$

is the estimated mean. Note the difference between Eqs. (4.2) and (4.3), where the $n-1$ in Eq. (4.3) is due to a lost degree of freedom by estimating the population mean.

## 4.2.5   An Application of the Variance Property

If $x_1, \ldots, x_n$ are $n$ observations (a sample with size $n$) from a variable $x$ whose expectation is $\mu$ and variance is $\text{var}(x) = \sigma^2$, the estimated value of the expectation is

$$\bar{x} = \widehat{\mu} = \frac{1}{n} \sum_{j=1}^{n} x_j$$

This is a sample mean and is still a variable. According to the properties of the expectation, we immediately know that

$$E(\bar{x}) = \frac{1}{n} \sum_{k=1}^{n} E(x_j) = \frac{1}{n} \sum_{k=1}^{n} \mu = \frac{1}{n} n\mu = \mu \qquad (4.4)$$

The sample mean is still a variable because it is not the true expectation. The variance of $\bar{x}$ can be derived based on the properties of variance,

$$\text{var}(\bar{x}) = \text{var}\left(\frac{1}{n} \sum_{j=1}^{n} x_j\right) = \frac{1}{n^2} \text{var}\left(\sum_{j=1}^{n} x_j\right) = \frac{1}{n^2} \sum_{j=1}^{n} \text{var}(x_j) = \frac{1}{n^2} \sum_{j=1}^{N} \sigma^2 = \frac{n\sigma^2}{n^2} = \frac{\sigma^2}{n} \qquad (4.5)$$

Note that $\text{var}(x_j) = \sigma^2$ for all $j = 1, \ldots, n$. The symbol "var()" is called the variance operator. It represents the variance of the population from which $x_j$ is sampled. It does not mean the variance of this single observation. A single observed number is just a number and there is no variance. Comparing Eqs. (4.4) and (4.5), we notice that the expectation of a sample mean equals the expectation of the variable, but the variance of the sample mean equals the variance of the variable divided by the sample size. If the sample size is extremely large, say $n \to \infty$, the sample mean will approach the expectation because the variance of the sample mean is virtually zero.

## 4.3 Covariance

### 4.3.1 Definition

Covariance describes the behavior of two variables varying together. Consider variables $x$ and $y$ with a joint distribution $f(x, y)$, the covariance is defined as

$$\sigma_{xy} = \text{cov}(x, y) = \int_{-\infty}^{\infty} \int_{-\infty}^{\infty} (x - \mu_x)(y - \mu_y)f(x, y)dxdy$$

$$= \int_{-\infty}^{\infty} \int_{-\infty}^{\infty} xyf(x, y)dxdy - \mu_x\mu_y$$

$$= \text{E}(xy) - \text{E}(x)\text{E}(y)$$

where $\mu_x = \text{E}(x)$ and $\mu_y = \text{E}(y)$, and

$$\text{E}(xy) = \int_{-\infty}^{\infty} \int_{-\infty}^{\infty} xyf(x, y)dxdy$$

For discrete variables, the definition is slightly different. Let $f(x_k)$ be the frequency of variable $x$ with $m$ categories, i.e., $x$ can take $m$ different values or $x = x_k$ for $k = 1, \ldots, m$. Let $f(y_l)$ be the frequency of variable $y$ with $n$ categories, i.e., $y$ can take $n$ different values or $y = y_l$ for $l = 1, \ldots, n$. The joint frequency of $x$ and $y$ is denoted by $f(x_k, y_l)$ for $k = 1, \ldots, m$ and $l = 1, \ldots, n$. Covariance between $x$ and $y$ is defined by

$$\sigma_{xy} = \text{cov}(x, y) = \sum_{k=1}^{m} \sum_{l=1}^{n} (x_k - \mu_x)(y_l - \mu_y)f(x_k, y_l)$$

$$= \sum_{k=1}^{m} \sum_{l=1}^{n} x_k y_l f(x_k, y_l) - \mu_x\mu_y$$

$$= \text{E}(xy) - \text{E}(x)\text{E}(y)$$

where

$$\text{E}(xy) = \sum_{k=1}^{m} \sum_{l=1}^{n} x_k y_l f(x_k, y_l)$$

Note that

$$\sigma_{xy} = \text{E}(xy) - \text{E}(x)\text{E}(y)$$

regardless whether the variables are discrete or continuous.

### 4.3.2 Properties of Covariance

1. If $x$ and $y$ are independent, then $\text{cov}(x, y) = 0$.
2. Covariance is symmetrical, i.e., $\text{cov}(x, y) = \text{cov}(y, x)$.
3. If $c$ is a constant, then $\text{cov}(c, x) = 0$. Also $\text{cov}(a, b) = 0$ if both $a$ and $b$ are constants.

4. The covariance of a variable with itself equals the variance, $\text{cov}(x, x) = \text{var}(x)$.
5. Let $z = x + y$ where $x$ and $y$ are two variables (correlated), then

$$\text{var}(z) = \text{var}(x + y) = \text{var}(x) + \text{var}(y) + 2\text{cov}(x, y)$$

6. If $x_1, \ldots, x_n$ are $n$ variables that are mutually correlated, then

$$\text{var}\left(\sum_{j=1}^{n} x_j\right) = \sum_{j=1}^{n} \text{var}(x_j) + 2\sum_{j=1}^{n-1} \sum_{j'=j+1}^{n} \text{cov}(x_j, x_{j'})$$

7. Let $x = a_1 x_1 + a_2 x_2$ and $y = b_1 y_1 + b_2 y_2$, then

$$\begin{aligned} \text{cov}(x, y) &= \text{cov}(a_1 x_1 + a_2 x_2, b_1 y_1 + b_2 y_2) \\ &= a_1 b_1 \text{cov}(x_1, y_1) + a_1 b_2 \text{cov}(x_1, y_2) + a_2 b_1 \text{cov}(x_2, y_1) + a_2 b_2 \text{cov}(x_2, y_2) \end{aligned}$$

8. If both $a$ and $b$ are constants, then $\text{cov}(ax, by) = ab\, \text{cov}(x, y)$.

Property (4) declares that variance is a special case of covariance. With this property, property (6) can also be expressed by

$$\text{var}\left(\sum_{j=1}^{n} x_j\right) = \sum_{j=1}^{n} \sum_{j'=1}^{n} \text{cov}(x_j, x_{j'})$$

### 4.3.3  Estimating Covariance from a Sample

Let $\{x_1, y_1\}, \ldots, \{x_n, y_n\}$ be a random sample of $n$ pairs of observations from the joint distribution of variables $x$ and $y$. The estimated covariance between $x$ and $y$ is

$$\widehat{\sigma}_{xy} = \frac{1}{n-1} \sum_{j=1}^{n} (x_j - \bar{x})(y_j - \bar{y}) = \frac{1}{n-1}\left(\sum_{j=1}^{n} x_j y_j - n\bar{x}\bar{y}\right)$$

If the means of both variables are known, then

$$\widehat{\sigma}_{xy} = \frac{1}{n} \sum_{j=1}^{n} (x_j - \mu_x)(y_j - \mu_y) = \frac{1}{n}\left(\sum_{j=1}^{n} x_j y_j - n\mu_x\mu_y\right)$$

The following example shows a joint distribution for $x$ and $y$, where $x$ has four categories (1, 2, 3, and 4) and $y$ has three categories ($-1$, 0, and 1). The joint distribution is given in Table 4.1 (a $4 \times 3$ table). The cells give the joint probabilities; the last row gives the marginal probabilities for variable $y$ and the last column gives the marginal probabilities of variable $x$. The marginal probabilities are the sum of all elements of each row or each column. You will see that $x$ and $y$ are independent, i.e., $\text{cov}(xy) = 0$.

The expectation and variance of variable $x$ are (only uses the marginal probabilities of $x$)

$$\mu_x = E(x) = \sum_{i=1}^{4} x_i f(x_i) = 1 \times 0.1 + 2 \times 0.2 + 3 \times 0.3 + 4 \times 0.4 = 3.0$$

$$E(x^2) = \sum_{i=1}^{4} x_i^2 f(x_i) = 1^2 \times 0.1 + 2^2 \times 0.2 + 3^2 \times 0.3 + 4^2 \times 0.4 = 10.0$$

$$\sigma_x^2 = var(x) = E(x^2) - E^2(x) = 10.0 - 3.0^2 = 1.0$$

The expectation and variance of variable $y$ are (only uses the marginal probabilities of $y$)

$$\mu_y = E(y) = \sum_{j=1}^{3} y_j f(y_j) = (-1) \times 0.3 + (0) \times 0.5 + 1 \times 0.2 = -0.1$$

$$E(y^2) = \sum_{j=1}^{3} y_j^2 f(y_j) = (-1)^2 \times 0.3 + (0)^2 \times 0.5 + 1^2 \times 0.2 = 0.5$$

$$\sigma_y^2 = var(y) = E(y^2) - E^2(y) = 0.5 - (-0.1)^2 = 0.49$$

The covariance between $x$ and $y$ is (uses the joint probabilities)

$$\sigma_{xy} = cov(x, y) = E(xy) - E(x)E(y) = -0.30 - 3.0 \times (-0.1) = 0.0$$

where

$$
\begin{aligned}
E(xy) &= \sum_{i=1}^{4} \sum_{j=1}^{3} x_i y_j f(x_i, y_j) \\
&= (1)(-1) \times 0.03 + (1)(0) \times 0.05 + (1)(1) \times 0.02 \\
&\quad + (2)(-1) \times 0.06 + (2)(0) \times 0.10 + (2)(1) \times 0.04 \\
&\quad + (3)(-1) \times 0.09 + (3)(0) \times 0.15 + (3)(1) \times 0.06 \\
&\quad + (4)(-1) \times 0.12 + (4)(0) \times 0.20 + (4)(1) \times 0.08 \\
&= -0.30
\end{aligned}
$$

### 4.3.4   Conditional Expectation and Conditional Variance

Consider two variables, $x$ and $y$, and they form a joint distribution. Let $f(y|x)$ be the conditional distribution of $y$ given $x$. The conditional expectation of $y$ given $x$ is denoted by $E(y|x)$ and is defined as a function of variable $x$. The conditional expectation of $y$, when specified at $x$, is

$$E(y|x) = \int_{-\infty}^{\infty} y f(y|x) dy$$

In other words, $E(y|x)$ is the mean for variable $y$ given $x$ is a fixed value. When the conditional distribution is discrete with $m$ categories, the conditional expectation is defined as

$$E(y|x) = \sum_{j=1}^{m} y_j f(y_j|x)$$

Conditional variance of $y$ given $x$ is defined as

**Table 4.1**  Joint distribution of two discrete variables, $x$ and $y$

| $xy$ | $-1$ | $0$ | $1$ | $f(x)$ |
|---|---|---|---|---|
| 1 | 0.03 | 0.05 | 0.02 | 0.1 |
| 2 | 0.06 | 0.10 | 0.04 | 0.2 |
| 3 | 0.09 | 0.15 | 0.06 | 0.3 |
| 4 | 0.12 | 0.20 | 0.08 | 0.4 |
| $f(y)$ | 0.3 | 0.5 | 0.2 | 1.0 |

$$\text{var}(y|x) = \int_{-\infty}^{\infty} [y - E(y|x)]^2 f(y|x)dy$$
$$= E(y^2|x) - E^2(y|x)$$

Note that var($y|x$) is also defined as a function of variable $x$.

For the joint distribution given in Table 4.1, let us try to find E($y|x = 2$) and var($y|x = 2$). First, we need to find the conditional distribution from the joint and marginal distributions using

$$f(y|x = 2) = f(x, y)/f(x = 2)$$

Therefore,

$$f(y = -1|x = 2) = f(x = 2, y = -1)/f(x = 2) = 0.06/0.2 = 0.30$$

$$f(y = 0|x = 2) = f(x = 2, y = 0)/f(x = 2) = 0.10/0.2 = 0.50$$

$$f(y = 1|x = 2) = f(x = 2, y = 1)/f(x = 2) = 0.04/0.2 = 0.20$$

The conditional expectation of $y$ given $x = 2$ is

$$E(y|x = 2) = \sum_{k=1}^{3} y_j f(y_j|x = 2)$$
$$= (-1)f(y = -1|x = 2) + (0)f(y = 0|x = 2) + (-1)f(y = 1|x = 2)$$
$$= (-1) \times 0.3 + (0) \times 0.5 + (1) \times 0.2$$
$$= -0.1$$

The conditional expectation of $y^2$ given $x = 2$ is

$$E(y^2|x = 2) = \sum_{k=1}^{3} y_j^2 f(y_j|x = 2)$$
$$= (-1)^2 f(y = -1|x = 2) + (0)^2 f(y = 0|x = 2) + (-1)^2 f(y = 1|x = 2)$$
$$= (-1)^2 \times 0.3 + (0)^2 \times 0.5 + (1)^2 \times 0.2$$
$$= 0.5$$

Therefore, the conditional variance of $y$ given $x = 2$ is

$$\text{var}(y|x = 2) = E(y^2|x) - E^2(y|x) = 0.5 - (-0.1)^2 = 0.49$$

Below is an important property of conditional expectation and conditional variance. If $x$ and $y$ are random variables for which the necessary expectations and variances exist, then

$$\text{var}(y) = E[\text{var}(y|x)] + \text{var}[E(y|x)] \tag{4.6}$$

Verbally, the property states that the variance of $y$ equals the expectation of the conditional variance of $y$ given $x$ plus the variance of the conditional expectation of $y$ given $x$. The expectation in the first term is taken with respect to $x$ because $\text{var}(y|x)$ is a function of $x$. The variance in the second term is also taken with respect to $x$ because $E(y|x)$ is a function of $x$. If $x$ is a class variable (discrete), property (6) is a partitioning of the total variance of $y$ into a within group variance, $E[\text{var}(y|x)]$, and a between group variance, $\text{var}[E(y|x)]$. Equation (4.6) is also called the Law of Total Variance.

## 4.4  Sample Estimates of Variance and Covariance

The method to be described here does not involve distribution. Later in the course, we will learn the maximum likelihood estimation of parameters, where distribution is required. Several simple rules need to be remembered, as described below,

1. No frequency or density distribution is involved.
2. A sample is required for each variable to calculate the variance.
3. A joint sample is required for both variables to calculate the covariance.
4. It is irrelevant whether a variable is continuous or discrete; they are treated equally.

Let $n$ be the sample size and $\{x_j, y_j\}$ be the $j$th data point of the sample for $j = 1, \ldots, n$. The average or mean of $x$ is

$$\widehat{\mu}_x = \bar{x} = \frac{1}{n} \sum_{j=1}^{n} x_j$$

The estimated variance of $x$ is

$$\widehat{\sigma}_x^2 = \frac{1}{n-1} \sum_{j=1}^{n} \left(x_j - \bar{x}\right)^2 = \frac{1}{n-1} \left( \sum_{j=1}^{n} x_j^2 - n(\bar{x})^2 \right)$$

The average or mean of $y$ is

$$\widehat{\mu}_y = \bar{y} = \frac{1}{n} \sum_{j=1}^{n} y_j$$

The estimated variance of $y$ is

$$\widehat{\sigma}_y^2 = \frac{1}{n-1} \sum_{j=1}^{n} \left(y_j - \bar{y}\right)^2 = \frac{1}{n-1} \left( \sum_{j=1}^{n} y_j^2 - n(\bar{y})^2 \right)$$

The estimated covariance between $x$ and $y$ is

$$\widehat{\sigma}_{xy} = \frac{1}{n-1} \sum_{j=1}^{n} \left(x_j - \bar{x}\right)\left(y_j - \bar{y}\right) = \frac{1}{n-1} \left( \sum_{j=1}^{n} x_j y_j - n(\bar{x})(\bar{y}) \right)$$

## 4.5     Linear Model

Consider two variables, $x$ and $y$. Assume that the change of $y$ depends on the change of $x$, then $y$ is called the dependent variable and $x$ is called the independent variable. No matter what the exact relationship is between $x$ and $y$, within a short range of value in $x$, variable $y$ can always be approximated by the following linear model,

$$y = a + bx + e$$

where $a$ is the intercept, $b$ is the regression coefficient (the slope of the linear function of $y$ on $x$), and $e$ is the residual error. The residual error is collectively contributed by all other unknown variables. The residual error is usually assumed to be normally distributed with mean zero and variance $\sigma^2$, i.e., $e \sim N(0, \sigma^2)$. The normal assumption often is not required for parameter estimation but it is required for significance test.

### 4.5.1    Regression

The regression coefficient $b$ represents the amount of change in $y$ corresponding to one unit change in $x$. Mathematically, the regression coefficient is defined as

$$b_{yx} = \frac{\mathrm{cov}(x, y)}{\mathrm{var}(x)} = \frac{\sigma_{xy}}{\sigma_x^2}$$

Because the covariance can take any real value, we have $-\infty < b < \infty$. Sometimes $b$ is denoted by $b_{yx}$, reflecting the direction of the regression process. The subscript in $b_{yx}$ says that $y$ is the response variable and $x$ is the independent variable, which is different from $b_{xy}$, the regression of $x$ on $y$,

$$b_{xy} = \frac{\mathrm{cov}(x, y)}{\mathrm{var}(y)} = \frac{\sigma_{xy}}{\sigma_y^2}$$

The predicted $y$ value is the conditional expectation of $y$ given $x$ and denoted by

$$\widehat{y} = \mathrm{E}(y|x) = a + bx$$

The line representing the predicted $y$ is shown in Fig. 4.1, where a scatter plot for $x$ and $y$ is presented.

### 4.5.2    Correlation

Correlation describes the association of two variables. The degree of association is measured by the so-called correlation coefficient, which is defined by

$$r_{xy} = \frac{\mathrm{cov}(x, y)}{\sqrt{\mathrm{var}(x)\mathrm{var}(y)}} = \frac{\sigma_{xy}}{\sigma_x \sigma_y}$$

where $\sigma_x = \sqrt{\sigma_x^2}$ and $\sigma_y = \sqrt{\sigma_y^2}$. In contrast to regression coefficient, correlation has no direction, i.e., it is symmetrical in the sense that $r_{xy} = r_{yx}$. In addition, correlation coefficient can only take value within the range defined by $-1 \leq r_{xy} \leq +1$.

The correlation coefficient can be considered as the covariance between two standardized variables. Define $z_x = (x - \mu_x)/\sigma_x$ and $z_y = (y - \mu_y)/\sigma_y$ as the standardized variables of $x$ and $y$, respectively. A standardized variable has a standardized normal distribution if the original variable is normally distributed. This can be shown by

$$E(z_x) = E\left(\frac{x - \mu_x}{\sigma_x}\right) = \frac{1}{\sigma_x}E(x) - \frac{1}{\sigma_x}\mu_x = \frac{1}{\sigma_x}\mu_x - \frac{1}{\sigma_x}\mu_x = 0$$

and

$$\text{var}(z_x) = \text{var}\left(\frac{x - \mu_x}{\sigma_x}\right) = \frac{1}{\sigma_x^2}\text{var}(x) - \frac{1}{\sigma_x^2}\text{var}(\mu_x) = \frac{1}{\sigma_x^2} \times \sigma_x^2 = 1$$

Note that $\mu_x$ is a constant parameter, not a variable, and thus the variance of $\mu_x$ is zero. Similarly, variable $z_y$ is also a standardized normal variable. The covariance between $z_x$ and $z_y$ is

$$\text{cov}(z_x, z_y) = E\{[z_x - E(z_x)][z_y - E(z_y)]\} = E(z_x z_y)$$

which is

$$E(z_x z_y) = E\left[\left(\frac{x - \mu_x}{\sigma_x}\right)\left(\frac{y - \mu_y}{\sigma_y}\right)\right] = \frac{E[(x - \mu_x)(y - \mu_y)]}{\sigma_x \sigma_y} = \frac{\text{cov}(x, y)}{\sigma_x \sigma_y} = r_{xy}$$

Therefore, $\text{cov}(z_x, z_y) = r_{xy}$. The relationship between the two regression coefficients and the correlation coefficient is

$$b_{yx} b_{xy} = \frac{\text{cov}(x, y)}{\text{var}(x)} \times \frac{\text{cov}(x, y)}{\text{var}(y)} = \frac{\text{cov}^2(x, y)}{\text{var}(x)\text{var}(y)} = r_{xy}^2$$

Figure 4.1 shows the scatter plots of a dataset generated with the same slope but with different correlation coefficients. A high correlation means that the data points concentrate along the regression line (panel A) while a low correlation means that the data points scatter away from the regression line (panel B).

### 4.5.3  Estimation of Regression and Correlation Coefficients

We have learned how to estimate variance and covariance given a random sample from the joint distribution of $x$ and $y$. The estimated regression coefficient is obtained by replacing the variance and covariance by their estimated values, as shown below,

$$\widehat{b}_{yx} = \frac{\sum_{j=1}^{n}(x_j - \widehat{\mu}_x)(y_j - \widehat{\mu}_y)}{\sum_{j=1}^{n}(x_j - \widehat{\mu}_x)^2}$$

Similarly, the correlation coefficient is estimated using

$$\widehat{r}_{xy} = \frac{\sum_{j=1}^{n}(x_j - \widehat{\mu}_x)(y_j - \widehat{\mu}_y)}{\sqrt{\sum_{j=1}^{n}(x_j - \widehat{\mu}_x)^2 \sum_{j=1}^{n}(y_j - \widehat{\mu}_y)^2}}$$

Note that the degree of freedom used to estimate the variance and covariance have been cancelled out because they appear in both the numerator and the denominator.

In regression analysis, we also need to estimate the intercept, denoted by $a$, because the regression model is $y = a + bx + e$. Without giving the derivation, we simply provide the estimated intercept as

$$\widehat{a} = \bar{y} - \widehat{b}\bar{x}$$

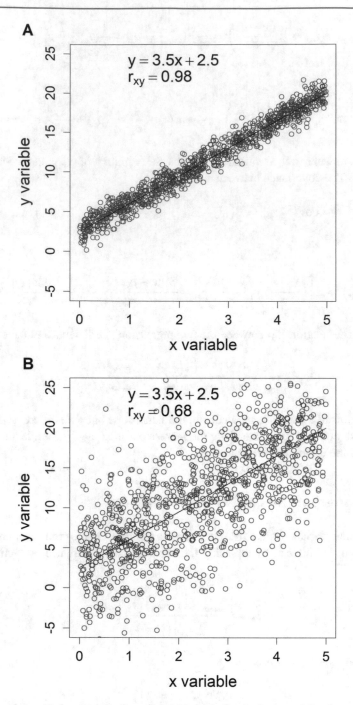

**Fig. 4.1** Scatter plots of *y* against *x* along with the regression lines. Panel (**a**) shows the plot for $b_{yx} = 3.5$ and $r_{xy} = 0.98$. Panel (**b**) shows the plot for $b_{yx} = 3.5$ and $r_{xy} = 0.68$. Both panels have the same slope but Panel (**b**) has a much lower correlation coefficient than Panel (**a**)

Suppose that a new *x* value is observed in the future, denoted by $x_{New}$, and we do not have the *y* value yet corresponding to this new *x*. We can predict *y* using

$$\widehat{y}_{New} = \widehat{a} + \widehat{b}x_{New}$$

We have only introduced the simple regression analysis where only one independent variable is available. We will learn multiple regression analysis that involves multiple independent variables after the review of matrix algebra.

## 4.6    Matrix Algebra

The lecture note of matrix algebra was adopted from Carey's Psychology 7291 (Multivariate Statistics) lectures (8/27/1998) (http://psych.colorado.edu/~carey/Courses/PSYC7291/handouts/matrix.algebra.pdf)

### 4.6.1    Definitions

A matrix is a collection of numbers ordered by rows and columns. It is customary to enclose the elements of a matrix in parentheses, brackets, or braces. For example, the following is a matrix,

$$X = \begin{bmatrix} 5 & 8 & 2 \\ -1 & 0 & 7 \end{bmatrix}$$

This matrix has two rows and three columns, so it is referred to as a $2 \times 3$ matrix. The elements of a matrix are numbered in the following way:

$$X = \begin{bmatrix} X_{11} & X_{12} & X_{13} \\ X_{21} & X_{22} & X_{23} \end{bmatrix}$$

That is, the first subscript of an element in a matrix refers to the row and the second subscript refers to the column. It is important to remember this convention when matrix algebra is performed.

A vector is a special type of matrix that has only one row (called a row vector) or one column (called a column vector). Below, $a$ is a column vector while $b$ is a row vector.

$$a = \begin{bmatrix} 7 \\ 2 \\ 3 \end{bmatrix}, \quad b = \begin{bmatrix} 2 & 7 & 1 \end{bmatrix}$$

A scalar is a matrix with only one row and one column. It is customary to denote scalars by italicized, lower case letters (e.g., $x$), to denote vectors by bold, lower case letters (e.g., $\mathbf{x}$), and to denote matrices with more than one row and one column by bold, upper case letters (e.g., $\mathbf{X}$). However, in the current literature, the bold face notation for matrix and vector is no longer required. In the lecture notes of this class, all variables are italicized in plain font, regardless whether they are matrices, vectors, or scalars.

A square matrix has as many rows as it has columns. Matrix $A$ is square but matrix $B$ is not,

$$A = \begin{bmatrix} 1 & 6 \\ 3 & 2 \end{bmatrix}, \quad B = \begin{bmatrix} 1 & 0 & 7 \\ 9 & 3 & -2 \end{bmatrix}$$

A symmetric matrix is a square matrix in which $x_{ij} = x_{ji}$ for all $i$ and $j$. Matrix $A$ is symmetric, but matrix $B$ is not,

$$A = \begin{bmatrix} 9 & 1 & 5 \\ 1 & 6 & 2 \\ 5 & 2 & 7 \end{bmatrix}, \quad B = \begin{bmatrix} 9 & 1 & 5 \\ 2 & 6 & 2 \\ 5 & 1 & 7 \end{bmatrix}$$

A diagonal matrix is a symmetric matrix where all of the off diagonal elements are 0. Matrix $A$ is diagonal.

$$A = \begin{bmatrix} 9 & 0 & 0 \\ 0 & 6 & 0 \\ 0 & 0 & 7 \end{bmatrix}$$

An identity matrix is a diagonal matrix with 1s and only 1s on the diagonal. The identity matrix is almost always denoted as $I$,

$$I = \begin{bmatrix} 1 & 0 & 0 \\ 0 & 1 & 0 \\ 0 & 0 & 1 \end{bmatrix}$$

Sometimes people use the $I_n$ notation for an $n \times n$ identity matrix. For example, the above identity matrix can be written as $I_3$.

### 4.6.2   Matrix Addition and Subtraction

To add two matrices, they both must have the same number of rows and the same number of columns, i.e., the two matrices must have the same dimensions. The elements of the two matrices are simply added together, element by element, to produce the results. That is, for $A + B = C$, we have

$$A_{ij} + B_{ij} = C_{ij}$$

for all $i$ and $j$. Thus,

$$\begin{bmatrix} 1 & 9 & -2 \\ 3 & 6 & 0 \end{bmatrix} + \begin{bmatrix} 8 & -4 & 3 \\ -7 & 1 & 6 \end{bmatrix} = \begin{bmatrix} 9 & 5 & 1 \\ -4 & 7 & 6 \end{bmatrix}$$

Matrix subtraction works in the same way, except that elements are subtracted instead of added. For example,

$$\begin{bmatrix} 1 & 9 & -2 \\ 3 & 6 & 0 \end{bmatrix} - \begin{bmatrix} 8 & -4 & 3 \\ -7 & 1 & 6 \end{bmatrix} = \begin{bmatrix} -7 & 13 & -5 \\ 10 & 5 & -6 \end{bmatrix}$$

### 4.6.3   Matrix Multiplication

There are several rules for matrix multiplication. The first concerns the multiplication between a matrix and a scalar. Here, each element in the product matrix is simply the scalar multiplied by the element in the matrix. That is, for $aB = C$, then

$$aB_{ij} = C_{ij}$$

for all $i$ and $j$. Thus,

$$8 \times \begin{bmatrix} 2 & 6 \\ 3 & 7 \end{bmatrix} = \begin{bmatrix} 8 \times 2 & 8 \times 6 \\ 8 \times 3 & 8 \times 7 \end{bmatrix} = \begin{bmatrix} 16 & 48 \\ 24 & 56 \end{bmatrix}$$

Matrix multiplication involving a scalar is commutative, that is, $aB = Ba$. The next rule involves the multiplication of a row vector by a column vector. To perform this function, the row vector must have as many columns as the column vector has rows. For example,

$$[1 \quad 7 \quad 5]\begin{bmatrix} 2 \\ 4 \\ 1 \end{bmatrix}$$

is legal. However,

$$[1 \quad 7 \quad 5]\begin{bmatrix} 2 \\ 4 \end{bmatrix}$$

is not legal because the row vector has three columns while the column vector has two rows. The product of a row vector multiplied by a column vector will be a scalar. This scalar is simply the sum of the first row vector element multiplied by the first column vector element plus the second row vector element multiplied by the second column vector element plus the product of the third elements, etc. In matrix algebra, if $ab = c$, then

$$\sum_{j=1}^{n} a_j b_j = c$$

Thus,

$$[2 \quad 6 \quad 3]\begin{bmatrix} 8 \\ 1 \\ 4 \end{bmatrix} = 2 \times 8 + 6 \times 1 + 3 \times 4 = 34$$

All other types of matrix multiplication involve the multiplication of a row vector and a column vector. Specifically, in the expression $AB = C$,

$$a_{i.}b_{.j} = \sum_{k=1}^{q} a_{ik}b_{kj} = C_{ij}$$

where $q$ is the number of columns of matrix $A$ (also the number of rows of matrix $B$), $a_{i.}$ is the $i$th row of matrix $A$ and $b_{.j}$ is the $j$th column of matrix $B$. Thus, if

$$A = \begin{bmatrix} 2 & 8 & -1 \\ 3 & 6 & 4 \end{bmatrix} \text{ and } B = \begin{bmatrix} 1 & 7 \\ 9 & -2 \\ 6 & 3 \end{bmatrix}$$

then

$$C_{11} = a_{1.}b_{.1} = [2 \quad 8 \quad -1]\begin{bmatrix} 1 \\ 9 \\ 6 \end{bmatrix} = 2 \times 1 + 8 \times 9 - 1 \times 6 = 68$$

and

$$C_{12} = a_{1.}b_{.2} = [2 \quad 8 \quad -1]\begin{bmatrix} 7 \\ -2 \\ 3 \end{bmatrix} = 2 \times 7 + 8 \times (-2) - 1 \times 3 = -5$$

and

$$C_{21} = a_2.b._1 = \begin{bmatrix} 3 & 6 & 4 \end{bmatrix} \begin{bmatrix} 1 \\ 9 \\ 6 \end{bmatrix} = 3 \times 1 + 6 \times 9 + 4 \times 6 = 81$$

and

$$C_{22} = a_2.b._2 = \begin{bmatrix} 3 & 6 & 4 \end{bmatrix} \begin{bmatrix} 7 \\ -2 \\ 3 \end{bmatrix} = 3 \times 7 + 6 \times (-2) + 4 \times 3 = 21$$

Hence,

$$C = AB = \begin{bmatrix} 2 & 8 & -1 \\ 3 & 6 & 4 \end{bmatrix} \begin{bmatrix} 1 & 7 \\ 9 & -2 \\ 6 & 3 \end{bmatrix} = \begin{bmatrix} 68 & -5 \\ 81 & 21 \end{bmatrix}$$

For matrix multiplication to be legal, the number of columns of the first matrix must be the same as the number of rows of the second matrix. This, of course, is the requirement for multiplying a row vector by a column vector. The resulting matrix will have as many rows as the first matrix and as many columns as the second matrix. Because $A$ has 2 rows and 3 columns while $B$ has 3 rows and 2 columns, the matrix multiplication can legally proceed and the resulting matrix will have 2 rows and 2 columns.

Because of these requirements, matrix multiplication is usually not commutative. That is, usually $AB \neq BA$. And even if $AB$ is a legal operation, there is no guarantee that $BA$ will also be legal. For these reasons, the terms pre-multiply and post-multiply are often encountered in matrix algebra while they are seldom encountered in scalar algebra.

One special case to be aware of is when a column vector is post-multiplied by a row vector. That is, what is

$$\begin{bmatrix} -3 \\ 4 \\ 7 \end{bmatrix} \begin{bmatrix} 8 & 2 \end{bmatrix} = ?$$

In this case, one simply follows the rules given above for the multiplication of two matrices. Note that the first matrix has one column and the second matrix has one row, so the matrix multiplication is legal. The resulting matrix will have as many rows as the first matrix (3) and as many columns as the second matrix (2). Hence, the result is

$$\begin{bmatrix} -3 \\ 4 \\ 7 \end{bmatrix} \begin{bmatrix} 8 & 2 \end{bmatrix} = \begin{bmatrix} -24 & -6 \\ 32 & 8 \\ 56 & 14 \end{bmatrix}$$

Similarly, multiplication of a matrix with a vector (or a vector times a matrix) will also conform to the multiplication of two matrices. For example,

$$\begin{bmatrix} 8 & 5 \\ 6 & 1 \\ 9 & 4 \end{bmatrix} \begin{bmatrix} 2 \\ 8 \\ 5 \end{bmatrix}$$

is an illegal operation because the number of columns in the first matrix (2) does not match the number of rows in the second matrix (3). However,

$$\begin{bmatrix} 8 & 5 \\ 6 & 1 \\ 9 & 4 \end{bmatrix} \begin{bmatrix} 3 \\ 7 \end{bmatrix} = \begin{bmatrix} 8 \times 3 + 5 \times 7 \\ 6 \times 3 + 1 \times 7 \\ 9 \times 3 + 4 \times 7 \end{bmatrix} = \begin{bmatrix} 59 \\ 25 \\ 55 \end{bmatrix}$$

and

$$\begin{bmatrix} 2 & 7 & 3 \end{bmatrix} \begin{bmatrix} 8 & 5 \\ 6 & 1 \\ 9 & 4 \end{bmatrix} = \begin{bmatrix} 2 \times 8 + 7 \times 6 + 3 \times 9 & 2 \times 5 + 7 \times 1 + 3 \times 4 \end{bmatrix} = \begin{bmatrix} 85 & 29 \end{bmatrix}$$

are legal matrix multiplications.

The last special case of matrix multiplication involves the identity matrix, $I$. The identity matrix operates as the number 1 does in scalar algebra. That is, any vector or matrix multiplied by an identity matrix is simply the original vector or original matrix. Hence, $aI = a$, $IX = X$, etc. Note, however, that a scalar multiplied by an identify matrix becomes a diagonal matrix with the scalars on the diagonal. That is,

$$4 \times \begin{bmatrix} 1 & 0 \\ 0 & 1 \end{bmatrix} = \begin{bmatrix} 4 & 0 \\ 0 & 4 \end{bmatrix}$$

not 4. This should be verified by reviewing the rules for multiplying a scalar and a matrix given above.

### 4.6.4 Matrix Transpose

The transpose of a matrix is denoted by a prime ($A^{'}$) or a superscript t or T ($A^{t}$ or $A^{T}$). The first row of a matrix becomes the first column of the transposed matrix, the second row of the matrix becomes the second column of the transposed matrix, etc. Thus, if

$$A = \begin{bmatrix} 2 & 7 & 1 \\ 8 & 6 & 4 \end{bmatrix}$$

then

$$A^{T} = \begin{bmatrix} 2 & 8 \\ 7 & 6 \\ 1 & 4 \end{bmatrix}$$

The transpose of a row vector will be a column vector, and the transpose of a column vector will be a row vector. The transpose of a symmetric matrix is simply the original matrix.

### 4.6.5 Matrix Inverse

In scalar algebra, the inverse of a number is that number which, when multiplied by the original number, gives a product of 1. Hence, the inverse of $x$ is simple $1/x$, or, in slightly different notation, $x^{-1}$. In matrix algebra, the inverse of a matrix is that matrix which, when multiplied by the original matrix, gives an identity matrix. The inverse of a matrix is denoted by the superscript "$-1$". Hence,

$$AA^{-1} = A^{-1}A = I$$

A matrix must be square to have an inverse, but not all square matrices have an inverse. In some cases, the inverse does not exist. For covariance and correlation matrices, an inverse always exists, provided that there are more subjects than there are variables and that every variable has a variance greater than 0.

It is important to know what an inverse is in multivariate statistics, but it is not necessary to know how to compute an inverse. For a small $2 \times 2$ matrix, the inverse is

$$A^{-1} = \begin{bmatrix} a_{11} & a_{12} \\ a_{21} & a_{22} \end{bmatrix}^{-1} = \frac{1}{a_{11}a_{22} - a_{12}a_{21}} \begin{bmatrix} a_{22} & -a_{12} \\ -a_{21} & a_{11} \end{bmatrix}$$

For example,

$$\begin{bmatrix} 4 & 2 \\ 3 & 5 \end{bmatrix}^{-1} = \frac{1}{4 \times 5 - 2 \times 3} \begin{bmatrix} 5 & -2 \\ -3 & 4 \end{bmatrix} = \frac{1}{14} \begin{bmatrix} 5 & -2 \\ -3 & 4 \end{bmatrix} = \begin{bmatrix} 5/14 & -2/14 \\ -3/14 & 4/14 \end{bmatrix} = \begin{bmatrix} 0.3571 & -0.1429 \\ -0.2143 & 0.2857 \end{bmatrix}$$

Below is a very important property of matrix inverse: The inverse of a diagonal matrix is also diagonal with each diagonal element taking the reciprocal of the diagonal element of the original matrix. For example, if $A$ is

$$A = \begin{bmatrix} a_{11} & 0 & 0 \\ 0 & a_{22} & 0 \\ 0 & 0 & a_{33} \end{bmatrix}$$

then

$$A^{-1} = \begin{bmatrix} a_{11} & 0 & 0 \\ 0 & a_{22} & 0 \\ 0 & 0 & a_{33} \end{bmatrix}^{-1} = \begin{bmatrix} 1/a_{11} & 0 & 0 \\ 0 & 1/a_{22} & 0 \\ 0 & 0 & 1/a_{33} \end{bmatrix}$$

Therefore, inverting a diagonal matrix is extremely fast and no computer program is needed.

### 4.6.6  Generalized Inverse

Let $A$ be a square matrix of less than full rank. The inverse of such a matrix does not exist. We can define a generalized inverse of $A$, denoted by $A^-$, which is defined as a matrix with the following property,

$$AA^-A = A$$

A generalized inverse is not unique. There are many different generalized inverses, all having the above property. In fact, a generalized inverse can be defined for a rectangular matrix. The following $4 \times 4$ matrix does not have an inverse, but a generalized inverse can be defined,

$$A = \begin{bmatrix} 5 & 3 & 1 & -4 \\ 8 & 5 & 2 & 3 \\ 21 & 13 & 5 & 2 \\ 3 & 2 & 1 & 7 \end{bmatrix} \quad \text{and} \quad A^- = \begin{bmatrix} 0.0132756 & 0.0073877 & 0.0280509 & -0.005888 \\ 0.0067766 & 0.0051658 & 0.0171083 & -0.001611 \\ 0.0002777 & 0.002944 & 0.0061656 & 0.0026662 \\ -0.066489 & 0.0285508 & -0.009387 & 0.0950397 \end{bmatrix}$$

You can verify that $AA^-A = A$.

### 4.6.7 Determinant of a Matrix

The determinant of a matrix is a scalar and is denoted by |A| or det(A). The determinant has very important mathematical properties, but it is very difficult to provide a substantive definition. For covariance and correlation matrices, the determinant is a number that is sometimes used to express the "generalized variance" of the matrix. That is, covariance matrices with small determinants denote variables that are redundant or highly correlated. Matrices with large determinants denote variables that are independent of one another. The determinant has several very important properties for some multivariate stats (e.g., change in $R^2$ in multiple regression can be expressed as a ratio of determinants). Only idiots calculate the determinant of a large matrix by hand. We will try to avoid them. For a small $2 \times 2$ matrix, the determinant is

$$| A |= \det(A) = \det \begin{bmatrix} a_{11} & a_{12} \\ a_{21} & a_{22} \end{bmatrix} = a_{11}a_{22} - a_{12}a_{21}$$

For example,

$$\det \begin{bmatrix} 4 & 2 \\ 3 & 5 \end{bmatrix} = 4 \times 5 - 2 \times 3 = 14$$

The scalar involved in matrix inverse is the determinant of the matrix. Therefore, the explicit form of the inverse of a $2 \times 2$ matrix is

$$A^{-1} = \begin{bmatrix} a_{11} & a_{12} \\ a_{21} & a_{22} \end{bmatrix}^{-1} = \frac{1}{| A |} \begin{bmatrix} a_{22} & -a_{12} \\ -a_{21} & a_{11} \end{bmatrix}$$

### 4.6.8 Trace of a Matrix

The trace of a matrix is sometimes, although not always, denoted by tr(A). The trace is used only for square matrices and equals the sum of the diagonal elements of the matrix. For example,

$$\text{tr} \begin{bmatrix} 3 & 7 & 2 \\ -1 & 6 & 4 \\ 9 & 0 & -5 \end{bmatrix} = 3 + 6 - 5 = 4$$

### 4.6.9 Orthogonal Matrices

Only square matrices may be orthogonal matrices although not all square matrices are orthogonal. An orthogonal matrix satisfies the following property,

$$AA^T = I$$

Thus, the inverse of an orthogonal matrix is simply the transpose of that matrix. Orthogonal matrices are very important in factor analysis. Matrices of eigenvectors (discussed below) are orthogonal matrices.

## 4.6.10 Eigenvalues and Eigenvectors

The eigenvalues and eigenvectors of a matrix play an important part in multivariate analysis. This discussion applies to correlation matrices and covariance matrices that (1) have more subjects than variables, (2) have non-zero variances, (3) are calculated from data having no missing values, and (4) no variable is a perfect linear combination of the other variables. Any such covariance matrix $V$ can be mathematically decomposed into a product,

$$V = UDU^{-1}$$

where $U$ is a square matrix of eigenvectors and $D$ is a diagonal matrix with the eigenvalues on the diagonal. If there are $n$ variables, both $U$ and $D$ will be $n \times n$ matrices. Eigenvalues are also called characteristic roots or latent roots. Eigenvectors are sometimes referred to as characteristic vectors or latent vectors. Each eigenvalue has its associated eigenvector. That is, the first eigenvalue in $D$ ($d_{11}$) is associated with the first column vector in $U$, the second diagonal element in $D$ (i.e., the second eigenvalue or $d_{22}$) is associated with the second column in $U$, and so on. Actually, the order of the eigenvalues is arbitrary from a mathematical viewpoint. However, if the diagonals of $D$ become switched around, then the corresponding columns in $U$ must also be switched appropriately. It is customary to order the eigenvalues so that the largest one is in the upper left ($d_{11}$) and then they proceed in descending order until the smallest one is in $d_{nn}$, or the extreme lower right. The eigenvectors in $U$ are then ordered accordingly so that column 1 in $U$ is associated with the largest eigenvalue and column $n$ is associated with the lowest eigenvalue.

Some important points about eigenvectors and eigenvalues are:

1. The eigenvectors are scaled so that $U$ is an orthogonal matrix. Thus, $U^T = U^{-1}$ and $UU^T = I$. Thus, each eigenvector is said to be orthogonal to all the other eigenvectors.
2. The eigenvalues will all be greater than 0.0, providing that the four conditions outlined above for $V$ are true.
3. For a covariance matrix, the sum of the diagonal elements of the covariance matrix equals the sum of the eigenvalues, or in math terms, $\text{tr}(V) = \text{tr}(D)$. For a correlation matrix, all the eigenvalues sum to $n$, the number of variables. Furthermore, in case you have a burning passion to know about it, the determinant of $V$ equals the product of the eigenvalues of $V$.
4. It is a royal pain to compute eigenvalues and eigenvectors, so don't let me catch you doing it.
5. VERY IMPORTANT: The decomposition of a matrix into its eigenvalues and eigenvectors is a mathematical/geometric decomposition. The decomposition literally rearranges the dimensions in an $n$ dimensional space ($n$ being the number of variables) in such a way that the axis (e.g., North-South, East-West) are all perpendicular. This rearrangement may, but is not guaranteed to, uncover an important biological construct or even to have a biologically meaningful interpretation.
6. ALSO VERY IMPORTANT: An eigenvalue tells us the proportion of total variability in a matrix associated with its corresponding eigenvector. Consequently, the eigenvector that corresponds to the highest eigenvalue tells us the dimension (axis) that generates the maximum amount of individual variability in the variables. The next eigenvector is a dimension perpendicular to the first that accounts for the second largest amount of variability, and so on.

Eigen decomposition is a very useful tool in genome-wide association studies (GWAS), which will be introduced in Chap. 19. Let $K$ be an $n \times n$ symmetric matrix with an existing inverse. Let $\lambda$ be a positive number (often a parameter in quantitative genetics). Let us define the following matrix,

$$H = K\lambda + I$$

where $I$ is an $n \times n$ identity matrix. The inverse of matrix $H$ is $H^{-1} = (K\lambda + I)^{-1}$, including parameter $\lambda$. If the parameter ($\lambda$) changes, the $H$ matrix also changes and it must be reinverted again. When the dimension of matrix $K$ is very large, repeatedly inverting a large matrix presents a huge cost in computational speed and computer memory. Let $K = UDU^T$ be decomposed into eigenvectors and eigenvalues. Since $UU^T = UU^{-1} = I$ and $D$ is diagonal. We can easily invert matrix $H$ using

$$H^{-1} = (K\lambda + I)^{-1} = \left(UDU^T\lambda + I\right)^{-1} = U\left(D\lambda + UU^T\right)^{-1}U^T = U(D\lambda + I)^{-1}U^T$$

Since $D$ is a diagonal matrix, we have

$$(D\lambda + I)^{-1} = \begin{bmatrix} \lambda d_1 + 1 & \cdots & 0 \\ \vdots & \ddots & \vdots \\ 0 & \cdots & \lambda d_n + 1 \end{bmatrix}^{-1} = \begin{bmatrix} 1/(\lambda d_1 + 1) & \cdots & 0 \\ \vdots & \ddots & \vdots \\ 0 & \cdots & 1/(\lambda d_n + 1) \end{bmatrix}$$

when $\lambda$ changes, we only need to reinvert this diagonal matrix $D\lambda + I$.

## 4.7 Linear Combination, Quadratic Form, and Covariance Matrix

1. Linear combination

Let $a$ be an $n \times 1$ vector of constants and $x$ be an $n \times 1$ vector of variables, then

$$L = \sum_{j=1}^{n} a_j x_j = a^T x$$

is a linear combination of $x$. For example,

$$L = 3x_1 + 4x_2 + 5x_3 = \begin{bmatrix} 3 & 4 & 5 \end{bmatrix} \begin{bmatrix} x_1 \\ x_2 \\ x_3 \end{bmatrix}$$

where

$$a = \begin{bmatrix} 3 \\ 4 \\ 5 \end{bmatrix} \quad \text{and} \quad x = \begin{bmatrix} x_1 \\ x_2 \\ x_3 \end{bmatrix}$$

2. Quadratic form

Let $A$ be an $n \times n$ constant matrix (usually symmetrical) and $x$ be an $n \times 1$ vector of variables, then

$$Q = x^T A x$$

is a quadratic form. For example, let

$$x = \begin{bmatrix} x_1 \\ x_2 \\ x_3 \end{bmatrix} \quad \text{and} \quad A = \begin{bmatrix} a_{11} & a_{12} & a_{13} \\ a_{21} & a_{22} & a_{23} \\ a_{31} & a_{32} & a_{33} \end{bmatrix}$$

then

$$x^T A x = \begin{bmatrix} x_1 & x_2 & x_3 \end{bmatrix} \begin{bmatrix} a_{11} & a_{12} & a_{13} \\ a_{21} & a_{22} & a_{23} \\ a_{31} & a_{32} & a_{33} \end{bmatrix} \begin{bmatrix} x_1 \\ x_2 \\ x_3 \end{bmatrix} = \sum_{i=1}^{n} \sum_{j=1}^{n} a_{ij} x_i x_j$$

is a quadratic form. A quadratic form is a $1 \times 1$ matrix, i.e., a scalar. However, many computer languages, e.g., R, treat a scalar differently from a $1 \times 1$ matrix. For example, if $a$ is a scalar, you can multiply a matrix, say $A_{3 \times 4}$, by this scalar to get $aA$, which is a $3 \times 4$ matrix. However, if $a$ is a $1 \times 1$ matrix, $aA$ is illegal because the number of columns of $a$ does not match the number of rows of $A$. The IML procedure in SAS (PROC IML) does not distinguish a scalar from a $1 \times 1$ matrix. A $1 \times 1$ matrix can be converted into a scalar using the `drop()` function in R. For example, if $a$ is a $1 \times 1$ matrix and $A$ is a $3 \times 4$ matrix, $a * A$ is illegal but $drop(a) * A$ is legal. In Chap. 17 when we learn the Smith-Hazel selection index, we need to multiply a scalar by a matrix, but the scalar will be calculated as a $1 \times 1$ matrix. The `drop()` function is required if the R language is used to calculate genetic gains from an index selection.

3. Variance-covariance matrix

Let $x = \begin{bmatrix} x_1 & x_2 & x_3 \end{bmatrix}^T$ be a vector of variables and denote the variance of $x_i$ by $\text{var}(x_i) = \sigma_i^2$ and the covariance between $x_i$ and $x_j$ by $\text{cov}(x_i, x_j) = \sigma_{ij}$. The variance-covariance matrix is

$$\text{var}(x) = \begin{bmatrix} \text{var}(x_1) & \text{cov}(x_1, x_2) & \text{cov}(x_1, x_3) \\ \text{cov}(x_2, x_1) & \text{var}(x_2) & \text{cov}(x_2, x_3) \\ \text{cov}(x_3, x_1) & \text{cov}(x_3, x_2) & \text{var}(x_3) \end{bmatrix} = \begin{bmatrix} \sigma_1^2 & \sigma_{12} & \sigma_{13} \\ \sigma_{21} & \sigma_2^2 & \sigma_{23} \\ \sigma_{31} & \sigma_{32} & \sigma_3^2 \end{bmatrix}$$

We often just call it the variance matrix. A variance matrix is always symmetric. Let $y = \begin{bmatrix} y_1 & y_2 \end{bmatrix}^T$ be a vector of two other variables. We can define the following $3 \times 2$ covariance matrix between vector $x$ and vector $y$,

$$\text{cov}(x, y^T) = \begin{bmatrix} \text{cov}(x_1, y_1) & \text{cov}(x_1, y_2) \\ \text{cov}(x_2, y_1) & \text{cov}(x_2, y_2) \\ \text{cov}(x_3, y_1) & \text{cov}(x_3, y_2) \end{bmatrix}$$

A covariance matrix is not necessarily square unless the number of elements in $x$ and the number of elements in $y$ are equal. If a covariance matrix is square, it is not necessarily symmetric.

4. Expectation and variance of a linear combination

Let $x = \begin{bmatrix} x_1 & x_2 & x_3 \end{bmatrix}^T$ be a vector of variables and $\mu = \begin{bmatrix} \mu_1 & \mu_2 & \mu_3 \end{bmatrix}^T$ be the expectations of the $x$ variables. Let $a = \begin{bmatrix} a_1 & a_2 & a_2 \end{bmatrix}^T$ be some constants. The expectation of a linear combination, $L = a^T x$, is

$$E(L) = E(a^T x) = a^T E(x) = a^T \mu = \sum_{j=1}^{3} a_j \mu_j$$

Let $y = \begin{bmatrix} y_1 & y_2 & y_3 \end{bmatrix}^T$ be another vector of variables with the same dimension as vector $x$, then

$$E(x + y) = E(x) + E(y) = \mu_x + \mu_y$$

where $\mu_x$ and $\mu_y$ are vectors of expectations of $x$ and $y$, respectively. The variance of a linear combination, $L = a^T x$, is

$$\text{var}(L) = \text{var}(a^T x) = a^T \text{var}(x) a = \begin{bmatrix} a_1 & a_2 & a_3 \end{bmatrix} \begin{bmatrix} \sigma_1^2 & \sigma_{12} & \sigma_{13} \\ \sigma_{21} & \sigma_2^2 & \sigma_{23} \\ \sigma_{31} & \sigma_{32} & \sigma_3^2 \end{bmatrix} \begin{bmatrix} a_1 \\ a_2 \\ a_3 \end{bmatrix} = \sum_{i=1}^{3} \sum_{j=1}^{3} a_i a_j \sigma_{ij}$$

where $\sigma_{ij} = \sigma_i^2$ if $i = j$.

5. Covariance between two linear combinations

Let $x = \begin{bmatrix} x_1 & x_2 & x_3 \end{bmatrix}^T$ be a vector of variables and $a = \begin{bmatrix} a_1 & a_2 & a_2 \end{bmatrix}^T$ be a constant vector. Let $y = \begin{bmatrix} y_1 & y_2 \end{bmatrix}^T$ be another set of variables and $b = \begin{bmatrix} b_1 & b_2 \end{bmatrix}^T$ be another vector of constants. Define $L_1 = a^T x$ and $L_2 = b^T y$ as two linear combinations. The covariance between $L_1$ and $L_2$ is

$$\text{cov}(L_1, L_2) = \text{cov}\left(a^T x, y^T b\right) = a^T \text{cov}\left(x, y^T\right) b$$

which is further expanded to

$$\text{cov}(L_1, L_2) = \begin{bmatrix} a_1 & a_2 & a_3 \end{bmatrix} \begin{bmatrix} \text{cov}(x_1, y_1) & \text{cov}(x_1, y_2) \\ \text{cov}(x_2, y_1) & \text{cov}(x_2, y_2) \\ \text{cov}(x_3, y_1) & \text{cov}(x_3, y_2) \end{bmatrix} \begin{bmatrix} b_1 \\ b_2 \end{bmatrix}$$

We now conclude the statistical review. Over 90% of the equations in quantitative genetics can be derived either directly or indirectly from the statistical techniques reviewed in this chapter. Therefore, for students lacking the necessary statistical background, this chapter should not be escaped by the instructor.

# Genetic Effects of Quantitative Traits

**5**

## 5.1 Phenotype, Genotype, and Environmental Error

A phenotypic value is the value observed when a character is measured on an individual. For example, the body weight of a particular pig is 300 lb. The value of 300 lb. is the phenotypic value of this pig for the trait called body weight. The genotypic value of a trait is the average of the phenotypic values of all individuals who have the same genotype. For example, there are 1000 individuals in a population who have the same genotype, $A_1A_2$. The phenotypic values of the 1000 individuals are $\{1.25, 0.89, \ldots, 2.10\}_{1000 \times 1}$. Because all the 1000 individuals have the same genotype, the average values of their phenotypic values,

$$\mathrm{E}(y) = G_{12} = \frac{1}{1000}(1.25 + 0.89 + \ldots + 2.10) = 1.75$$

is the genotypic value of genotype $A_1A_2$. Theoretically, the sample size should be infinitely large to make the average phenotypic value the genotypic value. When the sample size is finite, the average will be slightly different from the actual genotypic value.

An environmental error (deviation) is the difference between the phenotypic value and the genotypic value. Take the $A_1A_2$ genotype for example, let $y_{j(12)}$ be the phenotypic value of individual $j$ with genotype $A_1A_2$. We can write the following model to describe $y_{j(12)}$,

$$y_{j(12)} = G_{12} + E_{j(12)}$$

where $E_{j(12)}$ is the environmental deviation. Let $\bar{y}_{(12)}$ be the average of all the $y$'s with the same genotype, we have

$$\bar{y}_{(12)} = \frac{1}{n}\sum_{j=1}^{n} y_{j(12)} = \frac{1}{n}\sum_{j=1}^{n}\left(G_{12} - E_{j(12)}\right) = G_{12} + \frac{1}{n}\sum_{j=1}^{n} E_{j(12)} = G_{12} + \overline{E}_{(12)}$$

where

$$\lim_{n\to\infty} \overline{E}_{12} = \lim_{n\to\infty}\frac{1}{n}\sum_{j=1}^{n} E_{j(12)} = 0$$

Therefore,

$$\lim_{n\to\infty} \bar{y}_{12} = G_{12} + \lim_{n\to\infty} \overline{E}_{12} = G_{12}$$

Because environmental deviations can be positive and negative, the average of $E$'s for all individuals will be zero, and thus, the average phenotypic value is the genotypic value.

© Springer Nature Switzerland AG 2022
S. Xu, *Quantitative Genetics*, https://doi.org/10.1007/978-3-030-83940-6_5

## 5.2    The First Genetic Model

Let $P$ be the phenotype, $G$ be the genotype, and $E$ be the environmental error, the first genetic model is (Falconer and Mackay 1996; Lynch and Walsh 1998)

$$P = G + E$$

For the purpose of deduction, we must assign some values to the genotypes under discussion. This is done in the following way. Considering a single locus with two alleles, $A_1$ and $A_2$, we denote the genotypic value of $A_1A_1$ by $+a$, the genotypic value of $A_1A_2$ by $d$ and the genotypic value of $A_2A_2$ by $-a$. We shall adopt the convention that $A_1$ represents the allele that increases the genotypic value, called the "high allele." We thus have genotypic values defined in the first scale (scale 1), as demonstrated in Fig. 5.1.

The second scale (scale 2) of genotypic values are defined in the original genotypic values, denoted by $G_{11}$, $G_{12}$, and $G_{22}$, respectively, for the three genotypes, $A_1A_1$, $A_1A_2$, and $A_2A_2$, as illustrated in Fig. 5.1. The genotypic values defined in the first scale (scale 1) have special definitions: $a$ is called the additive genetic effect and $d$ is called the dominance genetic effect. If there is no dominance effect, the genotypic value of $A_1A_2$ should be half way between $A_1A_1$ and $A_2A_2$, i.e., $d = 0$. With dominance, $d \neq 0$, and the degree of dominance is usually expressed by the ratio $d/a$. Note that the value 0 reflects the mid-point in scale 1 (a relative mid-point value). The $\mu$ in Fig. 5.1 is the middle point value shown in scale 2 but it should belong to scale 1.

The relationships between genotypic values of scale 1 and scale 2 are shown below:

$$\mu = \frac{1}{2}(G_{11} + G_{22})$$

$$a = G_{11} - \frac{1}{2}(G_{11} + G_{22})$$

$$d = G_{12} - \frac{1}{2}(G_{11} + G_{22})$$

The genotypic values must be defined in scale 2 first before we can calculate the genotypic values in scale 1. We now show an example on how to define the genotypic values in the two different scales.

In Table 5.1, the plus sign (+) represents the wild allele and $pg$ represents the mutant allele (the dwarfing gene). First, we need to decide which allele should be labeled $A_1$ and which allele should be labeled $A_2$. Since the ++ genotype has a higher

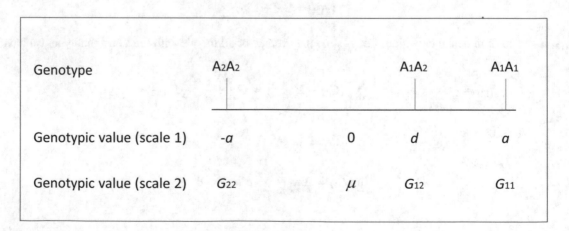

**Fig. 5.1** Genotypes and genotypic values defined in two different scales

**Table 5.1** Example showing the dwarfing gene (pg) effect in the mouse

| Genotype | ++ | +pg | pg pg |
|---|---|---|---|
| Average weight (g) | 14 | 12 | 6 |

**Table 5.2** Genotypic values in two different scales

| Genotype | + +($A_1A_1$) | + pg($A_1A_2$) | pg pg($A_2A_2$) |
|---|---|---|---|
| Genotypic value in scale 2 | 14 | 12 | 6 |
| Genotypic value in scale 1 | 4 | 2 | −4 |

value than the mutant genotype, $14 > 6$, we label the + allele as $A_1$ and thus the mutant allele as $A_2$. The three genotypes are then labeled $A_1A_1$ for ++, $A_1A_2$ for +pg and $A_2A_2$ for pgpg. The three genotypic values defined in scale 2 are $G_{11} = 14$, $G_{12} = 12$, and $G_{22} = 6$, respectively. The actual mid-point value ($\mu$), the additive effect ($a$), and the dominance effect ($d$) are

$$\mu = \frac{1}{2}(G_{11} + G_{22}) = \frac{1}{2}(14 + 6) = 10$$

$$a = G_{11} - \frac{1}{2}(G_{11} + G_{22}) = 14 - 10 = 4$$

$$d = G_{12} - \frac{1}{2}(G_{11} + G_{22}) = 12 - 10 = 2$$

The degree of dominance is $d/a = 2/4 = 0.5$. These genotypic values are shown in Table 5.2. Note that the mid-point is just a convenient way to express the position of the population. There are several other scales to express the position of the population. Remember to adjust the data by the mid-point to convert the genotypic values in scale 2 into genotypic values in scale 1. All subsequent analyses are conducted using the genotypic values defined in scale 1.

Before we proceed to the next section, let us use a matrix notation to define the relationship of the genotypes in the two scales. First, let us write the following equations to describe the relationships between scale 1 and scale 2,

$$\mu = \frac{1}{2}(G_{11} + G_{22}) = +\frac{1}{2}G_{11} + \frac{0}{2}G_{12} + \frac{1}{2}G_{22}$$

$$a = G_{11} - \frac{1}{2}(G_{11} + G_{22}) = +\frac{1}{2}G_{11} - \frac{0}{2}G_{12} - \frac{1}{2}G_{22}$$

$$d = G_{12} - \frac{1}{2}(G_{11} + G_{22}) = -\frac{1}{2}G_{11} + \frac{2}{2}G_{12} - \frac{1}{2}G_{22}$$

We then rewrite these relationships in matrix notation,

$$\begin{bmatrix} \mu \\ a \\ d \end{bmatrix} = \begin{bmatrix} +\frac{1}{2} & 0 & +\frac{1}{2} \\ +\frac{1}{2} & 0 & -\frac{1}{2} \\ -\frac{1}{2} & 1 & -\frac{1}{2} \end{bmatrix} \begin{bmatrix} G_{11} \\ G_{12} \\ G_{22} \end{bmatrix}$$

Define

$$\gamma = \begin{bmatrix} \mu \\ a \\ d \end{bmatrix}, \quad G = \begin{bmatrix} G_{11} \\ G_{12} \\ G_{22} \end{bmatrix} \quad \text{and} \quad H = \begin{bmatrix} +\frac{1}{2} & 0 & +\frac{1}{2} \\ +\frac{1}{2} & 0 & -\frac{1}{2} \\ -\frac{1}{2} & 1 & -\frac{1}{2} \end{bmatrix}$$

We then have a compact matrix expression for the relationship,

$$\gamma = HG$$

The reverse relationship is

$$G_{11} = \mu + a$$
$$G_{12} = \mu + d$$
$$G_{22} = \mu - a$$

In matrix notation,

$$\begin{bmatrix} G_{11} \\ G_{12} \\ G_{22} \end{bmatrix} = \begin{bmatrix} 1 & 1 & 0 \\ 1 & 0 & 1 \\ 1 & -1 & 0 \end{bmatrix} \begin{bmatrix} \mu \\ a \\ d \end{bmatrix}$$

Let

$$H^{-1} = \begin{bmatrix} 1 & 1 & 0 \\ 1 & 0 & 1 \\ 1 & -1 & 0 \end{bmatrix}$$

be the inverse matrix of $H$, we have $G = H^{-1}\gamma$. You can verify that $H^{-1}H = I$ (see Chap. 4). The relationship of the genotypic values in the two scales is verbally expressed by.

Genotypic value in scale 1 = Genotypic value in scale 2 – Actual mid-point value.

## 5.3    Population Mean

Let $p$ and $q$ be the frequencies of allele $A_1$ and $A_2$, respectively. The population mean (expectation) in scale 1 is calculated using Table 5.3. The population mean in scale 1 is the expectation of the genotypic value in scale 1,

$$M = \mathrm{E}(G) = ap^2 + d(2pq) + (-a)q^2 = a(p - q) + 2pqd$$

The population mean ($M$) has been expressed as a deviation from the mid-point value ($\mu$). Therefore, we should add the middle point value back to the population mean to get the actual population mean,

Actual population mean $= M + \mu$

We now use the dwarfing gene in the mouse as an example to show how to calculate the population mean. Assume that the frequency of the + allele ($A_1$) is $p = 0.9$ and that of the pg allele ($A_2$) is $q = 0.1$. We have

$$M = a(p - q) + 2pqd = 4 \times (0.9 - 0.1) + 2 \times 0.9 \times 0.1 \times 2 = 3.56$$

The actual population mean is $M + \mu = 3.56 + 10 = 13.56$.

**Table 5.3**  Genotypes, genotype frequencies, and genotypic values

| Genotype | Frequency $f(G)$ | Genotypic value ($G$) | $G \times f(G)$ |
|---|---|---|---|
| $A_1A_1$ | $p^2$ | $a$ | $a(p^2)$ |
| $A_1A_2$ | $2pq$ | $d$ | $d(2pq)$ |
| $A_2A_2$ | $q^2$ | $-a$ | $(-a)q^2$ |

## 5.4 Average Effect of Gene or Average Effect of Allele

Genotypes are not transmitted from generation to generation; it is the allele that is inherited from parent to offspring (Mendel's first law). The average effect of an allele is the mean deviation from the population mean of individuals which received that allele from one parent, the allele received from the other parent is randomly sampled from the population. This verbal description does not help. We now introduce the average effect of each allele.

First, let us look at the average effect of allele $A_1$. Consider one allele being $A_1$ and the other allele being randomly sampled from the population. The probability that the other allele is $A_1$ equals $p$. If this happens, the genotype will be $A_1A_1$ with a genotypic value $a$. On the other hand, the probability that the other allele is $A_2$ equals $q$. If the other allele is sampled, then the genotype will be $A_1A_2$ with a genotypic value $d$. So, in the population containing individuals with one $A_1$ allele being fixed, a proportion of $p$ will be $A_1A_1$ and a proportion of $q$ will be $A_1A_2$. Define the mean of this population by $M_{A_1}$, we have

$$M_{A_1} = ap + dq$$

We then express this mean by the deviation from the population mean, $M$, resulting in the defined *average effect of gene $A_1$*:

$$\alpha_1 = M_{A_1} - M = (ap + dq) - [a(p - q) + 2pqd] = q[a + d(q - p)]$$

The *average effect of $A_2$* is similarly defined by first obtaining

$$M_{A_2} = pd + q(-a) = pd - qa$$

and then

$$\alpha_2 = M_{A_2} - M = (pd - qa) - [a(p - q) + 2pqd] = -p[a + d(q - p)]$$

## 5.5 Average Effect of Gene Substitution

In a biallelic system (population with only two different alleles), the average effect of gene substitution is more informative than the average effect of gene. The average effect of gene substitution is defined by the difference between the two average effects of alleles, i.e.,

$$\alpha = \alpha_1 - \alpha_2 = a + d(q - p) \tag{5.1}$$

A biological interpretation of $\alpha$ is the change of population mean by replacing $A_2$ with $A_1$. If one allele is fixed at $A_2$, this population will contain $A_1A_2$ with probability $p$ and $A_2A_2$ with probability $q$. The mean of such a population is $pd - qa$. If we replace $A_2$ by $A_1$, the population will contain $A_1A_1$ and $A_1A_2$ instead. The mean value of such a population will be $pa + qd$. The average effect of gene substitution is

$$\alpha = (pa + qd) - (pd - qa) = a + (q - p)d$$

which is the same as Eq. (5.1). The average effects of genes (alleles) and the average effect of gene substitution can be derived using the conditional distribution and conditional mean approach.

## 5.6 Alternative Definition of Average Effect of Gene Substitution

Recall that the average effect of gene substitution is defined by Eq. (5.1). We can define the following $2 \times 2$ joint distribution table of gametes (Table 5.4).

**Table 5.4** Joint distribution of two gametes

| ♂\♀ | $A_1(p)$ | $A_2(q)$ | |
|---|---|---|---|
| $A_1(p)$ | $A_1A_1(a)\, p^2$ | $A_1A_2(d)\, pq$ | $\alpha_1$ |
| $A_2(q)$ | $A_2A_1(d)\, qp$ | $A_2A_2(-a)\, q^2$ | $\alpha_2$ |
| | $\alpha_1$ | $\alpha_2$ | |

The conditional mean of a population with one allele fixed at $A_1$ is

$$M_{A_1} = a\frac{p^2}{p^2+pq} + d\frac{pq}{p^2+pq} = a\frac{p^2}{p} + d\frac{pq}{p} = ap + dq$$

Subtracting the overall population mean from it leads to

$$\alpha_1 = M_{A_1} - M = ap + dq - [a(p-q) + 2pqd] = q[a + d(q-p)]$$

The conditional mean of a population with one allele fixed at $A_2$ is

$$M_{A_2} = d\frac{qp}{qp+q^2} + (-a)\frac{q^2}{qp+q^2} = d\frac{qp}{q} + (-a)\frac{q^2}{q} = dp - aq$$

Subtracting the overall population mean from it leads to

$$\alpha_2 = M_{A_2} - M = dp - aq - [a(p-q) + 2pqd] = -p[a + d(q-p)]$$

Therefore, the average effect of gene substitution is

$$\alpha = \alpha_1 - \alpha_2 = q[a + d(q-p)] - (-p)[a + d(q-p)] = a + d(q-p)$$

## 5.7    Breeding Value

The breeding value of an individual (or a genotype) is defined as the sum of effects of the two alleles carried by the individual (or the genotype). Table 5.5 shows the breeding values of the three genotypes in a Hardy-Weinberg equilibrium population. When the population is in H-W equilibrium, the average breeding value is zero.

$$E(A) = p^2(2q\alpha) + 2pq(q-p)\alpha + q^2(-2p\alpha) = 0$$

You can verify this using rules like $p^2 + 2pq + q^2 = (p+q)^2 = 1$ and $p + q = 1$.

## 5.8    Dominance Deviation

Dominance deviation is defined as the difference between the genotypic value and the breeding value, $D = G - A - M$. For example, the dominance deviation of genotype $A_1A_1$ is $-2q^2d$ (see Table 5.6). You can verify that, under H-W equilibrium, the average dominance deviation is zero.

$$E(D) = p^2(-2q^2d) + (2pq)(2pqd) + q^2(-2p^2d) = 0$$

**Table 5.5**  Genotype frequencies and breeding values

| Genotype | Frequency $f(A)$ | Breeding value $(A)$ |
|---|---|---|
| $A_1A_1$ | $p^2$ | $2\alpha_1 = 2q\alpha$ |
| $A_1A_2$ | $2pq$ | $\alpha_1 + \alpha_2 = (q - p)\alpha$ |
| $A_2A_2$ | $q^2$ | $2\alpha_2 = -2p\alpha$ |

**Table 5.6**  Distribution of dominance deviations

| Genotype | Frequency $f(x)$ | Breedingvalue $(A)$ | Genotypicvalue $(G)$ | Dominance deviation $(D)$ |
|---|---|---|---|---|
| $A_1A_1$ | $p^2$ | $2\alpha_1 = 2q\alpha$ | $a$ | $a - 2\alpha_1 - M = -2q^2d$ |
| $A_1A_2$ | $2pq$ | $\alpha_1 + \alpha_2 = (q - p)\alpha$ | $d$ | $d - (\alpha_1 + \alpha_2) - M = 2pqd$ |
| $A_2A_2$ | $q^2$ | $2\alpha_2 = -2p\alpha$ | $-a$ | $(-a) - 2\alpha_2 - M = -2p^2d$ |

Again, try to use the following rules, $p + q = 1$ and $p^2 + 2pq + q^2 = (p + q)^2 = 1$, as much as possible to simplify your derivation. The genetic model is now further partitioned into an additive effect and a dominance effect,

$$P = G + E = A + D + E$$

## 5.9  Epistatic Effects Involving Two Loci

What we have discussed so far are the genetic effects for a single locus, i.e., the additive and dominance effects. The dominance effect is also called the allelic interaction effect because it is due to the interaction between the two alleles within a locus. When we consider two or more loci simultaneously, we must consider the interaction effects between alleles of different loci. Such interaction effects are called epistatic effects, which are also called non-allelic interaction effects (Falconer and Mackay 1996; Lynch and Walsh 1998; Hartl and Clark 2007). Let us denote the first locus by $A$ and the second locus by $B$. Each locus has three genotypes (in a biallelic system) and thus the two locus model should deal with $3 \times 3 = 9$ genotypes (see Table 5.7).

Consider loci $A$ and $B$, the marginal genotypic values are defined separately for the two loci. They are expressed by

$$G_A = \mu_A + Z_A a_A + W_A d_A$$
$$G_B = \mu_B + Z_B a_B + W_B d_B$$

where $\mu_A$ and $\mu_B$ are the mid-point values for loci A and B, respectively. The other terms are defined by Tables 5.8 and 5.9, where $Z_A$ and $W_A$ are genotype indicator variables for locus $A$. Similarly, $Z_B$ and $W_B$ are the genotype indicator variables for locus $B$.

If epistatic effects are absent, the total genotypic value for the two loci is

$$G = G_A + G_B = \mu_A + \mu_B + Z_A a_A + Z_B a_B + W_A d_A + W_B d_B$$

We simply add the two locus-specific genotypic values together to obtain the total genotypic values. When epistatic effects are present, we need to add additional terms due to interactions between alleles of the two loci. Let us combine the means of the two loci into a single term, called $\mu = \mu_A + \mu_B$. This modified (epistatic effect) model is

$$G = \mu + Z_A a_A + Z_B a_B + W_A d_A + W_B d_B$$
$$+ Z_A Z_B \gamma_{aa} + Z_A W_B \gamma_{ad} + W_A Z_B \gamma_{da} + W_A W_B \gamma_{dd}$$

where $\gamma_{aa}$, $\gamma_{ad}$, $\gamma_{da}$, and $\gamma_{dd}$ are called the additive by additive (AA), additive by dominance (AD), dominance by additive (DA), and dominance by dominance (DD) effects, respectively. The AD and DA are different effects because AD is the additive effect of the first locus interacting with the dominance effect of the second locus while DA is the dominance effect of the first locus interacting with the additive effect of the second locus. However, AD and DA may not be separable in some

**Table 5.7** Nine genotypes for two loci of a random population

| $A \backslash B$ | $B_1B_1$ | $B_1B_2$ | $B_2B_2$ |
|---|---|---|---|
| $A_1A_1$ | $A_1A_1\ B_1B_1$ | $A_1A_1\ B_1B_2$ | $A_1A_1\ B_2B_2$ |
| $A_1A_2$ | $A_1A_2\ B_1B_1$ | $A_1A_2\ B_1B_2$ | $A_1A_2\ B_2B_2$ |
| $A_2A_2$ | $A_2A_2\ B_1B_1$ | $A_2A_2\ B_1B_2$ | $A_2A_2\ B_2B_2$ |

**Table 5.8** Genotypic values for locus $A$

| Genotype | Genotypic value ($G_A$) | $Z_A$ | $W_A$ |
|---|---|---|---|
| $A_1A_1$ | $\mu_A + a_A$ | 1 | 0 |
| $A_1A_2$ | $\mu_A + d_A$ | 0 | 1 |
| $A_2A_2$ | $\mu_A - a_A$ | −1 | 0 |

**Table 5.9** Genotypic values for locus $B$

| Genotype | Genotypic value ($G_B$) | $Z_B$ | $W_B$ |
|---|---|---|---|
| $B_1B_1$ | $\mu_B + a_B$ | 1 | 0 |
| $B_1B_2$ | $\mu_B + d_B$ | 0 | 1 |
| $B_2B_2$ | $\mu_B - a_B$ | −1 | 0 |

situations. They are definitely not separable if the genotypes are not observed. As a result, they are often combined into a single term called AD effect. Using the epistatic effect model and the definitions of the $Z$ and $W$ variables, we can find the genotypic effect for each particular genotype. For example, assume that the genotype is $A_1A_1B_1B_2$, we immediately know that $Z_A = 1$, $Z_B = 0$, $W_A = 0$, and $W_B = 1$. It is the combination of the first genotype ($A_1A_1$) of locus $A$ and the second genotype ($B_1B_2$) of locus $B$. Therefore,

$$G_{12} = (1)\mu + (1)a_A + (0)a_B + (0)d_A + (1)d_B$$
$$+ (1)(0)\gamma_{aa} + (1)(1)\gamma_{ad} + (0)(0)\gamma_{da} + (0)(1)\gamma_{dd}$$

Let us keep only the non-zero terms so that the above genotypic value becomes

$$G_{12} = \mu + a_A + d_B + \gamma_{ad}$$

Using matrix notation, we can write the above equation as

$$G_{12} = \begin{bmatrix} 1 & 1 & 0 & 0 & 1 & 0 & 1 & 0 & 0 \end{bmatrix} \begin{bmatrix} \mu \\ a_A \\ a_B \\ d_A \\ d_B \\ \gamma_{aa} \\ \gamma_{ad} \\ \gamma_{da} \\ \gamma_{dd} \end{bmatrix}$$

**Table 5.10** Coefficients of linear contrasts ($H^{-1}$ matrix) for two loci

| $G \backslash \beta$ | $\mu$ | $\alpha_A$ | $\alpha_B$ | $d_A$ | $d_B$ | $\gamma_{aa}$ | $\gamma_{ad}$ | $\gamma_{da}$ | $\gamma_{dd}$ |
|---|---|---|---|---|---|---|---|---|---|
| $G_{11}$ | 1 | 1 | 1 | 0 | 0 | 1 | 0 | 0 | 0 |
| $G_{12}$ | 1 | 1 | 0 | 0 | 1 | 0 | 1 | 0 | 0 |
| $G_{13}$ | 1 | 1 | -1 | 0 | 0 | -1 | 0 | 0 | 0 |
| $G_{21}$ | 1 | 0 | 1 | 1 | 0 | 0 | 0 | 1 | 0 |
| $G_{22}$ | 1 | 0 | 0 | 1 | 1 | 0 | 0 | 0 | 1 |
| $G_{23}$ | 1 | 0 | -1 | 1 | 0 | 0 | 0 | -1 | 0 |
| $G_{31}$ | 1 | -1 | 1 | 0 | 0 | -1 | 0 | 0 | 0 |
| $G_{32}$ | 1 | -1 | 0 | 0 | 1 | 0 | -1 | 0 | 0 |
| $G_{33}$ | 1 | -1 | -1 | 0 | 0 | 1 | 0 | 0 | 0 |

**Table 5.11** The $H$ matrix, the inverse matrix of $H^{-1}$

| $\beta \backslash G$ | $G_{11}$ | $G_{12}$ | $G_{13}$ | $G_{21}$ | $G_{22}$ | $G_{23}$ | $G_{31}$ | $G_{32}$ | $G_{33}$ |
|---|---|---|---|---|---|---|---|---|---|
| $\mu$ | 0.25 | 0 | 0.25 | 0 | 0 | 0 | 0.25 | 0 | 0.25 |
| $\alpha_A$ | 0.25 | 0 | 0.25 | 0 | 0 | 0 | -0.25 | 0 | -0.25 |
| $\alpha_B$ | 0.25 | 0 | -0.25 | 0 | 0 | 0 | 0.25 | 0 | -0.25 |
| $d_A$ | -0.25 | 0 | -0.25 | 0.5 | 0 | 0.5 | -0.25 | 0 | -0.25 |
| $d_B$ | -0.25 | 0.5 | -0.25 | 0 | 0 | 0 | -0.25 | 0.5 | -0.25 |
| $\gamma_{aa}$ | 0.25 | 0 | -0.25 | 0 | 0 | 0 | -0.25 | 0 | 0.25 |
| $\gamma_{ad}$ | -0.25 | 0.5 | -0.25 | 0 | 0 | 0 | 0.25 | -0.5 | 0.25 |
| $\gamma_{da}$ | -0.25 | 0 | 0.25 | 0.5 | 0 | -0.5 | -0.25 | 0 | 0.25 |
| $\gamma_{dd}$ | 0.25 | -0.5 | 0.25 | -0.5 | 1 | -0.5 | 0.25 | -0.5 | 0.25 |

Let $G_{ij}$ be the genotypic value for the $i$th genotype of locus $A$ and the $j$th genotype of locus $B$, for $i, j = 1, 2, 3$. We can write the coefficients for all $G_{ij}$ in Table 5.10.

Let $H^{-1}$ be the $9 \times 9$ coefficient matrix shown in Table 5.10. Define

$$\beta = \begin{bmatrix} \mu & a_A & a_B & d_A & d_B & \gamma_{aa} & \gamma_{ad} & \gamma_{da} & \gamma_{dd} \end{bmatrix}^T$$

as a column vector of the *genetic effects*, and

$$G = \begin{bmatrix} G_{11} & G_{12} & G_{13} & G_{21} & G_{22} & G_{23} & G_{31} & G_{32} & G_{33} \end{bmatrix}^T$$

as a column vector of the *genotypic values*. The superscript $^T$ means the transposition of the original vector. The transposition of a row vector is a column vector and the transposition of a column vector is a row vector. The relationship between $G$ (genotypic values) and $\beta$ (genetic effects) is

$$G = H^{-1}\beta$$

The reverse relationship is

$$\beta = HG$$

where $H^{-1}$ is the inverse of $H$, shown in Table 5.11.

When we learned the additive and dominance effects in the single locus model, we defined genotypic values in two different scales. For the epistatic model in two loci, we also define genotypic values in two different scales: scale 1 is the $\beta$ scale and scale 2 is the $G$ scale. The relationship of the two scales is $\beta = HG$. To estimate all kinds of genetic effects, we need a sampled dataset and perform a regression or linear model analysis, which will be introduced in Chap. 8. After we learned epistasis, the genetic model is further modified into

$$P = G + E = A + D + \underbrace{AA + AD + DA + DD}_{\text{Epistatic effects}} + E$$

Alternatively, we can make the following partitioning,

$$P = G + E = A + \underbrace{D + AA + AD + DA + DD}_{\text{Non-additive effects}} + E$$

## 5.10  An Example of Epistatic Effects

Two loci on the rice genome were selected for evaluating their effects on the grain weight trait, called 1000 grain weight (KGW). The experiment was replicated in years 1998 and 1999. For the combined data (pooled for the 2 years), Xu (2013) detected two loci (bin729 and bin1064) with epistatic effects for KGW using a mixed model procedure. For demonstration purpose in this chapter, we will use a simple method to evaluate epistasis with the 1998 data only. The nine genotypic values are listed in Table 5.12. For example, $G_{23} = 25.46367$ is the value of genotype $A_1A_2B_2B_2$. These genotypic values were estimated from the actual data. However, for the purpose of demonstration, we treat the estimated genotypic values as the true values. They are the genotypic values in scale 2 under the epistatic model. We now arrange the nine genotypic values in a column vector in the following order:

$$G = \begin{bmatrix} G_{11} \\ G_{12} \\ G_{13} \\ G_{21} \\ G_{22} \\ G_{23} \\ G_{31} \\ G_{32} \\ G_{33} \end{bmatrix} = \begin{bmatrix} 27.0185 \\ 27.0014 \\ 26.2599 \\ 26.0836 \\ 25.6126 \\ 25.4637 \\ 24.8704 \\ 24.1392 \\ 25.9409 \end{bmatrix}$$

We then use $\beta = HG$ to convert the nine genotypic values in scale 2 into nine genetic effects (including the grand mean) in scaled 1,

$$\beta = \begin{bmatrix} \mu \\ a_A \\ a_B \\ d_A \\ d_B \\ \gamma_{aa} \\ \gamma_{ad} \\ \gamma_{da} \\ \gamma_{dd} \end{bmatrix} = \begin{bmatrix} 0.25 & 0 & 0.25 & 0 & 0 & 0 & 0.25 & 0 & 0.25 \\ 0.25 & 0 & 0.25 & 0 & 0 & 0 & -0.25 & 0 & -0.25 \\ 0.25 & 0 & -0.25 & 0 & 0 & 0 & 0.25 & 0 & -0.25 \\ -0.25 & 0 & -0.25 & 0.5 & 0 & 0.5 & -0.25 & 0 & -0.25 \\ -0.25 & 0.5 & -0.25 & 0 & 0 & 0 & -0.25 & 0.5 & -0.25 \\ 0.25 & 0 & -0.25 & 0 & 0 & 0 & -0.25 & 0 & 0.25 \\ -0.25 & 0.5 & -0.25 & 0 & 0 & 0 & 0.25 & -0.5 & 0.25 \\ -0.25 & 0 & 0.25 & 0.5 & 0 & -0.5 & -0.25 & 0 & 0.25 \\ 0.25 & -0.5 & 0.25 & -0.5 & 1 & -0.5 & 0.25 & -0.5 & 0.25 \end{bmatrix} \begin{bmatrix} 27.0185 \\ 27.0014 \\ 26.2599 \\ 26.0836 \\ 25.6126 \\ 25.4637 \\ 24.8704 \\ 24.1392 \\ 25.9409 \end{bmatrix} = \begin{bmatrix} 26.02242 \\ 0.616774 \\ -0.077975 \\ -0.24878 \\ -0.45212 \\ 0.457275 \\ 0.814307 \\ 0.387989 \\ 0.291084 \end{bmatrix}$$

Table 5.12  Genotypic values of KGW for two loci in the rice population (year 1998)

| $A \backslash B$ | $B_1B_1$ | $B_1B_2$ | $B_2B_2$ |
|---|---|---|---|
| $A_1A_1$ | 27.01847 | 27.00138 | 26.25993 |
| $A_1A_2$ | 26.08362 | 25.6126 | 25.46367 |
| $A_2A_2$ | 24.87036 | 24.13922 | 25.94094 |

All genetic effects are very large except $a_B = -0.07801$ (the additive effect of locus $B$) is negligible. Note again, these genetic effects are not the estimated genetic effects; they are defined genetic effects in scale 1. We can verify that the reverse relationship $G = H^{-1}\beta$ also holds.

$$
G = \begin{bmatrix} G_{11} \\ G_{12} \\ G_{13} \\ G_{21} \\ G_{22} \\ G_{23} \\ G_{31} \\ G_{32} \\ G_{33} \end{bmatrix} = \begin{bmatrix} 1 & 1 & 1 & 0 & 0 & 1 & 0 & 0 & 0 \\ 1 & 1 & 0 & 0 & 1 & 0 & 1 & 0 & 0 \\ 1 & 1 & -1 & 0 & 0 & -1 & 0 & 0 & 0 \\ 1 & 0 & 1 & 1 & 0 & 0 & 0 & 1 & 0 \\ 1 & 0 & 0 & 1 & 1 & 0 & 0 & 0 & 1 \\ 1 & 0 & -1 & 1 & 0 & 0 & 0 & -1 & 0 \\ 1 & -1 & 1 & 0 & 0 & -1 & 0 & 0 & 0 \\ 1 & -1 & 0 & 0 & 1 & 0 & -1 & 0 & 0 \\ 1 & -1 & -1 & 0 & 0 & 1 & 0 & 0 & 0 \end{bmatrix} \begin{bmatrix} 26.02242 \\ 0.616774 \\ -0.0779575 \\ -0.24878 \\ -0.45212 \\ 0.457279 \\ 0.814307 \\ 0.387989 \\ 0.291084 \end{bmatrix} = \begin{bmatrix} 27.0185 \\ 27.0014 \\ 26.2599 \\ 26.0836 \\ 25.6126 \\ 25.4637 \\ 24.8704 \\ 24.1392 \\ 25.9409 \end{bmatrix}
$$

The $3 \times 3$ table given in Table 5.12 is often presented graphically in the following two forms of 2D plot (Figs. 5.2 and 5.3).

In the absence of epistasis, the three lines should be parallel. In this case, the three lines are not parallel, and thus the plot indicates strong epistasis. Another way of presenting epistasis is to use a 3D plot as shown in Fig. 5.4. The differences of the nine genotypic values are obvious from the 3D plot, indicating existence of epistasis. The 3D plot looks nicer than the 2D plot, but it is not as informative as the 2D plot.

We have not yet tested the significance of the epistatic effects, which will be studied in Chaps. 7 and 8.

**Fig. 5.2** The 2D plot of interaction (epistasis) between locus Bin726 (**A**) and Bin1064 (**B**) of the rice genome for trait KGW collected from year 1998

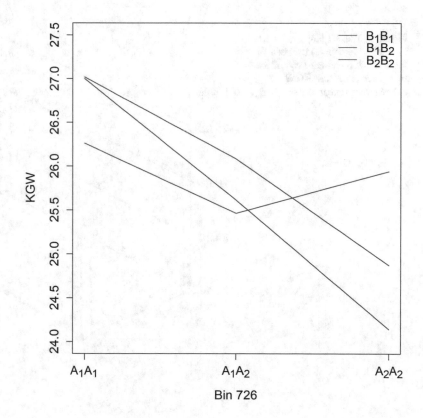

**Fig. 5.3** The 2D plot of interaction (epistasis) between locus Bin1064 (**B**) and Bin726 (**A**) of the rice genome for trait KGW collected from year 1998

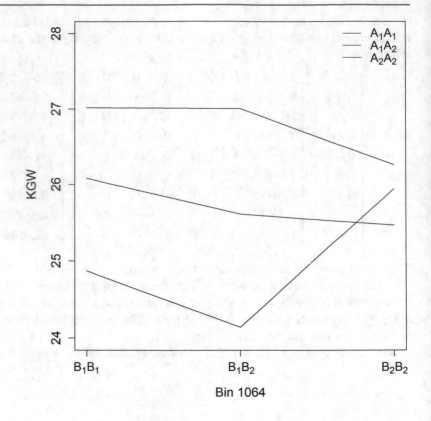

**Fig. 5.4** The 3D plot of interaction (epistasis) between locus Bin726 and Bin1064 of the rice genome for trait KGW collected from year 1998

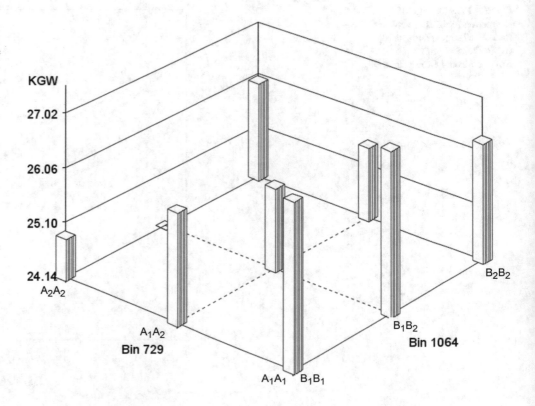

## 5.11 Population Mean of Multiple Loci

Recall that the genetic model for two loci in the presence of epistasis is

$$G = Z_A a_A + Z_B a_B + W_A d_A + W_B d_B$$
$$+ Z_A Z_B \gamma_{aa} + Z_A W_B \gamma_{ad} + W_A Z_B \gamma_{da} + W_A W_B \gamma_{dd}$$

where the mid-point value $\mu$ has been excluded. The population mean is the expectation of $G$, i.e.,

$$M = \mathrm{E}(G) = \mathrm{E}(Z_A)a_A + \mathrm{E}(Z_B)a_B + \mathrm{E}(W_A)d_A + \mathrm{E}(W_B)d_B$$
$$+ \mathrm{E}(Z_A Z_B)\gamma_{aa} + \mathrm{E}(Z_A W_B)\gamma_{ad} + \mathrm{E}(W_A Z_B)\gamma_{da} + \mathrm{E}(W_A W_B)\gamma_{dd}$$

To derive the population mean, we need to find the expectation terms of the genotype indicator variables. Assume that the population is in Hardy-Weinberg equilibrium and the two loci are not correlated (linkage equilibrium). Table 5.13 and Table 5.14 show the distributions of the genotype indicator variables.

From these tables, we can derive the expectations of various terms,

$$\mathrm{E}(Z_A) = p_A - q_A$$
$$\mathrm{E}(W_A) = 2p_A q_A$$
$$\mathrm{E}(Z_B) = p_B - q_B$$
$$\mathrm{E}(W_B) = 2p_B q_B$$

and

$$\mathrm{E}(Z_A Z_B) = \mathrm{E}(Z_A)\mathrm{E}(Z_B) = (p_A - q_A)(p_B - q_B)$$
$$\mathrm{E}(Z_A W_B) = \mathrm{E}(Z_A)\mathrm{E}(W_B) = 2p_B q_B(p_A - q_A)$$
$$\mathrm{E}(W_A Z_B) = \mathrm{E}(W_A)\mathrm{E}(Z_B) = 2p_A q_A(p_B - q_B)$$
$$\mathrm{E}(W_A W_B) = \mathrm{E}(Z_A)\mathrm{E}(W_B) = (2p_B q_B)(2p_A q_A)$$

Therefore,

$$M = (p_A - q_A)a_A + 2p_A q_A d_A + (p_B - q_B)a_B + 2p_B q_B d_B$$
$$+ (p_A - q_A)(p_B - q_B)\gamma_{aa} + 2p_B q_B(p_A - q_A)\gamma_{ad}$$
$$+ 2p_A q_A(p_B - q_B)\gamma_{da} + 4p_B q_B p_A q_A \gamma_{dd}$$

**Table 5.13** Genotype indicator variables for locus $A$

| Genotype | Frequency | $Z_A$ | $W_A$ |
| --- | --- | --- | --- |
| $A_1 A_1$ | $p_A^2$ | 1 | 0 |
| $A_1 A_2$ | $2p_A q_A$ | 0 | 1 |
| $A_2 A_2$ | $q_A^2$ | −1 | 0 |

**Table 5.14** Genotype indicator variables for locus $B$

| Genotype | Frequency | $Z_B$ | $W_B$ |
| --- | --- | --- | --- |
| $B_1 B_1$ | $p_B^2$ | 1 | 0 |
| $B_1 B_2$ | $2p_B q_B$ | 0 | 1 |
| $B_2 B_2$ | $q_B^2$ | −1 | 0 |

Let

$$M_A = (p_A - q_A)a_A + 2p_Aq_Ad_A$$

and

$$M_B = (p_B - q_B)a_B + 2p_Bq_Bd_B$$

be the locus-specific population means. If the epistatic effects are ignored, the population mean of the two locus model is

$$M = M_A + M_B$$

Extending to multiple loci ($k = 1, \ldots, m$), say $m$ loci, the above formula becomes

$$M = \sum_{k=1}^{m} [(p_k - q_k)a_k + 2p_kq_kd_k]$$

## References

Falconer DS, Mackay TFC. Introduction to quantitative genetics. Harlow, UK: Addison Wesley Longman; 1996.
Hartl DL, Clark AG. *Principles of population genetics*. Sunderland, MA: Sinauer associates; 2007.
Lynch M, Walsh B. Genetics and analysis of quantitative traits. Sunderland, MA: Sinauer Associates, Inc.; 1998.
Xu S. Mapping quantitative trait loci by controlling polygenic background effects. Genetics. 2013;195:1209–22.

# Genetic Variances of Quantitative Traits

Quantitative genetics centers around the study of genetic and environmental variation, for it is in terms of variation that the primary questions are formulated (Falconer and Mackay 1996). Following the partitioning of the genetic effects,

$$P = G + E = A + D + AA + AD + DA + DD + E$$

We now partition the total phenotypic variance into

$$V_P = V_G + V_E = V_A + \underbrace{V_D + V_{AA} + V_{AD} + V_{DA} + V_{DD}}_{\text{Non-additive variances}} + V_E$$

The relative importance of each effect can only be answered if it is expressed in terms of the variance attributable to the different sources of variation. Here, we introduce the concept of heritability. There are two types of heritability, one is the broad sense heritability defined by

$$H^2 = \frac{V_G}{V_P}$$

The other is the narrow sense heritability defined as

$$h^2 = \frac{V_A}{V_P}$$

The narrow sense heritability is very important in animal and plant breeding because it reflects the proportion of genetic variance that can be transmitted to the next generation.

## 6.1 Total Genetic Variance

The genetic variance can be derived using theTable 6.1. Recall that the population mean is

$$M = \mathrm{E}(G) = p^2 a + 2pqd + q^2(-a) = a(p - q) + 2pqd$$

The definition of the total genetic variance is

$$V_G = \mathrm{var}(G) = \mathrm{E}(G^2) - \mathrm{E}^2(G) = p^2 a^2 + 2pqd^2 + q^2(-a)^2 - M^2 = 2pq\alpha^2 + (2pqd)^2$$

where

© Springer Nature Switzerland AG 2022
S. Xu, *Quantitative Genetics*, https://doi.org/10.1007/978-3-030-83940-6_6

$$\alpha = a + d(q - p)$$

is the average effect of gene substitution.

## 6.2    Additive and Dominance Variances

The additive variance is the variance of the breeding values (see Table 6.1). Mathematically, it is defined as

$$V_A = \mathrm{var}(A) = \mathrm{E}(A^2) - \mathrm{E}^2(A)$$

where

$$\mathrm{E}(A) = p^2(2q\alpha) + 2pq(q - p)\alpha + q^2(-2p\alpha) = \ldots = 0$$

and

$$\mathrm{E}(A^2) = p^2(2q\alpha)^2 + 2pq[(q - p)\alpha]^2 + q^2(-2p\alpha)^2 = \ldots = 2pq\alpha^2$$

Therefore,

$$V_A = \mathrm{E}(A^2) - \mathrm{E}^2(A) = 2pq\alpha^2 - 0^2 = 2pq\alpha^2$$

The dominance variance is the variance of the dominance deviations (see Table 6.1). It is defined as

$$V_D = \mathrm{var}(D) = \mathrm{E}(D^2) - \mathrm{E}^2(D)$$

where

$$\mathrm{E}(D) = p^2(-2q^2 d) + 2pq(2pqd) + q^2(-2p^2 d) = \ldots = 0$$

and

$$\mathrm{E}(D^2) = p^2(-2q^2 d)^2 + 2pq(2pqd)^2 + q^2(-2p^2 d)^2 = \ldots = (2pqd)^2$$

Therefore,

$$V_D = \mathrm{E}(D^2) - \mathrm{E}^2(D) = (2pqd)^2 - 0^2 = (2pqd)^2$$

The covariance between the additive and dominance effects is defined as

$$\mathrm{cov}(A, D) = \mathrm{E}(A \times D) - \mathrm{E}(A)\mathrm{E}(D) = \mathrm{E}(A \times D)$$

because of the zero expectations of $A$ and $D$. The expectation of the product is

**Table 6.1** Distribution of genetic effects

| Genotype | Frequency | Breeding value ($A$) | Genotypic value ($G$) | Dominance deviation ($D$) |
|---|---|---|---|---|
| $A_1 A_1$ | $p^2$ | $2q\alpha$ | $a$ | $-2q^2 d$ |
| $A_1 A_2$ | $2pq$ | $(q - p)\alpha$ | $d$ | $2pqd$ |
| $A_2 A_2$ | $q^2$ | $-2p\alpha$ | $-a$ | $-2p^2 d$ |

$$E(A \times D) = p^2(2q\alpha)(-2q^2d) + 2pq[(q-p)\alpha](2pqd) + q^2(-2p\alpha)(-2p^2d)$$
$$= -4p^2q^3\alpha d + 4p^2q^2(p-q)\alpha d + 4q^2p^3\alpha d$$
$$= 4\alpha dp^2q^2[-q + (q-p) + p] = 0$$

Therefore,

$$\mathrm{cov}(A, D) = E(A \times D) - E(A)E(D) = 0 - 0 \times 0 = 0$$

The total genetic variance under the additive and dominance model is

$$V_G = V_A + V_D = 2pq\alpha^2 + (2pqd)^2$$

## 6.3 Epistatic Variances (General Definition)

Epistasis can happen when two or more loci are considered. Epistatic effects are defined as various interaction effects between alleles of different loci, in contrast to the dominance effect, which is defined as the interaction between alleles within a locus. The epistatic variance can be partitioned into different variance components,

$$V_I = V_{AA} + V_{AD} + V_{DA} + V_{DD} + V_{AAA} + \cdots$$

Note that $V_{AD}$ is the variance of interaction between the additive effect of the first locus and the dominance effect of the second locus. It is different from $V_{DA}$, which is the variance of interaction between the dominance effect of the first locus and the additive effect of the second locus. For more than two loci, higher order interactions may also be included, e.g., $V_{AAA}$ is the interaction of additive effects of all three loci and is called the additive×additive×additive variance. The list goes on forever. However, higher order interactions are often not as important as lower order interactions. In addition, they are difficult to estimate and test (requiring a very large sample size). Therefore, we often ignore higher order interactions and only consider pair-wise interactions (between two loci).

## 6.4 Epistatic Variance Between Two Loci

Consider locus $A$ and locus $B$ and assume that the two loci are independent (in linkage equilibrium). Using the genotype indicator variables, we have defined the epistatic genetic model in Chap. 5, which is rewritten here

$$G = \mu + Z_A a_A + Z_B a_B + W_A d_A + W_B d_B$$
$$+ Z_A Z_B \gamma_{aa} + Z_A W_B \gamma_{ad} + W_A Z_B \gamma_{da} + W_A W_B \gamma_{dd}$$

When none of the genotype indicator variables are correlated, the total genetic variance can be written as

$$V_G = \mathrm{var}(Z_A a_A) + \mathrm{var}(Z_B a_B) + \mathrm{var}(W_A d_A) + \mathrm{var}(W_B d_B)$$
$$+ \mathrm{var}(Z_A Z_B \gamma_{aa}) + \mathrm{var}(Z_A W_B \gamma_{ad}) + \mathrm{var}(W_A Z_B \gamma_{da}) + \mathrm{var}(W_A W_B \gamma_{dd})$$

According to the properties of variance, we can further express the above variance using

$$V_G = \mathrm{var}(Z_A)a_A^2 + \mathrm{var}(Z_B)a_B^2 + \mathrm{var}(W_A)d_A^2 + \mathrm{var}(W_B)d_B^2$$
$$+ \mathrm{var}(Z_A Z_B)\gamma_{aa}^2 + \mathrm{var}(Z_A W_B)\gamma_{ad}^2 + \mathrm{var}(W_A Z_B)\gamma_{da}^2 + \mathrm{var}(W_A W_B)\gamma_{dd}^2$$

**Table 6.2**  Genotypic values for locus $A$

| Genotype | Frequency | $Z_A$ | $W_A$ |
|---|---|---|---|
| $A_1A_1$ | $p_A^2 = 1/4$ | 1 | 0 |
| $A_1A_2$ | $2p_Aq_A = 1/2$ | 0 | 1 |
| $A_2A_2$ | $q_A^2 = 1/4$ | $-1$ | 0 |

Unfortunately, mutual independences of the coefficients are often violated. In some special populations, the assumption holds well. Kao and Zeng (2002) showed that this assumption holds exactly for an $F_2$ population where $p_A = p_B = q_A = q_B = 1/2$ and the population is in Hardy-Weinberg equilibrium. We now show the variance of each coefficient for the $F_2$ population. Let us consider locus $A$ with $p_A$ denoting the frequency of $A_1$ and $q_A$ denoting the frequency of allele $A_2$ (see Table 6.2). In an $F_2$ population, $p_A = q_A = 1/2$, which makes the derivation of the variances extremely easy. The variance of $Z_A$ is

$$\text{var}(Z_A) = \text{E}\left(Z_A^2\right) - \text{E}^2(Z_A) = \left(p_A^2 + q_A^2\right) - \left(p_A - q_A\right)^2 = 1/2$$

where $\text{E}(Z_A) = p_A^2 - q_A^2 = p_A - q_A = 0$ and $\text{E}\left(Z_A^2\right) = p_A^2 + q_A^2 = 1/2$. The variance of $W_A$ is

$$\text{var}(W_A) = \text{E}\left(W_A^2\right) - \text{E}^2(W_A) = 2p_Aq_A - (2p_Aq_A)^2 = 2p_Aq_A(1 - 2p_Aq_A) = 1/4$$

We can show that the covariance between $Z_A$ and $W_A$ is zero.

$$\text{cov}(Z_A, W_A) = \text{E}(Z_AW_A) - \text{E}(Z_A)\text{E}(W_A) = 0 - 2p_Aq_A(p_A - q_A) = 0$$

because $p_A = q_A = 1/2$. For the coefficients of the epistatic effects, the variances are derived using the properties of variance and covariance. When $Z_A$ and $Z_B$ are not correlated, we have

$$\text{var}(Z_AZ_B) = \text{E}\left[(Z_AZ_B)^2\right] - \text{E}^2(Z_AZ_B)$$

$$= \text{E}\left(Z_A^2\right)\text{E}\left(Z_B^2\right) - \text{E}^2(Z_A)\text{E}^2(Z_B)$$

$$= \left(p_A^2 + q_A^2\right)\left(p_B^2 + q_B^2\right) - \left(p_A - q_A\right)^2\left(p_B - q_B\right)^2$$

$$= \left(p_A^2 + q_A^2\right)\left(p_B^2 + q_B^2\right)$$

$$= 1/4$$

The remaining terms are

$$\text{var}(Z_AW_B) = \text{E}\left[(Z_AW_B)^2\right] - \text{E}^2(Z_AW_B)$$

$$= \text{E}\left(Z_A^2\right)\text{E}\left(W_B^2\right) - \text{E}^2(Z_A)\text{E}^2(W_B)$$

$$= 2p_Bq_B\left(p_A^2 + q_A^2\right) - (2p_Bq_B)^2\left(p_A - q_A\right)^2$$

$$= 2p_Bq_B\left(p_A^2 + q_A^2\right)$$

$$= 1/4$$

$$\text{var}(W_A Z_B) = \text{E}\left[(W_A Z_B)^2\right] - \text{E}^2(W_A Z_B)$$

$$= 2p_A q_A\left(p_B^2 + q_B^2\right) - (2p_A q_A)^2(p_B - q_B)^2$$

$$= 2p_A q_A\left(p_B^2 + q_B^2\right)$$

$$= 1/4$$

$$\text{var}(W_A W_B) = \text{E}\left[(W_A W_B)^2\right] - \text{E}^2(W_A W_B)$$

$$= \text{E}(W_A^2)\text{E}(W_B^2) - \text{E}^2(W_A)\text{E}^2(W_B)$$

$$= (2p_A q_A)(2p_B q_B) - (2p_A q_A)^2(2p_B q_B)^2$$

$$= (2p_A q_A)(2p_B q_B)[1 - (2p_A q_A)(2p_B q_B)]$$

$$= 3/16$$

Therefore, the total genetic variance in an $F_2$ population is defined as

$$V_G = \frac{1}{2}a_A^2 + \frac{1}{2}a_B^2 + \frac{1}{4}d_A^2 + \frac{1}{4}d_B^2 + \frac{1}{4}\gamma_{aa}^2 + \frac{1}{4}\gamma_{ad}^2 + \frac{1}{4}\gamma_{da}^2 + \frac{3}{16}\gamma_{dd}^2$$

Now, let us define $V_A = \frac{1}{2}a_A^2 + \frac{1}{2}a_B^2$, $V_D = \frac{1}{4}d_A^2 + \frac{1}{4}d_B^2$, $V_{AA} = \frac{1}{4}\gamma_{aa}^2$, $V_{AD} = \frac{1}{4}\gamma_{ad}^2$, $V_{DA} = \frac{1}{4}\gamma_{da}^2$ and $V_{DD} = \frac{3}{16}\gamma_{dd}^2$. Therefore,

$$V_G = V_A + V_D + V_{AA} + V_{AD} + V_{DA} + V_{DD}$$

which completes the partitioning of the total genetic variance. How to estimate these genetic variances will be dealt later in the course.

## 6.5   Average Effect of Gene Substitution and Regression Coefficient

The average effect of gene substitution turns out to be the regression coefficient of the genotypic value on the number of $A_1$ allele in the genotypes. We are going to prove this relationship here. Let us define $X$ as the number of high alleles ($A_1$) in the genotype. As a result, $X = 2$ in the $A_1 A_1$ genotype, $X = 1$ in the heterozygote and $X = 0$ in $A_2 A_2$. Let $Y$ be the genotypic value defined in scale 2 (see Fig. 5.1 in Chap. 5), i.e., $Y = G_{11} = \mu + a$ for $A_1 A_1$, $Y = G_{12} = \mu + d$ for $A_1 A_2$ and $Y = G_{22} = \mu - a$ for $A_2 A_2$. The joint distribution of $X$ and $Y$ is summarized in Table 6.3. We have derived the expectation of $X$ as $\text{E}(X) = 2p$ and the expectation of $Y$ as $\text{E}(Y) = M + \mu$. The centered $X$ and $Y$ variables are defined as $X - \text{E}(X)$ and $Y - \text{E}(Y)$, respectively. The joint distribution of the centered variables is given in Table 6.4. Let us define $\delta_{11} = -2q^2 d$, $\delta_{12} = 2pqd$, and $\delta_{22} = -2p^2 d$ as the dominance deviations. We can see the relationship of the centered $Y$ and $X$ as

$$Y - E(Y) = [X - E(X)]\alpha + \delta$$

Therefore,

$$\text{cov}(X, Y) = \text{E}\{[Y - \text{E}(Y)][X - \text{E}(X)]\}$$

$$= \text{E}\{[[X - \text{E}(X)]\alpha + \delta][X - \text{E}(X)]\}$$

$$= \alpha\text{E}[X - \text{E}(X)]^2 + \text{E}[\delta(X - \text{E}(X))]$$

where

**Fig. 6.1** The average effect of gene substitution ($\alpha$) is the slope of the regression line of the genotypic value ($Y$) on the number of "high alleles" ($X$) in the genotype. The left $y$-axis gives the genotypic values in scale 2 and the right $y$-axis gives the genotypic values in scale 1

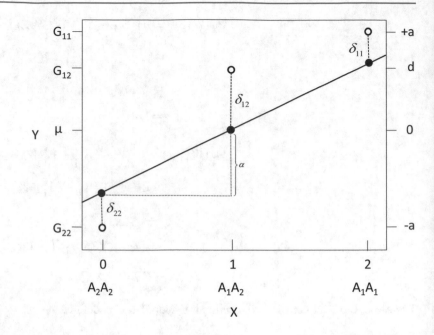

**Table 6.3** Joint distribution of genotypic value ($Y$) and the number of high-effect alleles in the genotype ($X$)

| Genotype | Frequency | X | Y |
|---|---|---|---|
| $A_1A_1$ | $p^2$ | 2 | $G_{11}$ |
| $A_1A_2$ | $2pq$ | 1 | $G_{12}$ |
| $A_2A_2$ | $q^2$ | 0 | $G_{22}$ |

**Table 6.4** Distribution of centered $X$ and $Y$ variables

| Genotype | Frequency | $X - E(X)$ | $Y - E(Y)$ |
|---|---|---|---|
| $A_1A_1$ | $p^2$ | $2q$ | $2q\alpha - 2q^2d$ |
| $A_1A_2$ | $2pq$ | $q - p$ | $(q - p)\alpha + 2pqd$ |
| $A_2A_2$ | $q^2$ | $-2p$ | $-2p\alpha - 2p^2d$ |

$$E\{\delta[X - E(X)]\} = p^2(2q)(-2q^2d) + 2pq(q-p)(2pqd) + q^2(-2p)(-2p^2d) = 0$$

and

$$E[X - E(X)]^2 = \text{var}(X) = 2pq$$

Therefore,

$$\text{cov}(X, Y) = \alpha\text{var}(X) = 2pq\alpha$$

We can prove that $b_{YX} = \alpha$, which states that the regression coefficient of the genotypic value on the number of high-effect alleles equals the average effect of gene substitution. From the definition of regression coefficient, we know that $b_{YX} = \text{cov}(X, Y)/\text{var}(X)$. Since we have already proved that $\text{var}(X) = 2pq$ and $\text{cov}(X, Y) = 2pq\alpha$, the proof is immediately obtained

$$b_{YX} = \frac{\text{cov}(X, Y)}{\text{var}(X)} = \frac{2pq\alpha}{2pq} = \alpha$$

The linear model of the genotypic value on the number of high alleles is

$$Y = b_0 + b_{YX}X + \delta$$

where

$$b_0 = \text{E}(Y) - b_{YX}\text{E}(X) = \mu + M - 2p\alpha = \alpha_0$$

Therefore, the above model is

$$Y = \underbrace{\alpha_0}_{\text{intercept}} + X\underbrace{\alpha}_{\text{slope}} + \underbrace{\delta}_{\text{residual}}$$

Figure 6.1 shows the regression line of $Y$ on $X$.

## References

Falconer DS, Mackay TFC. Introduction to quantitative genetics. Harlow, UK: Addison Wesley Longman; 1996.

Kao CH, Zeng ZB. Modeling epistasis of quantitative trait loci using Cockerham's model. Genetics. 2002;160:1243–61.

# Environmental Effects and Environmental Errors

Any non-genetic factors that cause the variation of a quantitative trait are environmental factors. Environmental variance is a source of error that reduces the precision of genetic studies. The aim of an experimenter is to reduce this error as much as possible through careful design of experiments. People rarely pay attention to the difference between environmental effects and environmental errors. In fact, in all textbooks and literature, environmental effects and environmental errors are all called environmental errors. I believe that it is fundamentally important to clarify the differences. This chapter mainly deals with the difference and investigates genotype by environment interaction in great detail. Some materials of this chapter are adopted from Professor Nyquist's lecture notes for his Quantitative Genetics class at Purdue University (AGR/ANSC611).

## 7.1 Environmental Effects

Environmental effects are recognizable or controllable non-genetic factors that cause the variation of a quantitative trait. People often call environmental effects systematic environmental errors. A typical example is the year (temporal) effect or location (spatial) effect on a quantitative trait. When you collect yields of plants, you know when and where the plants were grown. These effects are recognizable and must be removed from your genetic analysis. In human genetics studies, age is a non-genetic factor (you carry the same set of genes throughout your life, regardless your age) and it is recognizable. By removing these non-genetic effects, we mean that their effects on the phenotypic difference must be considered in the model, and this difference should not be counted as part of the phenotypic variance. This explains why genetic experiments are often conducted in highly controlled environments (to reduce the environmental effects and to emphasize the genetic effects). When these environmental effects are hard to control experimentally, we can deal with the effects by using different statistical models and data analysis procedures.

## 7.2 Environmental Errors

Environmental errors are defined as any unrecognizable or uncontrollable non-genetic factors that affect the phenotype of a quantitative trait. We often call them the random environmental errors. This type of environmental errors is hard or impossible to control. For example, the uneven rain drop in a field plot will affect plant growth. A plant happened to be bitten by an insect while the plants in the neighbor did not. When we said that the phenotypic variance is the sum of genetic variance and the environmental variance, that environmental variance is referred to the random environmental error variance. The phenotypic variance should include the random environment variance but not the variance caused by environmental effects (systematic environmental errors).

Environmental errors can be further partitioned into general (permanent) environmental errors and special (temporary) environmental errors. General environmental errors are errors that affect the performance of an organism for life (permanently). For example, starvation for a long period of time for a child will have a permanent damage on the child health, slow the child's growth, and affect the entire adult life of the child. Special environmental errors, however, only happen randomly and without any permanent consequence on the phenotype. For example, an outstanding high school student may get a very bad score on a math test due to sickness during the test day. When the student gets healthy again, he/she will perform

© Springer Nature Switzerland AG 2022
S. Xu, *Quantitative Genetics*, https://doi.org/10.1007/978-3-030-83940-6_7

normally. If we use $E$ to denote the environmental error, the partitioning of the environmental error can be written as $E = E_g + E_s$, where $E_g$ represents the general environmental error and $E_s$ represents the special environmental error.

## 7.3    Repeatability

In genetic studies, the environmental effects are explicitly put in the model as fixed effects and their effects on the phenotypic values should be removed. They should not contribute to the variation of the phenotype. Therefore, when you calculate the phenotypic variance, remember to exclude environmental effects. Recall that we partitioned the total phenotypic variance into genetic variance and environmental variance (see Chap. 6),

$$V_P = V_G + V_E$$

where $V_E$ only contains the variance of the environmental errors. The variance of the environmental effects (if any) should not be included in $V_E$. The random environmental error has been partitioned into general and special errors,

$$E = E_g + E_s$$

Corresponding to the partitioning of the environmental error, the environmental variance can also be partitioned this way,

$$V_E = V_{E_g} + V_{E_s}$$

Therefore,

$$V_P = V_G + V_E = V_G + V_{E_g} + V_{E_s}$$

The general and special environmental variances can be separated if a trait can be measured repeatedly in life. In that case, we can define a genetic parameter called repeatability, which measures the correlation between repeated measurements of the same character within the same individual. Mathematically, it is defined as a variance ratio and denoted by $\rho$ (Rho),

$$\rho = \frac{V_G + V_{E_g}}{V_P} = \frac{V_G}{V_P} + \frac{V_{E_g}}{V_P} = H^2 + \frac{V_{E_g}}{V_P}$$

where $H^2 = V_G/V_P$ is the broad sense heritability. Therefore, $\rho \geq H^2$, meaning that repeatability is the upper limit of the broad sense heritability. Theoretically, repeatability is always positive. In practice, the estimated repeatability can be negative when the sample size is small and the true repeatability is small.

## 7.3.1    Estimation of Repeatability

The way we estimate repeatability is to use a statistical method called the intra-class correlation analysis, which involves some calculation from the result of a one-way analysis of variances (ANOVA). The underlying assumptions of the ANOVA used for estimating repeatability are: (1) equal variance of different measurements; (2) different measurements reflect what is genetically the same character. We will know what they mean after we discuss the ANOVA model. Let $j$ index an individual for $j = 1, \ldots, n$ where $n$ is the number of individuals. Assume that each individual is measured $m_j$ times for a quantitative trait. For simplicity, let $m_j = m$ for all $j = 1, \ldots, n$, which means that all individuals are measured exactly the same number of times. Let $y_{jk}$ be the $k$th measurement from the $j$th individual. The data structure looks like the one given in Table 7.1.

We now introduce the following linear model

$$y_{jk} = \mu + B_j + \varepsilon_{jk}$$

**Table 7.1** Data structure for repeated measurement of a quantitative trait

| Individual | Measurement 1 | Measurement 2 | $\cdots$ | Measurement $m$ |
|---|---|---|---|---|
| 1 | $y_{11}$ | $y_{12}$ | $\cdots$ | $y_{1m}$ |
| 2 | $y_{21}$ | $y_{22}$ | $\cdots$ | $y_{2m}$ |
| 3 | $y_{31}$ | $y_{32}$ | $\cdots$ | $y_{3m}$ |
| $\vdots$ | $\vdots$ | $\vdots$ | $\vdots$ | $\vdots$ |
| $n$ | $y_{n1}$ | $y_{n2}$ | $\cdots$ | $y_{nm}$ |

where $\mu$ is the population mean of the trait (also called the grand mean), $B_j$ is the "true effect" of the $j$th individual (a theoretical value), and $\varepsilon_{jk}$ is the special environmental error. It is the deviation of the $j$th measurement from the "true effect." The true effect $B_j$ can be considered as the average value of an infinite number of measurements from individual $j$. The grand mean is treated as a fixed effect (a parameter). The "true effect" of individual $j$, $B_j$, is assumed to be a random variable sampled from a normal distribution with mean zero and variance $\sigma_B^2$, i.e., $B_j N\left(0, \sigma_B^2\right)$. Note that $B_j$ is the true value of an individual and it is partitioned into

$$B = G + E_g$$

Verbally, the true measurement of an individual is determined by its genetic value plus a *permanent* environmental effect. The corresponding partitioning of the variance also applies

$$\text{var}(B) = \text{var}(G) + \text{var}\left(E_g\right)$$

provided that $G$ and $E_g$ are independent. Using statistical and genetic notations, the above equation is rewritten as

$$\sigma_B^2 = V_G + V_{E_g}$$

Therefore, the variance of $B$ consists of the genetic variance and the permanent environmental variance. If somehow we can estimate the variance of $B$ and the total phenotypic variance, then we can take the ratio to get an estimated repeatability,

$$\widehat{\rho} = \frac{\widehat{\sigma}_B^2}{\widehat{\sigma}_P^2}$$

where $\sigma_P^2$ is just an alternative notation for $V_P$. The special environmental error is assumed to be normally distributed $\varepsilon_{jk} N\left(0, \sigma_\varepsilon^2\right)$, where the variance $\sigma_\varepsilon^2$ is denoted by $V_{E_s}$ here. The total variance of $y$ (excluding the variance due to classifiable environmental effects) is $\text{var}(y) = \sigma_P^2 = V_P$. Since

$$\underbrace{\sigma_Y^2}_{V_P} = \underbrace{\sigma_B^2}_{V_G + V_{E_g}} + \underbrace{\sigma_\varepsilon^2}_{V_{E_s}}$$

it can be seen that

$$\rho = \underbrace{\frac{V_G + V_{E_g}}{V_P}}_{\text{Genetic terminology}} = \underbrace{\frac{\sigma_B^2}{\sigma_B^2 + \sigma_\varepsilon^2}}_{\text{Statistical terminology}}$$

Statistically, $\sigma_B^2$ and $\sigma_\varepsilon^2$ can be estimated using the analysis of variances (ANOVA), as shown in Table 7.2. Note that

**Table 7.2** The analysis of variance (ANOVA) table

| Source of variation | Degree of freedom (df) | Sum of squares (SS) | Mean squares (MS) | Expected MS |
|---|---|---|---|---|
| Between groups ($B$) | $df_B = n - 1$ | $SS_B = m \sum_{j=1}^{n} \left( \bar{y}_{j.} - \bar{y}_{..} \right)^2$ | $MS_B = SS_B/df_B$ | $\sigma_\varepsilon^2 + m\sigma_B^2$ |
| Within groups ($E$) | $df_E = n(m - 1)$ | $SS_E = \sum_{j=1}^{n} \sum_{k=1}^{m} \left( y_{jk} - \bar{y}_{j.} \right)^2$ | $MS_E = SS_E/df_E$ | $\sigma_\varepsilon^2$ |

$$\bar{y}_{j.} = m^{-1} \sum_{k=1}^{m} y_{jk}$$

is the sample mean for individual $j$ and

$$\bar{y}_{..} = (nm)^{-1} \sum_{j=1}^{n} \sum_{k=1}^{m} y_{jk}$$

is the overall mean. The word "between" represents "between groups" or "between individuals," while the word "within" means "within groups" or "within individuals." The total variance of the trait is partitioned into "between individuals" and "within individuals" variances. Eventually, we will have two quantities calculated from the data, which are the two "mean squares," $MS_B$ and $MS_E$. Statisticians have already found that if our sample size is sufficiently large, these two calculated mean squares would equal the expected mean squares given in the last column of Table 7.2. Therefore,

$$\begin{cases} MS_E = \sigma_\varepsilon^2 \\ MS_B = \sigma_\varepsilon^2 + m\sigma_B^2 \end{cases}$$

We have two equations and two unknown parameters ($\sigma_B^2$ and $\sigma_\varepsilon^2$). The solutions of the unknown parameters are

$$\begin{cases} \hat{\sigma}_\varepsilon^2 = MS_E \\ \hat{\sigma}_B^2 = (MS_B - MS_E)/m \end{cases}$$

The estimated repeatability is

$$\hat{\rho} = \frac{\hat{\sigma}_B^2}{\hat{\sigma}_B^2 + \hat{\sigma}_\varepsilon^2} = \frac{MS_B - MS_E}{MS_B + (m - 1)MS_E}$$

The sums of squares in Table 7.2 are called the definition formulas. Programs are available to perform such an analysis of variances. However, if we want to perform ANOVA using excel spreadsheet, we should use computing formulas of the sums of squares. The computing formulas are

$$SS_T = \sum_{j=1}^{n} \sum_{k=1}^{m} \left( y_{jk} - \bar{y}_{..} \right)^2 = \sum_{j=1}^{n} \sum_{k=1}^{m} y_{jk}^2 - nm(\bar{y}_{..})^2$$

$$SS_B = m \sum_{j=1}^{n} \left( \bar{y}_{j.} - \bar{y}_{..} \right)^2 = m \sum_{j=1}^{n} \left( \bar{y}_{j.} \right)^2 - nm(\bar{y}_{..})^2$$

$$SS_E = SS_T - SS_B = \sum_{j=1}^{n} \sum_{k=1}^{m} y_{jk}^2 - m \sum_{j=1}^{n} \left( \bar{y}_{j.} \right)^2$$

where $SS_T$ is called the total sum of squares.

The ANOVA method for variance component estimation is called the moment method or Henderson's Type 1 method (Henderson 1985; Henderson 1986). The method has been replaced by some advanced methods, such as the minimum variance quadratic unbiased estimation (MIVQUE0)(Rao 1971b; Rao 1971a; Swallow and Searle 1978), the maximum likelihood (ML) method (Hartley and Rao 1967) and the restricted maximum likelihood (REML) method (Patterson and Thompson 1971). All these methods have been implemented in the MIXED Procedure in SAS (SAS Institute Inc. 2009).

If there are only two measurements per individual, say $y_{j1}$ and $y_{j2}$ for the two measurements from individual $j$. The repeatability is simply the Pearson correlation coefficient between $y_1$ and $y_2$, assuming that $\text{var}(y_1)$ and $\text{var}(y_2)$ are roughly the same. For three or more repeated measurements, the repeatability can be approximated by the average of all pair-wise correlation coefficients between all $m$ measurements. Again, the assumption is that all measurements have roughly the same variance.

### 7.3.2  Proof of the Intra-Class Correlation Coefficient

Here, we prove that the repeatability is indeed the intra-class correlation,

$$\rho = \frac{\sigma_B^2}{\sigma_B^2 + \sigma_\varepsilon^2}$$

Define $y_{jk}$ and $y_{jk'}$ for $k \neq k'$ as two repeated measurements on individual $j$. The intra-class correlation is defined as the correlation coefficient between any two repeated measurements within the same class (or group). The class here is the individual. Therefore, the intra-class correlation is

$$r_{y_{jk}y_{jk'}} = \frac{\text{cov}\left(y_{jk}, y_{jk'}\right)}{\sigma_{y_{jk}}\sigma_{y_{jk'}}}$$

The two repeated measurements can be described by the following linear models

$$\begin{cases} y_{jk} = \mu + B_j + \varepsilon_{jk} \\ y_{jk'} = \mu + B_j + \varepsilon_{jk'} \end{cases}$$

Using the properties of variance and covariance, we have

$$\sigma_{y_{jk}}^2 = \text{var}\left(y_{jk}\right) = \text{var}\left(B_j\right) + \text{var}\left(\varepsilon_{jk}\right) = \sigma_B^2 + \sigma_\varepsilon^2$$
$$\sigma_{y_{jk'}}^2 = \text{var}\left(y_{jk'}\right) = \text{var}\left(B_j\right) + \text{var}\left(\varepsilon_{jk'}\right) = \sigma_B^2 + \sigma_\varepsilon^2$$

The covariance between $y_{jk}$ and $y_{jk'}$ is

$$\begin{aligned} \text{cov}\left(y_{jk}, y_{jk'}\right) &= \text{cov}\left(\mu + B_j + \varepsilon_{jk}, \mu + B_j + \varepsilon_{jk'}\right) \\ &= \text{cov}\left(B_j, B_j\right) + \text{cov}\left(B_j, \varepsilon_{jk'}\right) + \text{cov}\left(\varepsilon_{jk}, B_j\right) + \text{cov}\left(\varepsilon_{jk'}, \varepsilon_{jk}\right) \\ &= \text{cov}\left(B_j, B_j\right) = \text{var}\left(B_j\right) = \sigma_B^2 \end{aligned}$$

Therefore,

$$r_{y_{jk}y_{jk'}} = \frac{\text{cov}\left(y_{jk}, y_{jk'}\right)}{\sigma_{y_{jk}}\sigma_{y_{jk'}}} = \frac{\sigma_B^2}{\sqrt{\sigma_B^2 + \sigma_\varepsilon^2}\sqrt{\sigma_B^2 + \sigma_\varepsilon^2}} = \frac{\sigma_B^2}{\sigma_B^2 + \sigma_\varepsilon^2} = \rho$$

Note that when we use the variance operator to write $\text{var}(B_j)$, we mean the variance of the population from which $B_j$ is sampled. Since all $B_j$ for $j = 1, \ldots, n$ are sample from the same normal distribution $N\left(0, \sigma_B^2\right)$, we have $\text{var}\left(B_j\right) = \sigma_B^2$. Similarly, $\text{var}\left(\varepsilon_{jk}\right) = \text{var}\left(\varepsilon_{jk'}\right) = \sigma_\varepsilon^2$ because both $\varepsilon_{jk}$ and $\varepsilon_{jk'}$ are assumed to be sampled from the same $N\left(0, \sigma_\varepsilon^2\right)$ distribution.

**Table 7.3** The analysis of variance (ANOVA) table for unbalanced data

| Source of variation | Degree of freedom (df) | Sum of squares (SS) | Mean squares (MS) | Expected MS |
|---|---|---|---|---|
| Between groups ($B$) | $df_B = n - 1$ | $SS_B = \sum_{j=1}^{n} m_j\left(\bar{y}_{j.} - \bar{y}_{..}\right)^2$ | $MS_B = SS_B/df_B$ | $\sigma_\varepsilon^2 + m_0\sigma_B^2$ |
| Within groups ($E$) | $df_E = \sum_{j=1}^{n}\left(m_j - 1\right)$ | $SS_E = \sum_{j=1}^{n}\sum_{k=1}^{m_j}\left(y_{jk} - \bar{y}_{j.}\right)^2$ | $MS_E = SS_E/df_E$ | $\sigma_\varepsilon^2$ |

### 7.3.3  Estimation of Repeatability with Variable Numbers of Repeats

In practice, the number of repeated measurements may vary across individuals. Such data are called unbalance data. There are many reasons causing the unbalanced data. For example, an individual may die early before all records are collected. A particular field plot may be destroyed by an unexpected event. In this case, we can still use the ANOVA intra-class correlation to estimate the repeatability of a trait, but we have to modify the formulas for calculation of the sums of squares. Let $n$ be the number of individuals (groups or classes) and $m_j$ be the number of repeated measurements for individual $j$ for $j = 1, \ldots, n$. The modifications are reflected in the following places of the ANOVA table (see Table 7.3). Note that

$$\bar{y}_{j.} = \frac{1}{m_j}\sum_{k=1}^{m_j} y_{jk}$$

is the sample mean for individual $j$ and

$$\bar{y}_{..} = \left[\sum_{j=1}^{n} m_j\right]^{-1}\left[\sum_{j=1}^{n}\sum_{k=1}^{m_j} y_{jk}\right]$$

is the grand mean (mean of the whole data). In addition, the $m_0$ appearing in the expected MS column is calculated using

$$m_0 = \frac{1}{n-1}\left[\sum_{j=1}^{n} m_j - \sum_{j=1}^{n} m_j^2 \Big/ \sum_{j=1}^{n} m_j\right]$$

You can verify that if $m_j = m$ for all $j = 1, \ldots, n$, then $m_0 = m$. The estimated repeatability is

$$\hat{\rho} = \frac{\hat{\sigma}_B^2}{\hat{\sigma}_B^2 + \hat{\sigma}_\varepsilon^2} = \frac{MS_B - MS_E}{MS_B + (m_0 - 1)MS_E}$$

For unbalanced data, the computing formulas for the SS terms are

$$SS_T = \sum_{j=1}^{n}\sum_{k=1}^{m_j}\left(y_{jk} - \bar{y}_{..}\right)^2 = \sum_{j=1}^{n}\sum_{k=1}^{m_j} y_{jk}^2 - (\bar{y}_{..})^2\sum_{j=1}^{n} m_j$$

$$SS_B = \sum_{j=1}^{n} m_j\left(\bar{y}_{j.} - \bar{y}_{..}\right)^2 = \sum_{j=1}^{n} m_j(\bar{y}_{j.})^2 - (\bar{y}_{..})^2\sum_{j=1}^{n} m_j$$

$$SS_E = SS_T - SS_B = \sum_{j=1}^{n}\sum_{k=1}^{m_j} y_{jk}^2 - \sum_{j=1}^{n} m_j(\bar{y}_{j.})^2$$

It is better to use the advanced methods introduced early to estimate the between-individual and within-individual variance components for the unbalanced data.

When a parameter is estimated, it should be associated with a standard error, although the parameter itself is a constant. The standard error for $\hat{\rho}$ is hard to calculate because it is defined as a ratio of variance components. We will use a Delta method to approximate the standard error. The Delta method will be introduced in Chap. 10, where we use the Delta method to calculate standard error of an estimated heritability. The same Delta method can be used to calculate the standard error of an estimated repeatability.

### 7.3.4 An Example for Estimating Repeatability

Table 7.4 shows a hypothetical sample of SAT test scores for high school students. A random sample of 20 high school juniors and seniors from a high school participated in the survey. The SAT test scores of three replicates are listed in the table. We want to estimate the repeatability of SAT test score (considered as a quantitative trait) for high school students. Before we perform the analysis, we need to understand the data. The classes or groups here are students (each student is a group). Therefore, the number of groups (individuals) is $n = 20$. The number of repeated measurements is $m = 3$, which is a constant number across students (balanced data). We first calculated the degrees of freedoms and the sums of squares, and then put these quantities in an ANOVA table (Table 7.5).

The following are the intermediate results, from which an ANOVA is constructed. The degrees of freedom are

$$df_B = n - 1 = 20 - 1 = 19$$

and

$$df_E = n(m - 1) = 20 \times (3 - 1) = 40$$

The grand mean is

**Table 7.4** Hypothetic SAT test scores of high school students in the old scale (sum of reading, mathematics, and writing)

| Student | Test 1 | Test 2 | Test 3 |
|---|---|---|---|
| 1 | 1810 | 1840 | 1880 |
| 2 | 1710 | 1820 | 1730 |
| 3 | 1990 | 2070 | 2210 |
| 4 | 1510 | 1790 | 1890 |
| 5 | 2260 | 2170 | 2230 |
| 6 | 2080 | 2150 | 2190 |
| 7 | 2270 | 2320 | 2420 |
| 8 | 1820 | 1750 | 1990 |
| 9 | 2060 | 2360 | 2300 |
| 10 | 2180 | 2330 | 2450 |
| 11 | 2010 | 2210 | 2250 |
| 12 | 1950 | 1880 | 2020 |
| 13 | 1850 | 1880 | 1920 |
| 14 | 1830 | 1940 | 2080 |
| 15 | 2170 | 2070 | 2220 |
| 16 | 2070 | 2240 | 2310 |
| 17 | 1890 | 1850 | 2070 |
| 18 | 1870 | 1850 | 1870 |
| 19 | 1830 | 2040 | 2110 |
| 20 | 2170 | 2150 | 2200 |

**Table 7.5** The ANOVA table for the SAT test score data

| Source | DF | SS | MS | EMS |
|---|---|---|---|---|
| Between (B) | 19 | 1,979,793 | 104199.6 | $\sigma_E^2 + m\sigma_B^2$ |
| Within (E) | 40 | 441,600 | 11,040 | $\sigma_E^2$ |

$$\bar{y}_{..} = \frac{1}{nm} \sum_{j=1}^{n} \sum_{k=1}^{m} y_{jk} = \frac{1}{20 \times 3} \times 122380 = 2039.667$$

which leads to

$$nm\bar{y}_{..}^2 = 20 \times 3 \times 2039.667^2 = 249614407$$

The total sum of squares is

$$SS_T = \sum_{j=1}^{n} \sum_{k=1}^{m} \left(y_{jk} - \bar{y}_{..}\right)^2 = \sum_{j=1}^{n} \sum_{k=1}^{m} y_{jk}^2 - nm\bar{y}_{..}^2 = 252035800 - 249614407 = 2421393.3$$

The between-student sum of squares is

$$SS_B = m \sum_{j=1}^{n} \left(\bar{y}_{j.} - \bar{y}_{..}\right)^2 = m \sum_{j=1}^{n} \left(\bar{y}_{j.}\right)^2 - nm \left(\bar{y}_{..}\right)^2 = 251594200 - 249614407 = 1979793.3$$

The within-student sum of squares is

$$SS_E = \sum_{j=1}^{n} \sum_{k=1}^{m} \left(y_{jk} - \bar{y}_{j.}\right)^2 = SS_T - SS_B = 2421393.3 - 1979793.3 = 441600$$

The between-student mean squares is

$$MS_B = SS_B/df_B = 1979793.3/19 = 104199.6$$

The within-student mean squares is

$$MS_E = SS_E/Df_E = 441600/40 = 11040$$

The estimated variance components are

$$\widehat{\sigma}_e^2 = MS_E = 11040$$

and

$$\widehat{\sigma}_B^2 = (MS_B - MS_E)/m = (104199.6 - 11040)/3 = 31053.216$$

Therefore, the estimated repeatability is

$$\rho = \frac{\widehat{\sigma}_B^2}{\widehat{\sigma}_B^2 + \widehat{\sigma}_E^2} = \frac{31053.216}{31053.216 + 11040} = 0.737725$$

The results are summarized in Table 7.5.

To analyze this data using either an R program or an SAS program, the input data must be rearranged into the following form (Table 7.6). In other words, the response variable must be placed in a single column. The VARCOMP procedure in SAS can be used to estimate the variance components. Below is the code to read the data and to estimate the variance components.

**Table 7.6**  Rearrangement of the SAT test score data (part of the full data)

| Student | Test | Score |
|---|---|---|
| 1 | 1 | 1810 |
| 1 | 2 | 1840 |
| 1 | 3 | 1880 |
| 2 | 1 | 1710 |
| 2 | 2 | 1820 |
| 2 | 3 | 1730 |
| 3 | 1 | 1990 |
| 3 | 2 | 2070 |
| 3 | 3 | 2210 |
| 4 | 1 | 1510 |
| 4 | 2 | 1790 |
| 4 | 3 | 1890 |
| 5 | 1 | 2260 |
| 5 | 2 | 2170 |
| 5 | 3 | 2230 |
| . | . | . |
| . | . | . |
| . | . | . |
| 20 | 1 | 2170 |
| 20 | 2 | 2150 |
| 20 | 3 | 2200 |

**Table 7.7**  The ANOVA table generated from `proc varcomp`

| Type 1 Analysis of Variance | | | | |
|---|---|---|---|---|
| Source | DF | Sum of Squares | Mean Square | Expected Mean Square |
| Student | 19 | 1,979,793 | 104,200 | Var(error) + 3 Var(student) |
| Error | 40 | 441,600 | 11,040 | Var(error) |
| Corrected Total | 59 | 2,421,393 | | |

**Table 7.8**  Estimated variance components from `proc varcomp`

| Type 1 Estimates | |
|---|---|
| Variance Component | Estimate |
| Var(student) | 31053.2 |
| Var(error) | 11040.0 |

```
filename aa "data-7-1.csv";
proc import datafile=aa out=one dbms=csv replace;
run;
proc varcomp data=one method=type1;
    class student;
    model score = student;
run;
```

The output includes the ANOVA table (Table 7.7) and the estimated variance components (Table 7.8). For such balanced data, all methods (TYPE1, MIVQUE0, ML, and REML) generate the same results, unless the true variance of between students is very small. The TYPE1 and MIVQUE0 methods allow negative estimate of the between-group variance but ML and REML do not.

### 7.3.5    Application of Repeatability

Repeatability can help us calculate the variance of the mean value of multiple measurements. Let $y_{jk}$ be the trait value of individual $j$ measured at the $k$th time point for $k = 1, \ldots, m$. Recall that the mean value for individual $j$ is

$$\bar{y}_{j.} = \frac{1}{m} \sum_{k=1}^{m} y_{jk}$$

We have already learned that $\mathrm{var}(y_{jk}) = \sigma_P^2 = V_P$ is the phenotypic variance of the trait. With multiple measurements, what is the variance of the mean values across individuals? The variance of the mean is denoted by $V_{\bar{P}} = \mathrm{var}(\bar{y}_{j.})$. Using the properties of variance and covariance, we can show

$$
\begin{aligned}
V_{\bar{P}} = \mathrm{var}(\bar{y}_{j.}) &= \mathrm{var}\left(\frac{1}{m}\sum_{k=1}^{m} y_{jk}\right) = \frac{1}{m^2}\mathrm{var}\left(\sum_{k=1}^{m} y_{jk}\right) \\
&= \frac{1}{m^2}\left[\sum_{k=1}^{m}\mathrm{var}(y_{jk}) + 2\sum_{k=1}^{m-1}\sum_{k'=k+1}^{m}\mathrm{cov}(y_{jk}, y_{jk'})\right] \\
&= \frac{1}{m^2}\left[\sum_{k=1}^{m}\sigma_{y_{jk}}^2 + 2\sum_{k=1}^{m-1}\sum_{k'=k+1}^{m} r_{y_{jk}y_{jk'}}\sqrt{\sigma_{y_{jk}}^2 \sigma_{y_{jk'}}^2}\right] \\
&= \frac{1}{m^2}\left[mV_P + 2\times\frac{1}{2}m(m-1)\rho V_P\right] \\
&= \frac{(m-1)\rho + 1}{m}V_P
\end{aligned}
$$

Note that

$$\mathrm{cov}(y_{jk}, y_{jk'}) = r_{y_{jk}y_{jk'}}\sqrt{\sigma_{y_{jk}}^2 \sigma_{y_{jk'}}^2} = \rho V_P$$

This relationship

$$V_{\bar{P}} = \frac{(m-1)\rho + 1}{m}V_P$$

is determined by both the number of repeated measurements and the repeatability. Let us define

$$\lambda = \frac{(m-1)\rho + 1}{m}$$

so that $V_{\bar{P}} = \lambda V_P$. Since $0 \le \rho \le 1$, the range of $\lambda$ is between $1/m$ and 1, i.e., $\frac{1}{m} \le \lambda \le 1$. When $m = 1$, we have $\lambda = 1$. The limit of $\lambda$ when $m \to \infty$ is

$$\lim_{m\to\infty}\lambda = \lim_{m\to\infty}\frac{(m-1)\rho + 1}{m} = \rho$$

Given the relationship $V_{\bar{P}} = \lambda V_P$, we define the heritability (heritability will be introduced formally in a later chapter) for the mean of multiple measurements as

$$h_{\mathrm{mean}}^2 = \frac{V_A}{V_{\bar{P}}} = \frac{V_A}{\lambda V_P} = \frac{m}{(m-1)\rho + 1}h^2 \tag{7.1}$$

This heritability is referred to as the heritability on the basis of mean, and it is always larger than or equal to the usual heritability $h^2$ defined on the basis of individual measurement. If we consider heritability as the accuracy of predicted breeding value, we should use the mean value as the trait if there are multiple measurements per individual.

Plants differ from animals in that multiple seeds are available from each inbred line or hybrid. Repeatability in plants is called heritability because each individual replicate is generated from a separate seed. A line can be measured multiple times for a quantitative trait. It is very rare to use a single measurement of a trait to represent the phenotype of an inbred variety. The environmental error always plays an important role in shaping the phenotype. Therefore, the phenotype of a plant is almost always the mean value of multiple repeated measurements. Therefore, when you report the heritability of a trait in plants, you must tell the number of repeated measurements. When a plant is measured an infinite number of times, the heritability of the trait is always unity. This may be hard to understand. However, we can look at the problem using a different approach. Let $V_A$ be the additive genetic variance (ignoring dominance and epistatic variances) and the phenotypic variance is $V_P = V_A + V_E$. When a trait is measured $m$ times, the phenotypic variance of the mean is

$$V_{\bar{P}} = V_A + \frac{1}{m} V_E$$

Therefore, the heritability of the mean is

$$h^2_{\text{mean}} = \frac{V_A}{V_{\bar{P}}} = \frac{V_A}{V_A + \frac{1}{m} V_E} = \frac{V_A/V_P}{V_A/V_P + \frac{1}{m} V_E/V_P} = \frac{h^2}{h^2 + \frac{1}{m}\left(1 - h^2\right)} \tag{7.2}$$

where $h^2 = V_A/V_P$ and $V_E/V_P = 1 - h^2$. It is clear that when $m \to \infty$, we have

$$\lim_{m \to \infty} h^2_{\text{mean}} = \lim_{m \to \infty} \frac{h^2}{h^2 + \frac{1}{m}\left(1 - h^2\right)} = 1$$

When there is no permanent environmental error, the repeatability is identical to the heritability, i.e., $\rho = h^2$, which makes Eq. (7.1) identical to Eq. (7.2). Note that we can define a term

$$e^2 = V_E/V_P = 1 - h^2$$

which has no official name but we can call it "environmentability," mimicking the name of "heritability." Therefore,

$$h^2_{\text{mean}} = \frac{h^2}{h^2 + \frac{1}{m}\left(1 - h^2\right)} = \frac{h^2}{h^2 + \frac{1}{m} e^2}$$

## 7.4 Genotype by Environment (G × E) Interaction

### 7.4.1 Definition of G × E Interaction

Genotype by environment interaction is a phenomenon that the difference in performance between two or more genotypes in one environment is different from the difference in another environment. G × E interaction can be illustrated in Fig. 7.1 (the lower panel), where genotype A performs better than genotype B in the North but it performs worse than genotype B in the South. When G × E interaction is present, the two lines are NOT parallel. On the contrary, the two lines are parallel if G × E interaction is absent (the upper panel of Fig. 7.1). G×E interaction is an important factor to consider by breeders when a new cultivar is released from nursery to a farm. We will learn how to test G×E interaction later using a real data set collected from a rice breeding experiment.

### Absence of G x E Interaction

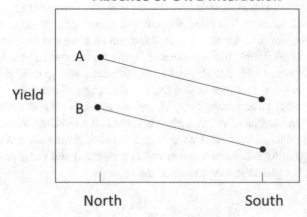

### Presence of G x E Interaction

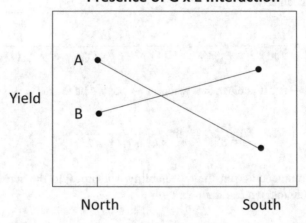

**Fig. 7.1** Illustration of G × E interactions: North and South represent two environments while A and B represent two genotypes (varieties or breeds). The upper panel shows the absence of G × E interaction and the lower panel shows the presence of G × E interaction

One important aspect of G × E interaction is that the E here is not the random environmental error; rather, it is an (recognizable, classifiable, or controllable) environmental effect. Recall that the model describing the phenotype is

$$P = G + E$$

The $E$ in the model is the random environmental error. When an environmental effect exists, the model should be modified by

$$P = G + E_X + E$$

where $E_X$ is the environmental effect, not environmental error. When the G × E interaction is absent, the above model is sufficient. When the G × E interaction is present, we should add another term in the model,

$$P = G + E_X + G \times E_X + E$$

where $G \times E$ is the usual notation for G by E interaction, but it is really $G \times E_X$. It is this $E_X$ that interacts with the genotype, not the random environmental error. Hereafter, we use $G \times E$ to denote the genotype by environment interaction although the $E$ in $G \times E$ is really $E_X$.

From the above models, we can derive the variance of the phenotypic value. In the absence of G × E interaction,

$$\text{var}(P) = \text{var}(G) + \text{var}(E_X) + \text{var}(E)$$

where $\text{var}(G) = V_G$ is the genetic variance, $\text{var}(E_X) = V_X$ is the variance caused by the environmental effect and $\text{var}(E) = V_E$ is the environmental error variance. However, $\text{var}(P) \neq V_P$ because the former contains variation caused by $E_X$. We have learned before that the phenotypic variance should not include variation caused by systematic environmental error (effect), i.e., $V_P = V_G + V_E$. This is very confusing but we have to live with it. Things are getting worse when the G × E interaction is present. The variance of the model becomes

$$\text{var}(P) = \text{var}(G) + \text{var}(E_X) + \text{var}(G \times E) + \text{var}(E) = V_G + V_X + V_{G \times E} + V_E$$

It is still in debate whether we should define $V_P = V_G + V_E$ or $V_P = V_G + V_{G \times E} + V_E$. I prefer including the G × E variance in the total phenotypic variance, i.e.,

$$V_P = V_G + V_{G \times E} + V_E$$

For this class, let us include the G×E variance in the total phenotypic variance. To prevent confusion, it is better to specify the name of the environmental effect, for example, year (Y) effect and G × Y interaction, location (L) effect, and G × L interaction.

## 7.4.2 Theoretical Evaluation of G × E Interaction

The data were obtained from Yu et al. (Yu et al. 2011) and reanalyzed by Xu (Xu 2013). The rice yield trait was measured in two different years (1998 and 1999). Three genotypes ($A_1A_1$, $A_1A_2$, and $A_2A_2$) of a locus (bin1005) were evaluated. So, we have two levels of E ($E_1$ = 1998 and $E_2$ = 1999) and three levels of G ($G_1 = A_1A_1$, $G_2 = A_1A_2$ and $G_3 = A_2A_2$). The counts and frequencies of the 2 × 3 cells are given in Table 7.9 and the average yields of the six genotype-year combinations are given in Table 7.10.

**Table 7.9** Joint distribution (counts and frequencies) of three genotypes and two environments of the rice yield trait analysis

| E G | $G_1$ | $G_2$ | $G_3$ | Marginal |
|---|---|---|---|---|
| $E_1$ | $n_{11} = 60$ | $n_{12} = 127$ | $n_{13} = 59$ | $n_{1.} = 246$ |
| | $p_{11} = 11.49\%$ | $p_{12} = 24.33\%$ | $p_{13} = 11.30\%$ | $p_{1.} = 47.13\%$ |
| $E_2$ | $n_{21} = 66$ | $n_{22} = 146$ | $n_{23} = 64$ | $n_{2.} = 276$ |
| | $p_{21} = 12.64\%$ | $p_{22} = 27.97\%$ | $p_{23} = 12.26\%$ | $p_{2.} = 52.87\%$ |
| Marginal | $n_{.1} = 126$ | $n_{.2} = 273$ | $n_{.3} = 123$ | $n_{..} = 522$ |
| | $p_{.1} = 24.14\%$ | $p_{.2} = 52.3\%$ | $p_{.3} = 23.56\%$ | $p_{..} = 100\%$ |

**Table 7.10** Average yield of rice for each of the six genotype and year combinations

| E G | $G_1$ | $G_2$ | $G_3$ | Marginal |
|---|---|---|---|---|
| $E_1$ | $Y_{11} = 43.6448$ | $Y_{12} = 46.4986$ | $Y_{13} = 45.7171$ | $\overline{Y}_{1.} = 45.6151$ |
| $E_2$ | $Y_{21} = 36.1015$ | $Y_{22} = 43.2745$ | $Y_{23} = 43.1836$ | $\overline{Y}_{2.} = 41.5381$ |
| Marginal | $\overline{Y}_{.1} = 39.6936$ | $\overline{Y}_{.2} = 44.7744$ | $\overline{Y}_{.3} = 44.3988$ | $\overline{Y}_{..} = 43.4595$ |

**Table 7.11** Interaction of the yield of rice for each of the six genotype and year combinations

| | $G_1$ | $G_2$ | $G_3$ |
|---|---|---|---|
| $E_1$ | $\Delta_{11} = 1.7955$ | $\Delta_{12} = -0.4314$ | $\Delta_{13} = -0.8374$ |
| $E_2$ | $\Delta_{21} = -1.6707$ | $\Delta_{22} = 0.4214$ | $\Delta_{23} = 0.706$ |

The marginal mean for a year is calculated using

$$\overline{Y}_{j\cdot} = \frac{1}{p_{j\cdot}} \sum_{k=1}^{3} p_{jk} Y_{jk}$$

The marginal mean for a genotype is

$$\overline{Y}_{\cdot k} = \frac{1}{p_{\cdot k}} \sum_{j=1}^{2} p_{jk} Y_{jk}$$

The grand mean is calculated using

$$\overline{Y}_{\cdot\cdot} = \sum_{j=1}^{2} \sum_{k=1}^{3} p_{jk} Y_{jk}$$

There are $2 \times 3 = 6$ interaction terms and the $j$th row and the $k$th column interaction is defined as

$$\Delta_{jk} = \left( Y_{jk} - \overline{Y}_{\cdot\cdot} \right) - \left( \overline{Y}_{j\cdot} - \overline{Y} \right) - \left( \overline{Y}_{\cdot k} - \overline{Y}_{\cdot\cdot} \right) = Y_{jk} - \overline{Y}_{j\cdot} - \overline{Y}_{\cdot k} + \overline{Y}_{\cdot\cdot}$$

In statistics, the marginal means are called the least squares means. The $G \times E$ interaction is present if at least one of the six delta's is not zero. The six interaction values are shown in Table 7.11. None of the six interaction terms are zero and thus $G \times E$ interaction is present. Figure 7.2 shows the plot of the average yield for the two years and the three genotypes. It is clear that the two lines are not parallel.

### 7.4.3 Significance Test of G × E Interaction

With the original data of the rice yield trait (data-7-2.csv), we performed ANOVA using the GLM procedure in SAS. The ANOVA table and the F test statistics are given in Table 7.12. Both G and E are significant due to their large F values. The $G \times E$ interaction term is also significant because the F value, 5.4775, is greater than 3.013192 (the critical value of an F distribution with 2 and 526 degrees of freedom).

The SAS program that reads the data and performs analysis of variances is.

```
filename aa "data-7-2.csv";
proc import datafile=aa out=one dbms=csv replace;
run;
proc glm data=one;
  class bin1005 year;
  model yd = bin1005|year;
quit;
```

The output contains several tables, but two tables are the ones we should look into, the overall ANOVA table (Table 7.13) and the Type I ANOVA table (Table 7.14). The last column of Tables 7.13 and 7.14 represents the p-values. The p-values are all less than 0.05 and thus all effects, including the genotype by year effect, are significant. When the p-value of an effect is too

**Fig. 7.2** Plot of the genotype by year interactions for the rice yield trait

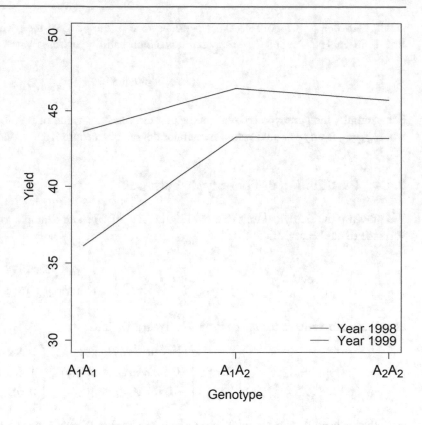

**Table 7.12** ANOVA table for testing G × E interaction for the yield trait of rice

| Source | Df | SS | MS | F value | Pr(>F) | Significance |
|---|---|---|---|---|---|---|
| G | 2 | 2367.5 | 1183.74 | 25.9339 | 1.85E-11 | *** |
| E | 1 | 2191.3 | 2191.35 | 48.0090 | 1.27E-11 | *** |
| G × E | 2 | 500.0 | 250.02 | 5.4775 | 0.004426 | ** |
| Residuals | 516 | 23552.6 | 45.64 | | | |

**Table 7.13** The overall ANOVA table containing the model and residual error terms

| Source | DF | Sum of Squares | Mean Square | F Value | Pr > F |
|---|---|---|---|---|---|
| Model | 5 | 5058.86603 | 1011.77321 | 22.17 | <0.0001 |
| Error | 516 | 23552.56467 | 45.64451 | | |
| Corrected Total | 521 | 28611.43070 | | | |

**Table 7.14** The Type I ANOVA table containing individual effects

| Source | DF | Type I SS | Mean Square | F Value | Pr > F |
|---|---|---|---|---|---|
| bin1005 | 2 | 2367.482719 | 1183.741360 | 25.93 | <0.0001 |
| Year | 1 | 2191.348823 | 2191.348823 | 48.01 | <0.0001 |
| bin1005*year | 2 | 500.034486 | 250.017243 | 5.48 | 0.0044 |

small, SAS does not report the exact p-value; instead, the place of the p-value is filled with <0.0001. The column header, Pr > F, in Tables 7.13 and 7.14 says that it is the probability that the $f$ variable is greater than your calculated F test statistic,

$$p\text{-value} = \Pr\left( f_{df_1, df_2} > F \right)$$

For example, the genotype by year interaction term has a $p$-value of 0.004, which means that $\Pr(f_{2, 516} > 5.48) = 0.0044$, where $f_{2, 516}$ is an $f$ variable with a numerator degree of freedom 2 and a denominator degree of freedom 516.

### 7.4.4 Partitioning of Phenotypic Variance

We now use the rice yield trait (Yu et al. 2011; Xu 2013) as an example to demonstrate the partitioning of the phenotypic variance (data-7-3.csv). Let us define

$$X_j = \begin{cases} +1 & \text{for year 1998} \\ -1 & \text{for year 1999} \end{cases}$$

as an indicator variable for the environment (year). Define.

$$Z_j = \begin{cases} +1 & \text{for } A_1A_1 \\ 0 & \text{for } A_1A_2 \\ -1 & \text{for } A_2A_2 \end{cases} \quad \text{and} \quad W_j = \begin{cases} 0 & \text{for } A_1A_1 \\ 1 & \text{for } A_1A_2 \\ 0 & \text{for } A_2A_2 \end{cases}.$$

as indicator variables for the additive and dominance coefficients. Let $y_j$ be the $j$th data point (yield) for $j = 1, \ldots, n$, where $n = 556$ is the total number of data points. We use the following linear model to describe the yield,

$$y_j = \beta_0 + X_j\beta_X + Z_j\beta_a + W_j\beta_d + (X_j \times Z_j)\beta_{aX} + (X_j \times W_j)\beta_{dX} + \varepsilon_j$$

where $\beta_0$ is the intercept (or population mean), $\beta_X$ is the environmental (or year) effect, $\beta_a$ is the additive effect, $\beta_d$ is the dominance effect, $\beta_{aX}$ is the additive by year interaction effect, $\beta_{dX}$ is the dominance by year interaction, and $\varepsilon_j$ is a random environmental error with an assumed $N(0, \sigma_\varepsilon^2)$ distribution. Note that the coefficient for $\beta_{aX}$ is $X_j$ times $Z_j$ and the coefficient for $\beta_{dX}$ is $X_j$ times $W_j$. Recall that the genotype by environment interaction model is

$$P = G + E_X + G \times E_X + E$$

Comparing this with the previous model, we can see that $P = y_j - \beta_0$, $E_X = X_j\beta_X$, $G = Z_j\beta_a + W_j\beta_d$, $G \times E_X = (X_j \times Z_j)\beta_{aX} + (X_j \times W_j)\beta_{dX}$ and $E = \varepsilon_j$. The phenotype is adjusted by the population mean. The data (phenotypes and genotypes) allow us to estimate all the effects in the model. The data with the (X, Z, W) coding are given in "data-7-2. xlsx". The estimated parameters are listed in Table 7.15. This detailed analysis shows that the year (environmental) effect, additive effect, and dominance effect are all significant. Interestingly, the significant G×E interaction is only caused by the additive by year interaction because the dominance by year interaction is not significant. The variance of $y$ is

**Table 7.15** Estimated parameters, their estimation errors, and the test statistics

| Parameter | Estimate | Standard Error | t Value | Pr > \|t\| |
|---|---|---|---|---|
| $\beta_0$ | 42.16179567 | 0.42859919 | | |
| $\beta_X$ | 2.51919741 | 0.42859919 | 5.88 | <0.0001 |
| $\beta_a$ | −2.28858915 | 0.42859919 | −5.34 | <0.0001 |
| $\beta_d$ | 2.72481110 | 0.59304895 | 4.59 | <0.0001 |
| $\beta_{aX}$ | 1.25241274 | 0.42859919 | 2.92 | 0.0036 |
| $\beta_{dX}$ | −0.90711914 | 0.59304895 | −1.53 | 0.1267 |

$$\text{var}(y_j) = \underbrace{\text{var}(Z_j)\beta_a^2 + \text{var}(W_j)\beta_d^2}_{V_G} + \underbrace{\text{var}(X_j)\beta_X^2}_{V_X}$$

$$+ \underbrace{\text{var}(X_j \times Z_j)\beta_{aX}^2 + \text{var}(X_j \times W_j)\beta_{dX}^2}_{V_{G \times E}}$$

$$+ 2\sum \text{cov} + \underbrace{\text{var}(\varepsilon_j)}_{V_E}$$

where $\sum$cov is the sum of all covariance terms. The above equation is rewritten as

$$\text{var}(y_j) = V_G + V_X + V_{G \times E} + 2\sum \text{cov} + V_E$$

The variance-covariance matrix for the five coefficients calculated from the data is listed in Table 7.16. The diagonal elements of the matrix in Table 7.16 are the variances and the off-diagonal elements are the covariance terms. For example, $\text{var}(Z_j) = 0.4721$ and $\text{cov}(Z_j, W_j) = -0.0019$. The year variance is

$$V_X = \text{var}(X_j)\beta_X^2 = 1.00 \times 2.5192^2 = 6.3464$$

The genetic variance is

$$V_G = \text{var}(Z_j)\beta_a^2 + \text{var}(W_j)\beta_d^2$$
$$= 0.4721 \times (-2.2886)^2 + 0.2496 \times 2.7248^2$$
$$= 4.3259$$

The G × E interaction variance is

$$V_{G \times E} = \text{var}(X_j \times Z_j)\beta_{aX}^2 + \text{var}(X_j \times W_j)\beta_{dX}^2$$
$$= 0.4721 \times 1.2524^2 + 0.5297 \times (-0.9071)^2$$
$$= 1.1764$$

The variance caused by all the covariance terms is $2\sum \text{cov} = -2.3745$. The residual error variance is $V_E = \sigma_\varepsilon^2 = 45.6445$. The variance of $y_j$ is

$$\text{var}(y_j) = V_G + V_X + V_{G \times E} + 2\sum \text{cov} + V_E$$
$$= 4.3259 + 6.3464 + 1.1764 - 2.3745 + 45.6445$$
$$= 55.1186$$

This variance is not $V_P$ because it contains variation caused by the year variance. The sum of all covariance terms may also be excluded. We now have two options to define the phenotypic variance. One option is

**Table 7.16**  Variance-covariance matrix for the five coefficients of the model

|     | X | Z | W | XZ | XW |
|-----|--------|---------|---------|--------|--------|
| X   | 1.0000 | 0.0000  | 0.0000  | 0.0036 | 0.5297 |
| Z   | 0.0000 | 0.4721  | −0.0019 | 0.0000 | 0.0000 |
| W   | 0.0000 | −0.0019 | 0.2496  | 0.0000 | 0.0000 |
| XZ  | 0.0036 | 0.0000  | 0.0000  | 0.4721 | 0.0000 |
| XW  | 0.5297 | 0.0000  | 0.0000  | 0.0000 | 0.5297 |

$$V_P = V_G + V_E = 4.3259 + 45.6445 = 49.9704$$

The second option is

$$V_P = V_G + V_{G \times E} + V_E = 4.3259 + 1.1764 + 45.6445 = 51.1468$$

Both options give smaller phenotypic variances than var($y_j$) = 55.1186. The broad sense heritability using option two is

$$H^2 = \frac{V_G}{V_G + V_{G \times E} + V_E} = \frac{4.3259}{4.3259 + 1.1764 + 45.6445} = \frac{4.3259}{51.1468} = 0.084578$$

A single locus explains 8.46% of the phenotypic variance, which represents a significant genetic contribution for such a complex trait (yield).

### 7.4.5 Tukey's One Degree of Freedom G × E Interaction Test

This method is to test the interaction of two factors in a two-way analysis of variance (ANOVA) without replication (Tukey 1949; Šimeček and Šimečková 2013). Let $n$ be the number of genotypes and $m$ be the number of blocks (environments). Since there is only one observation per cell, it is impossible to test the interaction. Tukey (Tukey 1949) proposed a one degree-of-freedom non-additivity restraint test for interaction, trying to "create something out of nothing." A one degree of freedom test is better than no test at all. The Tukey's constraint interaction model is

$$Y_{ij} = \mu + \alpha_i + \beta_j + \lambda \xi_{ij} + \varepsilon_{ij}$$

where $\xi_{ij} = \alpha_j \beta_j$ is the product of effects of the two factors, $\lambda$ is the regression coefficient of $y_{ij}$ on $\xi_{ij}$ that is treated as an independent variable. If we rewrite the model as

$$Y_{ij} = \mu + \alpha_i + \beta_j + \lambda (\alpha_i \beta_j) + \varepsilon_{ij} \tag{7.3}$$

we can see that the model is not linear on the parameters. Therefore, parameter estimation involves iterations. By testing the null hypothesis of $\lambda = 0$, we are able to detect some departures from additivity based only on the single parameter $\lambda$. Zhenyu Jia at UC Riverside (2018, personal communication) suggested that the nonlinear model can be fitted with data using the maximum likelihood method. The likelihood ratio test can be adopted here to test $H_0 : \lambda = 0$. The reduced (null) model is

$$Y_{ij} = \mu + \alpha_i + \beta_j + \varepsilon_{ij} \tag{7.4}$$

Let $\theta_1 = \{\alpha_i, \beta_j, \sigma^2, \lambda\}$ be the parameters from model (7.3) and $\theta_0 = \{\alpha_i, \beta_j, \sigma^2\}$ be the parameters from model (7.4). The likelihood ratio test is

$$\psi = -2 \left[ L\left(\widehat{\theta}_0\right) - L\left(\widehat{\theta}_1\right) \right]$$

The two models differ by one parameter, $\lambda$, and thus $\psi$ will follows a Chi-square distribution with one degree of freedom under the null model.

A two-step approach was proposed (Rasch et al. 2009; Šimeček and Šimečková 2013) to test the one degree of freedom interaction term. In the first step, model (7.4) is fitted and $\alpha_i$ and $\beta_j$ are estimated from this model. The estimated effects are denoted by $\widehat{\alpha}_i$ and $\widehat{\beta}_j$. In the second step, define $\widehat{\xi}_{ij} = \widehat{\alpha}_i \widehat{\beta}_j$ as a cofactor (a continuous independent variable) and refit the following model

$$Y_{ij} = \mu + \alpha_i + \beta_j + \lambda \widehat{\xi}_{ij} + \varepsilon_{ij}$$

which allows the test of $\lambda = 0$. This is called covariance analysis and $\lambda$ is simply a regression coefficient.

The original Tukey's test implemented a single-step approach. The method is carried out following the steps described below. But first, we need to define some commonly seen terminologies in a factorial design of experiment. Recall that $Y_{ij}$ is an element of an $n \times m$ matrix (table). From this table, we compute the row mean, denoted by $\overline{Y}_{i\cdot}$, and the column mean, denoted by $\overline{Y}_{\cdot j}$. We also compute the grand mean, denoted by $\overline{Y}_{\cdot\cdot}$. Such a notation system is called the dot notation in elementary statistics. Using these quantities, we compute sum of squares, degrees of freedom and mean squares, as described below.

(1) Calculate various sum of squares

$$SS_G = m \sum_{i=1}^{n} \left(\overline{Y}_{i\cdot} - \overline{Y}_{\cdot\cdot}\right)^2$$

$$SS_B = n \sum_{j=1}^{m} \left(\overline{Y}_{\cdot j} - \overline{Y}_{\cdot\cdot}\right)^2$$

$$SS_{G\times E} = \frac{\left[\sum_{i=1}^{n}\sum_{j=1}^{m} Y_{ij}\left(\overline{Y}_{i\cdot} - \overline{Y}_{\cdot\cdot}\right)\left(\overline{Y}_{\cdot j} - \overline{Y}_{\cdot\cdot}\right)\right]^2}{\sum_{i=1}^{n}\left(\overline{Y}_{i\cdot} - \overline{Y}_{\cdot\cdot}\right)^2 \sum_{j=1}^{m}\left(\overline{Y}_{\cdot j} - \overline{Y}_{\cdot\cdot}\right)^2}$$

$$SS_T = \sum_{i=1}^{n}\sum_{j=1}^{m}\left(Y_{ij} - \overline{Y}_{\cdot\cdot}\right)^2$$

and

$$SS_E = SS_T - SS_G - SS_B - SS_{G\times E}$$

(2) Calculate degrees of freedoms.

$$df_G = n - 1, df_B = m - 1, df_{G \times E} = 1.$$

and

$$df_E = (nm - 1) - (n - 1) - (m - 1) - 1 = nm - n - m$$

(3) Construct the ANOVA table (Table 7.17)
(4) Reject the null hypothesis if the $p$-value for G × E is smaller than 0.05.

There are no SAS procedures to do the Tukey's non-additivity test. There is an R package called "additivityTests" which consists of several R functions related to the Tukey's one degree of freedom non-additivity test. One function within that package has a function called tukey.test(). This function, however, only tests the G × E effect, not the main effects of G and E. I wrote an R function also called `tukey.test()` to replace the same R function within the "additivityTests" package. The R function is available on the course website and also given below.

**Table 7.17**  ANOVA table for Tukey's one degree of freedom additivity test

| Source | Df | SS | MS | F | P value |
|---|---|---|---|---|---|
| G | $df_A = n\text{-}1$ | $SS_G$ | $MS_A = SS_A/df_A$ | $F_A = MS_A/MS_E$ | Prob>$F_A$ |
| B | $df_B = m\text{-}1$ | $SS_B$ | $MS_B = SS_B/df_B$ | $F_B = MS_B/MS_E$ | Prob>$F_B$ |
| G×E | $Df_{G\times E} = 1$ | $SS_{G\times E}$ | $MS_{G\times E} = SS_{G\times E}/df_{G\times E}$ | $F_{G\times E} = MS_{G\times E}/MS_E$ | Prob>$F_{G\times E}$ |
| E | $df_E = nm\text{-}n\text{-}m$ | $SS_E$ | $MS_E = SS_E/df_E$ | | |

```
tukey.test<-function(Y=NULL){
  n<-nrow(Y)
  m<-ncol(Y)
  Yi.<-apply(Y,1,mean)
  Y.j<-apply(Y,2,mean)
  Y..<-mean(Y)
  SST<-sum((Y-Y..)^2)
  SSA<-m*sum((Yi.-Y..)^2)
  SSB<-n*sum((Y.j-Y..)^2)
  SSAB<-n*m*(t(as.matrix((Yi.-Y..)))%*%Y%*%as.matrix((Y.j-Y..)))^2/SSA/SSB
  SSE<-SST-SSA-SSB-SSAB
  dfA<-n-1
  dfB<-m-1
  dfAB<-1
  dfE<-n*m-n-m
  MSA<-SSA/dfA
  MSB<-SSB/dfB
  MSAB<-SSAB/dfAB
  MSE<-SSE/dfE
  FA<-MSA/MSE
  FB<-MSB/MSE
  FAB<-MSAB/MSE
  pA<-1-pf(FA,dfA,dfE)
  pB<-1-pf(FB,dfB,dfE)
  pAB<-1-pf(FAB,dfAB,dfE)
  Source<-c("A","B","AB","E")
  Df<-c(dfA,dfB,dfAB,dfE)
  SS<-round(c(SSA,SSB,SSAB,SSE),6)
  MS<-round(c(MSA,MSB,MSAB,MSE),6)
  F<-round(c(FA,FB,FAB,NA),6)
  p<-round(c(pA,pB,pAB,NA),6)
  ANOVA<-data.frame(Source,Df,SS,MS,F,p)
  return(ANOVA)
}
```

**Table 7.18** A hypothetical data with 10 genotypes and 5 environments (years) used for illustration for the Tukey's one-degree-of-freedom non-additivity test

| Genotype | Year1 | Year2 | Year3 | Year4 | Year5 |
|---|---|---|---|---|---|
| 1 | 3.129639 | 0.502099 | 0.648725 | −0.23966 | −2.21283 |
| 2 | 3.447046 | 0.857642 | 0.977676 | 0.097842 | −1.91319 |
| 3 | 5.211288 | 2.639684 | 2.760404 | 1.88609 | −0.10297 |
| 4 | 3.749767 | 1.138569 | 1.269763 | 0.410356 | −1.59578 |
| 5 | 3.796805 | 1.221267 | 1.327797 | 0.460502 | −1.53608 |
| 6 | 2.928469 | 2.079143 | 2.108889 | 1.828281 | 1.16302 |
| 7 | 1.682818 | 0.817779 | 0.859608 | 0.571313 | −0.09995 |
| 8 | −0.05124 | −0.8963 | −0.87694 | −1.15481 | −1.82423 |
| 9 | 0.52994 | −0.31826 | −0.26439 | −0.56248 | −1.25288 |
| 10 | 0.772169 | −0.07763 | −0.03281 | −0.33724 | −1.01222 |

Table 7.18 shows a sample data with 10 genotypes and 5 blocks (environments) without replicates within the genotype and block combinations. The following code reads the sample data and call the "tukey.test" function to do the test.

```
dir<-"C:\crUsers\crChapter 7"
setwd(dir)

YY<-read.csv(file="data-7-4.csv",header=T)
Y<-as.matrix(YY[,-1])
tukey.test(Y)

rm(tukey.test)
library(additivityTests)
tukey.test(Y)
```

The first call of tukey.test activates the tukey.test function written by the instructor and the second call of tukey.test activates the tukey.test function within an existing R package called "additivityTests." The output from the first tukey.test call is.

```
Source  Df       SS        MS          F        p
   G    9   49.252511   5.472501   14.290211  0.000000
   Y    4   65.379707  16.344927   42.681115  0.000000
  GY    1    2.980093   2.980093    7.781847  0.008482
   E   35   13.403409   0.382955         NA        NA
```

To improve the readability of the output, we present it in an ANOVA Table (Table 7.19). The *p*-value for the one degree of freedom test for interaction is 0.008482, smaller than 0.05, meaning that the G × Y interaction is significant. The output from the second call of the tukey.test function from the "additivityTests" package (Rasch et al. 2009; Šimeček and Šimečková 2013) is.

```
Tukey test on 5% alpha-level:
Test statistic: 7.782
Critival value: 4.121
The additivity hypothesis was rejected.
```

For an inbred rice data replicated in four environments (data-7-5.csv), we want to test whether G×E interaction is significant or not for two traits: 1000 grain weight (KGW) and yield (YD). The ANOVA table for trait KGW is shown in Table 7.20. The *p*-value for the one degree of freedom test for interaction is 0.038201, smaller than 0.05, meaning that the

Table 7.19  ANOVA table for Tukey's one degree of freedom additivity test

| Source | Df | SS | MS | F | P value |
|--------|----|----|----|----|---------|
| G | 9 | 49.252511 | 5.472501 | 14.290211 | 0.000000 |
| Y | 4 | 65.379707 | 16.344927 | 42.681115 | 0.000000 |
| G×Y | 1 | 2.980093 | 2.980093 | 7.781847 | 0.008482 |
| E | 35 | 13.403409 | 0.382955 | | |

Table 7.20  ANOVA table for Tukey's one degree of freedom additivity test (KGW)

| Source | Df | SS | MS | F | P value |
|--------|----|----|----|----|---------|
| G | 209 | 5281.0258 | 25.2681 | 22.27648 | 0.0000 |
| B | 3 | 87.0346 | 29.0115 | 25.576771 | 0.0000 |
| G×E | 1 | 4.8885 | 4.8885 | 4.309785 | 0.038301 |
| E | 626 | 710.0675 | 1.1343 | | |

**Table 7.21**  ANOVA table for Tukey's one degree of freedom additivity test (YD)

| Source | Df | SS | MS | F | P value |
|---|---|---|---|---|---|
| G | 209 | 16174.566 | 77.39027 | 5.447436 | 0.0000 |
| B | 3 | 18014.115 | 6004.70505 | 422.666151 | 0.0000 |
| G × E | 1 | 3508.139 | 3508.13870 | 246.934940 | 0.0000 |
| E | 626 | 8893.415 | 14.20673 | | |

G × E interaction is significant. Table 7.21 shows the test result for trait yield (YD). The $p$-value for the one degree of freedom test for interaction of YD is extremely small (0.0000), so the G × E interaction is extremely strong. If there are missing values in the two-way factorial design without replications, the missing values may be imputed prior to the Tukey's test. For example, if $Y_{ij}$ is missing, the imputed value, denoted by $\widehat{Y}_{ij}$, should take

$$\widehat{Y}_{ij} = \left( \overline{Y}_{i\cdot} - \overline{Y}_{\cdot\cdot} \right) + \left( \overline{Y}_{\cdot j} - \overline{Y}_{\cdot\cdot} \right) + \overline{Y}_{\cdot\cdot} = \overline{Y}_{i\cdot} + \overline{Y}_{\cdot j} - \overline{Y}_{\cdot\cdot}$$

The validity of this imputation depends on two assumptions: (1) missing is random and (2) the number of missing values cannot be higher than 10% of the total number of cells ($n \times m$).

Finally, there is an R package called "additivityTests" which consists of several R functions related to the Tukey's one degree of freedom non-additivity test. One function within this package has the same name as the function provided above "tukey.test" and this function only tests the G×E effect but does not test the main effects of G and E.

# References

Hartley HO, Rao JNK. Maximum-likelihood estimation for the mixed analysis of variance model. Biometrika. 1967;54:93–108.

Henderson CR. MIVQUE and REML estimation of additive and nonadditive genetic variances. J Anim Sci. 1985;61:113–21.

Henderson CR. Recent developments in variance and covariance estimations. J Anim Sci. 1986;63:208–16.

Patterson HD, Thompson R. Recovery of inter-block information when block sizes are unequal. Biometrika. 1971;58:545–54.

Rao CR. Estimation of variance and covariance components-MINQUE theory. J Multivar Anal. 1971a;1:257–75.

Rao CR. Minimum variance quadratic unbiased estimation of variance components. J Multivar Anal. 1971b;1:445–56.

Rasch D, Rusch T, Šimečková M, Kubinger KD, Moder K, Šimeˇcek P. Tests of additivity in mixed and fixed effect two-way ANOVA models with single sub-class numbers. Stat Papers. 2009;50:905–16.

SAS Institute Inc. 2009. *SAS/STAT: Users' Guide, Version 9.3*. SAS Institute Inc., Cary, NC.

Šimeček P, Šimečková M. Modification of Tukey's additivity test. J Stat Plann Inference. 2013;143:197–201.

Swallow WH, Searle SR. Minimum variance quadratic unbiased estimation (MIVQUE) of variance components. Technometrics. 1978;20:265–72.

Tukey JW. One degree of freedom for non-additivity. Biometrics. 1949;5:232–42.

Xu S. Mapping quantitative trait loci by controlling polygenic background effects. Genetics. 2013;195:1209–22.

Yu H, Xie W, Wang J, Xing Y, Xu C, Li X, Xiao J, Zhang Q. Gains in QTL detection using an ultra-high density SNP map based on population sequencing relative to traditional RFLP/SSR markers. PLoS One. 2011;6:e17595. https://doi.org/10.11371/journal.pone.0017595.

# Major Gene Detection and Segregation Analysis

**8**

As demonstrated in Chap. 2, the continuous distribution of a character does not have to be caused by many loci; one or two loci may be sufficient to cause a continuous distribution of a trait, as long as environment plays a major role in the expression of the trait. If a character is controlled by only a few major genes and the segregation of these genes account for a large proportion of the phenotypic variance, the genetic mechanism of such a character can be studied using a segregation analysis. In this chapter, we will investigate the statistical methods for estimating and testing such major genes, assuming that genotypes of the major genes are observable. If the genotypes are not observable and we believe that the trait is controlled by a single major gene, we can perform a segregation analysis (Morton 1958; Elston and Stewart 1971; Elston 1981; Zhang et al. 2003; Gai 2005). Segregation analysis will be studied towards the end of the chapter.

To detect a major gene, we often need to design a line crossing experiment, like Mendel did in his common pea hybridization experiments. Many line crossing experiments are available. The commonly used crossing designs are backcross (BC) design, $F_2$ design, double haploid (DH) design, recombinant inbred line (RIL) design, four-way (FW) mating design, and diallel cross (DC) design. Of course, detecting major gene can also be performed in random populations, e.g., human populations and populations of large animals (Elston and Stewart 1971). In this chapter, we will use the RIL and $F_2$ designs as examples to demonstrate the methods of major gene detection. We will introduce three methods for major gene detection: (1) two sample t or F test, (2) multiple sample F-test (ANOVA) and (3) regression analysis.

## 8.1 Two Sample t-Test or F-Test

The t-test applies to major gene detection for populations with two alternative genotypes, e.g., BC, DH, and RIL. Let us use an RIL population as an example. The RIL mating design is sketched in Fig. 8.1, where multiple $F_1$ plants are generated by crossing two inbred parents ($P_1$ and $P_2$). Each $F_1$ is then undergoing multiple generations of selfings until all progenies are fixed (pure inbred) for either one of the two genotypes, $A_1A_1$ or $A_2A_2$. Heterozygotes will not appear in the RIL population. A hypothetical sample dataset is given in Table 8.1.

We have two independent samples, one for each genotype. Let $Y_1$ represent the sample with genotype $A_1A_1$ and $Y_2$ the sample with genotype $A_2A_2$. The t-test statistic is defined by

$$t = \frac{|\overline{Y}_1 - \overline{Y}_2|}{s_{\overline{Y}_1 - \overline{Y}_2}}$$

where

$$s_{\overline{Y}_1 - \overline{Y}_2} = \sqrt{\frac{(n_1 - 1)s_1^2 + (n_2 - 1)s_2^2}{n_1 + n_2 - 2}\left(\frac{1}{n_1} + \frac{1}{n_2}\right)}$$

is the standard error of the difference between the means of the two genotypes. The sample means and sample variances are calculated using

© Springer Nature Switzerland AG 2022
S. Xu, *Quantitative Genetics*, https://doi.org/10.1007/978-3-030-83940-6_8

**Fig. 8.1** The recombinant inbred
line (RIL) mating design for major
gene detection

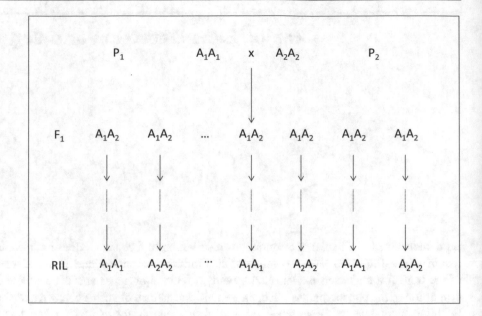

**Table 8.1** Phenotypic values of 15 individuals from two genotypes

| Genotype $A_1A_1$ $(Y_1)$ | Genotype $A_2A_2$ $(Y_2)$ |
|---|---|
| 10.92 | 6.82 |
| 16.09 | 14.67 |
| 13.57 | 15.77 |
| 12.60 | 7.52 |
| 13.59 | 7.99 |
| 17.59 | 9.85 |
| 17.90 | 1.69 |
|  | −1.09 |

$$\overline{Y}_k = \frac{1}{n_k} \sum_{j=1}^{n_k} Y_{jk}, \quad \forall k = 1, 2$$

and

$$s_k^2 = \frac{1}{n_k - 1} \sum_{j=1}^{n_k} \left( Y_{jk} - \overline{Y}_k \right)^2, \quad \forall k = 1, 2$$

In this example, $n_1 = 7$, $n_2 = 8$, $\overline{Y}_1 = 14.6086$, $\overline{Y}_2 = 7.9025$, $s_1^2 = 6.9455$, and $s_2^2 = 33.1822$. The degree of freedom is $df = n_1 + n_2 - 2 = 13$. Therefore, the standard error of the difference between the means of the two genotypes is

$$s_{\overline{Y}_1 - \overline{Y}_2} = \sqrt{\frac{(7-1) \times 6.9455 + (8-1) \times 33.1822}{7+8-2} \left( \frac{1}{7} + \frac{1}{8} \right)} = 2.3758$$

The t-test statistic is

$$t = \frac{\mid 14.6086 - 7.9025 \mid}{2.3758} = 2.8226$$

The critical value at Type I error rate of 0.05 (two-tail test) is 2.1604. Since $t > 2.1604$, we claim that this gene is significant, e.g., we detected a major gene. The $p$-value is calculated using

$$p = 2 \times [1 - T_{13}(t < 2.8226)] = 2 \times (1 - 0.9928) = 0.0144$$

where $T_{13}(t < 2.8226)$ is the cumulative distribution function of a $t$ variable with 13 degrees of freedom. If the $p$-value is smaller than 0.05, the test is significant. When we perform a t-test, it is always a two-tailed t-test in major gene detection because we do not know a priori which genotype is better than the other. It is very confusing in t-test because the one-tailed test vs two-tailed test issue. In practice, we may convert the t-test into an F-test and the result of the F-test is always one-tailed (equivalent to two-tailed t-test). The relationship between the t-test and the F-test is

$$F = t^2 = \frac{\left(\overline{Y}_1 - \overline{Y}_2\right)^2}{s^2_{\overline{Y}_1 - \overline{Y}_2}} = 2.8226^2 = 7.9671$$

Now, the $p$-value can be easily calculated using

$$p = 1 - F_{1,13}(f < 7.9671) = 1 - 0.9856 = 0.0144$$

where $F_{1,13}(f < 7.9671)$ is the cumulative distribution function of an $f$ variable with 1 and 13 degrees of freedom. From now on, let us forget t-test and always use the F-test for major gene detection.

The TTEST and ANOVA procedures in SAS can do the t and F tests. Below is the SAS code to read the data and do the tests.

```
data one;
    input genotype $ y;
cards;
A1A1   10.92
A1A1   16.09
A1A1   13.57
A1A1   12.6
A1A1   13.59
A1A1   17.59
A1A1   17.9
A2A2    6.82
A2A2   14.67
A2A2   15.77
A2A2    7.52
A2A2    7.99
A2A2    9.85
A2A2    1.69
A2A2   -1.09
;
run;
ods graphics off;
proc ttest data=one;
    class genotype;
run;
proc anova data=one;
    class genotype;
    model y = genotype;
run;
```

**Table 8.2** Output from the TTEST procedure in SAS for data from Table 8.1

| Method | Variances | DF | t value | Pr > \|t\| |
|---|---|---|---|---|
| Pooled | Equal | 13 | 2.82 | 0.0144 |
| Satterthwaite | Unequal | 10.077 | 2.96 | 0.0142 |

**Table 8.3** Output from the ANOVA procedure in SAS for data from Table 8.1

| Source | DF | Sum of squares | Mean square | F value | Pr > F |
|---|---|---|---|---|---|
| Model | 1 | 167.8932043 | 167.8932043 | 7.97 | 0.0144 |
| Error | 13 | 273.9484357 | 21.0729566 | | |
| Corrected Total | 14 | 441.8416400 | | | |

**Table 8.4** Output from the upper tailed t-test using the TTEST procedure in SAS for data from Table 8.1

| Method | Variances | DF | t value | Pr > t |
|---|---|---|---|---|
| Pooled | Equal | 13 | 2.82 | 0.0072 |
| Satterthwaite | Unequal | 10.077 | 2.96 | 0.0071 |

The data (data-8-1.xlsx) are read through the data step. Since the dataset is very small, they can be read instream (reading raw data with the input statement). The `ods graphics off` statement turns off the graphic presentation. The output from `proc. ttest` is shown in Table 8.2.

The pooled method generated the same result as the one reported early. The assumption for the pooled method is that the two groups ($A_1A_1$ and $A_2A_2$) have the same variances. The Satterwaite method does not require this assumption and the result is not much different from the pooled method for this example. The output from `proc anova` is shown in Table 8.3. The F value is indeed the square of the t value (pooled or equal variance) and the p-value is identical to the p-value of the t-test. The pooled t-test and the F-test for two genotypes are exactly the same. Why do we need the t-test? The t-test is needed in one particular situation— one-tailed test. If the alternative hypothesis is that the genotypic value of $A_1A_1$ is higher than that of $A_2A_2$, we must use a one-tailed t-test. Since the default sorting order is $A_1A_1$ appears first and $A_2A_2$ appears second, $A_1A_1$ is considered as group 1 and $A_2A_2$ is considered as group 2. The t-test is an upper tailed t-test. The SAS code for such a one-tailed t-test is

```
proc ttest data=one side=u;
    class genotype;
run;
```

where the `side = u` option indicates upper tail. The result is shown in Table 8.4. The p-value is 0.0072, half the p-value of the two-tailed test. If we set the lower tail for the test, i.e., `side = l`, in the data step, the p-value would be 0.99, and we cannot reject the null hypothesis. Again, for major gene detection, use the default two-tailed test, unless there is a particular reason to pursue a one-tailed t-test.

## 8.2   F-Test for Multiple Samples (ANOVA)

For populations that contain more than two genotypes, t-test cannot be applied because it only applies to two genotypes (samples). The $F_2$ mating design generates three possible genotypes. The F-test is designed to handle multiple samples. Therefore, we will use the F-test to detect major genes in $F_2$ populations. A sketch of the $F_2$ mating design is shown in Fig. 8.2. The three genotypes of an $F_2$ population are $A_1A_1$, $A_1A_2$, and $A_2A_2$. A sample dataset with 30 individuals are listed in Table 8.5 (data-8-2.xlsx). We will use ANOVA to test the differences over the three genotypes. The groups in the ANOVA refer to the genotypes and thus there are three groups. The data are unbalanced with 7 observations in the first group, 15 observations in

**Fig. 8.2** The $F_2$ mating design for major genes detection

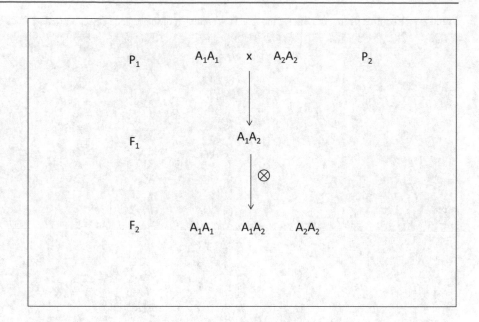

**Table 8.5** Genotypes and phenotypes of 30 individuals

| Genotype $A_1A_1$ | Genotype $A_1A_2$ | Genotype $A_2A_2$ |
|---|---|---|
| 13.92 | 9.62 | 6.82 |
| 19.09 | 7.37 | 14.67 |
| 16.57 | 22.00 | 15.77 |
| 15.60 | 6.87 | 7.52 |
| 16.59 | 8.83 | 7.99 |
| 19.59 | 6.24 | 9.85 |
| 19.90 | 15.21 | 1.69 |
| | 14.61 | -1.09 |
| | 16.82 | |
| | 15.31 | |
| | 8.61 | |
| | 19.76 | |
| | −0.37 | |
| | 16.03 | |
| | −2.66 | |

**Table 8.6** ANOVA table for major genes detection in an $F_2$ population

| Source | Df | SS | MS | F value | Pr(>F) | |
|---|---|---|---|---|---|---|
| Between | 2 | 344.95 | 172.47 | 4.9013 | 0.01528 | * |
| Within | 27 | 950.13 | 35.19 | | | |

the second group, and 8 observations in the third group. The ANOVA table is shown in Table 8.6. The critical value for this test is 3.3541 and the actual F value is 4.9013. Therefore, the major genes detection is significant. We also look at the p-value, which is 0.01528, less than 0.05 and thus the test is significant. To perform the analysis of variances, we must rearrange the data in the format shown in Table 8.7. The SAS code for the ANOVA procedure is.

**Table 8.7** Required format of data for the ANOVA approach of major gene detection (data-8-2.xlsx)

| Plant | Genotype | $y$ |
|---|---|---|
| 1 | $A_1A_1$ | 13.92 |
| 2 | $A_1A_1$ | 19.09 |
| 3 | $A_1A_1$ | 16.57 |
| 4 | $A_1A_1$ | 15.6 |
| 5 | $A_1A_1$ | 16.59 |
| 6 | $A_1A_1$ | 19.59 |
| 7 | $A_1A_1$ | 19.9 |
| 8 | $A_1A_2$ | 9.62 |
| 9 | $A_1A_2$ | 7.37 |
| 10 | $A_1A_2$ | 22 |
| 11 | $A_1A_2$ | 6.87 |
| 12 | $A_1A_2$ | 8.83 |
| 13 | $A_1A_2$ | 6.24 |
| 14 | $A_1A_2$ | 15.21 |
| 15 | $A_1A_2$ | 14.61 |
| 16 | $A_1A_2$ | 16.82 |
| 17 | $A_1A_2$ | 15.31 |
| 18 | $A_1A_2$ | 8.61 |
| 19 | $A_1A_2$ | 19.76 |
| 20 | $A_1A_2$ | −0.37 |
| 21 | $A_1A_2$ | 16.03 |
| 22 | $A_1A_2$ | −2.66 |
| 23 | $A_2A_2$ | 6.82 |
| 24 | $A_2A_2$ | 14.67 |
| 25 | $A_2A_2$ | 15.77 |
| 26 | $A_2A_2$ | 7.52 |
| 27 | $A_2A_2$ | 7.99 |
| 28 | $A_2A_2$ | 9.85 |
| 29 | $A_2A_2$ | 1.69 |
| 30 | $A_2A_2$ | −1.09 |

**Table 8.8** ANOVA table for major gene detection in an $F_2$ population generated from the ANOVA procedure in SAS

| Source | DF | Sum of squares | Mean square | F value | Pr > F |
|---|---|---|---|---|---|
| Model | 2 | 344.949444 | 172.474722 | 4.90 | 0.0153 |
| Error | 27 | 950.126293 | 35.189863 | | |
| Corrected Total | 29 | 1295.075737 | | | |

```
filename aa "data-8-2.xlsx";
proc import datafile=aa out=one dbms=xlsx replace;
run;
ods graphics off;
proc anova data=one;
   class genotype;
   model y=genotype;
run;
```

The output is shown in Table 8.8, where the p-value is 0.0153, same as the p-value shown in Table 8.6.

## 8.3 Regression Analysis

### 8.3.1 Two Genotypes

The regression method is sufficiently general to perform major gene detection for populations with arbitrary number of genotypes, including RIL and $F_2$ populations. We will use matrix algebra to describe the regression method and then reanalyze the two sample data (introduced before) using the regression method. Eventually, we will demonstrate the equivalence of the regression analysis to the F-test in ANOVA. First, we must recode the genotype from a character variable into a numerical variable. The coding is somehow arbitrary, but I prefer the following coding system,

$$X_{j1} = \begin{cases} +1 & \text{for} \quad A_1A_1 \\ -1 & \text{for} \quad A_2A_2 \end{cases}$$

Let us also define a variable $X_{j0}$ to store a constant number 1; it is really not a variable, but defined as a variable to simplify the notation of the regression analysis. The numerically coded data for the RIL example are listed in Table 8.9. Let $y_j$ be the phenotypic value of the $j$th plant and the following linear model can be used to describe the phenotype,

$$y_j = X_{j0}b_0 + X_{j1}b_1 + \varepsilon_j$$

where $b_0$ is the intercept and $X_{j0}b_0 = b_0$ because $X_{j0} = 1$ for all $j = 1, \ldots, n$. The regression coefficient $b_1 = a$ is the additive genetic effect defined in Chap. 5 (half the distance between the values of the two genotypes). If $G_{11}$ and $G_{22}$ represent the genotypic values for $A_1A_1$ and $A_2A_2$, respectively, then

$$a = \frac{1}{2}(G_{11} - G_{22})$$

Let us define the data and the residual errors in matrix forms as given below,

$$y = \begin{bmatrix} 6.82 \\ 10.92 \\ \vdots \\ 17.9 \end{bmatrix}_{n \times 1}, X_0 = \begin{bmatrix} 1 \\ 1 \\ \vdots \\ 1 \end{bmatrix}_{n \times 1}, X_1 = \begin{bmatrix} -1 \\ +1 \\ \vdots \\ +1 \end{bmatrix}_{n \times 1} \text{and } \varepsilon = \begin{bmatrix} \varepsilon_1 \\ \varepsilon_2 \\ \vdots \\ \varepsilon_{15} \end{bmatrix}_{n \times 1}$$

.

**Table 8.9** Reformatted dataset for the hypothetical RIL population (data-8-3.csv)

| Phenotype($y$) | Genotype | $X_0$ | $X_1$ |
|---|---|---|---|
| 6.82 | $A_2A_2$ | 1 | −1 |
| 10.92 | $A_1A_1$ | 1 | +1 |
| 14.67 | $A_2A_2$ | 1 | −1 |
| 15.77 | $A_2A_2$ | 1 | −1 |
| 16.09 | $A_1A_1$ | 1 | +1 |
| 7.52 | $A_2A_2$ | 1 | −1 |
| 13.57 | $A_1A_1$ | 1 | +1 |
| 7.99 | $A_2A_2$ | 1 | −1 |
| 12.6 | $A_1A_1$ | 1 | +1 |
| 9.85 | $A_2A_2$ | 1 | −1 |
| 1.69 | $A_2A_2$ | 1 | −1 |
| 13.59 | $A_1A_1$ | 1 | +1 |
| 17.59 | $A_1A_1$ | 1 | +1 |
| −1.09 | $A_2A_2$ | 1 | −1 |
| 17.9 | $A_1A_1$ | 1 | +1 |

where $n = 15$ is the sample size. The last vector is a vector of residual errors. If we collect all the $n$ models given in the above equation together and put them into a single matrix, the model is rewritten as

$$
\begin{bmatrix} y_1 \\ y_2 \\ \vdots \\ y_n \end{bmatrix} = \begin{bmatrix} X_{10} \\ X_{20} \\ \vdots \\ X_{n0} \end{bmatrix} b_0 + \begin{bmatrix} X_{11} \\ X_{21} \\ \vdots \\ X_{n1} \end{bmatrix} b_1 + \begin{bmatrix} \varepsilon_1 \\ \varepsilon_2 \\ \vdots \\ \varepsilon_{15} \end{bmatrix}
$$

An even more compact form of the model is

$$
y = X_0 b_0 + X_1 b_1 + \varepsilon
$$

The residual error vector is assumed to be normally distributed, using notation such as $\varepsilon \sim N(0, I_n \sigma^2)$. This is a matrix notation for the distribution of a vector, where $I_n$ is an identity matrix with order $n$ and $I_n \sigma^2$ is a diagonal matrix with all diagonal elements equal to $\sigma^2$. This special matrix notation is equivalent to $\varepsilon_j \sim N(0, \sigma^2)$ for all $j = 1, \ldots, n$. The most compact notation for the above linear model is

$$
y = Xb + \varepsilon
$$

where $X_{n \times 2} = X_0 \| X_1$ is an $n \times 2$ matrix that combines the two vectors horizontally, called horizontal concatenation, and $b_{2 \times 1} = b_0 /\!/ b_1$ is a $2 \times 1$ vertical concatenation matrix of $b_0$ and $b_1$. We want to estimate $b$ and $\sigma^2$ using the least squares method and perform a hypothesis test for the estimated genetic effect. The estimates of the parameters are

$$
\widehat{b} = \left( X^T X \right)^{-1} X^T y
$$

and

$$
\widehat{\sigma}^2 = \frac{1}{n-2} \left( y - X\widehat{b} \right)^T \left( y - X\widehat{b} \right)
$$

The estimated $b$ is a $2 \times 1$ vector,

$$
\widehat{b} = \begin{bmatrix} \widehat{b}_0 \\ \widehat{b}_1 \end{bmatrix}
$$

The variance-covariance matrix of the estimated $b$ is

$$
\mathrm{var}\left( \widehat{b} \right) = \left( X^T X \right)^{-1} \widehat{\sigma}^2
$$

which is a $2 \times 2$ matrix whose elements are shown as

$$
\mathrm{var}\left( \widehat{b} \right) = \mathrm{var}\begin{bmatrix} \widehat{b}_0 \\ \widehat{b}_1 \end{bmatrix} = \begin{bmatrix} \mathrm{var}\left( \widehat{b}_0 \right) & \mathrm{cov}\left( \widehat{b}_0, \widehat{b}_1 \right) \\ \mathrm{cov}\left( \widehat{b}_1, \widehat{b}_0 \right) & \mathrm{var}\left( \widehat{b}_1 \right) \end{bmatrix}
$$

The null hypothesis for major gene detection is $H_0 : a = 0$ which is formulated as $H_0 : b_1 = 0$. The F-test statistic is

$$
F = \frac{\widehat{b}_1^2}{\mathrm{var}\left( \widehat{b}_1 \right)}
$$

The numerator and denominator degrees of freedom are 1 and $n - 2$, respectively. The estimated regression coefficients for the sample data are

$$\begin{bmatrix} \widehat{b}_0 \\ \widehat{b}_1 \end{bmatrix} = \begin{bmatrix} 11.2555 \\ 3.3530 \end{bmatrix}$$

The estimated residual error variance is $\widehat{\sigma}^2 = 21.07296$. The variance-covariance matrix of $\widehat{b}$ is

$$\text{var}\left(\widehat{b}\right) = \left(X^T X\right)^{-1}\widehat{\sigma}_\varepsilon^2 = \begin{bmatrix} 0.06696 & 0.00446 \\ 0.00446 & 0.06696 \end{bmatrix} \times 21.07296 = \begin{bmatrix} 1.41114 & 0.09408 \\ 0.09408 & 1.41114 \end{bmatrix}$$

Therefore, the F-test statistic is

$$F = \frac{\widehat{b}_1^2}{\text{var}\left(\widehat{b}_1\right)} = \frac{3.3530^2}{1.41114} = 7.9672$$

The p-value is

$$p = 1 - F_{1,13}(F < 7.9671) = 1 - 0.9856 = 0.0144$$

These results are the same as those given in the t-test described in the previous section. The REG procedure in SAS can be used to perform the regression analysis. The SAS code and the output are given below.

```
filename aa "data-8-3.csv";
proc import datafile=aa out=one dbms=csv replace;
run;
ods graphics off;
proc reg data=one;
   model y=x1;
quit;
proc reg data=one;
   model y=x0 x1/noint;
quit;
```

The two `proc reg` statements produce exactly the same result. The first one only includes `x1` in the model statement because, by default, SAS includes the intercept automatically. The second one includes `x0` explicitly in the model statement, but we have to add the `noint` (no intercept) option to prevent the model from adding another intercept. The REG procedure generates the output shown in Table 8.10. The REG procedure only reports the results of two-sided t-tests. However, the squares of the t-tests are the F-tests. The p-values, however, remain the same, regardless whether the tests are t-tests or F-tests. For example, the t-value for the regression coefficient is 2.82 and the square of 2.82 is the F-value 7.96. The p-value for the regression coefficient is 0.0144, identical to the p-value reported earlier. PROC REG always reports the p-value for the intercept, but we are not interested in testing whether the intercept is significantly different from zero or not, and do not care about the p-value of the intercept.

**Table 8.10**  Result from the REG procedure in SAS for the data from Table 8.9

| Variable | DF | Parameter estimate | Standard error | t value | Pr > |t| |
|----------|----|--------------------|----------------|---------|----------|
| X0 | 1 | 11.25554 | 1.18791 | 9.48 | <0.0001 |
| X1 | 1 | 3.35304 | 1.18791 | 2.82 | 0.0144 |

### 8.3.2   Three Genotypes

We now demonstrate regression analysis for major gene detection in $F_2$ populations. We define two genotype indicator variables, one for the additive effect and one for the dominance effect,

$$X_{j1} = \begin{cases} +1 & \text{for } A_1A_1 \\ 0 & \text{for } A_1A_2 \\ -1 & \text{for } A_2A_2 \end{cases} \quad \text{and } X_{j2} = \begin{cases} 0 & \text{for } A_1A_1 \\ 1 & \text{for } A_1A_2 \\ 0 & \text{for } A_2A_2 \end{cases}$$

The linear model is

$$y_j = X_{j0}b_0 + X_{j1}b_1 + X_{j2}b_2 + \varepsilon_j$$

where $b_1 = a$ is the additive effect and $b_2 = d$ is the dominance effect. In a compact matrix notation, the model is

$$y = Xb + \varepsilon$$

where $X = X_0 \| X_1 \| X_2$ is an $n \times 3$ design matrix and $b = b_0 / / b_1 / / b_2$ is a $3 \times 1$ vector for the regression coefficients (including the intercept). The reformatted dataset are given in Table 8.11 (data-8-4.csv).

All formulas remain the same as what we learned for the RIL population except that the matrix dimensions have been increased by one because we now have an additional regression coefficient. The results of the regression analysis are

$$\widehat{b} = \begin{bmatrix} b_0 \\ b_1 \\ b_2 \end{bmatrix} = \begin{bmatrix} 12.6127 \\ 4.7102 \\ -1.6627 \end{bmatrix}$$

The residual error variance is $\widehat{\sigma}^2 = 35.1899$. The variance-covariance matrix of $\widehat{b}$ is

$$\text{var}\begin{bmatrix} \widehat{b}_0 \\ \widehat{b}_1 \\ \widehat{b}_2 \end{bmatrix} = \begin{bmatrix} \text{var}\left(\widehat{b}_0\right) & \text{cov}\left(\widehat{b}_0, \widehat{b}_1\right) & \text{cov}\left(\widehat{b}_0, \widehat{b}_2\right) \\ \text{cov}\left(\widehat{b}_1, \widehat{b}_0\right) & \text{var}\left(\widehat{b}_1\right) & \text{cov}\left(\widehat{b}_1, \widehat{b}_2\right) \\ \text{cov}\left(\widehat{b}_2, \widehat{b}_0\right) & \text{cov}\left(\widehat{b}_2, \widehat{b}_1\right) & \text{var}\left(\widehat{b}_2\right) \end{bmatrix} = \begin{bmatrix} 2.3564 & 0.15709 & -2.3564 \\ 0.15709 & 2.3564 & -0.1571 \\ -2.3564 & -0.1571 & 4.7024 \end{bmatrix}$$

The null hypothesis is $H_0 : b_1 = b_2 = 0$, which is equivalent to $H_0 : a = d = 0$. The corresponding F-test statistic is

$$F = \frac{1}{2}\begin{bmatrix} \widehat{b}_1 & \widehat{b}_2 \end{bmatrix} \begin{bmatrix} \text{var}\left(\widehat{b}_1\right) & \text{cov}\left(\widehat{b}_1, \widehat{b}_2\right) \\ \text{cov}\left(\widehat{b}_2, \widehat{b}_1\right) & \text{var}\left(\widehat{b}_2\right) \end{bmatrix}^{-1} \begin{bmatrix} \widehat{b}_1 \\ \widehat{b}_2 \end{bmatrix}$$

$$= \frac{1}{2} \times \begin{bmatrix} 4.7102 & -1.6627 \end{bmatrix} \begin{bmatrix} 2.3565 & -0.1571 \\ -0.1571 & 4.7025 \end{bmatrix}^{-1} \begin{bmatrix} 4.7102 \\ -1.6627 \end{bmatrix}$$

$$= 4.9013$$

The p-value is

$$p = 1 - F_{2,27}(f < 4.9013) = 1 - 0.9847 = 0.0153$$

**Table 8.11** Reformatted dataset of the $F_2$ population for major gene detection

| Plant | Genotype | $y$ | $X_0$ | $X_1$ | $X_2$ |
|---|---|---|---|---|---|
| 1 | $A_1A_1$ | 13.92 | 1 | 1 | 0 |
| 2 | $A_1A_1$ | 19.09 | 1 | 1 | 0 |
| 3 | $A_1A_1$ | 16.57 | 1 | 1 | 0 |
| 4 | $A_1A_1$ | 15.6 | 1 | 1 | 0 |
| 5 | $A_1A_1$ | 16.59 | 1 | 1 | 0 |
| 6 | $A_1A_1$ | 19.59 | 1 | 1 | 0 |
| 7 | $A_1A_1$ | 19.9 | 1 | 1 | 0 |
| 8 | $A_1A_2$ | 9.62 | 1 | 0 | 1 |
| 9 | $A_1A_2$ | 7.37 | 1 | 0 | 1 |
| 10 | $A_1A_2$ | 22 | 1 | 0 | 1 |
| 11 | $A_1A_2$ | 6.87 | 1 | 0 | 1 |
| 12 | $A_1A_2$ | 8.83 | 1 | 0 | 1 |
| 13 | $A_1A_2$ | 6.24 | 1 | 0 | 1 |
| 14 | $A_1A_2$ | 15.21 | 1 | 0 | 1 |
| 15 | $A_1A_2$ | 14.61 | 1 | 0 | 1 |
| 16 | $A_1A_2$ | 16.82 | 1 | 0 | 1 |
| 17 | $A_1A_2$ | 15.31 | 1 | 0 | 1 |
| 18 | $A_1A_2$ | 8.61 | 1 | 0 | 1 |
| 19 | $A_1A_2$ | 19.76 | 1 | 0 | 1 |
| 20 | $A_1A_2$ | −0.37 | 1 | 0 | 1 |
| 21 | $A_1A_2$ | 16.03 | 1 | 0 | 1 |
| 22 | $A_1A_2$ | −2.66 | 1 | 0 | 1 |
| 23 | $A_2A_2$ | 6.82 | 1 | −1 | 0 |
| 24 | $A_2A_2$ | 14.67 | 1 | −1 | 0 |
| 25 | $A_2A_2$ | 15.77 | 1 | −1 | 0 |
| 26 | $A_2A_2$ | 7.52 | 1 | −1 | 0 |
| 27 | $A_2A_2$ | 7.99 | 1 | −1 | 0 |
| 28 | $A_2A_2$ | 9.85 | 1 | −1 | 0 |
| 29 | $A_2A_2$ | 1.69 | 1 | −1 | 0 |
| 30 | $A_2A_2$ | −1.09 | 1 | −1 | 0 |

Therefore, we reject the null hypothesis and claim that a major gene has been detected. Note that the F test result is the same as that given in Table 8.6 of the ANOVA procedure. The regression analysis can go beyond the overall null hypothesis test. We can actually tell which particular effect is significant. To test the additive effect, we use

$$F_1 = \frac{\widehat{b}_1^2}{\text{var}\left(\widehat{b}_1\right)} = \frac{4.7102^2}{2.3565} = 9.4148$$

The corresponding p-value is 0.00485, indicating significant additive effect. The F test for dominance is

$$F_2 = \frac{\widehat{b}_2^2}{\text{var}\left(\widehat{b}_2\right)} = \frac{(-1.6627)^2}{4.7025} = 0.5879$$

with a p-value of 0.44989, which is larger than 0.05 and thus the dominance effect is not significant. The major gene is due to the significant additive effect, not due to the dominance effect.

We can use the REG procedure in SAS to perform the major gene detection. The SAS code is nearly the same as that of the two-genotype case.

**Table 8.12** ANOVA table generated from PROC REG for the $F_2$ population

| Source | DF | Sum of squares | Mean square | F value | Pr > F |
|---|---|---|---|---|---|
| Model | 2 | 344.94944 | 172.47472 | 4.90 | 0.0153 |
| Error | 27 | 950.12629 | 35.18986 | | |
| Corrected Total | 29 | 1295.07574 | | | |

**Table 8.13** Regression coefficients estimated from PROC REG for the $F_2$ population

| Variable | DF | Parameter estimate | Standard error | t value | Pr > |t| |
|---|---|---|---|---|---|
| Intercept | 1 | 12.61268 | 1.53508 | 8.22 | <0.0001 |
| X1 | 1 | 4.71018 | 1.53508 | 3.07 | 0.0049 |
| X2 | 1 | −1.66268 | 2.16851 | −0.77 | 0.4499 |

```
filename aa "data-8-4.csv";
proc import datafile=aa out=one dbms=csv replace;
run;
ods graphics off;
proc reg data=one;
   model y=x1 x2;
quit;
```

The output includes two tables, the first one (Table 8.12) showing the F-test for the model (both the additive and the dominance) and the second one (Table 8.13) showing individual tests for additive and dominance effects separately.

## 8.4   Major Gene Detection Involving Epistatic Effects

We can use the linear model (regression analysis) to detect two major genes simultaneously along with their potential interactions (epistatic effects). We now use the 1000 grain weight (KGW) trait in rice as an example to demonstrate the procedure. The two loci under investigation are Bin729 located on chromosome 5 and Bin1064 located on chromosome 8. The experiment was replicated in years 1998 and 1999 (Hua et al. 2002; Xu et al. 2016). In Chap. 5, we have seen the summarized data for KGW in year 1998 and learned the concept of epistasis. We did not show how to test statistical significance of the epistatic effects. We now perform statistical test for epistasis using the phenotypes measured from 1998. Partial records of the data are illustrated in Table 8.14, where $A$ indicates the $A_1A_1$ homozygote, $H$ indicates the heterozygote $A_1A_2$, and $B$ indicates the other homozygote $A_2A_2$.

### 8.4.1   Test of Epistasis

Using the analysis of variances (ANOVA) procedure, we can test epistasis for KGW. The results are given in Table 8.15 for testing the overall model effect and Table 8.16 for the two major genes and their interaction effects. The overall effect is significant because the p-value is less than 0.0001. Bin729 is very significant ($p < 0.001$) but Bin1064 is not. The interaction between the two loci has a p-value of 0.0442, less than the 0.05 threshold. The SAS code to read the data and perform analysis of variance is shown here.

```
filename aa "data-8-5.csv";
proc import datafile=aa out=one dbms=csv replace;
run;
ods graphics off;
proc glm data=one;
   class bin729 bin1064;
   model kgw98=bin729|bin1064;
quit;
```

**Table 8.14** Genotypes of two markers (Bin729 and Bin1064) and phenotypes of the KGW trait measured in 1998 for an immortalized $F_2$ (IMF2) population of rice (data-8-5.csv)

| Line | IMF2 | Bin729[a] | Bin1064[a] | KGW |
|------|------|-----------|------------|-----|
| 1 | F001 | A | H | 28.459 |
| 2 | F002 | B | A | 25.657 |
| 3 | F003 | A | H | 27.046 |
| 4 | F005 | A | B | . |
| 5 | F006 | H | H | 25.324 |
| 6 | F008 | B | H | 27.862 |
| 7 | F009 | A | H | 27.687 |
| 8 | F010 | H | A | 27.231 |
| 9 | F012 | A | A | 24.932 |
| 10 | F014 | H | H | 25.504 |
| 11 | F015 | A | H | 27.103 |
| 12 | F017 | B | H | 23.464 |
| 13 | F018 | H | H | 25.144 |
| 14 | F019 | H | A | 26.977 |
| 15 | F020 | H | B | 24.351 |
| 16 | F022 | H | A | 26.638 |
| 17 | F025 | H | B | 28.459 |
| ... | ... | ... | ... | ... |
| 276 | F358 | H | H | 23.155 |
| 277 | F359 | A | B | 29.273 |
| 278 | F360 | B | A | . |

[a]Genotypes coded by A, H, and B represent one of the homozygotes, the heterozygote, and the other homozygote

**Table 8.15** ANOVA tables for the overall model effect of KGW generated from PROC GLM

| Source | DF | Sum of squares | Mean square | F value | Pr > F |
|--------|----|----------------|-------------|---------|--------|
| Model | 8 | 190.8489022 | 23.8561128 | 7.96 | <0.0001 |
| Error | 237 | 710.2260691 | 2.9967345 | | |
| Corrected total | 245 | 901.0749713 | | | |

**Table 8.16** ANOVA tables for the main effects and epistatic effects of KGW generated from PROC GLM

| Source | DF | Type I SS | Mean square | F value | Pr > F |
|--------|----|-----------|-------------|---------|--------|
| Bin729 | 2 | 150.5794351 | 75.2897176 | 25.12 | <0.0001 |
| Bin1064 | 2 | 10.4540035 | 5.2270018 | 1.74 | 0.1770 |
| Bin729*Bin1064 | 4 | 29.8154636 | 7.4538659 | 2.49 | 0.0442 |

## 8.4.2 Epistatic Variance Components and Significance Test for each Type of Effects

Let us adopt the following coding system to indicate genotypes. Define.

$$Z_{j1} = \begin{cases} +1 & \text{for } A \\ 0 & \text{for } H \\ -1 & \text{for } B \end{cases} \quad \text{and} \quad W_{j1} = \begin{cases} 0 & \text{for } A \\ 1 & \text{for } H \\ 0 & \text{for } B \end{cases}.$$

for the first locus (Bin729) and.

$$Z_{j2} = \begin{cases} +1 & \text{for } A \\ 0 & \text{for } H \\ -1 & \text{for } B \end{cases} \quad \text{and} \quad W_{j2} = \begin{cases} 0 & \text{for } A \\ 1 & \text{for } H \\ 0 & \text{for } B \end{cases}.$$

for the second locus (Bin1064). The epistatic effect model for the phenotypic value of individual $j$ is

$$y_j = \mu + Z_{j1}a_1 + W_{j1}d_1 + Z_{j2}a_2 + W_{j2}d_2$$
$$+ Z_{j1}Z_{j2}\gamma_{aa} + Z_{j1}W_{j2}\gamma_{ad} + W_{j1}Z_{j2}\gamma_{da} + W_{j1}W_{j2}\gamma_{dd} + \varepsilon_j$$

The sample data show that $Z_{j1}W_{j2} = W_{j1}Z_{j2}$ and thus the two different epistatic effects cannot be estimated separately. Therefore, we leave one of the redundant coefficients out and use

$$y_j = \mu + Z_{j1}a_1 + W_{j1}d_1 + Z_{j2}a_2 + W_{j2}d_2$$
$$+ Z_{j1}Z_{j2}\gamma_{aa} + (Z_{j1}W_{j2})\gamma_{(ad)} + W_{j1}W_{j2}\gamma_{dd} + \varepsilon_j$$

where $(Z_{j1}W_{j2}) = \frac{1}{2}(Z_{j1}W_{j2} + W_{j1}Z_{j2})$ and $\gamma_{(ad)} = \gamma_{ad} + \gamma_{da}$. For the KGW trait measured in 1998, the multiple linear regression analysis generated the following results (see Table 8.17). The estimated residual variance is $\hat{\sigma}^2 = V_E = 3.005$. The phenotypic variance is $\hat{\sigma}_y^2 = V_P = 3.6779$. Therefore, $V_G = V_P - V_E = 3.6779 - 3.005 = 0.6729$. The estimated broad sense heritability is $H^2 = V_G/V_P = 0.6729/3.6779 = 0.1829$. The two loci jointly contribute 18% of the phenotypic variance. The next step is to partition the total genetic variance into different variance components, which requires the estimated genetic effects (Table 8.17) and the variance-covariance matrix of the genotype indicator variables (Table 8.18). The total phenotypic variance is

**Table 8.17**  Estimated genetic effects and their tests for the KGW trait measured in 1998

| Effect | Estimate | Std. error | t-value | p-value | Significance |
|---|---|---|---|---|---|
| $\mu$ | 26.0842 | 0.2538 | | | |
| $a_1$ | 0.7991 | 0.2427 | 3.293 | 0.00114 | ** |
| $d_1$ | −0.3456 | 0.3403 | −1.016 | 0.31081 | |
| $a_2$ | 0.1641 | 0.1689 | 0.972 | 0.33226 | |
| $d_2$ | −0.5481 | 0.341 | −1.608 | 0.10927 | |
| $\gamma_{aa}$ | 0.5174 | 0.2535 | 2.041 | 0.04235 | * |
| $\gamma_{(ad)}$ | 0.7094 | 0.3328 | 2.132 | 0.03407 | * |
| $\gamma_{dd}$ | 0.4107 | 0.4582 | 0.896 | 0.37105 | |

**Table 8.18**  The variance-covariance matrix of seven genotype indicator variables

| | $Z_1$ | $W_1$ | $Z_2$ | $W_2$ | $ZZ$ | $(ZW)$ | $WW$ |
|---|---|---|---|---|---|---|---|
| $Z_1$ | 0.4582 | −0.0098 | −0.0784 | −0.0055 | 0.0014 | 0.2202 | −0.0052 |
| $W_1$ | −0.0098 | 0.2490 | −0.0247 | 0.0123 | 0.0431 | −0.0020 | 0.1319 |
| $Z_2$ | −0.0784 | −0.0247 | 0.4917 | 0.0275 | 0.0102 | 0.0002 | 0.0156 |
| $W_2$ | −0.0055 | 0.0123 | 0.0275 | 0.2509 | 0.0403 | 0.0018 | 0.1423 |
| $ZZ$ | 0.0014 | 0.0431 | 0.0102 | 0.0403 | 0.2320 | 0.0003 | 0.0229 |
| $(ZW)$ | 0.2202 | −0.0020 | 0.0002 | 0.0018 | 0.0003 | 0.2202 | −0.0010 |
| $WW$ | −0.0052 | 0.1319 | 0.0156 | 0.1423 | 0.0229 | −0.0010 | 0.2057 |

$$\text{var}(y) = \sigma_{Z_1}^2 a_1^2 + \sigma_{W_1}^2 d_1^2 + \sigma_{Z_2}^2 a_2^2 + \sigma_{W_2}^2 d_2^2 + \sigma_{ZZ}^2 \gamma_{aa}^2 + \sigma_{(ZW)}^2 \gamma_{(ad)}^2 + \sigma_{WW}^2 \gamma_{dd}^2 + 2 \sum \text{cov} + \sigma_\varepsilon^2$$

Therefore, the different types of genetic variances are

$$V_A = \sigma_{Z_1}^2 a_1^2 + \sigma_{Z_2}^2 a_2^2 = 0.4582 \times 0.7991^2 + 0.4917 \times 0.1641^2 = 0.8112$$

$$V_D = \sigma_{W_1}^2 d_1^2 + \sigma_{W_2}^2 d_2^2 = 0.249033 \times (-0.3456)^2 + 0.250851 \times (-0.5481)^2 = 0.1051$$

$$V_{AA} = \sigma_{ZZ}^2 \gamma_{aa}^2 = 0.2320 \times 0.5174^2 = 0.0621$$

$$V_{(AD)} = \sigma_{(ZW)}^2 \gamma_{(ad)}^2 = 0.2202 \times 0.7094^2 = 0.1108$$

$$V_{DD} = \sigma_{WW}^2 \gamma_{dd}^2 = 0.2057 \times 0.4107^2 = 0.0347$$

The sum of all the covariance terms is

$$2 \sum \text{cov} = V_G - \left( V_A + V_D + V_{AA} + V_{(AD)} + V_{DD} \right) = -0.45097$$

The narrow sense heritability is

$$h^2 = \frac{V_A}{V_P} = \frac{0.811154}{3.6779} = 0.220548$$

You may wonder why the narrow sense heritability is greater than the broad sense heritability,

$$H^2 = \frac{V_G}{V_P} = \frac{0.6729}{3.6779} = 0.1829$$

The reason is that different genetic effects in our population are correlated and the sum of the covariance terms is negative. This is contradicted to what we have learned before where $H^2$ is always greater than or equal to $h^2$. The traditional model we learned before assumes that different types of genetic effects are not correlated, and thus there are no covariance terms. The assumption of uncorrelated effects was violated in this example, which explains why we got this peculiar result. The proportion of phenotypic variance contributed by other genetic variance components are also calculated and listed in Table 8.19. The additive variance contributes the most, followed by the additive×dominance variance, the dominance variance, and the additive×additive variance. The contribution from the dominance×dominance variance is negligible. Result of this analysis is consistent with common knowledge about this trait (KGW), i.e., KGW is a highly heritable trait contributed largely through additive effect. An extra data step is required to partition the genetic effects into additive, dominance and epsitatic effects. The SAS program to code the genotype indicator variables is shown below,

**Table 8.19** Estimated genetic variance components and the proportions contributed to the phenotypic variances

| Type of effect | Variance | Proportion |
| --- | --- | --- |
| A | 0.811154 | 0.220548 |
| D | 0.105103 | 0.028577 |
| A × A | 0.062102 | 0.016885 |
| (A × D) | 0.110817 | 0.030131 |
| D × D | 0.034696 | 0.009434 |

**Table 8.20** Output from the GLM procedure in SAS for detailed epistatic effects

| Parameter | Estimate | Standard error | t value | Pr > |t| |
|---|---|---|---|---|
| Intercept | 26.08422529 | 0.25375715 | 102.79 | <0.0001 |
| z1 | 0.79913522 | 0.24268579 | 3.29 | 0.0011 |
| w1 | −0.34560042 | 0.34026389 | −1.02 | 0.3108 |
| z2 | 0.16410152 | 0.16890633 | 0.97 | 0.3323 |
| w2 | −0.54812410 | 0.34097429 | −1.61 | 0.1093 |
| Zz | 0.51738000 | 0.25348626 | 2.04 | 0.0423 |
| Zw | 0.70943026 | 0.33281702 | 2.13 | 0.0341 |
| Ww | 0.41066824 | 0.45822834 | 0.90 | 0.3710 |

```
filename aa "data-8-5.csv";
proc import datafile=aa out=one dbms=csv replace;
run;
data two;
   set one;
   z1=0;
   w1=0;
   z1=(bin729='A')-(bin729='B');
   w1=(bin729='H');
   z2=0;
   w2=0;
   z2=(bin1064='A')-(bin1064='B');
   w2=(bin1064='H');
   zz=z1*z2;
   zw=(z1*w2+w2*z1)/2;
   ww=w1*w2;
run;
ods graphics off;
proc glm data=two;
   model kgw98=z1 w1 z2 w2 zz zw ww;
quit;
proc corr data=two cov;
   var z1 w1 z2 w2 zz zw ww;
run;
```

The output of PROC GLM is shown in Table 8.20. The output of PROC CORR includes a covariance matrix of the genotype indicator variables, which is identical to Table 8.18.

## 8.5    Segregation Analysis

### 8.5.1    Qualitative Traits

Segregation analysis differs from major gene detection in that the genotypes of the locus of interest are not observed. For a given disease phenotype (a binary trait), we assume that the trait is controlled by a single locus under a particular mode of inheritance, e.g., dominance or recessive. We then test the hypothesis using a particular statistical method.

We first consider segregation analysis for diseases controlled by a dominant allele. Let $A_1A_1$, $A_1A_2$, and $A_2A_2$ be the three genotypes of a hypothetic locus for the disease trait. Assume that $A_1$ is dominant over $A_2$ so that individuals with genotypes $A_1A_1$ and $A_1A_2$ are affected and individuals with genotype $A_2A_2$ are normal. A dominant disease allele often has very low frequency, and thus homozygotes of the dominant allele are very rare. A typical example of this type of disease is achondroplasia—a form of short-limbed dwarfism (Wikipedia contributors 2019). The dominant allele is due to a single mutation in the FGFR3 gene. Due to the dominant nature of the disease-associated mutations in the FGFR3 gene, achondroplasia is caused by only one copy of a mutant allele in the FGFR3 gene. To test whether a disease is caused by a single dominant allele, we need to collect family data. In human, a nuclear family (parents and children) is often very small.

Therefore, we must collect data from many nuclear families, typically a few hundred families. Since the single mutation is very rare, a dwarf is often assumed to be a heterozygote ($A_1A_2$). If a dwarf marries a normal person ($A_1A_2 \times A_2A_2$), a child will have a 50% probability being dwarf. If both parents are dwarfs ($A_1A_2 \times A_1A_2$), a child will have a 75% probability of being dwarf. Let us call families with a single dwarf parent Type 1 families and families with both parents being dwarfs Type 2 families. Let $n_D$ and $n_N$ be the numbers of dwarf and normal children, respectively, and $n = n_D + n_N$. The Chi-square test is

$$X_1^2 = \frac{\left(n_D - \frac{1}{2}n\right)^2}{\frac{1}{2}n} + \frac{\left(n_N - \frac{1}{2}n\right)^2}{\frac{1}{2}n}$$

for the Type 1 families and

$$X_2^2 = \frac{\left(n_D - \frac{3}{4}n\right)^2}{\frac{3}{4}n} + \frac{\left(n_N - \frac{1}{4}n\right)^2}{\frac{1}{4}n}$$

for the Type 2 families. Under the true model, the Chi-square tests for both types of families follow a Chi-square distribution with one degree of freedom. The combined test is

$$X^2 = X_1^2 + X_2^2$$

The overall p-value is

$$p = 1 - \Pr\left(\chi_2^2 \leq X^2\right) \tag{8.1}$$

where $\chi_2^2$ is a Chi-square variable with two degrees of freedom. If we are only provided the p-values from the two types of families, say $p_1$ and $p_2$, and do not know the Chi-square test statistics from the two types of families, we can still combine the two p-values using a meta-analysis approach. The combined single test is

$$X^2 = \left[\Phi^{-1}(1 - 0.5p_1)\right]^2 + \left[\Phi^{-1}(1 - 0.5p_2)\right]^2$$

Let $z_k = \Phi^{-1}(1 - 0.5p_k)$ be the probit (quantile) of $1 - 0.5p_k$ for $k = 1, 2$. The 0.5 in front of $p_k$ is due to the two-tailed nature of the normal test. Under the true model, the $p_k$ value is uniformly distributed between 0 and 1. In this case, $z_k$ should be the positive half of the standardized normal variable (truncated standardized normal distribution). As a result,

$$X^2 = \sum_{k=1}^{2} z_k^2 = \sum_{k=1}^{2} \left[\Phi^{-1}(1 - 0.5p_k)\right]^2$$

should follow a Chi-square distribution with two degrees of freedom. The new p-value should be calculated using Eq. (8.1). This method is more intuitive than the Fisher's method for meta-analysis because there is no need to prove the distribution of $X^2$. However, the calculation is complicated because you need the probit function, which is not always available in some computing platforms. The combined p-value calculated using this way is slightly different from that of the Fisher's method.

We now demonstrate the method using a hypothetical example. For the Type 1 families, let $n_D = 35$ and $n_N = 40$. The Chi-square test statistic is

$$X_1^2 = \frac{\left(n_D - \frac{1}{2}n\right)^2}{\frac{1}{2}n} + \frac{\left(n_N - \frac{1}{2}n\right)^2}{\frac{1}{2}n} = \frac{(35 - 37.5)^2}{37.5} + \frac{(40 - 37.5)^2}{37.5} = 0.3333$$

The associated p-value is

$$p_1 = 1 - \Pr\left(\chi_1^2 < 0.3333\right) = 0.5637$$

For the Type 2 families, let $n_D = 150$ and $n_N = 45$. The Chi-square test statistic is

$$X_2^2 = \frac{(150 - 146.25)^2}{146.25} + \frac{(45 - 48.75)^2}{48.75} = 0.3846$$

The corresponding p-value is

$$p_2 = 1 - \Pr\left(\chi_1^2 < 0.3846\right) = 0.5351$$

The combined test is

$$X^2 = \left[\Phi^{-1}(1 - 0.5p_1)\right]^2 + \left[\Phi^{-1}(1 - 0.5p_2)\right]^2 = 0.3333 + 0.3846 = 0.7179$$

The p-value corresponding to the overall test is

$$p = 1 - \Pr\left(\chi_2^2 < 0.7179\right) = 0.6984$$

The conclusion of this analysis is that we cannot reject the hypothesis that the dwarfism is controlled by a dominant allele.

We now consider traits controlled by two recessive alleles. Most single-gene controlled genetic diseases are caused by two copies of a recessive allele. Using the $A_1A_1$, $A_1A_2$, and $A_2A_2$ notation for the three genotypes with $A_2$ being the recessive disease allele, individuals with genotype $A_2A_2$ are affected. Collecting family data for statistical test is hard because affected individuals often do not survive to the adulthood and thus have no children. If two carriers (heterozygotes) marry ($A_1A_2 \times A_1A_2$), their children will have a 25% probability to be affected. If a normal individual marries a carrier, they will not give affected children. Therefore, we must collect families from mating type $A_1A_2 \times A_1A_2$. The problem is that we do not know the mating type of the parents being $A_1A_2 \times A_1A_2$ unless we observe an affected child from this family. If we only collect families with at least one affected child, the data are ascertained and there is an ascertainment bias. Without the bias correction, the 3:1 ratio will not hold, even if the disease is indeed controlled by two recessive alleles. The bias correction is very important when the family sizes are very small. For large family sizes, all families with $A_1A_2 \times A_1A_2$ mating type will be collected and the ascertainment bias can be ignored.

Let us use the genetic basis of "colored lamb" as an example to illustrate how to test the recessive allele hypothesis. When I was an undergraduate student, I choose a senior project in the area of Mendelian genetics. We went to several government-owned sheep farms of China to study sheep genetics. We first tried to identify problems the farmers had with their sheep populations. One problem caught my attention: the genetic basis of colored lambs. The local farmers called the colored lambs flower lambs. A colored lamb (Fig. 8.3) is defined as a mostly white lamb with just a few spots of black, gray, or brown. Natural dyes can dye white wool just about any color in the rainbow, but dark wool is difficult to dye with natural dyes. Because of this, white wool brings higher prices than dark wool and most farmers do not want sheep that are anything but solid white. Through generations of selection to eliminate colored sheep, the color allele frequency has been very low, but colored lambs still appear sporadically from families of normal (solid white) parents. A colored lamb will be raised for the dinner table but will not be kept for the next year's breeding cycle. This becomes a problem to the fiber sheep industry. A hypothesis is that the colored phenotype is controlled by two recessive color alleles. Phenotypic selection cannot eliminate heterozygotes which look normal but carry a color allele. To test the hypothesis, we need to test the 3:1 ratio of normal lambs to colored lambs from families of heterozygous parents ($A_1A_2 \times A_1A_2$).

For the senior project, we visited about ten farms in the west of Liaoning Province of China to collect data. Whenever we found a colored lamb in a family, we searched the records of that family to see how many lambs produced from that family. We collected about 300 such families and recorded the number of colored lambs and the number of white lambs for each family. I did not keep the data but recall that the proportion of colored lambs was way higher than the expected 25%, perhaps 40–50%. Therefore, I gave up the project and choose a different senior project. In 1987, I came to the USA to start my Ph.D. training at Purdue University. I took many statistical courses, including a course entitled "Statistical Inference." I learned the concept of conditional probability and the maximum likelihood method for parameter estimation. I revisited my failed senior project on the genetic basis of colored lambs. Using the probability theory and the maximum likelihood method I learned from that class, I figured out why the ratio was so high. It was due to an ascertainment bias because I failed to collect families of both heterozygous parents if they did not produce colored lambs. Unfortunately, I did not keep the original data and cannot analyze the data using the correct method.

**Fig. 8.3** A colored lamb (also called flower lambs): mostly white lamb with a few spots of black and brown (Jeremiah Gard, https://redriverzoo.org/rare-jacob-sheep-born-red-river-zoo/)

Let $n$ be the total number of families and $n_k$ be the number of lambs collected from the $k$th family for $k = 1, \cdots, n$. Let $c_k$ be the number of colored lambs and $w_k$ be the number of white lambs from the $k$th family, where $c_k + w_k = n_k$. The colored lambs per family ($c_k$) follows a binomial distribution with size $n_k$ and probability $\theta$. Theoretically, if the color trait is indeed controlled by two recessive alleles and there is no ascertainment bias, $\theta = 0.25$. To test the hypothesis, we estimate $\theta$ from the data to see how close the estimate $\theta$ to the expected 0.25. The binomial likelihood is defined as

$$\Pr(\theta) = \prod_{k=1}^{n} \frac{n_k!}{c_k! w_k!} \theta^{c_k} (1 - \theta)^{w_k}$$

We want to find the $\theta$ value that maximizes this probability. Such a $\theta$ value is called the maximum likelihood estimate of $\theta$ and is denoted by $\widetilde{\theta}$. It is tedious to find $\widetilde{\theta}$ using the above probability. Therefore, we try to maximize the natural log of the probability, and this log probability is called the log likelihood function,

$$L(\theta) = \ln\left[\Pr(\theta)\right] = \sum_{k=1}^{n} \ln \frac{n_k!}{c_k! w_k!} + \sum_{k=1}^{n} c_k \ln(\theta) + \sum_{k=1}^{n} w_k \ln(1 - \theta)$$

Ignoring the constant which does not include $\theta$, we have

$$L(\theta) = \sum_{k=1}^{n} c_k \ln(\theta) + \sum_{k=1}^{n} w_k \ln(1 - \theta) \tag{8.2}$$

If we can find the derivative of the log likelihood function and let it equal zero, the solution should be the maximum likelihood estimate of $\theta$.

$$\frac{\partial L(\theta)}{\partial \theta} = \sum_{k=1}^{n} \frac{c_k}{\theta} - \sum_{k=1}^{n} \frac{w_k}{1-\theta} = \frac{\sum_{k=1}^{n} c_k}{\theta} - \frac{\sum_{k=1}^{n} w_k}{1-\theta} = 0$$

Solving the above equation leads to

$$\widetilde{\theta} = \frac{\sum_{k=1}^{n} c_k}{\sum_{k=1}^{n} c_k + \sum_{k=1}^{n} w_k}$$

which is simply the ratio of the colored lambs to the total number of lambs. A more formal expression for the maximum likelihood estimate is

$$\widetilde{\theta} = \arg \max_{\theta} \sum_{k=1}^{n} \{c_k \ln(\theta) + w_k \ln(1-\theta)\}$$

We can use the likelihood ratio test to test the hypothesis, $H_0 : \theta = 0.25$, which is defined as

$$\lambda_{\text{Biased}} = -2\left[L(0.25) - L\left(\widetilde{\theta}\right)\right]$$

where $L\left(\widetilde{\theta}\right)$ is the likelihood function in Eq. (8.2) evaluated at $\theta = \widetilde{\theta}$ and $L(0.25)$ is the same likelihood function evaluated at $\theta = 0.25$. Under the true model, $\lambda$ follows a Chi-square distribution with one degree of freedom. The estimated parameter and the likelihood ratio test are subject to ascertainment bias.

Not all families are counted! The families who do not produce a colored lamb have been ignored. The probability that a family is selected is the probability that there is at least one colored lamb observed from this family, which is defined as

$$\psi_k = 1 - (1-\theta)^{w_k + c_k}$$

Therefore, the conditional probability that we can observe the data from the $k$th family is

$$\Pr(c_k|\theta) = \psi_k^{-1} \frac{n_k!}{c_k! w_k!} \theta^{c_k} (1-\theta)^{w_k}$$

The probability of observing the data from the sample takes the products of all family specific probabilities,

$$\Pr(\theta) = \prod_{k=1}^{n} \psi_k^{-1} \frac{n_k!}{c_k! w_k!} \theta^{c_k} (1-\theta)^{w_k}$$

We want to find the $\theta$ value that maximizes this probability. Such a $\theta$ value is called the maximum likelihood estimate of $\theta$ and is denoted by $\widehat{\theta}$. The log likelihood function is

$$L(\theta) = \sum_{k=1}^{n} c_k \ln(\theta) + \sum_{k=1}^{n} w_k \ln(1-\theta) - \sum_{k=1}^{n} \ln\left[1 - (1-\theta)^{w_k+c_k}\right] \tag{8.3}$$

where a constant has been ignored. The problem is complicated and there is perhaps no explicit solution. Therefore, a numeric approach, such as the Newton method or the grid search, must be used to find the maximum likelihood estimate of $\theta$, denoted by $\widehat{\theta}$. The likelihood ratio test for $H_0 : \theta = 0.25$ is

$$\lambda_{\text{Unbiased}} = -2\left[L(0.25) - L\left(\widehat{\theta}\right)\right]$$

**Table 8.21** Simulated data for colored lambs used for illustration

| Family | Size | White | Colored |
|--------|------|-------|---------|
| 1 | 2 | 1 | 1 |
| 2 | 1 | 0 | 1 |
| 3 | 2 | 1 | 1 |
| 4 | 2 | 1 | 1 |
| 5 | 2 | 1 | 1 |
| 6 | 4 | 2 | 2 |
| 7 | 2 | 1 | 1 |
| 8 | 4 | 3 | 1 |
| 9 | 1 | 0 | 1 |
| 10 | 2 | 0 | 2 |

where $L(0.25)$ and $L(\widehat{\theta})$ are the likelihood functions defined in Eq. (8.3) evaluated at $\theta = 0.25$ and $\theta = \widehat{\theta}$, respectively. Under the true model, $\lambda_{\text{Unbiased}}$ follows a Chi-square distribution with one degree of freedom. The $p$-value is calculated from

$$p\text{-value} = 1 - \Pr\left(\chi_1^2 \leq \lambda_{\text{Unbiased}}\right)$$

If the p-value is greater than 0.05, we will not reject $\theta = 0.25$ and conclude that the colored lamb phenotype is controlled by two recessive alleles.

To illustrate the test for the hypothesis of two recessive alleles causing the colored lamb phenotype, we simulated 10 families from heterozygous parents with an average number of progenies being 1.5 per family. The 10 families were recorded because there was at least one colored lamb from each family. In fact, many more families were simulated but many of them did not have colored lambs and thus were deleted from the sample. The number of progenies was simulated from a Poisson distribution with mean 1.5. The number of colored lambs was generated from probability $\theta = 0.25$ from each family. The sampled data are shown in Table 8.21 (data-8-6.csv).

The average family size from the ascertained sample is 2.2, much higher than 1.5. The total number of white lambs is 10 and the total number of colored lambs is 12, leading to an estimated proportion of colored lambs of $\widetilde{\theta} = 12/(12 + 10) = 0.5454$, way higher than the expected 0.25. The likelihood ratio test for $H_0 : \theta = 0.25$ is

$$\lambda_{\text{Biased}} = -2\left[L(0.25) - L\left(\widetilde{\theta}\right)\right] = -2(-19.51235 + 15.1582) = 8.7083$$

with a $p$-value of

$$p\text{-value} = 1 - \Pr\left(\chi_1^2 \leq 8.7083\right) = 0.003168$$

Although the data were simulated from the true model, the test actually rejected the true model due to ascertainment bias.

We now use the ascertainment bias corrected method to do the test. The estimate proportion of colored lambs is $\widehat{\theta} = 0.275417$, much closer to the expected 0.25. The likelihood ratio test is

$$\lambda_{\text{Unbiased}} = -2\left[L(0.25) - L\left(\widehat{\theta}\right)\right] = -2(-11.01891 + 11.00443) = 0.02896882$$

The corresponding $p$-value is

$$p\text{-value} = 1 - \Pr\left(\chi_1^2 \leq 0.02896882\right) = 0.8648511$$

As a result, we cannot reject the true model because the $p$-value is much greater than 0.05.

### 8.5.2  Quantitative Traits

Segregation analysis for quantitative traits is often different from that for qualitative traits. The method is closely related to that of major gene detection but does not require observed genotypes of the major gene. The key of such a segregation analysis is the mixture model analysis (McLachlan and Peel 2000; Zhang et al. 2003; Gai 2005). Let the three unobserved genotypes be $G_1 = A_1A_1$, $G_2 = A_1A_2$, and $G_3 = A_2A_2$. In an $F_2$ population, the frequencies are 0.25, 0.5, and 0.25, respectively, for the three genotypes. However, we can estimate the genotypic frequencies if the population is a random population. Let $y_j$ be the phenotype of individual $j$ for $j = 1, \ldots, n$, where $n$ is the sample size. Let $\mu_k$ be the genotypic value of $G_k$ for $k = 1, 2, 3$. Given the genotype, the density of $y_j$ is

$$f(y_j|G_k) = N(y_j|\mu_k, \sigma^2)$$

where $\sigma^2$ is the residual error variance and independent of the genotype. The marginal density is

$$f(y_j) = \sum_{k=1}^{3} \pi_k N(y_j|\mu_k, \sigma^2)$$

where $\pi_k$ is the frequency of genotype $G_k$, also called the mixing probability or mixing proportion. Let $\theta = \{\mu_1, \mu_2, \mu_3, \pi_1, \pi_2, \pi_3, \sigma^2\}$ be the unknown parameters subject to estimation, where $\pi_3 = 1 - \pi_1 - \pi_2$ is not a parameter given $\pi_1$ and $\pi_2$. The log likelihood function is

$$L(\theta) = \sum_{j=1}^{n} \ln f(y_j) = \sum_{j=1}^{n} \ln \sum_{k=1}^{3} \pi_k N(y_j|\mu_k, \sigma^2) \tag{8.4}$$

The maximum likelihood method is often used to estimate the parameters. The likelihood ratio test is used to test the null hypothesis $H_0 : \mu_1 = \mu_2 = \mu_3 = \mu$. Rejection of the null hypothesis indicates a major gene segregating in the population for the quantitative trait. The expectation and maximization (EM) algorithm (Dempster et al. 1977) is often used to obtain the maximum likelihood estimates of the parameters. However, the Bayesian method implemented via the Markov chain Monte Carlo (MCMC) is the best method for the mixture model analysis (Geman and Geman 1984; Gilks et al. 1996). We now introduce the EM algorithm without providing the derivation. Let $\delta_j = [\delta_{j1} \quad \delta_{j2} \quad \delta_{j3}]$ be three dummy variables representing one of the three genotypes, where $\delta_{jk} = 1$ and $\delta_{jk'} = 0$ if the individual's genotype is $G_k$ and $k' \neq k$. The prior probability for $\delta_{jk} = 1$ is $\mathrm{E}(\delta_{jk}) = \mathrm{Pr}(\delta_{jk} = 1) = \pi_k$ for all $j = 1, \ldots, n$ and $k = 1, 2, 3$. If $\mu_k$ and $\sigma^2$ are known, the posterior probability of $\delta_{jk} = 1$ is

$$\mathrm{E}(\delta_{jk}|y_k) = \mathrm{Pr}(\delta_{jk} = 1|y_j) = \frac{\pi_k f(y_j|G_k)}{\sum_{k'=1}^{3} \pi_k f(y_j|G_{k'})} \tag{8.5}$$

Given the initial values of $\theta^{(t)} = \left\{\mu_k^{(t)}, \pi_k^{(t)}, \sigma^{2(t)}\right\}$ where $t = 0$, we can calculate $\mathrm{E}(\delta_{jk}|y_k)$, which represents the expectation step. The maximization step is represented by updating the parameters using the following equations,

$$\mu^{(t+1)} = \left[\sum_{j=1}^{n} \mathrm{E}(\delta_{jk}|y_j)\right]^{-1} \sum_{j=1}^{n} \mathrm{E}(\delta_{jk}|y_j) y_j$$

is the updated mean for the $k$th genotype,

$$\pi_k^{(t+1)} = \frac{1}{n} \sum_{j=1}^{n} \mathrm{E}(\delta_{jk}|y_j)$$

is the updated mixing proportion for genotype $k$ and

$$\sigma^{2(t+1)} = \frac{1}{n} \sum_{j=1}^{n} \sum_{k=1}^{3} E\left(\delta_{jk}|y_j\right)\left(y_j - \mu_k^{(t)}\right)^2$$

is the updated residual error variance. Once the parameters are updated, we then go back to Eq. (8.5) to recalculate the conditional expectation. The iteration process continues until a certain criterion of convergence is reached, say

$$\left(\theta^{(t+1)} - \theta^{(t)}\right)^T \left(\theta^{(t+1)} - \theta^{(t)}\right) \leq \varepsilon$$

where $\varepsilon = 10^{-8}$, a small positive number set by the investigator, the smaller the criterion, the higher the precision. The parameter values in the last iteration are the maximum likelihood estimates, denoted by $\widehat{\theta} = \left\{\widehat{\mu}_k, \widehat{\pi}_k, \widehat{\sigma}^2\right\}$. Substituting $\theta$ in the likelihood function given in Eq. (8.4) by $\widehat{\theta}$ yields

$$L\left(\widehat{\theta}\right) = \sum_{j=1}^{n} \ln \sum_{k=1}^{3} \widehat{\pi}_k N\left(y_j|\widehat{\mu}_k, \widehat{\sigma}^2\right)$$

which is called the likelihood value under the full model. To perform a likelihood ratio test for the mixture model, we also need to analyze the data under the null model, i.e., $H_0 : \mu_1 = \mu_2 = \mu_3 = \mu$. Under the null model, the likelihood function is

$$L(\theta) = \sum_{j=1}^{n} \ln N\left(y_j|\mu, \sigma^2\right)$$

where $\theta = \{\mu, \sigma^2\}$ is redefined under the null model. The maximum likelihood estimates of the parameters are

$$\widetilde{\mu} = \frac{1}{n} \sum_{j=1}^{n} y_j$$

and

$$\widetilde{\sigma}^2 = \frac{1}{n-1} \sum_{j=1}^{n} \left(y_j - \widetilde{\mu}\right)^2$$

The likelihood value under the null model is

$$L\left(\widetilde{\theta}\right) = \sum_{j=1}^{n} \ln N\left(y_j|\widetilde{\mu}, \widetilde{\sigma}^2\right)$$

where $\widetilde{\theta} = \left\{\widetilde{\mu}, \widetilde{\sigma}^2\right\}$. The likelihood ratio test is

$$\lambda = -2\left[L\left(\widetilde{\theta}\right) - L\left(\widehat{\theta}\right)\right]$$

Under the null model, $\lambda$ follows a Chi-square distribution with $df = 3 - 1 = 2$ degrees of freedom. The full model has three groups and the null model has only one group. The difference is $3 - 1 = 2$, explaining why $df = 2$. The EM algorithm appears to be simple and efficient. The problem is that it is very sensitive to the initial values. Different initial values sometime lead to different solutions of the parameters. Therefore, it is suggested to start the EM with different sets of initial values. Alternatively, placing some constraints on the parameter space can also improve the probability of finding the global solution of the parameters. Therefore, the MCMC implemented Bayesian method often performs better than the EM algorithm. Bayesian statistics is beyond the scope of the course and will not be studied here.

A simulated dataset consisting of $n = 1000$ observations are stored in "data-8-7.csv". The data were generated from three genotypes with 250 observations from genotype $G_1 = A_1A_1$, 500 observations from $G_2 = A_1A_2$ and 250 observations from genotype $G_3 = A_2A_2$. The three means are $\mu_1 = 10$, $\mu_2 = 15$ and $\mu_3 = 20$, respectively, and the residual variance is $\sigma^2 = 10$. The mixing proportions are $\pi_1 = 0.25$, $\pi_2 = 0.50$, and $\pi_3 = 0.25$. These parameter values were used to simulate the data. The finite mixture model (FMM) procedure in SAS was used to analyze this data. The SAS code for PROC FMM is shown below,

```
proc import datafile="data-8-7.csv" out=one dbms=csv replace;
run;

proc fmm data=one;
    model y = /dist=normal k=3 equate=scale parms(5 5,10 5,15 5);
    output out=outpost posterior=post;
run;
```

The **proc import** statement reads the data and the **proc export** statement writes the posterior probabilities of all individuals to an external file. The **proc fmm** statement performs the mixture model analysis, where the distribution of $y_j$ within each group is normal; there are three groups, the three groups have the same residual error variance, and the initial values of parameters for the three genotypes are $(\mu_1, \sigma^2) = (5, 5)$, $(\mu_2, \sigma^2) = (10, 5)$, and $(\mu_3, \sigma^2) = (15, 5)$. The output includes a plot of distribution for the mixture model and various tables for the result. Figure 8.4 shows the distribution of the mixture model.

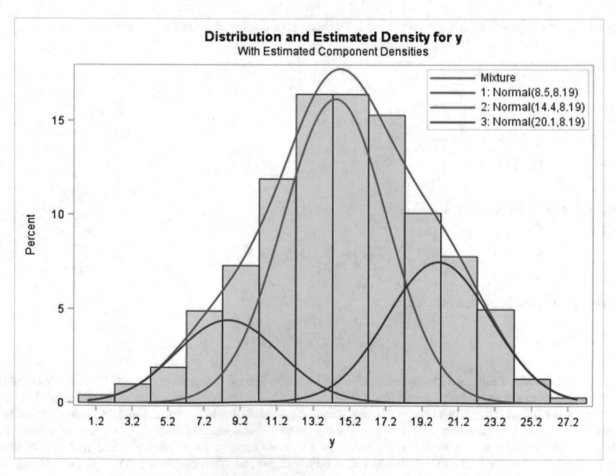

**Fig. 8.4** Distribution and estimated densities of Y for the three groups

**Table 8.22**  Fit statistics of the full model

| Fit Statistics | |
|---|---|
| −2 Log Likelihood | 5918.2 |
| AIC (smaller is better) | 5930.2 |
| AICC (smaller is better) | 5930.3 |
| BIC (smaller is better) | 5959.7 |
| Pearson statistic | 1000.0 |
| Effective parameters | 6 |
| Effective components | 3 |

**Table 8.23**  Parameter estimates of the mixture model

| Parameter estimates for normal model | | | | | |
|---|---|---|---|---|---|
| Component | Parameter | Estimate | Standard error | z value | Pr > |z| |
| 1 | Intercept | 8.5003 | 0.5601 | 15.18 | <0.0001 |
| 2 | Intercept | 14.3936 | 0.4163 | 34.58 | <0.0001 |
| 3 | Intercept | 20.0915 | 0.4851 | 41.41 | <0.0001 |
| 1 | Variance | 8.1908 | 1.1079 | | |
| 2 | Variance | 8.1908 | 1.1079 | | |
| 3 | Variance | 8.1908 | 1.1079 | | |

**Table 8.24**  Estimated mixing probabilities of the three groups

| Parameter estimates for mixing probabilities | | | | | |
|---|---|---|---|---|---|
| Component | Mixing probability | Linked scale | | | |
| | | GLogit(Prob) | Standard error | z value | Pr > |z| |
| 1 | 0.1571 | −0.5310 | 0.3445 | −1.54 | 0.1233 |
| 2 | 0.5756 | 0.7672 | 0.2051 | 3.74 | 0.0002 |
| 3 | 0.2673 | 0 | | | |

The output includes the fit statistics (Table 8.22), estimated parameters (Table 8.23), and the mixing probabilities (Table 8.24). The three intercepts in Table 8.23 are the estimated means of the three genotypes and they are close to the true values, 10, 15, and 20. The estimated residual variance is 8.19, also close to the true value of 10. The three mixing probabilities (Table 8.24) are slightly off the true segregation ratio 1:2:1, but reasonably close. The data were also analyzed under the null model using PROC FMM,

```
proc fmm data=one;
    model y = /dist=normal k=1 parms(5 5);
run;
```

The distribution of the null model is illustrated in Fig. 8.5.

We are not interested in the parameter estimates under the null model (Table 8.26), but the likelihood value in Table 8.25 is of interest, $-2L\left(\widetilde{\theta}\right) = 5927.5$. The corresponding likelihood value under the full model (Table 8.22) is 5918.2. Therefore, the likelihood ratio test is

$$\lambda = -2L\left(\widetilde{\theta}\right) - 2L\left(\widehat{\theta}\right) = 5927.5 - 5918.2 = 9.317477$$

with a p-value is

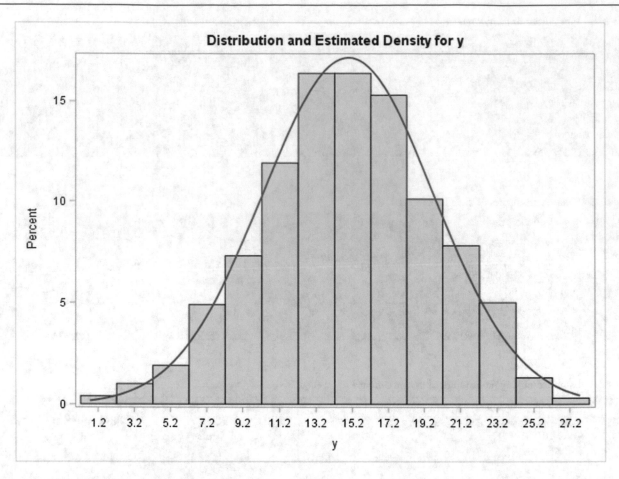

**Fig. 8.5** Distribution and estimated densities for Y under the null model (single group)

**Table 8.25** The fit statistics under the null model

| Fit statistics | |
|---|---|
| −2 Log likelihood | 5927.5 |
| AIC (smaller is better) | 5931.5 |
| AICC (smaller is better) | 5931.5 |
| BIC (smaller is better) | 5941.3 |
| Pearson statistic | 1000.0 |
| Unscaled Pearson Chi-Square | 21969.4 |

**Table 8.26** Estimated parameters from the null model

| Parameter estimates for normal model | | | | |
|---|---|---|---|---|
| Effect | Estimate | Standard error | z value | Pr > \|z\| |
| Intercept | 14.9903 | 0.1482 | 101.13 | <0.0001 |
| Variance | 21.9694 | 0.9825 | | |

$$p = 1 - \Pr\left(\chi_2^2 < 9.317477\right) = 0.009478414$$

The conclusion is positive for the presence of a major gene segregating for the quantitative trait of interest.

# References

Dempster AP, Laird MN, Rubin DB. Maximum likelihood from incomplete data via the EM algorithm. J Royal Stat Soc Series B. 1977;39:1–38.

Elston RC. Segregation analysis. In: Harris H, Hirschhorn K, editors. Advances in human genetics 11. Boston, MA: Springer; 1981.

Elston RC, Stewart J. A general model for the genetic analysis of pedigree data. Hum Hered. 1971;21:523–42.

Gai J. Segregation analysis on genetic system of quantitative traits in plants. Front Biol China. 2005;1:85–92.

Geman S, Geman D. Stochastic relaxation, Gibbs distributions and the Bayesian restoration of images. IEEE Trans Pattern Anal Mach Intell. 1984;6: 721–41.

Gilks WR, Richardson S, Spiegelhalter DJ. Markov chain Monte Carlo in practice. London: Chapman & Hall; 1996.

Hua JP, Xing YZ, Xu CG, Sun XL, Yu SB, Zhang Q. Genetic dissection of an elite rice hybrid revealed that heterozygotes are not always advantageous for performance. Genetics. 2002;162:1885–95.

McLachlan G, Peel D. Finite mixture models. New York: Wiley; 2000.

Morton NE. Segregation analysis in human genetics. Science. 1958;127:79–80.

Wikipedia Contributors. Achondroplasia. Vol 2019. Wikipedia, The Free Encyclopedia. 2019.

Xu S, Xu Y, Gong L, Zhang Q. Metabolomic prediction of yield in hybrid rice. Plant J. 2016;88:219–27.

Zhang Y-M, Gai J-Y, Yang Y-H. The EIM algorithm in the joint segregation analysis of quantitative traits. Genet Res. 2003;81:157–63.

# Resemblance between Relatives

One property of metric character is the resemblance between relatives. Relatives are more similar than genetically unrelated individuals because relatives share common genetic material. For a quantitative trait, resemblance is measured by the covariance between relatives for this trait. More closely related individuals tend to have a higher covariance than distantly related or no related individuals. Assume that relatives do not share common environmental effects (this assumption may often be violated), the covariance is solely caused by gene sharing. The covariance reflects a fraction of the genetic variance. This fraction depends on the degree of the relatedness or the relationship between the relatives. A common environment is defined as an environment shared by relatives. For example, full-siblings often share the same maternal effect. If maternal effect is very important in children's growth, the covariance between full-siblings in growth will be higher than the genetic covariance between the full-siblings.

Let $a_{12}$ be the proportion of common genetic material shared by two relatives (relative 1 and relative 2). We can write the genetic models for the two relatives by

$$\begin{cases} P_1 = \mu + A_1 + E_1 \\ P_2 = \mu + A_2 + E_2 \end{cases}$$

where $A_1$ and $A_2$ are the additive genetic effects (not the $A_1$ and $A_2$ alleles defined early in the course) for the two relatives, $E_1$ and $E_2$ are the environmental effects for the two relatives, respectively. We now express $A_2$ as a function of $A_1$ using

$$A_2 = a_{12}A_1 + (1 - a_{12})A_{-1}$$

where $A_{-1}$ is a special notation to indicate the genetic material (additive genetic effect) independent of $A_1$. The phenotypic covariance between the two relatives is

$$\begin{aligned} \mathrm{cov}(P_1, P_2) &= \mathrm{cov}(\mu + A_1 + E_1, \mu + A_2 + E_2) \\ &= \mathrm{cov}(A_1, A_2) \\ &= \mathrm{cov}(A_1, a_{12}A_1 + (1 - a_{12})A_{-1}) \\ &= a_{12}\mathrm{cov}(A_1, A_1) \\ &= a_{12}\mathrm{var}(A_1) \\ &= a_{12}V_A \end{aligned}$$

Deleting the intermediate steps of the derivation, we have

$$\mathrm{cov}(P_1, P_2) = \mathrm{cov}(A_1, A_2) = a_{12}V_A$$

This relationship means that when two relatives do not share a common environmental effect and all other genetic effects, except the additive effect, are ignored, their phenotypic covariance equals their genetic covariance. This genetic covariance is a proportion of the additive genetic variance. This relationship is general and applies to all types of relatives. For example,

© Springer Nature Switzerland AG 2022
S. Xu, *Quantitative Genetics*, https://doi.org/10.1007/978-3-030-83940-6_9

full-siblings share half of their genetic material, and thus $a_{12} = 0.5$. Two unrelated individuals share no common genetic material and thus $a_{12} = 0$, leading to $\mathrm{cov}(P_1, P_2) = \mathrm{cov}(A_1, A_2) = 0$.

The above discussion is only for an additive model, i.e., the phenotypic value is only controlled by the sum of allelic effects, not interaction effects (dominance) between different alleles. When dominance is present, the models will be

$$\begin{cases} P_1 = \mu + A_1 + D_1 + E_1 \\ P_2 = \mu + A_2 + D_2 + E_2 \end{cases}$$

Let $d_{12}$ be the probability that the two individuals share both alleles from their ancestors. We can then write

$$D_2 = d_{12}D_1 + (1 - d_{12})D_{-1}$$

The covariance between $D_1$ and $D_2$ is

$$\begin{aligned} \mathrm{cov}(D_1, D_2) &= \mathrm{cov}(D_1, d_{12}D_1 + (1 - d_{12})D_{-1}) \\ &= d_{12}\mathrm{cov}(D_1, D_1) \\ &= d_{12}\mathrm{var}(D_1) \\ &= d_{12}V_D \end{aligned}$$

Incorporating the dominance covariance, we have

$$\mathrm{cov}(P_1, P_2) = \mathrm{cov}(G_1, G_2) = \mathrm{cov}(A_1, A_2) + \mathrm{cov}(D_1, D_2) = a_{12}V_A + d_{12}V_D$$

where different types of relatives have different $d_{12}$ probabilities. Extending this formula to multiple loci with epistatic effects, the model becomes general

$$\mathrm{cov}(P_1, P_2) = \mathrm{cov}(G_1, G_2) = a_{12}V_A + d_{12}V_D + a_{12}^2 V_{AA} + a_{12}d_{12}V_{AD} + d_{12}a_{12}V_{DA} + d_{12}^2 V_{DD}$$

Although $V_{AD}$ and $V_{DA}$ can be different, $a_{12}d_{12} = d_{12}a_{12}$ always holds such that we cannot separate $V_{AD}$ and $V_{DA}$. As a result, we define $V_{(AD)} = V_{AD} + V_{DA}$ and rewrite the covariance by

$$\mathrm{cov}(P_1, P_2) = \mathrm{cov}(G_1, G_2) = a_{12}V_A + d_{12}V_D + a_{12}^2 V_{AA} + a_{12}d_{12}V_{(AD)} + d_{12}^2 V_{DD}$$

In the quantitative genetics literature, people often present one term, $V_{AD}$, which is equivalent to $V_{(AD)}$ defined in this class.

Remark: When we say covariance between relative 1 and relative 2, this covariance is the covariance between two relatives who have the same relationship as that between relatives 1 and 2. Alternatively, we can interpret the covariance between relatives 1 and 2 as the covariance over an infinite number of independent loci with equal effects (under the infinitesimal model).

## 9.1  Genetic Covariance Between Offspring and One Parent

### 9.1.1  Short Derivation

Let $G_F$ and $G_O$ be the genetic value of a parent and an offspring, respectively. They can be written in the following model,

$$\begin{cases} G_F = A_F + D_F \\ G_O = A_O + D_O = \dfrac{1}{2}(A_F + A_M) + A_W + D_O \end{cases}$$

where $A_F$ and $A_M$ are the additive genetic values of the father and the mother, respectively. The extra term $A_W$ is called the within-family segregation effect and is assumed to be sampled from an $N\left(0, \frac{1}{2}V_A\right)$ distribution. In other words, $E(A_W) = 0$ and $\mathrm{var}(A_W) = \frac{1}{2}V_A$. Later in the chapter, we will learn how to derive the within-family segregation variance. The genetic covariance between the father and the offspring is

$$
\begin{aligned}
\mathrm{cov}(G_F, G_O) &= \mathrm{cov}(A_F + D_F, A_O + D_O) \\
&= \mathrm{cov}\left[A_F + D_F, \frac{1}{2}(A_F + A_M) + A_W + D_O\right] \\
&= \mathrm{cov}\left(A_F, \frac{1}{2}A_F\right) \\
&= \frac{1}{2}\mathrm{cov}(A_F, A_F) \\
&= \frac{1}{2}\mathrm{var}(A_F) \\
&= \frac{1}{2}V_A
\end{aligned}
$$

In general, the genetic covariance between an offspring ($O$) and a parent ($P$) equals half the additive genetic variance, i.e.,

$$
\mathrm{cov}(O, P) = \frac{1}{2}V_A
$$

## 9.1.2 Long Derivation

This way of derivation requires a joint distribution table between the offspring and the single parent. Now, the genetic values of the offspring and the parent are treated as two variables. From the joint distribution, we can use the definition of covariance to derive the covariance between the two variables. This derivation requires what we have learned in Chap. 6, such as the additive variance $V_A = 2pq\alpha^2$, the average effect of gene substitution $\alpha = a + d(q - p)$ and the population mean $M = a(p - q) + 2pqd$.

We need the joint distribution of the genetic values between the parent and the progeny, but the joint distribution is difficult to derive. We must use the conditional and marginal distributions to derive the joint distribution, i.e., $f(x, y) = f(y|x)f(x)$. In this table, the probability that a child is $A_1A_1$ given that the father is $A_1A_2$ is

$$
\Pr(O = A_1A_1 | P = A_1A_2) = \frac{1}{2}p
$$

The reason is that the parent provides two possible gametes, $A_1$ and $A_2$, each with 0.5 probability. The gamete from the other parent is a random sample of the gene pool with probability $p$ for $A_1$ and probability $q$ for $A_2$. Table 9.2 shows the conditional probabilities of the offspring genotypes given the father is of $A_1A_2$ type. From this table, we can see that $\frac{1}{2}p$ of the offspring are $A_1A_1$. Similarly, $\frac{1}{2}q$ of the offspring are $A_1A_2$. Finally, $\frac{1}{2}p + \frac{1}{2}q = \frac{1}{2}(p + q) = \frac{1}{2}$ of the offspring are $A_1A_2$. We have now understood the conditional distribution (Table 9.1). Assuming that the population is in Hardy-Weinberg equilibrium, the marginal distributions for the parent are given in Table 9.3. Multiplying each element of the first row of the conditional table (Table 9.1) by the first element of the marginal table (Table 9.3) produces the joint probability table (Table 9.4).

**Table 9.1** The conditional distribution table of offspring given parent $f(O|P)$

| Parent | Offspring | | |
|---|---|---|---|
| | $A_1A_1$ | $A_1A_2$ | $A_2A_2$ |
| $A_1A_1$ | $p$ | $q$ | $0$ |
| $A_1A_2$ | $\frac{1}{2}p$ | $\frac{1}{2}p + \frac{1}{2}q = \frac{1}{2}$ | $\frac{1}{2}q$ |
| $A_2A_2$ | $0$ | $p$ | $q$ |

**Table 9.2** Condition distribution of an offspring genotype given father's genotype being $A_1A_2$

| | Other parent (mother) | | |
|---|---|---|---|
| Father | $A_1$ ($p$) | | $A_2$ ($q$) |
| $A_1$ ($\frac{1}{2}$) | $A_1A_1$ ($\frac{1}{2}p$) | | $A_1A_2$ ($\frac{1}{2}q$) |
| $A_2$ ($\frac{1}{2}$) | $A_2A_1$ ($\frac{1}{2}p$) | | $A_2A_2$ ($\frac{1}{2}q$) |

**Table 9.3** Marginal probabilities of genotypes for the parent

| Parent | Probability |
|---|---|
| $A_1A_1$ | $p^2$ |
| $A_1A_2$ | $2pq$ |
| $A_2A_2$ | $q^2$ |

**Table 9.4** The joint distribution table of the parent and the offspring $f(O, P)$

| | Offspring | | |
|---|---|---|---|
| Parent | $A_1A_1$ | $A_1A_2$ | $A_2A_2$ |
| $A_1A_1$ | $p^3$ | $p^2q$ | 0 |
| $A_1A_2$ | $p^2q$ | $pq$ | $pq^2$ |
| $A_2A_2$ | 0 | $pq^2$ | $q^3$ |

**Table 9.5** The joint distribution table of the parent and the offspring merged with the marginal probability

| | Offspring | | | |
|---|---|---|---|---|
| Parent | $A_1A_1(a)$ | $A_1A_2(d)$ | $A_2A_2(-a)$ | Marginal |
| $A_1A_1(a)$ | $p^3$ | $p^2q$ | 0 | $p^2$ |
| $A_1A_2(d)$ | $p^2q$ | $pq$ | $pq^2$ | $2pq$ |
| $A_2A_2(-a)$ | 0 | $pq^2$ | $q^3$ | $q^2$ |
| Marginal | $p^2$ | $2pq$ | $q^2$ | |

One property of the joint probability table is that the sum of all elements equals 1. Another property is that the sum of each row equals the marginal probability for the parent and the sum of each column is the marginal probability of the child. For example, the sum of the first column is $p^3 + p^2q = p^2(p + q) = p^2$, which is the probability of $A_1A_1$ for the offspring. Now, let us merge the joint probability with the marginal probability, adding the genotypic effect in the table and generate Table 9.5. The covariance between the parent and the offspring is

$$\mathrm{cov}(P, O) = \mathrm{cov}(G_P, G_O) = \mathrm{E}(G_P \times G_O) - \mathrm{E}(G_P)\mathrm{E}(G_O)$$

The marginal expectations are derived using

$$\mathrm{E}(G_P) = p^2 a + 2pqd + q^2(-a) = a(p - q) + 2pq = M$$

and

$$\mathrm{E}(G_O) = p^2 a + 2pqd + q^2(-a) = a(p - q) + 2pq = M$$

which only involve the marginal distributions. The expectation of the product is

$$\begin{aligned}
E(G_P \times G_O) =\; &p^3(a)(a) + p^2q(a)(d) + 0 \times (a)(-a) \\
&+ p^2q(d)(a) + pq(d)(d) + pq^2(d)(-a) \\
&+ 0 \times (-a)(a) + pq^2(-a)(d) + q^3(-a)(-a) \\
=\; &p^3a^2 + p^2qad + p^2qad + pqd^2 - pq^2ad - pq^2ad + q^3a^2 \\
=\; &(p^3 + q^3)a^2 + 2pq(p-q)ad + pqd^2
\end{aligned}$$

Therefore, the covariance is

$$\mathrm{cov}(P,O) = (p^3 + q^3)a^2 + 2pq(p-q)ad + pqd^2 - [a(p-q) + 2pqd]^2$$

After several steps of simplification, we get

$$\mathrm{cov}(P,O) = pq[a + d(q-p)]^2 = pq\alpha^2 = \frac{1}{2} \times 2pq\alpha^2 = \frac{1}{2}V_A$$

because the additive variance is $V_A = 2pq\alpha^2$ (see Chap. 6).

## 9.2 Genetic Covariance between Offspring and Mid-Parent

Mid-parent is defined as the average of the two parental values, $\overline{P} = (P_1 + P_2)/2$, where $P_1$ and $P_2$ are the phenotypic values of the two parents. Let $O$ be the phenotypic value of the offspring. The covariance between the offspring and the middle-parent is

$$\begin{aligned}
\mathrm{cov}(O, \overline{P}) &= \mathrm{cov}\left[O, \frac{1}{2}(P_1 + P_2)\right] \\
&= \frac{1}{2}\mathrm{cov}(O, P_1) + \frac{1}{2}\mathrm{cov}(O, P_2) \\
&= \frac{1}{2} \times \frac{1}{2}V_A + \frac{1}{2} \times \frac{1}{2}V_A \\
&= \frac{1}{2}V_A
\end{aligned}$$

## 9.3 Genetic Covariance between Half-Sibs

Half-sibs are individuals that have one parent in common and the other parents are not related. A father had one child in the first marriage, got devoiced, then married again to a different woman, and eventually had another child from the second marriage. The two children are half-sibs because they share the same father but have different mothers. The genetic models for half-sibs are

$$\begin{aligned}
G_1 &= A_1 + D_1 = \frac{1}{2}A_F + \frac{1}{2}A_{M_1} + A_{W_1} + D_1 \\
G_2 &= A_2 + D_2 = \frac{1}{2}A_F + \frac{1}{2}A_{M_2} + A_{W_2} + D_2
\end{aligned}$$

where $A_{M_1}$ and $A_{M_2}$ are the additive genetic effects of the mothers of the two children, one from each marriage. The $A_{W_1}$ and $A_{W_2}$ are the within-family segregation effects for the first and second marriages, respectively, and each one follows an $N\left(0, \frac{1}{2}V_A\right)$ distribution. The covariance between the genetic values of the half-sibs are

$$\begin{aligned}
\operatorname{cov}(G_1, G_2) &= \operatorname{cov}\left(\frac{1}{2}A_F + \frac{1}{2}A_{M_1} + A_{W_1} + D_1, \frac{1}{2}A_F + \frac{1}{2}A_{M_2} + A_{W_2} + D_2\right) \\
&= \operatorname{cov}\left(\frac{1}{2}A_F, \frac{1}{2}A_F\right) \\
&= \frac{1}{2} \times \frac{1}{2} \times \operatorname{cov}(A_F, A_F) \\
&= \frac{1}{4}\operatorname{var}(A_F) \\
&= \frac{1}{4}V_A
\end{aligned}$$

Using a generic notation for the covariance of relatives, we can write

$$\operatorname{cov}(HS) = \frac{1}{4}V_A$$

where $HS$ stands for half-sibs.

## 9.4    Genetic Covariance between Full-Sibs

Full-sibs are individuals who share both the father and the mother. We will also derive the covariance using both the short way and the long way.

### 9.4.1    Short Way of Derivation

The models for the two sibs are

$$G_1 = \frac{1}{2}A_F + \frac{1}{2}A_M + A_{W_1} + D_1$$
$$G_2 = \frac{1}{2}A_F + \frac{1}{2}A_M + A_{W_2} + D_2$$

The covariance is

$$\begin{aligned}
\operatorname{cov}(G_1, G_2) &= \operatorname{cov}\left(\frac{1}{2}A_F + \frac{1}{2}A_M + A_{W_1} + D_1, \frac{1}{2}A_F + \frac{1}{2}A_M + A_{W_2} + D_2\right) \\
&= \operatorname{cov}\left(\frac{1}{2}A_F + \frac{1}{2}A_M + D_1, \frac{1}{2}A_F + \frac{1}{2}A_M + D_2\right) \\
&= \frac{1}{4}\operatorname{cov}(A_F + A_M, A_F + A_M) + \operatorname{cov}(D_1, D_2) \\
&= \frac{1}{4}\operatorname{var}(A_F + A_M) + \frac{1}{4}\operatorname{var}(D_1) \\
&= \frac{1}{4}(V_A + V_A) + \frac{1}{4}V_D \\
&= \frac{1}{2}V_A + \frac{1}{4}V_D
\end{aligned}$$

Full-sibs share the same dominance effect only if they share the same genotype, but they do not always share the same genotype. They share identical genotype with a probability 1/4. This explains why $\operatorname{cov}(D_1, D_2) = \frac{1}{4}V_D$. Table 9.6 shows all possible sib pairs and their probabilities resulted from the mating of $A_1A_2$ and $A_3A_4$. You can see that among the 16 possible sib pairs, four cases show exactly the same genotypes for both sibs. Therefore, the probability is $4/16 = 1/4$. The generic notation for covariance between full-sibs (FS) is

**Table 9.6**  Genotypic array of full-sib pairs resulted from mating type $A_1A_2 \times A_3A_4$

| Sib 1 | Sib 2 | | | |
|---|---|---|---|---|
| | $A_1A_3$ ($\frac{1}{4}$) | $A_1A_4$ ($\frac{1}{4}$) | $A_2A_3$ ($\frac{1}{4}$) | $A_2A_4$ ($\frac{1}{4}$) |
| $A_1A_3$ ($\frac{1}{4}$) | $A_1A_3$-$A_1A_3$ ($\frac{1}{16}$) | $A_1A_3$-$A_1A_4$ ($\frac{1}{16}$) | $A_1A_3$-$A_2A_3$ ($\frac{1}{16}$) | $A_1A_3$-$A_2A_4$ ($\frac{1}{16}$) |
| $A_1A_4$ ($\frac{1}{4}$) | $A_1A_4$-$A_1A_3$ ($\frac{1}{16}$) | $A_1A_4$-$A_1A_4$ ($\frac{1}{16}$) | $A_1A_4$-$A_2A_3$ ($\frac{1}{16}$) | $A_1A_4$-$A_2A_4$ ($\frac{1}{16}$) |
| $A_2A_3$ ($\frac{1}{4}$) | $A_2A_3$-$A_1A_3$ ($\frac{1}{16}$) | $A_2A_3$-$A_1A_4$ ($\frac{1}{16}$) | $A_2A_3$-$A_2A_3$ ($\frac{1}{16}$) | $A_2A_3$-$A_2A_4$ ($\frac{1}{16}$) |
| $A_2A_4$ ($\frac{1}{4}$) | $A_2A_4$-$A_1A_3$ ($\frac{1}{16}$) | $A_2A_4$-$A_1A_4$ ($\frac{1}{16}$) | $A_2A_4$-$A_2A_3$ ($\frac{1}{16}$) | $A_2A_4$-$A_2A_4$ ($\frac{1}{16}$) |

**Table 9.7**  The joint distribution table of full-sib pairs merged with the marginal probabilities

| Sib 1 | Sib 2 | | | Marginal |
|---|---|---|---|---|
| | $A_1A_1(a)$ | $A_1A_2(d)$ | $A_2A_2(-a)$ | |
| $A_1A_1(a)$ | $p^2(p+\frac{1}{2}q)^2$ | $\frac{1}{2}p^2q(p+1)$ | $\frac{1}{4}p^2q^2$ | |
| $A_1A_2(d)$ | $\frac{1}{2}p^2q(p+1)$ | $pq(1+pq)$ | $\frac{1}{2}pq^2(q+1)$ | |
| $A_2A_2(-a)$ | $\frac{1}{4}p^2q^2$ | $\frac{1}{2}pq^2(q+1)$ | $q^2(q+\frac{1}{2}p)^2$ | |
| Marginal | $p^2$ | $2pq$ | $q^2$ | |

$$\mathrm{cov}(FS) = \frac{1}{2}V_A + \frac{1}{4}V_D$$

## 9.4.2  Long Way of Derivation

Ignoring the derivation of the joint distribution, I directly provide the joint distribution table between full-sibs (Table 9.7). The covariance between full-sibs

$$\mathrm{cov}(FS) = \mathrm{cov}(G_1, G_2) = \mathrm{E}(G_1 \times G_2) - \mathrm{E}(G_1)\mathrm{E}(G_2)$$

The marginal expectations are derived using

$$\mathrm{E}(G_1) = \mathrm{E}(G_2) = p^2a + 2pqd + q^2(-a) = a(p-q) + 2pq = M$$

which only involves the marginal distributions. The expectation of the product is

$$
\begin{aligned}
\mathrm{E}(G_1 \times G_2) =\ & p^2\left(p+\frac{1}{2}q\right)^2(a)(a) + \frac{1}{2}p^2q(p+1)(a)(d) + \frac{1}{4}p^2q^2 \times (a)(-a) \\
& + \frac{1}{2}p^2q(p+1)(d)(a) + pq(1+pq)(d)(d) + \frac{1}{2}pq^2(q+1)(d)(-a) \\
& + \frac{1}{4}p^2q^2 \times (-a)(a) + \frac{1}{2}pq^2(q+1)(-a)(d) + q^2\left(q+\frac{1}{2}p\right)^2(-a)(-a) \\
=\ & p^2\left(p+\frac{1}{2}q\right)^2a^2 + \frac{1}{2}p^2q(p+1)ad - \frac{1}{4}p^2q^2a^2 \\
& + \frac{1}{2}p^2q(p+1)ad + pq(1+pq)d^2 - \frac{1}{2}pq^2(q+1)ad \\
& - \frac{1}{4}p^2q^2a^2 - \frac{1}{2}pq^2(q+1)ad + q^2\left(q+\frac{1}{2}p\right)^2a^2 \\
=\ & \left[p^2\left(p+\frac{1}{2}q\right)^2 - \frac{1}{2}p^2q^2 + q^2\left(q+\frac{1}{2}p\right)^2\right]a^2 \\
& + [p^2q(p+1) - pq^2(q+1)]ad + [pq(1+pq)]d^2
\end{aligned}
$$

Therefore, the covariance is

$$\text{cov}(FS) = \left[ p^2 \left( p + \frac{1}{2}q \right)^2 - \frac{1}{2}p^2q^2 + q^2 \left( q + \frac{1}{2}p \right)^2 \right] a^2$$
$$+ \left[ p^2 q (p+1) - pq^2 (q+1) \right] ad + \left[ pq(1+pq) \right] d^2$$
$$- \left[ a(p-q) + 2pqd \right]^2$$

After several steps of simplification, we get

$$\text{cov}(FS) = pq[a + d(q-p)]^2 + p^2 q^2 d^2$$
$$= \frac{1}{2} \left( 2pq\alpha^2 \right) + \frac{1}{4} \left( 2pqd \right)^2$$
$$= \frac{1}{2} V_A + \frac{1}{4} V_D$$

## 9.5    Genetic Covariance between Monozygotic Twins (Identical Twins)

Monozygotic twins are also called identical twins, which are generated by splitting the same fertilized egg into two individuals. The twins are genetically the same individuals. Therefore, $\text{cov}(MZ) = V_G$. In contrast, dizygotic twins (also called fraternal twins) are generated via fertilizing two eggs by two sperms. Genetically, dizygotic twins are full-sibs. Therefore, $\text{cov}(DZ) = \text{cov}(FS) = \frac{1}{2}V_A + \frac{1}{4}V_D$.

## 9.6    Summary

The general formula of covariance between relatives is

$$\text{cov}(P_i, P_j) = \text{cov}(G_i, G_j) = a_{ij}V_A + d_{ij}V_D + a_{ij}^2 V_{AA} + a_{ij}d_{ij}V_{(AD)} + d_{ij}^2 V_{DD}$$

where the coefficients are listed below for some common relatives. These relationships in Table 9.8 are graphically represented in Fig. 9.1, called an extended pedigree. A pedigree containing only two parents and their children is called a nuclear family or conjugal family. In Chap. 11, we will learn the concepts of inbreeding coefficient and coancestry. The common relationships listed in Table 9.8 are special cases where inbreeding coefficients for all individuals in the pedigree are zero. When inbreeding (mating between relatives) is involved, Table 9.8 will be modified to reflect the general nature of pedigree relationships.

**Table 9.8** The additive and dominance coefficients of relationship for common relatives

| Degree of relationship | Relationship | Illustration in Fig. 9.1 | $a_{12}$ | $d_{12}$ |
|---|---|---|---|---|
| Zero-degree | Monozygotic twins (MZ) | H-H | 1 | 1 |
| First-degree | Parent-offspring | A-F | 1/2 | 0 |
| First-degree | Full-sibs (FS) | G-H | 1/2 | 1/4 |
| Second-degree | Half-sibs (HS) | F-H | 1/4 | 0 |
| Second-degree | Grandchildren-grandparent | B-M | 1/4 | 0 |
| Second-degree | Uncle(aunt)–nephew(niece) | I-N,H-M | 1/4 | 0 |
| Second-degree | Double first cousin | M-N | 1/4 | 1/16 |
| Third-degree | Great-grandparent-offspring | B-Q | 1/8 | 0 |
| Third-degree | Single first cousin | N-O | 1/8 | 0 |

**Fig. 9.1** A hypothetical pedigree showing most common types of relatives, including (1) Parent-offspring: A-F; (2) Full-sibs: G-H; (3) Half-sibs: F-H; (4) Grandparent-grandchild: B-M; (5) Uncle-nephew: I-N; (6) Aunt-niece: H-M; (7) Double-first cousins: M-N; (8) Single-first cousins: N-O; (9) Great-grandparent vs. great-grand offspring: B-Q. Circles and squares denote females and males, respectively

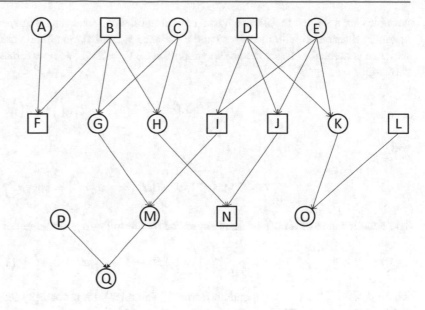

## 9.7 Environmental Covariance

Sometimes, a particular environmental variance may contribute to the variation between means of families but not to the variance within families. Such a variance contributes to the covariance of the related individuals. This between-family environmental component, for which we shall use the symbol $E_c$, is usually called the common or shared environment. Therefore, we can also partition the environmental error as $E = E_c + E_w$. The between-family environmental effect is denoted by $E_c$ and the within-family environmental effect is denoted by $E_w$. The corresponding environmental error variance is partitioned similarly, $V_E = V_{E_c} + V_{E_w}$. This type of partitioning of error variance is similar to the partitioning of the environmental error variance into the general (permanent) and the special (temporary) environmental error variances we have learned in Chap. 7, i.e., $V_E = V_{E_g} + V_{E_s}$.

## 9.8 Phenotypic Resemblance

What we have learned so far is that relatives do not share common environmental effects. As a result, the phenotypic covariance between relatives is the same as the genetic covariance. When the assumption is violated, i.e., relatives do share common environment, the phenotypic covariance between relatives is no long the same as their genetic covariance. The relationship becomes

$$\text{cov}(P_1, P_2) = \text{cov}(G_1, G_2) + \text{cov}(E_c, E_c) = a_{12}V_A + d_{12}V_D + V_{E_c}$$

where epistatic variances have been ignored in the above formula. Full-siblings tend to share more common environmental effect than other relatives because they share the same uterus condition and the same milk of their mother (maternal effect). Throughout this course, we assume that the common environmental effect is absent (by default); otherwise, it will be specifically indicated so.

## 9.9 Derivation of within Family Segregation Variance

The within-family segregation variance (within-family variance in short) is defined as the expectation of the variance of all members within a family. Mathematically, we should write the genetic effects of two parents using alleles as the basic units. Let $A_m = a_1^m + a_2^m$ and $A_f = a_1^f + a_2^f$, where $a_1^m$ and $a_2^m$ are the values for the first and second alleles of the mother, respectively, $a_1^f$ and $a_2^f$ are the values for the first and second alleles of the father, respectively. By definition, all random variables are assumed to follow a normal distribution with mean zero. Thus, each parental allelic effect (a random variable) is normally distributed with mean zero and a variance $V_a$, called the allelic variance. In diploid organisms, the additive genetic variance is defined as twice the allelic variance, i.e., $V_A = 2V_a$. We now reiterate the above definition for the allelic effects by showing

$$\mathrm{E}\left(a_1^m\right) = \mathrm{E}\left(a_2^m\right) = \mathrm{E}\left(a_1^f\right) = \mathrm{E}\left(a_2^f\right) = 0$$

and

$$\mathrm{var}\left(a_1^m\right) = \mathrm{var}\left(a_2^m\right) = \mathrm{var}\left(a_1^f\right) = \mathrm{var}\left(a_2^f\right) = V_a = \frac{1}{2}V_A$$

The genetic value of an offspring is expressed by the following model under the Mendelian inheritance,

$$A = \delta_m a_1^m + (1 - \delta_m)a_2^m + \delta_f a_1^f + \left(1 - \delta_f\right)a_2^f \tag{9.1}$$

where $\delta_m$ and $\delta_f$ are two independent Bernoulli variables with probability 0.5, and they represent the Mendelian segregation (inheritance) pattern. They are defined as

$$\delta_m = \begin{cases} 1 & \text{if } a_1^m \text{ is passed to the progeny with probability 0.5} \\ 0 & \text{if } a_2^m \text{ is passed to the progeny with probability 0.5} \end{cases}$$

for the mother and

$$\delta_f = \begin{cases} 1 & \text{if } a_1^f \text{ is passed to the progeny with probability 0.5} \\ 0 & \text{if } a_2^f \text{ is passed to the progeny with probability 0.5} \end{cases}$$

for the father. A Bernoulli variable with probability 0.5 has an expectation of $\mathrm{E}(\delta_m) = \mathrm{E}(\delta_f) = 1/2$ and a variance of var $(\delta_m) = \mathrm{var}(\delta_f) = 1/4$. The genetic value of the offspring $A_o$ is a composite quantity involving two Bernoulli variables, denoted by $\delta = \{\delta_m, \delta_f\}$, and four parental allelic values, denoted by $a = \left\{a_1^m, a_2^m, a_1^f, a_2^f\right\}$. With a few algebraic steps, Eq. (9.1) can be rewritten as

$$A = \delta_m\left(a_1^m - a_2^m\right) + \delta_f\left(a_1^f - a_2^f\right) + a_2^m + a_2^f$$

The conditional variance of $A_o$ given $a$ is taken with respect to $\delta$ and has the following expression:

$$\begin{aligned}
\mathrm{var}(A|a) &= \mathrm{var}\left[\delta_m\left(a_1^m - a_2^m\right) + \delta_f\left(a_1^f - a_2^f\right) + a_2^m + a_2^f\right] \\
&= \mathrm{var}(\delta_m)\left(a_1^m - a_2^m\right)^2 + \mathrm{var}(\delta_f)\left(a_1^f - a_2^f\right)^2 \\
&= \frac{1}{4}\left(a_1^m - a_2^m\right)^2 + \frac{1}{4}\left(a_1^f - a_2^f\right)^2
\end{aligned}$$

The within-family variance is the expectation of the above conditional variance taken with respect to $a$, which has the following final expression,

$$\begin{aligned}
\mathrm{E}[\mathrm{var}(A|a)] &= \frac{1}{4}\mathrm{E}\left[\left(a_1^m - a_2^m\right)^2\right] + \frac{1}{4}\mathrm{E}\left[\left(a_1^f - a_2^f\right)^2\right] \\
&= \frac{1}{4}\mathrm{var}\left(a_1^m - a_2^m\right) + \frac{1}{4}\mathrm{var}\left(a_1^f - a_2^f\right) \\
&= \frac{1}{4}\left[\mathrm{var}\left(a_1^m\right) + \mathrm{var}\left(a_2^m\right)\right] + \frac{1}{4}\left[\mathrm{var}\left(a_1^f\right) + \mathrm{var}\left(a_2^f\right)\right] \\
&= \frac{1}{2}V_A
\end{aligned}$$

Therefore, the within-family variance is half the additive genetic variance if the families are defined as full-sib families. The between-family variance is defined as

$$\begin{aligned}
\mathrm{var}[\mathrm{E}(A|a)] &= \mathrm{var}\left\{\mathrm{E}\left[\delta_m a_1^m + (1 - \delta_m)a_2^m + \delta_f a_1^f + (1 - \delta_f)a_2^f\right]\right\} \\
&= \mathrm{var}\left[\mathrm{E}(\delta_m)a_1^m + \mathrm{E}(1 - \delta_m)a_2^m + \mathrm{E}(\delta_f)a_1^f + \mathrm{E}(1 - \delta_f)a_2^f\right] \\
&= \mathrm{var}\left(\frac{1}{2}a_1^m + \frac{1}{2}a_2^m + \frac{1}{2}a_1^f + \frac{1}{2}a_2^f\right) \\
&= \frac{1}{4}\mathrm{var}\left(a_1^m + a_2^m + a_1^f + a_2^f\right) \\
&= \frac{1}{4}\left[\mathrm{var}(a_1^m) + \mathrm{var}(a_2^m) + \mathrm{var}\left(a_1^f\right) + \mathrm{var}\left(a_2^f\right)\right] \\
&= \frac{1}{4}(4V_a) \\
&= \frac{1}{2}V_A
\end{aligned}$$

The total genetic variance of the entire population is partitioned into the within-family segregation variance and the between-family variance,

$$\mathrm{var}(A) = \mathrm{var}[\mathrm{E}(A|a)] + \mathrm{E}[\mathrm{var}(A|a)] = \frac{1}{2}V_A + \frac{1}{2}V_A = V_A$$

This equation is called the law of total variance in statistics (Weiss 2005).

An alternative derivation of the within-family segregation variance is to take the difference between the total variance and the between-family variance (Foulley and Chevalet 1981; Wang and Xu 2019),

$$\mathrm{E}[\mathrm{var}(A|a)] = \mathrm{var}(A) - \mathrm{var}[\mathrm{E}(A|a)]$$

where $\mathrm{var}(A) = V_A$ and

$$\begin{aligned}
\mathrm{var}[\mathrm{E}(A|a)] &= \mathrm{var}\left(\frac{1}{2}A_F + \frac{1}{2}A_M\right) \\
&= \frac{1}{4}\mathrm{var}(A_F + A_M) \\
&= \frac{1}{4}[\mathrm{var}(A_F) + \mathrm{var}(A_M)] \\
&= \frac{1}{4}(2V_A) \\
&= \frac{1}{2}V_A
\end{aligned}$$

Therefore,

$$\mathrm{E}[\mathrm{var}(A|a)] = \mathrm{var}(A) - \mathrm{var}[\mathrm{E}(A|a)] = V_A - \frac{1}{2}V_A = \frac{1}{2}V_A$$

This alternative derivation involves no Mendelian variables and thus is not general.

When the two parents are genetically related and they are inbred themselves, i.e., in the presence of inbreeding, the within family variance is no longer $\frac{1}{2}V_A$; rather, it is

$$\mathrm{E}[\mathrm{var}(A|a)] = \frac{1}{2}\left[1 - \frac{1}{2}(f_S + f_D)\right]V_A \tag{9.2}$$

where $f_S$ and $f_D$ are the inbreeding coefficients of the sire and the dam, respectively. The between family variance is

$$\mathrm{var}[\mathrm{E}(A|a)] = \frac{1}{2}\left[1 + \frac{1}{2}(f_S + f_D) + 2f\right]V_A \tag{9.3}$$

where $f$ is the inbreeding coefficient of the progeny (identical to the coancestry between the two parents). Therefore, the total genetic variance is

$$\begin{aligned}
\mathrm{var}(A) &= \mathrm{var}[\mathrm{E}(A|a)] + \mathrm{E}[\mathrm{var}(A|a)]\\
&= \frac{1}{2}\left[1 + \frac{1}{2}(f_S + f_D) + 2f\right]V_A + \frac{1}{2}\left[1 - \frac{1}{2}(f_S + f_D)\right]V_A\\
&= (1 + f)V_A
\end{aligned}$$

Derivations of Eqs. (9.2) and (9.3) were provided by (Foulley and Chevalet 1981; Wang and Xu 2019). The inbreeding coefficient of an individual and the coancestry between individuals will be formally studied in Chap. 11.

## References

Foulley JL, Chevalet C. Méthode de prise en compte de la consanguinité dans un modèle simple de simulation de performances. Ann Genet Sel Anim. 1981;13:189–96.

Wang M, Xu S. Statistics of Mendelian segregation - a mixture model. J Anim Breed Genet. 2019;136(5):341–50. https://doi.org/10.1111/jbg.12394.

Weiss NA. A course in probability. Boston, MA: Addison–Wesley; 2005.

# Estimation of Heritability

The narrow sense heritability is defined as the ratio of the additive genetic variance to the phenotypic variance, as shown below,

$$h^2 = \frac{V_A}{V_P}$$

Heritability can also be interpreted as the regression coefficient of the breeding value on the phenotypic value from the same individual. The covariance between the breeding value and the phenotypic value (both from the same individual) is

$$\text{cov}(A, P) = \text{cov}(A, A + E) = \text{cov}(A, A) + \text{cov}(A, E) = \text{var}(A) = V_A$$

where $P$ is the phenotypic value and $A$ is the additive genetic value. Therefore, the regression coefficient is

$$b_{AP} = \frac{\text{cov}(A, P)}{\text{var}(P)} = \frac{V_A}{V_P} = h^2$$

As a regression coefficient, the heritability can be used to predict the breeding value from the phenotype value. Let $\widehat{A}$ be the predicted breeding value, we have,

$$\widehat{A} - \overline{A} = b_{AP}(P - \overline{P}) = h^2(P - \overline{P})$$

which leads to the following predicted breeding value:

$$\widehat{A} = \overline{A} + h^2(P - \overline{P})$$

Heritability highly depends on traits, populations, and environments. Different traits often have very different heritability. The same trait may also have different heritability in different populations, as different populations may have very different allele frequencies. Heritability also varies cross different environments. Heritability is defined as a ratio of variance components. Therefore, we can estimate the variance components first and then convert the estimated variance components into heritability. Sometimes, heritability can be estimated directly, e.g., the parent-offspring regression analysis. There are two types of data used for this purpose: (1) data from inbred lines and line crosses and (2) data sampled from out-bred populations. The methods are quite different for different types of data. Using outbred populations to estimate heritability is very common and thus we will study methods of estimating heritability in outbred populations in great details after briefly introducing methods of estimating heritability in inbred lines and line crosses.

© Springer Nature Switzerland AG 2022
S. Xu, *Quantitative Genetics*, https://doi.org/10.1007/978-3-030-83940-6_10

## 10.1   F$_2$ Derived from a Cross of Two Inbred Parents

Consider an F$_2$ mating design of experiments. Assume that you collect $n_{F_1}$ plants of F$_1$ derived from the cross of two inbred lines. You then collect $n_{F_2}$ plants of F$_2$ derived from selfing of the F$_1$ plants. For the $j$th F$_1$ plant, the model for a phenotypic value is

$$x_j = \mu + G + e_j$$

where $G$ is the genotypic value for the heterozygote. All F$_1$ plants have the same genotypic value. For the $j$th F$_2$ plant, the model is

$$y_j = \mu + G_j + e_j$$

where $G_j$ is the genotypic value for the $j$th plant and it varies across $j = 1, \ldots n_{F_2}$. The variance for $x_j$ is

$$V_{F_1} = \mathrm{var}(x_j) = \mathrm{var}(e_j) = V_E$$

The variance for $y_j$ is

$$V_{F_2} = \mathrm{var}(y_j) = \mathrm{var}(G_j) + \mathrm{var}(e_j) = V_G + V_E = V_P$$

Therefore, the difference of variances between the two generations of plants is

$$V_{F_2} - V_{F_1} = V_G + V_E - V_E = V_G$$

The estimated broad sense heritability is

$$\widehat{H}^2 = \frac{\widehat{V}_G}{\widehat{V}_P} = \frac{\widehat{V}_{F_2} - \widehat{V}_{F_1}}{\widehat{V}_{F_2}}$$

where

$$\widehat{V}_{F_1} = \frac{1}{n_{F_1} - 1} \sum_{j=1}^{n_{F_1}} (x_j - \bar{x})^2$$

and

$$\widehat{V}_{F_2} = \frac{1}{n_{F_2} - 1} \sum_{j=1}^{n_{F_2}} (y_j - \bar{y})^2$$

are the sample variances obtained from the data.

## 10.2   Multiple Inbred Lines or Multiple Hybrids

In plant breeding, people often estimate (broad sense) heritability with a population of inbred lines or a population of hybrids. This type of experiments must be evaluated in multiple environments. Within each genotype and environment combination, there should be a few replicates in order to separate potential G $\times$ E interaction from random residual errors. If there are no replications, the G $\times$ E interaction effects are confounded with the residual errors.

## 10.2.1 With Replications

Let $y_{ijk}$ be the phenotypic value of the $k$th replicate of genotype $i$ evaluated in block $j$ for $i = 1, \ldots, n$, $j = 1, \ldots, m$ and $k = 1, \ldots, r$, where $n$ is the number of genotypes (genetically different plants), $m$ is the number of blocks, and $r$ is the number of replicates. The $m$ block effects are systematic effects, such as year effects, location effects, and so on. Sometimes we call the blocks systematic environments. They are not random environmental errors! The model to describe the data is

$$y_{ijk} = \mu + \alpha_i + \beta_j + \gamma_{ij} + \varepsilon_{ijk}$$

where $\mu$ is the grand mean, $\alpha_i$ is the effect of the $i$th genotype, $\beta_j$ is the effect of the $j$th environment (block), $\gamma_{ij}$ is the genotype by environment interaction, and $\varepsilon_{ijk}$ is the residual error. The genotypic effects are random with an assumed $N(0, \sigma_G^2)$ distribution, the block effects are assumed to be fixed (no distribution is assigned to fixed effects), the interaction effects are normally distributed with mean zero and variance $\sigma_{G \times E}^2$ and the residual errors are normally distributed with mean 0 and variance $\sigma^2$. Because the model contains both fixed and random effects, it is called a linear mixed model. The ANOVA table, including the expected mean squares, is given in Table 10.1. The G × E effect is the genotype by block interaction effect. Since blocks are defined as systematic environmental effects, we use E instead of B in the ANOVA table. The mean squares (MS) equals the sum of squares (SS) divided by the degree of freedom (DF). Let the expected MS equal the observed (calculated) MS, we can find the estimated variance components. The following three equations and three unknowns allow us to solve for the unknown variance components.

$$MS_G = \sigma^2 + r\sigma_{G \times E}^2 + rm\sigma_G^2$$
$$MS_{G \times E} = \sigma^2 + r\sigma_{G \times E}^2$$
$$MS_E = \sigma^2$$

The solutions are

$$\hat{\sigma}^2 = MS_E$$
$$\hat{\sigma}_{G \times E}^2 = (MS_{G \times E} - MS_E)/r$$
$$\hat{\sigma}_G^2 = (MS_G - MS_{G \times E})/(mr)$$

The estimated broad sense heritability is

$$\hat{H}^2 = \frac{\hat{\sigma}_G^2}{\hat{\sigma}_G^2 + \hat{\sigma}_{G \times E}^2 + \hat{\sigma}^2}$$

The denominator is the phenotypic variance ($\hat{\sigma}_P^2 = \hat{\sigma}_G^2 + \hat{\sigma}_{G \times E}^2 + \hat{\sigma}^2$) and the numerator is the genotypic variance. There is a debate regarding whether $\sigma_{G \times E}^2$ should be included in the phenotypic variance or not. I prefer to include $\sigma_{G \times E}^2$ in the phenotypic variance.

A simulated data with $n = 10$ genotypes, $m = 5$ blocks (years), and $r = 3$ replicates per genotype and year combination are stored in "data-10-1.csv". The MIXED procedure in SAS is used to perform the analysis of variances. The SAS code to read and analyze the data are shown below.

**Table 10.1** ANOVA table for estimation of variance components

| Source | DF | SS | MS | Expected mean square |
|---|---|---|---|---|
| Block | $m - 1$ | $SS_B$ | $MS_B$ | $\sigma^2 + r\sigma_{G \times E}^2 + Q_B$ |
| Genotype | $n - 1$ | $SS_G$ | $MS_G$ | $\sigma^2 + r\sigma_{G \times E}^2 + rm\sigma_G^2$ |
| G × E | $(n - 1)(m - 1)$ | $SS_{G \times E}$ | $MS_{G \times E}$ | $\sigma^2 + r\sigma_{G \times E}^2$ |
| Residual | $nm(r - 1)$ | $SS_E$ | $MS_E$ | $\sigma^2$ |

```
filename aa "data-10-1.csv";
proc import datafile=aa out=one dbms=csv replace;
run;
proc mixed data=one method=type1;
    class genotype year;
    model y = year;
    random genotype genotype*year;
run;
```

The output of the analysis of variances are illustrated in Table 10.2, which is a modified version of the ANOVA table delivered by PROC MIXED. The estimated variance components are

$$\widehat{\sigma}^2_G = 34.1810$$
$$\widehat{\sigma}^2_{G \times E} = 22.2549$$
$$\widehat{\sigma}^2 = 9.1282$$

The estimated broad sense heritability is

$$\widehat{H}^2 = \frac{\widehat{\sigma}^2_G}{\widehat{\sigma}^2_G + \widehat{\sigma}^2_{G \times E} + \widehat{\sigma}^2} = \frac{34.1810}{34.1810 + 22.2549 + 9.1282} = 0.5213371$$

The denominator of the F test for G × E is the residual MS and thus $F_{G \times E} = 75.89/9.13 = 8.31$ with 36 as the numerator degree of freedom and 100 as the denominator degree of freedom. The denominator of the F test for genotype is the G × E MS and thus $F_G = 588.61/75.89 = 7.76$ with 9 as the numerator degree of freedom and 36 as the denominator degree of freedom. Both the genotype and genotype by year interaction effects are significant. The denominator of the F test for the year effect is the G × E MS and thus $F_Y = 24.58/75.89 = 0.32$ with 4 as the numerator degree of freedom and 36 as the denominator degree of freedom. The year effect is not significant.

## 10.2.2  Without Replications

Let $y_{ij}$ be the phenotypic value of genotype $i$ evaluated in block $j$ for $i = 1, \ldots, n$ and $j = 1, \ldots, m$, where $n$ is the number of genotypes and $m$ is the number of blocks. You can imagine that $y = \{y_{ij}\}$ is an $n \times m$ matrix. The $m$ blocks are systematic environmental effects, not environmental errors. The model to describe the data is

$$y_{ij} = \mu + \alpha_i + \beta_j + \gamma_{ij} + \varepsilon_{ij}$$

where $\mu$ is the grand mean, $\alpha_i$ is the effect of the $i$th genotype, $\beta_j$ is the effect of the $j$th block, $\gamma_{ij}$ is the genotype by block interaction, and $\varepsilon_{ij}$ is the residual error. Since there is only one observation per genotype and block combination, the G × E interaction is completely confounded with the residual error. Therefore, we cannot test the genotype by block interaction. We must revise the model as follows:

**Table 10.2** ANOVA table for estimation of variance components for the simulated data

| Source | DF | SS | MS | Expected Mean Square | F value | Pr > F |
|---|---|---|---|---|---|---|
| Year | 4 | 98.3 | 24.58 | $\sigma^2 + 3\sigma^2_{G \times E} + Q_B$ | 0.32 | 0.8601 |
| Genotype | 9 | 5297.5 | 588.61 | $\sigma^2 + 3\sigma^2_{G \times E} + 3 \times 5\sigma^2_G$ | 7.76 | <0.0001 |
| G × E | 36 | 2732.1 | 75.89 | $\sigma^2 + 3\sigma^2_{G \times E}$ | 8.31 | <0.0001 |
| Residual | 100 | 912.8 | 9.13 | $\sigma^2$ | | |

**Table 10.3** ANOVA table for estimation of variance components

| Source | DF | SS | MS | Expected mean squares |
|---|---|---|---|---|
| Block | $m - 1$ | $SS_B$ | $MS_B$ | $\sigma^2 + Q_B$ |
| Genotype | $n - 1$ | $SS_G$ | $MS_G$ | $\sigma^2 + m\sigma_G^2$ |
| Residual | $(n - 1)(m - 1)$ | $SS_E$ | $MS_E$ | $\sigma^2$ |

**Table 10.4** ANOVA table for estimation of variance components

| Source | DF | Sum of Squares | Mean Square | Expected Mean Square |
|---|---|---|---|---|
| Block | 3 | 87.034665 | 29.011555 | Var(residual) + Q(block) |
| Genotype | 209 | 5281.025830 | 25.268066 | Var(residual) + 4 Var(genotype) |
| Residual | 627 | 714.956061 | 1.140281 | Var(residual) |

$$y_{ij} = \mu + \alpha_i + \beta_j + e_{ij}$$

where $e_{ij} = \gamma_{ij} + \varepsilon_{ij}$ is a composite residual including both G × E interaction and the pure residual error. The ANOVA table is shown in Table 10.3. Let the expected MS equal the observed mean squares and solve for the variance components, we have the estimated variance components and thus the estimated broad sense heritability,

$$\widehat{\sigma}^2 = MS_E \quad \text{and} \quad \widehat{H}^2 = \frac{\widehat{\sigma}_G^2}{\widehat{\sigma}_G^2 + \widehat{\sigma}^2}$$
$$\sigma_G^2 = (MS_G - MS_E)/m$$

.

A real-life example in rice is used to show how to estimate the broad sense heritability of the thousand grain weight (KGW) trait. The population consists of 210 recombinant inbred lines. The trait was evaluated in four blocks (years). The data are stored in the course website with a filename "data-10-2.xlsx". The trait value of each genotype and block combination is the average of about $r = 10$ plants. Technically, the data were replicated 10 times, but we do not have the original records and were only provided the average values. Therefore, we treated the average value of the 10 plants as a data point and estimate the heritability on the basis of mean. Below is the SAS code for the MIXED procedure.

```
filename aa "data-10-2.xlsx";
proc import datafile=aa out=one dbms=xlsx replace;
run;
proc mixed data=one method=type1;
    class Genotype Block;
    model KGW = Block;
    random Genotype;
run;
```

The ANOVA table generated by PROC MIXED is given in Table 10.4. The estimated variance components are $\widehat{\sigma}_G^2 = 6.0319$ and $\widehat{\sigma}^2 = 1.1403$. The estimated broad sense heritability is

$$\widehat{H}^2 = \frac{\widehat{\sigma}_G^2}{\widehat{\sigma}_G^2 + \widehat{\sigma}^2} = \frac{6.0319}{6.0319 + 1.1403} = 0.8410$$

Note that the residual variance is a composite term containing both the G × E interaction variance and the pure environmental error variance. In addition, the above estimated heritability is based on the mean. This heritability can be converted back to the heritability on the basis of individual plant, which is

$$\widehat{H}^2 = \frac{\widehat{\sigma}_G^2}{\widehat{\sigma}_G^2 + r\widehat{\sigma}^2} = \frac{6.0319}{6.0319 + 10 \times 1.1403} = 0.345967$$

In plants, the estimated heritability is often the one on the basis of mean. This presents a big problem because the number of replicates used to calculate the mean value determines the mean-based heritability. For the same trait collected from the same population in the same environment, different people may report quite different estimated heritability if the numbers of replicates are different. Therefore, it is recommended to (1) report the number of replicates along with the estimated heritability or (2) convert the mean-based heritability into the individual-based heritability. Let $h_{mean}^2$ and $h_{individual}^2$ be the heritability on the basis of mean and on the basis of individual, respectively. Let $r$ be the number of replicates, $\sigma_{individual}^2$ be the residual variance when each data point represents a single plant, and $\sigma_{mean}^2$ be the residual variance when each data point represents the mean value of $r$ plants. The relationship can be described as

$$h_{individual}^2 = \frac{\sigma_G^2}{\sigma_G^2 + \sigma_{individual}^2} = \frac{\sigma_G^2}{\sigma_G^2 + r\sigma_{mean}^2}$$

and

$$h_{mean}^2 = \frac{\sigma_G^2}{\sigma_G^2 + \sigma_{mean}^2} = \frac{\sigma_G^2}{\sigma_G^2 + \sigma_{individual}^2/r}$$

## 10.3  Parent-Offspring Regression

There are two commonly used designs of experiments for estimating heritability in out-bred populations: (1) parent-offspring regression and (2) sib analysis. We will discuss the regression analysis first and then the sib analysis. In non-experimental populations, e.g., human populations, all types of relatives are included in data analysis. The data structure is complicated, so a complicated statistical analysis, e.g., maximum likelihood, is required, which will be studied in a later chapter. Parent-offspring regression analysis is the simplest method. There are two parents and one or more siblings in each family, and thus, there are many different ways to perform the regression analysis.

### 10.3.1  Single Parent Vs. Single Offspring

The pair of observations from each family are the phenotypic values of a single parent and a single offspring. The data structure looks like (Table 10.5). The phenotypic values of each pair of observations are described by

$$\begin{cases} P_P = A_P + D_P + E_P \\ P_O = A_O + D_O + E_O \end{cases}$$

Assuming that there is no environmental covariance between parent and offspring, the phenotypic covariance is equivalent to the genetic covariance. The genetic covariance between a parent and an offspring is half the additive genetic variance, i.e.,

**Table 10.5**  Data structure for parent-offspring regression analysis

| Family | Parent | Offspring |
|---|---|---|
| 1 | $P_1$ | $O_1$ |
| 2 | $P_2$ | $O_2$ |
| $\vdots$ | $\vdots$ | $\vdots$ |
| $n$ | $P_n$ | $O_n$ |

$$\text{cov}(P_O, P_P) = \text{cov}(A_O, A_P) = \frac{1}{2}V_A$$

It is known that $\text{var}(P_P) = V_P$, i.e., the variance of the phenotypic value of the parent is the phenotypic variance. The regression coefficient of the phenotypic value of an offspring on the phenotypic value of a single parent is

$$b_{OP} = \frac{\text{cov}(P_O, P_P)}{\text{var}(P_P)} = \frac{\frac{1}{2}V_A}{V_P} = \frac{1}{2}h^2$$

This simple relationship allows us to estimate the heritability because the regression coefficient can be estimated from sampled data,

$$\widehat{b}_{OP} = \frac{\sum_{j=1}^{n}(P_j - \overline{P})(O_j - \overline{O})}{\sum_{j=1}^{n}(P_j - \overline{P})^2}$$

Therefore,

$$\widehat{h}^2 = 2\widehat{b}_{OP}$$

Note that the regression coefficient is directional with $O$ being the response variable and $P$ being the independent variable.

## 10.3.2 Middle Parent Vs. Single Offspring

The mid-parent value is defined as the average of the phenotypic values of the two parents, i.e.,

$$\overline{P} = \frac{1}{2}\left(P_f + P_m\right)$$

The pair of observations are the mid-parent value and a single offspring in each family. The data structure looks like (Table 10.6). The phenotypic covariance between the mid-parent and an offspring is

$$\text{cov}\left(O, \overline{P}\right) = \text{cov}\left[O, \frac{1}{2}\left(P_m + P_f\right)\right] = \frac{1}{2}\text{cov}(O, P_m) + \frac{1}{2}\text{cov}(O, P_f) = \frac{1}{4}V_A + \frac{1}{4}V_A = \frac{1}{2}V_A$$

This is because $\text{cov}(O, P_m) = \text{cov}(O, P_f) = \frac{1}{2}V_A$. The variance of the mid-parent value is

$$\text{var}\left(\overline{P}\right) = \text{var}\left[\frac{1}{2}\left(P_m + P_f\right)\right] = \frac{1}{4}\left[\text{var}(P_m) + \text{var}(P_f)\right] = \frac{1}{4}(V_P + V_P) = \frac{1}{2}V_P$$

Therefore, the regression coefficient of the phenotypic value of an offspring on the mid-parent value is

$$b_{O\overline{P}} = \frac{\text{cov}\left(O, \overline{P}\right)}{\text{var}\left(\overline{P}\right)} = \frac{\frac{1}{2}V_A}{\frac{1}{2}V_P} = h^2$$

**Table 10.6** Data structure for mid-parent vs. offspring regression analysis

| Family | Father | Mother | Mid-parent | Offspring |
|---|---|---|---|---|
| 1 | $P_{1f}$ | $P_{1m}$ | $\overline{P_1}$ | $O_1$ |
| 2 | $P_{2f}$ | $P_{2m}$ | $\overline{P_2}$ | $O_2$ |
| $\vdots$ | $\vdots$ | $\vdots$ | $\vdots$ | $\vdots$ |
| $n$ | $P_{nf}$ | $P_{nm}$ | $\overline{P_n}$ | $O_n$ |

This simple relationship allows us to estimate the heritability because the regression coefficient can be estimated from sampled data,

$$\hat{b}_{O\bar{P}} = \frac{\sum_{j=1}^{n} \left( \bar{P}_j - \bar{\bar{P}} \right) \left( O_j - \bar{O} \right)}{\sum_{j=1}^{n} \left( \bar{P}_j - \bar{\bar{P}} \right)^2}$$

Therefore,

$$\hat{h}^2 = \hat{b}_{O\bar{P}}$$

### 10.3.3  Single Parent Vs. Mean Offspring

When there are $m$ progenies per family, we usually take the average of the phenotypic values of the $m$ progenies and regress the mean phenotypic value of the offspring on the phenotypic value of a parent. The data structure will look like (Table 10.7). The mean offspring value is

$$\bar{O}_j = \frac{1}{m} \sum_{k=1}^{m} O_{jk}$$

where $m$ is the number of offspring per family. First, let us look at the covariance between the mean offspring value and the phenotypic value of the parent,

$$\mathrm{cov}\left( \bar{O}, P \right) = \frac{1}{m} \mathrm{cov}\left( \sum_{k=1}^{m} O_k, P \right) = \frac{1}{m} \sum_{k=1}^{m} \mathrm{cov}(O_k, P) = \frac{1}{m} \sum_{k=1}^{m} \left( \frac{1}{2} V_A \right) = \frac{1}{2} V_A$$

Therefore,

$$b_{\bar{O}P} = \frac{\mathrm{cov}\left( \bar{O}, P \right)}{\mathrm{var}(P)} = \frac{\frac{1}{2} V_A}{V_P} = \frac{1}{2} h^2$$

The estimated regression coefficient is

$$\hat{b}_{\bar{O}P} = \frac{\sum_{j=1}^{n} \left( P_j - \bar{P} \right) \left( \bar{O}_j - \bar{\bar{O}} \right)}{\sum_{j=1}^{n} \left( P_j - \bar{P} \right)^2}$$

Therefore,

$$\hat{h}^2 = 2\hat{b}_{\bar{O}P}$$

**Table 10.7**  Data structure for parent vs. mean offspring regression analysis

| Family | Parent | Mean offspring |
|--------|--------|----------------|
| 1      | $P_1$  | $\bar{O}_1$    |
| 2      | $P_2$  | $\bar{O}_2$    |
| ⋮      | ⋮      | ⋮              |
| $n$    | $P_n$  | $\bar{O}_n$    |

**Table 10.8** Data structure for mid-parent vs. mean offspring regression analysis

| Family | Mid-parent | Mean offspring |
|---|---|---|
| 1 | $\overline{P_1}$ | $\overline{O_1}$ |
| 2 | $\overline{P_2}$ | $\overline{O_2}$ |
| $\vdots$ | | |
| $n$ | $\overline{P_n}$ | $\overline{O_n}$ |

**Table 10.9** Summary of the relationship of the estimated heritability and the parent-offspring regression

| $X$ | $Y$ | Heritability |
|---|---|---|
| Single parent | Single offspring | $2\widehat{b}_{YX}$ |
| Single parent | Mean offspring | $2\widehat{b}_{YX}$ |
| Mid-parent | Single offspring | $\widehat{b}_{YX}$ |
| Mid-parent | Mean offspring | $\widehat{b}_{YX}$ |

### 10.3.4  Middle Parent Vs. Mean Offspring

Table 10.8 shows the data structure for mid-parent analysis. It has been verified that $\mathrm{cov}\left(\overline{P},\overline{O}\right) = \frac{1}{2}V_A$ and $\mathrm{var}\left(\overline{P}\right) = \frac{1}{2}V_P$. Therefore,

$$b_{\overline{OP}} = \frac{\mathrm{cov}\left(\overline{O},\overline{P}\right)}{\mathrm{var}\left(\overline{P}\right)} = \frac{\frac{1}{2}V_A}{\frac{1}{2}V_P} = h^2$$

and

$$\widehat{h}^2 = \widehat{b}_{\overline{OP}}$$

Table 10.9 summarizes the parent-offspring regression analysis for heritability estimation. Whether or not the estimated heritability takes the regression coefficient or twice the regression coefficient depends on whether or not the parent is the mid-parent or a single parent.

### 10.3.5  Estimate Heritability Using Parent-Offspring Correlation

We have learned how to estimate heritability using parent-offspring regression. In fact, heritability can be estimated equally well using parent-offspring correlation, assuming that parent and offspring have the same phenotypic variance. Let $O$ be the phenotypic value of an offspring for a quantitative trait and $P$ be the phenotypic value of the corresponding trait measured from a parent. The correlation coefficient between the parent and the offspring is

$$r_{OP} = \frac{\mathrm{cov}(O,P)}{\sqrt{\mathrm{var}(O)\mathrm{var}(P)}}$$

where

$$\mathrm{cov}(O,P) = \frac{1}{2}V_A$$

and

$$\text{var}(O) = \text{var}(P) = V_P$$

Therefore,

$$r_{OP} = \frac{\text{cov}(O, P)}{\sqrt{\text{var}(O)\text{var}(P)}} = \frac{\frac{1}{2}V_A}{V_P} = \frac{1}{2}h^2$$

The reason that we do not often use this correlation to estimate heritability is that when the mid-parent and mean offspring values are used, the correlation coefficient is not a simple function of the heritability anymore. For example, assume that we use the mid-parent value to perform the correlation analysis. The correlation coefficient would be

$$r_{O\overline{P}} = \frac{\text{cov}\left(O, \overline{P}\right)}{\sqrt{\text{var}(O)\text{var}\left(\overline{P}\right)}} = \frac{\text{cov}\left(O, \overline{P}\right)}{\sqrt{\frac{1}{2}\text{var}(P)\text{var}(O)}} \frac{\frac{1}{2}V_A}{\sqrt{\frac{1}{2}V_P}} = \frac{1}{\sqrt{2}}h^2$$

Although we can still multiply $r_{O\overline{P}}$ by $\sqrt{2}$ to obtain an estimated heritability, it looks strange to use such a weird multiplier. If the offspring value is an average of multiple sibs, the correlation coefficient is even more complicated. Let $\overline{O} = \frac{1}{2}(O_1 + O_2)$ be the mean of two siblings. The parent-offspring correlation would be

$$r_{\overline{OP}} = \frac{\text{cov}\left(\overline{O}, \overline{P}\right)}{\sqrt{\text{var}\left(\overline{O}\right)\text{var}\left(\overline{P}\right)}}$$

where $\text{cov}\left(\overline{O}, \overline{P}\right) = \frac{1}{2}V_A$ and $\text{var}\left(\overline{P}\right) = \frac{1}{2}V_P$. However,

$$\begin{aligned}
\text{var}\left(\overline{O}\right) &= \frac{1}{4}\text{var}(O_1 + O_2) \\
&= \frac{1}{4}[\text{var}(O_1) + \text{var}(O_1) + 2\text{cov}(O_1, O_2)] \\
&= \frac{1}{4}\left(V_P + V_P + 2 \times \frac{1}{2}V_A\right) \\
&= \frac{1}{4}\left(2 + 2 \times \frac{1}{2}h^2\right)V_P \\
&= \frac{1}{2}\left(1 + \frac{1}{2}h^2\right)V_P
\end{aligned}$$

The correlation coefficient between the mid-parent and the mean offspring would be

$$r_{\overline{OP}} = \frac{\text{cov}\left(\overline{O}, \overline{P}\right)}{\sqrt{\text{var}\left(\overline{O}\right)\text{var}\left(\overline{P}\right)}} = \frac{\frac{1}{2}V_A}{\frac{1}{2}\sqrt{1 + \frac{1}{2}h^2}V_P} = \frac{h^2}{\sqrt{1 + \frac{1}{2}h^2}}$$

The estimated heritability must be obtained by solving the following quadratic equation for $h^2$,

$$h^4 - \frac{1}{2}r_{\overline{OP}}^2 h^2 - r_{\overline{OP}}^2 = 0$$

There will be two solutions and we have to take the positive one as the estimated heritability. The positive solution is

$$\hat{h}^2 = \frac{-\left(-\frac{1}{2}r_{\overline{OP}}^2\right) + \sqrt{\left(-\frac{1}{2}r_{\overline{OP}}^2\right)^2 - 4(1)\left(-r_{\overline{OP}}^2\right)}}{2(1)} = \frac{1}{4}r_{\overline{OP}}^2 + \frac{1}{4}\sqrt{r_{\overline{OP}}^4 + 16r_{\overline{OP}}^2}$$

It will be very strange to use parent-offspring correlation to estimate heritability if the mean offspring value is used. Therefore, only correlation between a single parent and a single offspring should be considered for estimation of heritability.

In animal and plant breeding, the parents are often selected based on the phenotypic values of the trait under study, e.g., only the top 50% parents are selected to breed. The correlation coefficient between the selected parents and their offspring will give a biased estimate of heritability, but the estimated heritability based on the regression coefficient is unbiased, regardless whether the parents are selected or not (Falconer and Mackay 1996; Lynch and Walsh 1998). To prove the above claims, we first need to show the variance of $P$ and the covariance between $O$ and $P$ after selection for the parental phenotypic value. The changes of variance and covariance after selection is called the Bulmer effect (Bulmer 1971). Let $P^*$ be the phenotypic value of the selected parents. Denote the selection intensity by $i$ and the standardized truncation point of selection by $t$ (topics related to selection will be studied in Chap. 14). The variance of $P^*$ is

$$\mathrm{var}(P^*) = [1 - i(i - t)]\mathrm{var}(P) = [1 - i(i - t)]V_P$$

The covariance between $P$ and $O$ after selection is

$$\mathrm{cov}(O^*, P^*) = \mathrm{cov}(O, P) - i(i - t)\frac{\mathrm{cov}(P, P)\mathrm{cov}(O, P)}{\mathrm{var}(P)}$$

$$= \frac{1}{2}V_A - \frac{1}{2}i(i - t)\frac{V_P V_A}{V_P}$$

$$= \frac{1}{2}[1 - i(i - t)]V_A$$

The regression coefficient of $O$ on $P$ after selection is

$$b_{O^*P^*} = \frac{\mathrm{cov}(O^*, P^*)}{\mathrm{var}(P^*)} = \frac{\frac{1}{2}[1 - i(i - t)]V_A}{[1 - i(i - t)]V_P} = \frac{1}{2}\frac{V_A}{V_P} = \frac{1}{2}h^2$$

which is still half of the heritability, i.e., selection does not cause bias in estimated heritability. To show that the correlation coefficient between $O$ and $P$ after selection is biased, we need to derive the variance of $O$ after selection. This variance is

$$\mathrm{var}(O^*) = \left[1 - i(i - t)r_{OP}^2\right]\mathrm{var}(O) = \left[1 - i(i - t)h^4\right]V_P$$

The correlation is

$$r_{O^*P^*} = \frac{\mathrm{cov}\left(P_0^*, P_1^*\right)}{\sqrt{\mathrm{var}\left(P_0^*\right)\mathrm{var}\left(P_1^*\right)}}$$

$$= \frac{\frac{1}{2}[1 - i(i - t)]V_A}{\sqrt{[1 - i(i - t)]V_P\left[1 - i(i - t)h^4\right]V_P}}$$

$$= \frac{1}{2\left[1 - i(i - t)h^4\right]}h^2$$

So, the estimated heritability using the correlation would be

$$h^2 = 2\left[1 - i(i - t)h^4\right]r_{O^*P^*} \tag{10.1}$$

which is a function the heritability. Since $0 \le 1 - i(i - t)h^4 \le 1$, the bias is always downwards. One can solve the quadratic Eq. (10.1) for $h^2$ to get an unbiased estimate of heritability. Clearly, correlation coefficient cannot be used to estimate heritability when selection is present. Although regression analysis is unbiased, the estimate is not stable (with a large standard error) when selection is present compared to the estimate without selection.

## 10.4   Sib Analysis

### 10.4.1 Full-Sib Analysis

Assume $n$ full-sib families each with $m$ siblings. The data structure looks like the one shown in Table 10.10. The covariance between full-sibs is

$$\text{cov}(FS) = \text{cov}(y_{jk}, y_{jk'}) = \frac{1}{2} V_A + \frac{1}{4} V_D$$

The total phenotypic variance is

$$\text{var}(P) = \text{var}(y_{jk}) = V_P$$

Therefore,

$$\rho_{FS} = \frac{\text{cov}(FS)}{\text{var}(P)} = \frac{\frac{1}{2} V_A + \frac{1}{4} V_D}{V_P} = \frac{1}{2} h^2 + \frac{1}{4} \frac{V_D}{V_P} \approx \frac{1}{2} h^2$$

provided that $V_D$ is absent or negligibly small. The ratio, $\text{cov}(FS)/\text{var}(P)$, can be estimated as the intra-class correlation in the analysis of variance (ANOVA). The analysis is exactly the same as what we learned for estimating repeatability except that the estimated intra-class correlation only represents half of the heritability,

$$\widehat{h}^2 = 2\widehat{\rho}_{FS} \tag{10.2}$$

This estimated heritability is biased because we have ignored the dominance variance. In addition, full-siblings tend to share common family environmental effect (e.g., maternal effect), which has been ignored as well. The intra-class correlation $\rho_{FS}$ is similar to the repeatability $\rho$. Both are intra-class correlation but $\rho_{FS}$ is the correlation between different siblings within the same family, whereas $\rho$ is the correlation between different measurements within the same individual.

Statistically, we can write the following model for the trait value of the $k$th sibling in the $j$th full-sib family,

$$y_{jk} = \mu + f_j + w_{jk}$$

where $\mu$ is the grand mean, $f_j$ is the effect of the $j$th family, $w_{jk}$ is the residual (not an error) of the $k$th sibling (deviation of the sibling from the family mean). Both $f_j$ and $w_{jk}$ are random variables with normal distributions, $f_j \sim N(0, \sigma_B^2)$ and $w_{jk} \sim N(0, \sigma_W^2)$. Let us look at the variance of the model,

$$\text{var}(y_{jk}) = \text{var}(f_j) + \text{var}(w_{jk}) = \underset{\text{between family}}{\sigma_B^2} + \underset{\text{within family}}{\sigma_W^2} = \sigma_y^2$$

The variance of $y$, however, is the phenotypic variance, i.e., $\text{var}(y_{jk}) = \sigma_y^2 = V_P$. The between family variance is equivalent to the covariance between full-siblings, i.e., $\text{cov}(FS) = \sigma_B^2$. This can be shown as follows. Let $k$ and $k'$ be two siblings within the $j$th family and their phenotypic values are described by

**Table 10.10**  Data structure for full-sib analysis

| Family | Sib$_1$ | Sib$_2$ | . . . | Sib$_m$ |
|---|---|---|---|---|
| 1 | $y_{11}$ | $y_{12}$ | | $y_{1m}$ |
| 2 | $y_{21}$ | $y_{22}$ | | $y_{2m}$ |
| ⋮ | | | | |
| $n$ | $y_{n1}$ | $y_{n2}$ | | $y_{nm}$ |

$$\begin{cases} y_{jk} = \mu + f_j + w_{jk} \\ y_{jk'} = \mu + f_j + w_{jk'} \end{cases}$$

The variances are

$$\mathrm{var}(y_{jk}) = \mathrm{var}(f_j) + \mathrm{var}(w_{jk}) = \sigma_B^2 + \sigma_W^2$$

and

$$\mathrm{var}(y_{jk'}) = \mathrm{var}(f_j) + \mathrm{var}(w_{jk'}) = \sigma_B^2 + \sigma_W^2$$

The covariance between $y_{jk}$ and $y_{jk'}$ is

$$\begin{aligned} \mathrm{cov}(y_{jk}, y_{jk'}) &= \mathrm{cov}(\mu + f_j + w_{jk}, \mu + f_j + w_{jk'}) \\ &= \mathrm{cov}(f_j, f_j) + \mathrm{cov}(f_j, w_{jk}) + \mathrm{cov}(w_{jk'}, f_j) + \mathrm{cov}(w_{jk}, w_{jk'}) \\ &= \mathrm{cov}(f_j, f_j) \\ &= \mathrm{var}(f_j) \\ &= \sigma_B^2 \end{aligned}$$

Therefore,

$$\mathrm{cov}(FS) = \mathrm{cov}(y_{jk}, y_{jk'}) = \sigma_B^2$$

We have now associated the covariance between full-siblings with the between family variance. Let us compare the statistical model with the genetic model so that an equivalence can be deduced.

$$\text{Statistical model}: \sigma_y^2 = \sigma_B^2 + \sigma_W^2$$
$$\text{Genetic model}: \quad V_P = V_A + V_D + V_E$$

As shown earlier, we have the following equivalence:

$$\sigma_y^2 = V_P$$
$$\sigma_B^2 = \mathrm{cov}(FS) = \frac{1}{2}V_A + \frac{1}{4}V_D$$
$$\sigma_W^2 = \sigma_y^2 - \sigma_B^2 = \frac{1}{2}V_A + \frac{3}{4}V_D + V_E$$

The residual variance (within-family variance) $\sigma_W^2$ is not just the error variance; it contains part of the genetic variance. The intra-class correlation (correlation between full-siblings) now can be expressed as the ratio of variance components,

$$\rho_{FS} = r_{y_{jk}y_{jk'}} = \frac{\mathrm{cov}(y_{jk}, y_{jk'})}{\sigma_{y_{jk}}\sigma_{y_{jk'}}} = \frac{\sigma_B^2}{\sqrt{\sigma_B^2 + \sigma_W^2}\sqrt{\sigma_B^2 + \sigma_W^2}} = \frac{\sigma_B^2}{\sigma_B^2 + \sigma_W^2}$$

Because the variance components can be estimated from analysis of variance, we have

$$\widehat{\rho}_{FS} = \frac{\widehat{\sigma}_B^2}{\widehat{\sigma}_B^2 + \widehat{\sigma}_W^2}$$

**Table 10.11**  The ANOVA table for full-sib analysis

| Source of variation | Degree of freedom (df) | Sum of squares (SS) | Mean squares (MS) | Expected MS |
|---|---|---|---|---|
| Between groups (B) | $df_B = n - 1$ | $SS_B = m \sum_{j=1}^{n} \left(\bar{y}_{j.} - \bar{y}_{..}\right)^2$ | $MS_B = SS_B/df_B$ | $\sigma_W^2 + m\sigma_B^2$ |
| Within groups (W) | $df_W = n(m - 1)$ | $SS_W = \sum_{j=1}^{n} \sum_{k=1}^{m} \left(y_{jk} - \bar{y}_{j.}\right)^2$ | $MS_W = SS_W/df_E$ | $\sigma_W^2$ |

**Table 10.12**  The ANOVA table of sib analysis for unbalanced data

| Source of variation | Degree of freedom (df) | Sum of squares (SS) | Mean squares (MS) | Expected MS |
|---|---|---|---|---|
| Between groups (B) | $df_B = n - 1$ | $SS_B = \sum_{j=1}^{n} m_j \left(\bar{y}_{j.} - \bar{y}_{..}\right)^2$ | $MS_B = SS_B/df_B$ | $\sigma_W^2 + m_0\sigma_B^2$ |
| Within groups (W) | $df_W = \sum_{j=1}^{n} (m_j - 1)$ | $SS_W = \sum_{j=1}^{n} \sum_{k=1}^{m_j} \left(y_{jk} - \bar{y}_{j.}\right)^2$ | $MS_W = SS_W/df_W$ | $\sigma_W^2$ |

When $V_D$ is absent or negligibly small, we have $h^2 \approx 2\hat{\rho}_{FS}$, as given in Eq. (10.2). The variance components can be estimated via analysis of variance by treating families as groups. The ANOVA table is listed in Table 10.11. The estimated variance components are

$$\begin{cases} \hat{\sigma}_W^2 = MS_W \\ \hat{\sigma}_B^2 = (MS_B - MS_W)/m \end{cases}$$

The estimated intra-class correlation is

$$\hat{\rho}_{FS} = \frac{\hat{\sigma}_B^2}{\hat{\sigma}_B^2 + \hat{\sigma}_W^2} = \frac{MS_B - MS_W}{MS_B + (m - 1)MS_W}$$

If the data are unbalanced, i.e., the number of siblings per family varies across different families, the ANOVA table is slightly different, as shown in Table 10.12, In the ANOVA table (Table 10.12), $m_j$ is the number of siblings in the $j$th family and $m_0$ appearing in the expected MS column is calculated using

$$m_0 = \frac{1}{n - 1}\left[\sum_{j=1}^{n} m_j - \sum_{j=1}^{n} m_j^2 / \sum_{j=1}^{n} m_j\right]$$

The estimated intra-class correlation is

$$\hat{\rho}_{FS} = \frac{\hat{\sigma}_B^2}{\hat{\sigma}_B^2 + \hat{\sigma}_W^2} = \frac{MS_B - MS_W}{MS_B + (m_0 - 1)MS_W}$$

The method is exactly the same as the ANOVA for estimating repeatability except that the groups defined here are families while the groups defined in the repeatability analysis are individuals.

## 10.4.2  Half-Sib Analysis

The covariance between half-sibs is

$$\mathrm{cov}(HS) = \frac{1}{4}V_A$$

Therefore, the intra-class correlation for half-sib analysis is

$$\rho_{HS} = \frac{\frac{1}{4} V_A}{V_P} = \frac{1}{4} h^2$$

The analysis of variance and intra-class correlation estimation of heritability for full-sib families also applies to half-sib families,

$$\widehat{\rho}_{HS} = \frac{\widehat{\sigma}_B^2}{\widehat{\sigma}_B^2 + \widehat{\sigma}_W^2}$$

The only difference is that $\widehat{h}^2 = 4\widehat{\rho}_{HS}$ for half-sib analysis instead of twice the intra-class correlation for full-sib analysis.

### 10.4.3 Nested or Hierarchical Mating Design

A common form of mating design in large animals is the nested or hierarchical mating design. The population contains several male parents. Each male (sire) is mated to several females (dams). The males and females are randomly selected from the population. The mating is also random. A number of offspring from each female are measured to provide the phenotypic data. The individuals measured thus form a mixed population of half-sib and full-sib families. Assume that the population is made of $s$ sires each mated to $d$ dams and each dam reproduces $n$ progenies. The total number of individuals is $sdn$ but there are $s$ half-sib families and $d$ full-sib families within each half-sib family. The data structure is shown in Table 10.13, where the number of sires is $s = 3$, the number of dams per sire is $d = 5$, and the number of sibs per dam is $n = 3$. It is called nested design because dams are nested within sires, which means that the dam 1 mated with sire 1 is not the same as the dam 1 mated with sire 2. In addition, siblings are nested within dams, which means that the order of siblings within the dam families are irrelevant. The statistical model for the nested design is

$$y_{ijk} = \mu + s_i + d_{j(i)} + w_{ijk}$$

where $y_{ijk}$ is the phenotypic value of the $k$th sibling in the $j$th dam within the $i$th sire. In the above model, $\mu$ is the grand mean, $s_i$ is the $i$th sire effect with an assumed $N(0, \sigma_S^2)$ distribution, $d_{j(i)}$ is the effect of the $j$th dam mated with the $i$th sire with an assumed $N(0, \sigma_D^2)$ distribution, and $w_{ijk}$ is the effect of the $k$th sibling produced prom the $j$th dam within the $i$th sire family, which is also called the residual with an assumed $N(0, \sigma_W^2)$ distribution. The total phenotypic valiance ($V_P$) is

**Table 10.13** Data structure of a nested mating design

| Sire | Dam | Sibling1 | Sibling2 | Sibling3 |
|------|-----|----------|----------|----------|
| 1 | 1 | $y_{111}$ | $y_{112}$ | $y_{113}$ |
| 1 | 2 | $y_{121}$ | $y_{122}$ | $y_{123}$ |
| 1 | 3 | $y_{131}$ | $y_{132}$ | $y_{133}$ |
| 1 | 4 | $y_{141}$ | $y_{142}$ | $y_{143}$ |
| 1 | 5 | $y_{151}$ | $y_{152}$ | $y_{153}$ |
| 2 | 1 | $y_{211}$ | $y_{212}$ | $y_{213}$ |
| 2 | 2 | $y_{221}$ | $y_{222}$ | $y_{223}$ |
| 2 | 3 | $y_{231}$ | $y_{232}$ | $y_{233}$ |
| 2 | 4 | $y_{241}$ | $y_{242}$ | $y_{243}$ |
| 2 | 5 | $y_{251}$ | $y_{252}$ | $y_{253}$ |
| 3 | 1 | $y_{311}$ | $y_{312}$ | $y_{313}$ |
| 3 | 2 | $y_{321}$ | $y_{322}$ | $y_{323}$ |
| 3 | 3 | $y_{331}$ | $y_{332}$ | $y_{333}$ |
| 3 | 4 | $y_{341}$ | $y_{342}$ | $y_{343}$ |
| 3 | 5 | $y_{341}$ | $y_{342}$ | $y_{343}$ |

$$\text{var}(y_{ijk}) = \text{var}(s_i) + \text{var}(d_{j(i)}) + \text{var}(w_{ijk}) = \sigma_S^2 + \sigma_D^2 + \sigma_W^2$$

The connection between the genetic variances and the variances in the statistical model are

$$\sigma_S^2 = \frac{1}{4} V_A$$

$$\sigma_D^2 = \frac{1}{4} V_A + \frac{1}{4} V_D$$

$$\sigma_W^2 = \frac{1}{2} V_A + \frac{3}{4} V_D + V_E$$

Note that $\sigma_D^2$ is the variance of the dam effects and is not the dominance variance $V_D$ introduced earlier. In addition, $\sigma_W^2$ is not just the error variance and it also contains part of the genetic variances. The variance components can be estimated from the nested design ANOVA, as listed in Table 10.14. You need to equate E(MS) to the measured MS to solve for the three variance components. The three equations are

$$MS_S = \sigma_W^2 + n\sigma_D^2 + dn\sigma_S^2$$
$$MS_D = \sigma_W^2 + n\sigma_D^2$$
$$MS_W = \sigma_W^2$$

Solutions of the three variance components are

$$\widehat{\sigma}_W^2 = MS_W$$

$$\widehat{\sigma}_D^2 = \frac{1}{n}(MS_D - MS_W)$$

$$\widehat{\sigma}_S^2 = \frac{1}{dn}(MS_S - MS_D)$$

There are three ways you can estimate the heritability using the three variance components, which are listed below,

(1) Paternal half-sibs

$$\widehat{h}^2 = \frac{4\widehat{\sigma}_S^2}{\widehat{\sigma}_S^2 + \widehat{\sigma}_D^2 + \widehat{\sigma}_W^2}$$

(2) Maternal half-sibs

$$\widehat{h}^2 = \frac{4\widehat{\sigma}_D^2}{\widehat{\sigma}_S^2 + \widehat{\sigma}_D^2 + \widehat{\sigma}_W^2}$$

(3) Full-sibs

$$\widehat{h}^2 = \frac{2(\widehat{\sigma}_S^2 + \widehat{\sigma}_D^2)}{\widehat{\sigma}_S^2 + \widehat{\sigma}_D^2 + \widehat{\sigma}_W^2}$$

**Table 10.14** The ANOVA table for the nested mating design

| Source of variation. | Df | SS | MS | E(MS) |
|---|---|---|---|---|
| Between sires | $df_S = s - 1$ | $SS_S$ | $MS_S = SS_S/df_S$ | $\sigma_W^2 + n\sigma_D^2 + dn\sigma_S^2$ |
| Between dams (sires) | $df_D = s(d - 1)$ | $SS_D$ | $MS_D = SS_D/df_D$ | $\sigma_W^2 + n\sigma_D^2$ |
| Between siblings (dams) | $df_W = sd(n - 1)$ | $SS_W$ | $MS_W = SS_W/df_W$ | $\sigma_W^2$ |

In real data analysis, all three methods should be tried. If you believe that maternal effect can be very strong to bias the estimated $\sigma_D^2$, then the paternal half-sib approach should be used. Under the assumption of no common environment effect for the full-sib families, the third method (full-sibs) should be taken. This method should have the minimum estimation error.

Although computer programs are required to perform the analysis of variances, as a researcher or teacher, we should understand how the sum of squares are calculated. Below are the computing formulas for different sums of squares (SS) in the ANOVA table (Table 10.14). First, we need to define the following means:

$$\bar{y}_{\cdots} = \frac{1}{sdn} \sum_{i=1}^{s} \sum_{j=1}^{d} \sum_{k=1}^{n} y_{ijk}$$

$$\bar{y}_{i\cdot\cdot} = \frac{1}{dn} \sum_{j=1}^{d} \sum_{k=1}^{n} y_{ijk}, \forall i = 1, \cdots, s$$

$$\bar{y}_{ij\cdot} = \frac{1}{n} \sum_{k=1}^{n} y_{ijk}, \forall i = 1, \cdots, s \text{ and } j = 1, \cdots, d$$

We then use the following computing formulas to calculate the sums of squares:

$$SS_S = dn \sum_{i=1}^{s} \bar{y}_{i\cdot\cdot}^2 - sdn \bar{y}_{\cdots}^2$$

$$SS_D = n \sum_{i=1}^{s} \sum_{j=1}^{d} \bar{y}_{ij\cdot}^2 - dn \sum_{i=1}^{s} \bar{y}_{i\cdot\cdot}^2$$

$$SS_W = \sum_{i=1}^{s} \sum_{j=1}^{d} \sum_{k=1}^{n} y_{ijk}^2 - n \sum_{i=1}^{s} \sum_{j=1}^{d} \bar{y}_{ij\cdot}^2$$

In real-life experiments, we often have unbalanced data. The number of sibs per dam often varies across different full-sib families. The number of dams mated with each sire also varies across the half-sib families. Let $s$ be the number of sires, $d_i$ be the number of dams mated with the $i$th sire and $n_{ij}$ be the number of siblings produced by the $j$th dam mated to the $i$th sire. The ANOVA table for the unbalanced data is given in Table 10.15. The estimated variance components are obtained via

$$\hat{\sigma}_W^2 = MS_W$$

$$\sigma_D^2 = \frac{1}{k_1}(MS_D - MS_W)$$

$$\sigma_S^2 = \frac{1}{k_3}[MS_S - (k_2/k_1)MS_D + (k_2/k_1 - 1)MS_W]$$

Recall that $n_{ij}$ is the number of full-siblings produced by the $j$th dam within the $i$th sire family. Let us define $n_{i\cdot} = \sum_{j=1}^{d_j} n_{ij}$ as the number of progeny of the $i$th sire and $n_{\cdot\cdot} = \sum_{i=1}^{s} \sum_{j=1}^{d_j} n_{ij}$ as the total number of progeny in the entire experiment. The coefficients involved in the E(MS) terms of the ANOVA table are

**Table 10.15** The ANOVA table for the nested mating design with unbalanced data

| Variation | Df | SS | MS | E(MS) |
|---|---|---|---|---|
| Between sires | $df_S = s - 1$ | $SS_S$ | $MS_S = SS_S/df_S$ | $\sigma_W^2 + k_2\sigma_D^2 + k_3\sigma_S^2$ |
| Between dams | $df_D = \sum_{i=1}^{s} (d_i - 1)$ | $SS_D$ | $MS_D = SS_D/df_D$ | $\sigma_W^2 + k_1\sigma_D^2$ |
| Between sibs | $df_W = \sum_{i=1}^{s} \sum_{j=1}^{d_i} (n_{ij} - 1)$ | $SS_W$ | $MS_W = SS_W/df_W$ | $\sigma_W^2$ |

$$k_1 = \frac{1}{df_D}\left(n.. - \sum_{i=1}^{s}\frac{1}{n_{i.}}\sum_{j=1}^{d_i}n_{ij}^2\right)$$

$$k_2 = \frac{1}{df_S}\left(\sum_{i=1}^{s}\frac{1}{n_{i.}}\sum_{j=1}^{d_i}n_{ij}^2 - \frac{1}{n..}\sum_{i=1}^{s}\sum_{j=1}^{d_i}n_{ij}^2\right)$$

$$k_3 = \frac{1}{df_S}\left(n.. - \frac{1}{n..}\sum_{i=1}^{s}n_{i.}^2\right)$$

Define the following means,

$$\bar{y}... = \frac{1}{n..}\sum_{i=1}^{s}\sum_{j=1}^{d_i}\sum_{k=1}^{n_{ij}}y_{ijk}$$

$$\bar{y}_{i..} = \frac{1}{n_{i.}}\sum_{j=1}^{d_i}\sum_{k=1}^{n_{ij}}y_{ijk}, \forall i-1,\cdots,s$$

$$\bar{y}_{ij.} = \frac{1}{n_{ij}}\sum_{k=1}^{n_{ij}}y_{ijk}, \forall i=1,\cdots,s \text{ and } j=1,\cdots,d_i$$

which allow us to calculate the sums of squares using

$$SS_S = \sum_{i=1}^{s}n_{i.}\bar{y}_{i..}^2 - n..\bar{y}_{...}^2$$

$$SS_D = \sum_{i=1}^{s}\sum_{j=1}^{d_i}n_{ij}\bar{y}_{ij.}^2 - \sum_{i=1}^{s}n_{i.}\bar{y}_{i..}^2$$

$$SS_W = \sum_{i=1}^{s}\sum_{j=1}^{d_i}\sum_{k=1}^{n_{ij}}y_{ijk}^2 - \sum_{i=1}^{s}\sum_{j=1}^{d_i}n_{ij}\bar{y}_{ij.}^2$$

The variance components estimated from the ANOVA tables are called the Henderson's Type 1 estimates (also called the moment methods)(Henderson 1985; Henderson 1986). Other methods are also available for estimation of variance components. They include the MIVQUE method (Rao 1971b; Rao 1971a), the maximum likelihood method (Hartley and Rao 1967; Harville 1977), the restricted maximum likelihood method (Patterson and Thompson 1971), and the Bayesian method (Harville 1974). The most powerful software package for mixed model analysis is the MIXED procedure in SAS (SAS Institute Inc. 2009), which will be introduced towards the end of this chapter.

## 10.5   Standard Error of an Estimated Heritability

Heritability is a parameter, like the repeatability we learned before. A parameter is a constant and thus there is no error. However, an estimated parameter is subject to error because it is obtained from a sample. The estimation error is jointly determined by the sample size and the method of estimation. When an estimated heritability is reported, it should be associated with a standard error, in a format like $\widehat{h}^2 \pm S_{\widehat{h}^2}$, where $S_{\widehat{h}^2}$ is the standard error.

### 10.5.1 Regression Method (Parent Vs. Progeny Regression)

We first address the standard error when the regression method is used to estimate heritability. Recall that the regression coefficient is estimated via

$$\widehat{b}_{YX} = \frac{\sum_{j=1}^{n}(X_j - \overline{X})(Y_j - \overline{Y})}{\sum_{j=1}^{n}(X_j - \overline{X})^2}$$

and

$$\widehat{\sigma}^2 = \frac{1}{n-2}\sum_{j=1}^{n}\left(Y_j - \widehat{b}_0 - X_j\widehat{b}_{YX}\right)^2$$

where $X_j$ is the phenotypic value from the parents of the $j$th family, and $Y_j$ is the phenotypic value measured from the progeny of the $j$th family. The variance of the estimated regression coefficient is

$$\text{var}\left(\widehat{b}_{YX}\right) = \frac{\widehat{\sigma}^2}{\sum_{j=1}^{n}(X_j - \overline{X})^2}$$

The standard error of $\widehat{b}_{YX}$ is

$$S_{\widehat{b}_{YX}} = \sqrt{\text{var}\left(\widehat{b}_{YX}\right)} = \frac{\widehat{\sigma}}{\sqrt{\sum_{j=1}^{n}(X_j - \overline{X})^2}}$$

Let $r = 2$ if $X$ is the phenotypic value of mid-parent and $r = 1$ if $X$ is the single parent value. The relationship between the heritability and the regression coefficient is

$$\widehat{h}^2 = \frac{2}{r}\widehat{b}_{YX}$$

As a result,

$$\text{var}\left(\widehat{h}^2\right) = \frac{4}{r^2}\text{var}\left(\widehat{b}_{YX}\right)$$

We now evaluate the theoretical value of $\text{var}\left(\widehat{h}^2\right)$ to compare the precisions of the four regression methods. First, let us assume that $n$ is sufficiently large so that

$$\sum_{j=1}^{n}(X_j - \overline{X})^2 = n\text{var}(X) = nV_P/r$$

Secondly, let $\widehat{\sigma}^2 = V_E/m$, where $m$ is the number of siblings per family with $m = 1$ being the special case for a single progeny per family. Therefore,

$$\text{var}\left(\widehat{h}^2\right) = \frac{4}{r^2}\text{var}\left(\widehat{b}_{YX}\right) = \frac{4}{r^2}\frac{\widehat{\sigma}^2}{\sum_{j=1}^{n}(X_j - \overline{X})^2} = \frac{4}{r^2}\frac{V_E/m}{nV_P/r} = \frac{4}{nmr}\frac{V_E}{V_P}$$

where $V_E/V_P = 1 - V_A/V_P = 1 - h^2$. This leads to

$$\text{var}\left(h^2\right) = \frac{4}{nmr}\frac{V_E}{V_P} = \frac{4}{nmr}\left(1 - h^2\right)$$

It is clear that the variance decreases as the heritability increases. So, low heritability is more difficult to estimate. The variance is inversely proportional to the sample size ($n$), meaning that large samples are required if a high precision is preferred. Using multiple progeny ($m$) per family can decrease the variance and thus increase the precision. Finally, mid-parent ($r = 2$) data can improve the precision of heritability estimation.

## 10.5.2  Analysis of Variances (Sib Analysis)

The estimated heritability is proportional to the intra-class correlation, which is expressed as variance ratio. As a ratio of estimated variance components, there is no analytical expression for the variance of a ratio. Therefore, some approximate methods must be adopted here. One powerful approximation is the delta method (Doob 1935; Dorfman 1938). We now show the delta method to calculate the variance of an estimated heritability as an intra-class correlation. We first introduce the variance of an estimated heritability using the nested mating design as an example. Recall that the estimated heritability is

$$\widehat{h}^2 = \frac{2(\widehat{\sigma}_S^2 + \widehat{\sigma}_D^2)}{\widehat{\sigma}_S^2 + \widehat{\sigma}_D^2 + \widehat{\sigma}_W^2}$$

Substituting the estimated variance components by linear function of all mean squares (MS), we have

$$\widehat{h}^2 = \frac{X}{Y}$$

where $X$ and $Y$ are defined as linear functions of all the mean squares in the ANOVA table (Table 10.14). The numerator is

$$X = 2\left[\frac{1}{dn}(MS_S - MS_D) + \frac{1}{n}(MS_D - MS_W)\right]$$
$$= \frac{2}{dn}MS_S + \frac{2(d-1)}{dn}MS_D - \frac{2}{n}MS_W$$
$$= a_S MS_S + a_D MS_D + a_W MS_W$$

where $a_S = 2/(dn)$, $a_D = 2(d-1)/(dn)$ and $a_W = 2/n$. The denominator is

$$Y = \frac{1}{dn}(MS_S - MS_D) + \frac{1}{n}(MS_D - MS_W) + MS_W$$
$$= \frac{1}{dn}MS_S + \frac{d-1}{dn}MS_D + \frac{n-1}{n}MS_W$$
$$= b_S MS_S + b_D MS_D + b_W MS_W$$

where $b_S = 1/(dn)$, $b_D = (d-1)/(dn)$ and $b_W = (n-1)/n$. The variance of $X$ is

$$\mathrm{var}(X) = a_S^2 \mathrm{var}(MS_S) + a_D^2 \mathrm{var}(MS_D) + a_W^2 \mathrm{var}(MS_W)$$
$$= a_S^2 \frac{2(MS_S)^2}{df_S + 2} + a_D^2 \frac{2(MS_D)^2}{df_D + 2} + a_W^2 \frac{2(MS_W)^2}{df_W + 2}$$

The variance of $Y$ is

$$\mathrm{var}(Y) = b_S^2 \mathrm{var}(MS_S) + b_D^2 \mathrm{var}(MS_D) + b_W^2 \mathrm{var}(MS_W)$$
$$= b_S^2 \frac{2(MS_S)^2}{df_S + 2} + b_D^2 \frac{2(MS_D)^2}{df_D + 2} + b_W^2 \frac{2(MS_W)^2}{df_W + 2}$$

The covariance between $X$ and $Y$ is

$$\mathrm{cov}(X, Y) = a_S b_S \mathrm{var}(MS_S) + a_D b_D \mathrm{var}(MS_D) + a_W b_W \mathrm{var}(MS_W)$$
$$= a_S b_S \frac{2(MS_S)^2}{df_S + 2} + a_D b_D \frac{2(MS_D)^2}{df_D + 2} + a_W b_W \frac{2(MS_W)^2}{df_W + 2}$$

Once the variances and covariance for $X$ and $Y$ are defined, the variance of the estimated heritability is defined as follows:

$$\text{var}\left(\hat{h}^2\right) = \text{var}\left(\frac{X}{Y}\right) \approx \left(\frac{X}{Y}\right)^2 \left[\frac{\text{var}(X)}{X^2} - \frac{2\text{cov}(X,Y)}{XY} + \frac{\text{var}(Y)}{Y^2}\right] \tag{10.3}$$

Equation (10.3) is called the delta method for the variance of estimated heritability (Lynch and Walsh 1998). We now provide a simple proof for Eq. (10.3). Let $X$ and $Y$ be two variables with variance-covariance matrix

$$\Sigma = \begin{bmatrix} \text{var}(X) & \text{cov}(X,Y) \\ \text{cov}(Y,X) & \text{var}(Y) \end{bmatrix}$$

Define a new variable $Z$ as

$$Z = \frac{X}{Y}$$

Let the partial derivatives of $Z$ with respect to $X$ and $Y$ be

$$\nabla = \begin{bmatrix} \frac{\partial Z}{\partial X} & \frac{\partial Z}{\partial Y} \end{bmatrix} = \begin{bmatrix} \frac{1}{Y} & -\frac{X}{Y^2} \end{bmatrix}$$

The delta method for variance of $Z$ is $\text{var}(Z) \approx \nabla \Sigma \nabla^T$ (Doob 1935; Dorfman 1938), where

$$\nabla \Sigma \nabla^T = \begin{bmatrix} \frac{1}{Y} & -\frac{X}{Y^2} \end{bmatrix} \begin{bmatrix} \text{var}(X) & \text{cov}(X,Y) \\ \text{cov}(Y,X) & \text{var}(Y) \end{bmatrix} \begin{bmatrix} \frac{1}{Y} \\ -\frac{X}{Y^2} \end{bmatrix}$$

$$= \left(\frac{1}{Y}\right)^2 \text{var}(X) - \frac{X}{Y^2}\frac{1}{Y}\text{cov}(X,Y) - \frac{1}{Y}\frac{X}{Y^2}\text{cov}(Y,X) + \left(-\frac{X}{Y^2}\right)^2 \text{var}(Y)$$

$$= \left(\frac{X}{Y}\right)^2 \left[\frac{\text{var}(X)}{X^2} - \frac{2\text{cov}(X,Y)}{XY} + \frac{\text{var}(Y)}{Y^2}\right]$$

## 10.6 Examples

Table 10.16 shows a sample data of 15 families with two parents and four offspring in each family. The data will be analyzed in both the regression analysis and the analysis of variances for estimating heritability.

### 10.6.1 Regression Analysis

Let $X$ be the phenotypic value of the male parent (Sire, column 2 in Table 10.16) and $Y$ be the phenotypic value of the first sibling (Sib1, column 4 in Table 10.16). We want to estimate the heritability of body weight of mice using the regression analysis of $Y$ on $X$. Below is the SAS code to read the data and perform the regression analysis.

```
filename aa "data-10-5.xlsx";
proc import datafile=aa out=one dbms=xlsx replace;
run;
ods graphics off;
proc reg data=one;
    model sib1 = sire;
quit;
```

**Table 10.16** Body weights (gram) of mice for both parents and progeny of 15 full-sib families of laboratory mice

| Family | Sire | Dam | Sib1 | Sib2 | Sib3 | Sib4 |
|--------|------|------|------|------|------|------|
| 1 | 14.7 | 21.4 | 19.9 | 21.1 | 21.0 | 19.2 |
| 2 | 19.2 | 18.1 | 18.7 | 20.8 | 19.6 | 21.4 |
| 3 | 13.5 | 15.7 | 20.8 | 15.5 | 11.5 | 12.7 |
| 4 | 29.4 | 27.0 | 26.1 | 25.0 | 22.0 | 36.5 |
| 5 | 15.4 | 25.3 | 22.5 | 17.8 | 23.7 | 15.7 |
| 6 | 30.5 | 21.7 | 24.9 | 23.0 | 17.9 | 25.7 |
| 7 | 30.4 | 29.8 | 15.8 | 19.3 | 26.7 | 24.5 |
| 8 | 24.4 | 5.0 | 16.4 | 20.2 | 16.2 | 21.5 |
| 9 | 25.7 | 13.4 | 19.9 | 16.9 | 19.6 | 29.8 |
| 10 | 17.3 | 16.8 | 22.9 | 5.9 | 13.4 | 15.2 |
| 11 | 25.6 | 17.7 | 21.7 | 22.5 | 15.3 | 15.3 |
| 12 | 9.7 | 22.2 | 20.9 | 22.5 | 20.8 | 18.2 |
| 13 | 23.9 | 20.3 | 31.5 | 32.2 | 19.6 | 18.6 |
| 14 | 26.7 | 20.4 | 22.9 | 20.0 | 9.2 | 17.0 |
| 15 | 17.2 | 20.8 | 17.6 | 20.5 | 33.1 | 23.8 |

**Table 10.17** Output of the REG procedure in SAS for the data in Table 10.16

| Parameter estimates | | | | | | |
|---------|-------|-----|--------------------|----------------|---------|---------|
| Variable | Label | DF | Parameter estimate | Standard error | t value | Pr > |t| |
| Intercept | Intercept | 1 | 19.37852 | 3.69464 | 5.25 | 0.0002 |
| Sire | Sire | 1 | 0.09834 | 0.16406 | 0.60 | 0.5592 |

The output is given in Table 10.17, where you can find the estimated regression coefficient and the standard error of the estimate. The t-test shows that it is not significantly different from zero due to the small sample size. The regression coefficient is $\widehat{b}_{YX} = 0.09834$ and thus

$$\widehat{h}^2 = 2\widehat{b}_{YX} = 2 \times 0.09834 = 0.19668$$

The standard error of the estimate is

$$S_{\widehat{h}^2} = 2S_{\widehat{b}_{YX}} = 2 \times 0.16406 = 0.32812$$

The estimated heritability should be reported as $0.1967 \pm 0.3281$. The standard error is so large compared to the estimate so that you can tell that the estimate should not be significant.

### 10.6.2 Analysis of Variances

For the same data listed in Table 10.16, we deleted the phenotypic values of the parents. The data then became the sib data with 15 families and four full-siblings per family. The sib data are rearranged in a format as illustrated in Table 10.18 (partial record) and stored in "data-10-4.xlsx". The SAS code to read the data and perform analysis of variance is shown below.

```
filename aa "data-10-4.xlsx";
proc import datafile=aa out=one dbms=xlsx replace;
run;
proc mixed data=one method=type1;
   class family;
   model weight = ;
   random family;
run;
```

**Table 10.18** Partial records of reformatted body weights (gram) of mice measured from 15 full-sib families of laboratory mice

| Family | Sib | Weight |
|---|---|---|
| 1 | 1 | 19.9 |
| 1 | 2 | 21.1 |
| 1 | 3 | 21 |
| 1 | 4 | 19.2 |
| 2 | 1 | 18.7 |
| 2 | 2 | 20.8 |
| 2 | 3 | 19.6 |
| 2 | 4 | 21.4 |
| 3 | 1 | 20.8 |
| 3 | 2 | 15.5 |
| 3 | 3 | 11.5 |
| 3 | 4 | 12.7 |
| 4 | 1 | 26.1 |
| 4 | 2 | 25 |
| 4 | 3 | 22 |
| 4 | 4 | 36.5 |
| 5 | 1 | 22.5 |
| 5 | 2 | 17.8 |
| 5 | 3 | 23.7 |
| 5 | 4 | 15.7 |
| . | . | . |
| . | . | . |
| . | . | . |
| . | . | . |
| 15 | 1 | 17.6 |
| 15 | 2 | 20.5 |
| 15 | 3 | 33.1 |
| 15 | 4 | 23.8 |

**Table 10.19** The ANOVA table of full-sib analysis for the sample data

| Source | Df | SS | MS | E(MS) |
|---|---|---|---|---|
| Between-family | 15−1 = 14 | 701.61 | 701.61/14 = 50.11 | $\sigma_W^2 + m\sigma_B^2$ |
| Within-family | 15(4−1) = 45 | 1054.55 | 1054.55/45 = 23.43 | $\sigma_W^2$ |

The output is shown in Table 10.19 where $m = 4$ and $n = 15$ (the data are balanced). The estimated variance components are

$$\widehat{\sigma}_W^2 = MS_W = 23.4344$$

and

$$\widehat{\sigma}_B^2 = \frac{1}{m}(MS_B - MS_W) = \frac{1}{4}(50.11 - 23.43) = 6.67$$

Therefore,

$$\widehat{\rho}_{FS} = \frac{\widehat{\sigma}_B^2}{\widehat{\sigma}_B^2 + \widehat{\sigma}_W^2} = \frac{6.67}{6.67 + 23.4344} = 0.2216$$

The estimated heritability is

$$\widehat{h}^2 = 2\widehat{\rho}_{FS} = 2 \times 0.2216 = 0.4432$$

We now calculate the standard error of the estimated heritability using the delta method introduced earlier. The estimated intra-class correlation is

$$\widehat{\rho}_{FS} = \frac{\widehat{\sigma}_B^2}{\widehat{\sigma}_B^2 + \widehat{\sigma}_W^2} = \frac{\frac{1}{4}MS_B - \frac{1}{4}MS_W}{\frac{1}{4}MS_B - \frac{1}{4}MS_B + MS_W} = \frac{\frac{1}{4}MS_B - \frac{1}{4}MS_W}{\frac{1}{4}MS_B + \frac{3}{4}MS_W} = \frac{X}{Y}$$

The variances of the MS are

$$\text{var}(MS_B) = \frac{2MS_B^2}{df_B + 2} = \frac{2 \times 50.11^2}{15 + 2} = 313.8765$$

$$\text{var}(MS_W) = \frac{2MS_W^2}{df_W + 2} = \frac{2 \times 23.43^2}{45 + 2} = 23.36021$$

The variances $X$ is

$$\begin{aligned}
\text{var}(X) &= \left(\frac{1}{4}\right)^2 \text{var}(MS_G) + \left(-\frac{1}{4}\right)^2 \text{var}(MS_E) \\
&= \left(\frac{1}{4}\right)^2 \times 313.8765 + \left(-\frac{1}{4}\right)^2 \times 23.36021 \\
&= 21.0773
\end{aligned}$$

The variance of $Y$ is

$$\begin{aligned}
\text{var}(Y) &= \left(\frac{1}{4}\right)^2 \text{var}(MS_G) + \left(1 - \frac{1}{4}\right)^2 \text{var}(MS_E) \\
&= \left(\frac{1}{4}\right)^2 \times 313.8765 + \left(1 - \frac{1}{4}\right)^2 \times 23.36021 \\
&= 32.7574
\end{aligned}$$

The covariance between $X$ and $Y$ is

$$\begin{aligned}
\text{cov}(X, Y) &= \left(\frac{1}{4}\right)^2 \text{var}(MS_G) + \left(-\frac{1}{4}\right)\left(1 - \frac{1}{4}\right) \text{var}(MS_E) \\
&= \left(\frac{1}{4}\right)^2 \times 313.8765 - \left(\frac{1}{4}\right) \times \left(1 - \frac{1}{4}\right) \times 23.36021 \\
&= 23.99732
\end{aligned}$$

The variance of the estimated intra-class correlation is

$$\text{var}(\widehat{\rho}_{FS}) = \left(\frac{X}{Y}\right)^2 \left(\frac{\text{var}(X)}{X^2} - \frac{2\text{cov}(X, Y)}{XY} + \frac{\text{var}(Y)}{Y^2}\right)$$

$$= \left(\frac{6.67}{30.1}\right)^2 \left(\frac{21.0773}{6.67^2} - \frac{2 \times 23.99732}{6.67 \times 30.1} + \frac{32.7574}{30.1^2}\right)$$

$$= 0.01330059$$

The standard error for the estimated heritability is

$$S_{\widehat{h}^2} = \sqrt{4\text{var}(\widehat{\rho}_{FS})} = \sqrt{4 \times 0.01330059} = 0.2306563$$

The estimated heritability should be reported as $0.4432 \pm 0.2306563$.

### 10.6.3  A Nested Mating Design

The following data were collected from a swine breeding experiment with 3 sires and 12 dams. The trait is the growth rate of piglets. We want to estimate the heritability of the growth rate. This is an unbalanced nested design of experiment. Table 10.20 shows the original data (盛志廉 and 陈瑶生 1999). To analyze the data using the MIXED procedure in SAS, the data must be reformatted into a form like Table 10.21 (partial record). The reformatted data is stored in "data-10-5.csv". Here is the SAS code to read the data and analyze the data.

```
filename aa "data-10-5.csv";
proc import datafile=aa out=one dbms=csv replace;
run;
proc mixed data=one method=type1;
    class sire dam;
    model growth =;
    random sire dam(sire);
run;
/*or*/
proc varcomp data=one method=type1;
    class sire dam;
    model growth=sire dam(sire);
run;
```

**Table 10.20**  Daily growth rate of piglets (gram/day) collected from a nested mating design of a swine (big) breeding experiment (盛志廉 and 陈瑶生 1999)

| Sire | Dam | Sib1 | Sib2 | Sib3 | Sib4 | Sib5 | $n_{ij}$ | $n_{i.}$ |
|------|-----|------|------|------|------|------|-----|-----|
| 1 | 1 | 403 | 392 | 382 | 395 |     | 4 |    |
| 1 | 2 | 404 | 395 |     |     |     | 2 |    |
| 1 | 3 | 397 | 406 | 425 | 418 | 394 | 5 | 11 |
| 2 | 1 | 395 | 412 | 407 |     |     | 3 |    |
| 2 | 2 | 384 | 395 |     |     |     | 2 |    |
| 2 | 3 | 410 | 382 | 394 | 390 |     | 4 |    |
| 2 | 4 | 405 | 396 | 387 |     |     | 3 |    |
| 2 | 5 | 382 | 395 | 406 | 392 | 403 | 5 | 17 |
| 3 | 1 | 404 | 415 | 369 | 418 | 407 | 5 |    |
| 3 | 2 | 423 | 397 | 409 | 411 |     | 4 |    |
| 3 | 3 | 405 | 418 | 392 |     |     | 3 |    |
| 3 | 4 | 385 | 402 | 394 |     |     | 3 | 15 |

**Table 10.21**   Reformatted data for the growth rate of piglets

| Sire | Dam | Progeny | Growth |
|------|-----|---------|--------|
| 1 | 1 | 1 | 403 |
| 1 | 1 | 2 | 392 |
| 1 | 1 | 3 | 382 |
| 1 | 1 | 4 | 395 |
| 1 | 2 | 1 | 404 |
| 1 | 2 | 2 | 395 |
| 1 | 3 | 1 | 397 |
| 1 | 3 | 2 | 406 |
| 1 | 3 | 3 | 425 |
| 1 | 3 | 4 | 418 |
| 1 | 3 | 5 | 394 |
| . | . | . | . |
| . | . | . | . |
| . | . | . | . |
| 3 | 1 | 1 | 404 |
| 3 | 1 | 2 | 415 |
| 3 | 1 | 3 | 369 |
| 3 | 1 | 4 | 418 |
| 3 | 1 | 5 | 407 |
| 3 | 2 | 1 | 423 |
| 3 | 2 | 2 | 397 |
| 3 | 2 | 3 | 409 |
| 3 | 2 | 4 | 411 |
| 3 | 3 | 1 | 405 |
| 3 | 3 | 2 | 418 |
| 3 | 3 | 3 | 392 |
| 3 | 4 | 1 | 385 |
| 3 | 4 | 2 | 402 |
| 3 | 4 | 3 | 394 |

**Table 10.22**   The ANOVA Table for the nested mating design with unbalanced data

| Source | Df | SS | MS | E(MS) |
|--------|-----|-----|-----|-------|
| Sires | $df_S = 2$ | $SS_S = 419.0$ | $MS_S = 209.51$ | $\sigma_W^2 + 3.9232\sigma_D^2 + 14.1163\sigma_S^2$ |
| Dams | $df_D = 9$ | $SS_D = 1300.7$ | $MS_D = 144.52$ | $\sigma_W^2 + 3.4744\sigma_D^2$ |
| Sibs | $df_W = 31$ | $SS_W = 4492.7$ | $MS_W = 144.93$ | $\sigma_W^2$ |

The `dam(sire)` notation in the `random` statement indicates that dams are nested within sires. In other words, dam number 1 within sire 1 is different from dam 1 in sire 2. The ANOVA table resulted from the analysis is given in Table 10.22. The mixed procedure already delivers the estimated variance components (Table 10.23), but we show the detailed steps of estimating the variance components from the intermediate results of the ANOVA table any way.

In this nested mating design, $s = 3$ and thus $df_S = s - 1 = 3 - 1 = 2$. The numbers of dams mated to the first sire is $d_1 = 3$, to the second sire is $d_2 = 5$ and to the third sire is $d_3 = 4$. Therefore, the degree of freedom for the dam component is

$$df_D = \sum_{i=1}^{s} (d_i - 1) = (3 - 1) + (5 - 1) + (4 - 1) = 9$$

The degree of freedom for the sib component requires information for each family size ($n_{ij}$), which is given in Table 10.24. Therefore,

**Table 10.23** Estimated variance components from the MIXED procedure in SAS

| Covariance parameter estimates | |
| --- | --- |
| Cov Parm | Estimate |
| Sire | 4.6075 |
| Dam(sire) | −0.1174 |
| Residual | 144.93 |

**Table 10.24** Number of siblings per family

| Sire. | Dam | $n_{ij}$ | $n_{i.}$ |
| --- | --- | --- | --- |
| 1 | 1 | 4 | |
| 1 | 2 | 2 | |
| 1 | 3 | 5 | 11 |
| 2 | 1 | 3 | |
| 2 | 2 | 2 | |
| 2 | 3 | 4 | |
| 2 | 4 | 3 | |
| 2 | 5 | 5 | 17 |
| 3 | 1 | 5 | |
| 3 | 2 | 4 | |
| 3 | 3 | 3 | |
| 3 | 4 | 3 | 15 |

$$df_W = \sum_{j=1}^{3} \sum_{j=1}^{d_i} (n_{ij} - 1) = (4 - 1) + (2 - 1) + (5 - 1) + (3 - 1) + \cdots + (3 - 1) = 31$$

The size of each sire family is given in Table 10.24 also (the last column), denoted by $n_{i.}$ for the $i$th sire. The total number of progeny in the experiment is $n_{..} = 43$. The following intermediate results are required to calculate the $k$ values.

$$\sum_{i=1}^{s} \frac{1}{n_{i.}} \sum_{j=1}^{d_i} n_{ij}^2 = \frac{1}{11}\left(4^2 + 2^2 + 5^2\right) + \frac{1}{17}\left(3^2 + 2^2 + 4^2 + 3^2 + 5^2\right) + \frac{1}{15}\left(5^2 + 4^2 + 3^2 + 3^2\right) = 11.73012$$

$$\frac{1}{n_{..}} \sum_{i=1}^{s} \sum_{j=1}^{d_i} n_{ij}^2 = \frac{1}{43}\left(4^2 + 2^2 + 5^2 + 3^2 + 2^2 + 4^2 + 3^2 + 5^2 + 5^2 + 4^2 + 3^2 + 3^2\right) = \frac{1}{43} \times 167 = 3.883721$$

$$\frac{1}{n_{..}} \sum_{i=1}^{s} n_{i.}^2 = \frac{1}{43}\left(11^2 + 17^2 + 15^2\right) = \frac{1}{43} \times 635 = 14.76744$$

Therefore, the three $k$ values are

$$k_1 = \frac{1}{df_D}\left(n_{..} - \sum_{i=1}^{s} \frac{1}{n_{i.}} \sum_{j=1}^{d_i} n_{ij}^2\right) = \frac{1}{9}(43 - 11.73012) = 3.474431$$

$$k_2 = \frac{1}{df_S}\left(\sum_{i=1}^{s} \frac{1}{n_{i.}} \sum_{j=1}^{d_i} n_{ij}^2 - \frac{1}{n_{..}} \sum_{i=1}^{s} \sum_{j=1}^{d_i} n_{ij}^2\right) = \frac{1}{2}(11.73012 - 3.883721) = 3.9232$$

$$k_3 = \frac{1}{df_S}\left(n_{..} - \frac{1}{n_{..}} \sum_{i=1}^{s} n_{i.}^2\right) = \frac{1}{2}(43 - 14.76744) = 14.11628$$

The three estimated variance components are

$$\widehat{\sigma}_W^2 = MS_W = 144.93$$

$$\sigma_D^2 = \frac{1}{k_1}(MS_D - MS_W) = \frac{1}{3.4744}(144.52 - 144.93) = -0.1175 \approx 0$$

$$\sigma_S^2 = \frac{1}{k_3}[MS_S - (k_2/k_1)MS_D + (k_2/k_1 - 1)MS_W]$$

$$= \frac{1}{14.1163}[209.51 - (3.9232/3.4744) \times 144.52 + (3.9232/3.4744 - 1) \times 144.93]$$

$$= 4.6075$$

The negative estimate $\sigma_D^2$ must be set to zero because variance cannot be negative. The estimated heritability via the three methods are.

Paternal half-sib:  $\widehat{h}^2 = \frac{4\widehat{\sigma}_S^2}{\widehat{\sigma}_S^2 + \widehat{\sigma}_D^2 + \widehat{\sigma}_W^2} = \frac{4 \times 4.6075}{144.93 + 0 + 4.6075} = \frac{4 \times 4.6075}{149.5375} = 0.123247$

Maternal half-sib:  $\widehat{h}^2 = \frac{4\widehat{\sigma}_D^2}{\widehat{\sigma}_S^2 + \widehat{\sigma}_D^2 + \widehat{\sigma}_W^2} = \frac{4 \times 0}{144.93 + 0 + 4.6075} = \frac{4 \times 0}{149.5375} = 0$

Full-sib:  $\widehat{h}^2 = \frac{2\left(\widehat{\sigma}_S^2 + \widehat{\sigma}_D^2\right)}{\widehat{\sigma}_S^2 + \widehat{\sigma}_D^2 + \widehat{\sigma}_W^2} = \frac{2 \times (4.6075 + 0)}{144.93 + 0 + 4.6075} = \frac{2 \times 4.6075}{149.5375} = 0.061623$

The three heritability estimates are not consistent mainly because the sample size is too small.

Let us show the steps in calculating the standard error for the last estimated heritability (full-sib). Recall that the intra-class correlation for the full-sib under the nested design is $\widehat{\rho}_{FS} = X/Y = 0.03081167$, where

$$X = \frac{1}{k_3}[MS_S - (k_2/k_1)MS_D + (k_2/k_1 - 1)MS_W] + \frac{1}{k_1}(MS_D - MS_W)$$

$$= \frac{1}{k_3}MS_S + \frac{k_3 - k_2}{k_1 k_3}MS_D + \frac{k_2 - k_1 - k_3}{k_1 k_3}MS_W$$

$$= 0.07084019 MS_S + 0.20782676 MS_D - 0.27866696 MS_W$$

and

$$Y = \frac{1}{k_3}[MS_S - (k_2/k_1)MS_D + (k_2/k_1 - 1)MS_W] + \frac{1}{k_1}(MS_D - MS_W) + MS_W$$

$$= \frac{1}{k_3}MS_S + \frac{k_3 - k_2}{k_1 k_3}MS_D + \frac{k_2 - k_1 - k_3 + k_1 k_3}{k_1 k_3}MS_W$$

$$= 0.07084019 MS_S + 0.20782676 MS_D + 0.72133304 MS_W$$

The variance of $X$ is

$$\text{var}(X) = 0.0708^2 \text{var}(MS_S) + 0.2078^2 MS_D + (-0.2787)^2 \text{var}(MS_W)$$

$$= 0.0708^2 \frac{2(MS_S)^2}{df_S + 2} + 0.2078^2 \frac{2(MS_D)^2}{df_D + 2} + (-0.2787)^2 \frac{2(MS_W)^2}{df_W + 2}$$

$$= 0.0708^2 \times 10973.61 + 0.2078^2 \times 26.27636 + (-0.2787)^2 \times 8.783636$$

$$= 111.9555$$

The variance of $Y$ is

$$\mathrm{var}(Y) = 0.0708^2 \mathrm{var}(MS_S) + 0.2078^2 MS_D + 0.721333^2 \mathrm{var}(MS_W)$$

$$= 0.0708^2 \frac{2(MS_S)^2}{df_S + 2} + 0.2078^2 \frac{2(MS_D)^2}{df_D + 2} + 0.721333^2 \frac{2(MS_W)^2}{df_W + 2}$$

$$= 0.0708^2 \times 10973.61 + 0.2078^2 \times 26.27636 + 0.721333^2 \times 8.783636$$

$$= 115.8437$$

The covariance between $X$ and $Y$ is

$$\mathrm{cov}(X, Y) = 0.0708^2 \mathrm{var}(MS_S) + 0.2078^2 MS_D + (-0.2787) \times 0.7213 \mathrm{var}(MS_W)$$

$$= 0.0708^2 \frac{2(MS_S)^2}{df_S + 2} + 0.2078^2 \frac{2(MS_D)^2}{df_D + 2} - 0.2787 \times 0.7213 \frac{2(MS_W)^2}{df_W + 2}$$

$$= 0.0708^2 \times 10973.61 + 0.2078^2 \times 26.27636 - 0.2787 \times 0.7213 \times 8.783636$$

$$= 109.5078$$

The variance of the estimated intra-class correlation is

$$\mathrm{var}(\widehat{\rho}_{FS}) = \left(\frac{X}{Y}\right)^2 \left(\frac{\mathrm{var}(X)}{X^2} - \frac{2\mathrm{cov}(X, Y)}{XY} + \frac{\mathrm{var}(Y)}{Y^2}\right)$$

$$= \left(\frac{4.489651}{149.4197}\right)^2 \left(\frac{111.9555}{4.489651^2} - \frac{2 \times 109.5078}{4.489651 \times 149.4197} + \frac{115.8437}{149.4197^2}\right)$$

$$= 0.004724454$$

The standard error for the estimated heritability is

$$S_{\widehat{h}^2} = \sqrt{4\mathrm{var}(\widehat{\rho}_{FS})} = \sqrt{4 \times 0.004724454} = 0.09720549$$

The estimated heritability should be reported as $0.061623 \pm 0.09720549$.

## References

Bulmer MG. The effect of selection on genetic variability. Am Nat. 1971;105:201–11.

Doob JL. The limiting distributions of certain statistics. Ann Math Stat. 1935;6:160–9.

Dorfman R. A note on the δ-method for finding variance formulae. Biomet Bull. 1938;1:129–37.

Falconer DS, Mackay TFC. Introduction to quantitative genetics. Harlow, UK: Addison Wesley Longman; 1996.

Hartley HO, Rao JNK. Maximum-likelihood estimation for the mixed analysis of variance model. Biometrika. 1967;54:93–108.

Harville DA. Bayesian inference for variance components using only error contrasts. Biometrika. 1974;61:383–5.

Harville DA. Maximum likelihood approaches to variance component estimation and to related problems. J Am Stat Assoc. 1977;72:320–38.

Henderson CR. MIVQUE and REML estimation of additive and nonadditive genetic variances. J Anim Sci. 1985;61:113–21.

Henderson CR. Recent developments in variance and covariance estimations. J Anim Sci. 1986;63:208–16.

Lynch M, Walsh B. Genetics and analysis of quantitative traits. Sunderland, MA: Sinauer Associates, Inc.; 1998.

Patterson HD, Thompson R. Recovery of inter-block information when block sizes are unequal. Biometrika. 1971;58:545–54.

Rao CR. Estimation of variance and covariance components-MINQUE theory. J Multivar Anal. 1971a;1:257–75.

Rao CR. Minimum variance quadratic unbiased estimation of variance components. J Multivar Anal. 1971b;1:445–56.

SAS Institute Inc. *SAS/STAT: Users' Guide, Version 9.3*. SAS Institute Inc., Cary, NC. 2009.

盛志廉, 陈瑶生. 1999. 数量遗传学. 科学出版社, 北京.

The coefficient of relationship between two relatives, previously denoted by $a_{ij}$, describes the closeness between two relatives, individuals $i$ and $j$. To fully understand the actual meaning of $a_{ij}$, we must first understand some fundamentally important concepts in population and quantitative genetics that describe the relationship between two genes (alleles).

## 11.1 Allelic Variance

First, let us define allelic variance and its relationship to the additive genetic variance. As described earlier, additive genetic variance is the variance of breeding values. However, the breeding value is made of two allelic effects, one from the male parent and the other from the female parent. Each allele is a haploid, i.e., a single copy of a gene. The variance of the allelic effect is called the allelic variance. To demonstrate the relationship between the allelic variance and the additive genetic variance, we must use the following model,

$$y = \mu + \alpha^s + \alpha^d + \delta + \varepsilon$$

where $\alpha^s$ and $\alpha^d$ are the effects of the paternal allele (allele from the sire, i.e., father) and the maternal allele (allele from the dam, i.e., mother), respectively, $\delta$ is the interaction (dominance) effect between the two alleles and $\varepsilon$ is the residual error. Each allelic effect is assumed to be sampled from a normal distribution with mean zero and variance $\sigma_a^2$, i.e., $\alpha^s \sim N(0, \sigma_a^2)$ and $\alpha^d \sim N(0, \sigma_a^2)$. We may use a genetic notation, $V_a = \sigma_a^2$, for the allelic variance. The dominance effect is assumed to be sampled from a $\delta \sim N(0, \sigma_\delta^2)$ distribution. Similarly, we may use the genetic notations $V_D = \sigma_\delta^2$ and $V_E = \sigma_\varepsilon^2$. Under the assumption that the parents are genetically unrelated, we have

$$\text{var}(y) = \text{var}(\alpha^s) + \text{var}(\alpha^d) + \text{var}(\delta) + \text{var}(\varepsilon)$$

The same model can be written in two different forms,

$$\begin{cases} \sigma_y^2 = 2\sigma_a^2 + \sigma_\delta^2 + \sigma_\varepsilon^2 \\ V_P = 2V_a + V_D + V_E \end{cases}$$

The upper one is the statistical model and the lower one is the genetic model. We also know that $V_P = V_A + V_D + V_E$. It is now obvious that

$$V_A = 2V_a$$

Verbally, it says that *the additive genetic variance equals twice the allelic variance.*

## 11.2    Genetic Covariance Between Relatives

Define the genetic models of two relatives by

$$
\begin{cases}
G_1 = A_1 + D_1 = \alpha_1^s + \alpha_1^d + \delta_1 \\
G_2 = A_2 + D_2 = \alpha_2^s + \alpha_2^d + \delta_2
\end{cases}
$$

The genetic covariance between the two individuals is

$$
\begin{aligned}
\mathrm{cov}(G_1, G_2) &= \mathrm{cov}\left(\alpha_1^s + \alpha_1^d + \delta_1, \alpha_2^s + \alpha_2^d + \delta_2\right) \\
&= \mathrm{cov}\left(\alpha_1^s, \alpha_2^s\right) + \mathrm{cov}\left(\alpha_1^s, \alpha_2^d\right) + \mathrm{cov}\left(\alpha_1^d, \alpha_2^s\right) + \mathrm{cov}\left(\alpha_1^d, \alpha_2^d\right) + \mathrm{cov}(\delta_1, \delta_2) \\
&= \left[\Pr\left(\alpha_1^s \equiv \alpha_2^s\right) + \Pr\left(\alpha_1^s \equiv \alpha_2^d\right) + \Pr\left(\alpha_1^d \equiv \alpha_2^s\right) + \Pr\left(\alpha_1^d \equiv \alpha_2^d\right)\right] V_a \\
&\quad + \left[\Pr\left(\alpha_1^s \equiv \alpha_2^s\right)\Pr\left(\alpha_1^d \equiv \alpha_2^d\right) + \Pr\left(\alpha_1^s \equiv \alpha_2^d\right)\Pr\left(\alpha_1^d \equiv \alpha_2^s\right)\right. \\
&\quad \left. - \Pr\left(\alpha_1^s \equiv \alpha_2^s\right)\Pr\left(\alpha_1^d \equiv \alpha_2^d\right)\Pr\left(\alpha_1^d \equiv \alpha_2^s\right)\Pr\left(\alpha_1^s \equiv \alpha_2^d\right)\right] V_D \\
&= 4\theta_{12} V_a + d_{12} V_D \\
&= 2\theta_{12} V_A + d_{12} V_D
\end{aligned}
$$

We now interpret some new notations involved in the above equation. The symbol $\alpha_1^s \equiv \alpha_2^d$ indicates the event that the paternal allele of individual 1 is identical-by-descent (IBD) to the maternal allele of individual 2. $\Pr\left(\alpha_1^s \equiv \alpha_2^d\right)$ is the probability that $\alpha_1^s \equiv \alpha_2^d$.

$$
\theta_{12} = \frac{1}{4}\left[\Pr\left(\alpha_1^s \equiv \alpha_2^s\right) + \Pr\left(\alpha_1^s \equiv \alpha_2^d\right) + \Pr\left(\alpha_1^d \equiv \alpha_2^s\right) + \Pr\left(\alpha_1^d \equiv \alpha_2^d\right)\right]
$$

is called the coancestry (coefficient) between individuals 1 and 2.

$$
\begin{aligned}
d_{12} &= \Pr\left(\alpha_1^s \equiv \alpha_2^s\right)\Pr\left(\alpha_1^d \equiv \alpha_2^d\right) + \Pr\left(\alpha_1^s \equiv \alpha_2^d\right)\Pr\left(\alpha_1^d \equiv \alpha_2^s\right) \\
&\quad - \Pr\left(\alpha_1^s \equiv \alpha_2^s\right)\Pr\left(\alpha_1^d \equiv \alpha_2^d\right)\Pr\left(\alpha_1^d \equiv \alpha_2^s\right)\Pr\left(\alpha_1^s \equiv \alpha_2^d\right)
\end{aligned}
$$

is called the fraternity. The coefficient of additive relationship is defined as twice the coancestry, i.e.,

$$
a_{12} = 2\theta_{12}
$$

Since $0 \leq \theta_{12} \leq 1$, we have $0 \leq a_{12} \leq 2$, which is different from what was previously defined. Therefore, the covariance between the additive genetic effects of two relatives can be as much as twice the additive genetic variance. The fraternity still takes value within the normal range, i.e., $0 \leq d_{12} \leq 1$. In general, the additive relationship between individuals $i$ and $j$ is

$$
a_{ij} = 2\theta_{ij}
$$

## 11.3    Genetic Covariance of an Individual with Itself

It is known that the covariance of a variable with itself is always equal to the variance of the variable. However, this is not necessarily so for the covariance between the breeding value of an individual and the breeding value of the same individual. Breeding value of an individual is the sum of the two allelic values from both parents. The breeding value is a composite term, not a simple variable.

$$\begin{aligned}
\mathrm{cov}(G_i, G_i) &= \mathrm{cov}\big(\alpha_i^s + \alpha_i^d + \delta_i, \alpha_i^s + \alpha_i^d + \delta_i\big) \\
&= \mathrm{cov}\big(\alpha_i^s, \alpha_i^s\big) + \mathrm{cov}\big(\alpha_i^s, \alpha_i^d\big) + \mathrm{cov}\big(\alpha_i^d, \alpha_i^s\big) + \mathrm{cov}\big(\alpha_i^d, \alpha_i^d\big) + \mathrm{cov}(\delta_i, \delta_i) \\
&= \sigma_a^2 + \mathrm{Pr}\big(\alpha_i^s \equiv \alpha_i^d\big)\sigma_a^2 + \mathrm{Pr}\big(\alpha_i^d \equiv \alpha_i^s\big)\sigma_a^2 + \sigma_a^2 + \sigma_\delta^2 \\
&= \big[2 + 2\mathrm{Pr}\big(\alpha_i^s \equiv \alpha_i^d\big)\big]\sigma_a^2 + \sigma_\delta^2 \\
&= \big[2 + 2\mathrm{Pr}\big(\alpha_i^s \equiv \alpha_i^d\big)\big]V_a + V_D \\
&= \big[1 + \mathrm{Pr}\big(\alpha_i^s \equiv \alpha_i^d\big)\big](2V_a) + V_D \\
&= (1 + f_i)V_A + V_D
\end{aligned}$$

where

$$f_i = \mathrm{Pr}\big(\alpha_i^s \equiv \alpha_i^d\big)$$

is called the inbreeding coefficient of individual $i$. The covariance of the genetic value of the same individual can be expressed in a general formula like

$$\mathrm{cov}(G_i, G_i) = a_{ii}V_A + d_{ii}V_D = (1 + f_i)V_A + V_D$$

Therefore, $a_{ii} = 1 + f_i$ and $d_{ii} = 1$. Because $a_{ii} = 2\theta_{ii}$, we have the definition of the coancestry of an individual with itself,

$$\theta_{ii} = \frac{1}{2}(1 + f_i)$$

## 11.4 Terminology

*Inbreeding*: The mating together of individuals that are related to each other by sharing common ancestry. It is a phenomenon of mating between relatives.

*Identity-by-descent*: Two genes that have originated from the replicate of a single gene in a "previous generation."

*Identity-by-state*: Two genes that have the same form (chemically), regardless of their origins. Please see Fig. 11.1 for the difference between IBD and IBS.

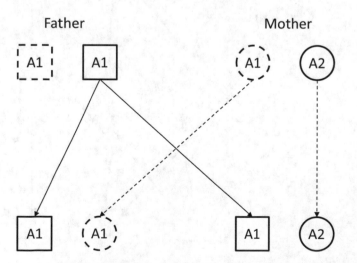

**Fig. 11.1** Difference between identity-by-descent (IBD) and identity-by-state (IBS). The first alleles of the two siblings are IBD because they are copies of the second allele of their father. The second allele of the first sibling and the first allele of the second sibling are IBS because they are copies of different founder alleles although both have the same form $A_1$. This figure is adopted from Lynch and Walsh (Lynch and Walsh 1998)

**Fig. 11.2** A hypothetical
complex pedigree with
22 individuals

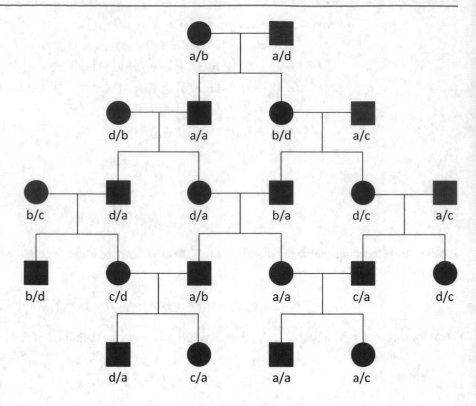

**Fig. 11.3** A descent graph of the
complex pedigree shown in
Fig. 11.2

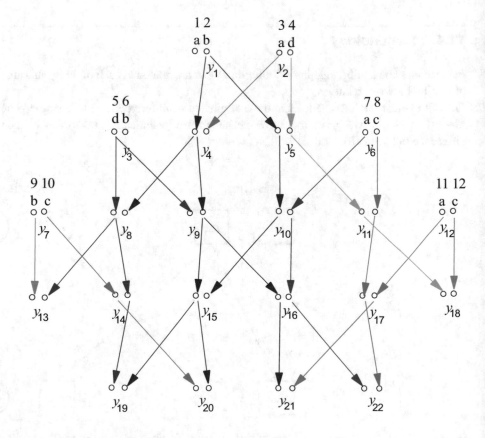

*Base population*: Identity-by-descent is essentially a comparison between the population in question and some specific or implied base population. In the base population, all genes are treated as independent. So the "previous generation" defined earlier must be a generation at or after the base population.

*Coancestry*: Coancestry is also called coefficient of kinship. It is the probability that two genes taken at random, one from each individual, are identical-by-descent. Coancestry describes the relationships of genes between individuals.

*Inbreeding coefficient*: The probability that the two genes at a locus in an individual are identical-by-descent. Inbreeding coefficient describes the relationship of two genes within an individual.

*Descent graph*: A pedigree showing the paths of gene flow from the founders to non-founders is called a descent graph (Sobel and Lange 1996).

Per definition of the coancestry and inbreeding coefficient, the inbreeding coefficient of an individual equals the coancestry of the two parents. A very complicated pedigree with 22 members is shown in Fig. 11.2. The genotype of each individual is given under the symbol of that individual. Females are indicated by circles and males are indicated by squares. The males in blue and the females in red are founders. The individuals in purple are non-founders. Founders provide new alleles to the pedigree while non-founders do not contribute new alleles. Figure 11.3 shows one possible way that alleles from the founders pass to the non-founders. A pedigree showing the paths of gene flow from the founders to non-founders is called a descent graph (Sobel and Lange 1996). Descent graphs allow people to use stochastic techniques to analyze pedigree data via the Markov chain Monte Carlo algorithm. In the descent graph (Fig. 11.3), the two alleles carried by $y_{15}$ are IBD because they are the copies of the founder allele number 2. The first allele of $y_{20}$ and the second allele of $y_{22}$ are IBS because they both have the same allelic form (labeled as c), but one is a copy of founder allele number 10 and the other is a copy of founder allele number 8.

## 11.5 Computing Coancestry Coefficients

Coancestry and inbreeding coefficients are very important concepts in quantitative and population genetics. There are two ways to calculate coancestry between pairs of individuals, one is called the path analysis and the other is called the tabular method. The path analysis requires a graphical presentation of a pedigree and it can only be used when you are only interested in the coancestry between two members of the pedigree. The tabular method is more general and used when the pedigree is very complicated. With the tabular method, coancestry coefficients must be calculated between all pairs of relatives, even if you are only interested in one pair of relatives. The two methods are available for calculating the coancestry between two individuals. There is no algorithm required to calculate inbreeding coefficient of an individual because the inbreeding coefficient of an individual is the coancestry between the two parents. So, you simply calculate the coancestry between the parents to obtain the inbreeding coefficient of the individual.

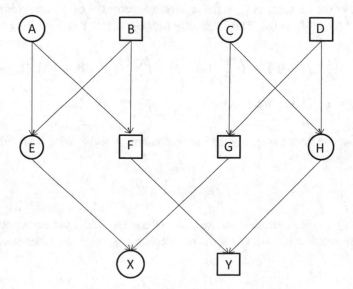

**Fig. 11.4** Pedigree of double first cousins

## 11.5.1 Path Analysis

Path analysis was developed by geneticist Sewall Wright (Wright 1921), one of the three founders of population genetics. The method first identifies the number of paths that connect the two relatives in question and then assigns a value to each path. The coancestry between the two relatives takes the sum of values of all paths. Figure 11.4 shows a pedigree involving double first cousins. Let us use the pedigree of double first cousins as an example to illustrate the path analysis for calculation of coancestry. First, we need to identify all possible paths to connect $X$ and $Y$. A path can only contain one ancestor. A path can only change direction once. A path cannot contain a node like "$\rightarrow$node$\leftarrow$". If a path involves change of direction, it can only occur as "backward, backward, ..., backward and forward, forward, ..., forward." The double first cousin pedigree has four paths to connect $X$ and $Y$.

$$\text{Path 1}: X \leftarrow E \leftarrow A \rightarrow F \rightarrow Y, \text{Value} = \left(\frac{1}{2}\right)^5 (1 + f_A)$$

$$\text{Path 2}: X \leftarrow E \leftarrow B \rightarrow F \rightarrow Y, \text{Value} = \left(\frac{1}{2}\right)^5 (1 + f_B)$$

$$\text{Path 3}: X \leftarrow G \leftarrow C \rightarrow H \rightarrow Y, \text{Value} = \left(\frac{1}{2}\right)^5 (1 + f_C)$$

$$\text{Path 4}: X \leftarrow G \leftarrow D \rightarrow H \rightarrow Y, \text{Value} = \left(\frac{1}{2}\right)^5 (1 + f_D)$$

The coancestry between $X$ and $Y$ is

$$\theta_{XY} = \left(\frac{1}{2}\right)^5 (1 + f_A) + \left(\frac{1}{2}\right)^5 (1 + f_B) + \left(\frac{1}{2}\right)^5 (1 + f_C) + \left(\frac{1}{2}\right)^5 (1 + f_D)$$

$$= 4 \times \left(\frac{1}{2}\right)^5 = 0.125$$

where $f_A = f_B = f_C = f_D = 0$ is assumed in this particular pedigree. In general,

$$\theta_{XY} = \sum_{k=1}^{n} \left(\frac{1}{2}\right)^{m_k - 1} \theta_{kk} = \sum_{k=1}^{n} \left(\frac{1}{2}\right)^{m_k - 1} \left(\frac{1}{2}\right)(1 + f_k) = \sum_{k=1}^{n} \left(\frac{1}{2}\right)^{m_k} (1 + f_k)$$

where $n$ is the number of paths, $m_k$ is the number of individuals (nodes) in the $k$th path (including $X$ and $Y$), $\theta_{kk} = \frac{1}{2}(1 + f_k)$ is the coancestry of the ancestor of the $k$th path, and $f_k$ is the inbreeding coefficient of the ancestor in the $k$th path. Let us assume that $f_A = 0.1, f_B = 0.2, f_C = 0.3,$ and $f_D = 0.4$. The coancestry between $X$ and $Y$ is

$$\theta_{XY} = \left(\frac{1}{2}\right)^5 (1 + 0.1) + \left(\frac{1}{2}\right)^5 (1 + 0.2) + \left(\frac{1}{2}\right)^5 (1 + 0.3) + \left(\frac{1}{2}\right)^5 (1 + 0.4)$$

$$= \left(\frac{1}{2}\right)^5 (1.1 + 1.2 + 1.3 + 1.4) = \left(\frac{1}{2}\right)^5 \times 5 = 0.15625$$

Assume that individual $Z$ is resulted from the mating between individuals $X$ and $Y$, what is the inbreeding coefficient of $Z$? We do not need to calculate that because it is

$$f_Z = \theta_{XY} = 0.15625$$

The method (path analysis) can be extended to situations where the founders are genetically related, but it is very complicated. It is not recommended to use path analysis to calculate coancestry between relatives drawn from a pedigree with related founders.

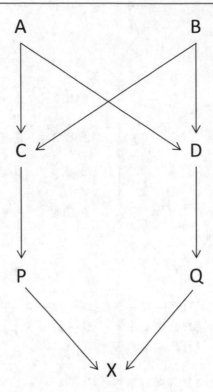

**Fig. 11.5** A sample pedigree with seven members

The most difficult part of path analysis is to correctly identify all possible legal paths. You cannot miss any paths and all paths must be legal. A legal path must have one and only one ancestor. A path can change direction at most once. The legal change of direction must be "from backward to forward," not "from forward to backward."

We now show an example to demonstrate the path analysis for calculating coancestry between two individuals. Figure 11.5 shows a pedigree with seven members. By default, all founders are non-inbred and also not related to each other. We want to find the inbreeding coefficient of individual $X$, i.e., $f_X = ?$. Since we know that $f_X = \theta_{PQ}$ because $P$ and $Q$ are the parents of $X$, we use the coancestry formula to calculate $\theta_{PQ}$. Here are the legal paths between $P$ and $Q$,

$$(1)\ P \leftarrow C \leftarrow A \rightarrow D \rightarrow Q$$
$$(2)\ P \leftarrow C \leftarrow B \rightarrow D \rightarrow Q$$

where $A$ and $B$ are ancestors (founders) of the pedigree, one for each path. The coancestry between $P$ and $Q$ is

$$\theta_{PQ} = \left(\frac{1}{2}\right)^5 (1 + f_A) + \left(\frac{1}{2}\right)^5 (1 + f_B) = \left(\frac{1}{2}\right)^5 (1 + 0) + \left(\frac{1}{2}\right)^5 (1 + 0) = 0.0625$$

Therefore, the inbreeding coefficient of $X$ is $f_X = \theta_{PQ} = 0.0625$. What is the coancestry of $X$ with itself, i.e., $\theta_{XX} = ?$. This time, we do not use the path analysis but simply use

$$\theta_{XX} = \frac{1}{2}(f_X + 1) = \frac{1}{2}(\theta_{PQ} + 1) = \frac{1}{2}(0.0625 + 1) = 0.53125$$

to obtain the coancestry of an individual with itself. What is the coancestry between $P$ and $X$? We found three paths connecting $P$ and $X$,

**Fig. 11.6** Pedigree with
genetically related ancestors

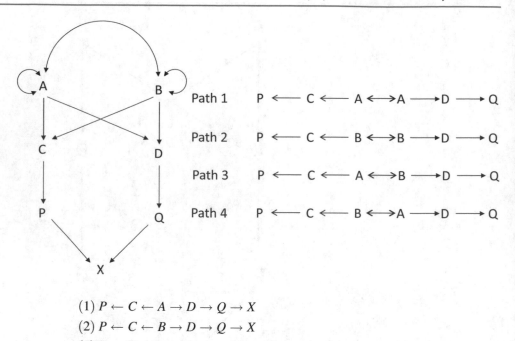

$$(1)\ P \leftarrow C \leftarrow A \rightarrow D \rightarrow Q \rightarrow X$$
$$(2)\ P \leftarrow C \leftarrow B \rightarrow D \rightarrow Q \rightarrow X$$
$$(3)\ P \rightarrow X$$

Do not forget the last path! The coancestry between $P$ and $X$ is

$$
\theta_{PX} = \left(\frac{1}{2}\right)^6 (1 + f_A) + \left(\frac{1}{2}\right)^6 (1 + f_B) + \left(\frac{1}{2}\right)^2 (1 + f_P)
$$
$$
= \left(\frac{1}{2}\right)^6 (1 + 0) + \left(\frac{1}{2}\right)^6 (1 + 0) + \left(\frac{1}{2}\right)^2 (1 + 0)
$$
$$
= 0.28125
$$

The ancestor of the last path is $P$.

By default, the inbreeding coefficients of all ancestors are assumed to be zero and the coancestry coefficients between ancestors are assumed to be zero as well. However, if any ancestors are inbred, they will be said so. If the coancestry between any pair of founders is not zero, it will be said also. We now assume that $f_A = 0.15$ and $f_B = 0.25$. We did not mention the coancestry between $A$ and $B$. Therefore, $\theta_{AB} = 0$ is assumed by default. Now let us see what the inbreeding coefficient of $X$ will be under $f_A = 0.15$ and $f_B = 0.25$. The answer is

$$
f_X = \theta_{PQ} = \left(\frac{1}{2}\right)^5 (1 + f_A) + \left(\frac{1}{2}\right)^5 (1 + f_B) = \left(\frac{1}{2}\right)^5 (1 + 0.15) + \left(\frac{1}{2}\right)^5 (1 + 0.25) = 0.075
$$

Suppose that $f_A = 0.15$, $f_B = 0.25$, and $\theta_{AB} = 0.25$, the situation is very complicated because the ancestors are related. We can still use the path analysis, but the tabular method is recommended. The path analysis for related ancestors involves an algorithm that is different from the one learned so far. A general formula for situations like this (with genetically related ancestors) is (Boucher 1988; Lynch and Walsh 1998)

$$
\theta_{XY} = \sum_{k=1}^{n_1} \left(\frac{1}{2}\right)^{m_k} \theta_{kk} + \sum_{i \neq j}^{n_2} \left(\frac{1}{2}\right)^{m_{ij}} \theta_{ij}
$$

where $n_1$ is the number of paths with one ancestor, $m_k$ is the number of segments with one-headed arrow (do not include the double-headed arrows) in the $k$th one-ancestor path, $\theta_{kk}$ is the coancestry of the ancestor with itself in the $k$th path, $n_2$ is the number of paths connecting $X$ and $Y$ through two ancestors ($i$ and $j$) per path, $m_{ij}$ is the number of segments with one-headed arrow connecting $X$ and $Y$ for the path identified by the $i$th and the $j$th ancestors, and $\theta_{ij}$ is the coancestry between the $i$th and the $j$th ancestors. Figure 11.6 shows a modified pedigree and the four paths connecting $P$ and $Q$. The first and second paths are one-ancestor paths ($n_1 = 2$). The third and fourth paths are two-ancestor paths ($n_2 = 2$). Each path has 4 segments of one-headed arrows and 1 curve with a double-headed arrow. Each one-headed segment is assigned a value of 1/2 and each double-headed curve is assigned a value of $\theta_{kk}$ or $\theta_{ij}$. The value of each path is the product of all the values of segments and curves. The coancestry between $P$ and $Q$ in the modified pedigree (Fig. 11.6) is

$$
\begin{aligned}
\theta_{PQ} &= \left(\frac{1}{2}\right)^4 \theta_{AA} + \left(\frac{1}{2}\right)^4 \theta_{BB} + \left(\frac{1}{2}\right)^4 \theta_{AB} + \left(\frac{1}{2}\right)^4 \theta_{BA} \\
&= \left(\frac{1}{2}\right)^4 \left(\frac{1}{2}\right)(1 + f_A) + \left(\frac{1}{2}\right)^4 \left(\frac{1}{2}\right)(1 + f_B) + \left(\frac{1}{2}\right)^4 \theta_{AB} + \left(\frac{1}{2}\right)^4 \theta_{AB} \\
&= \left(\frac{1}{2}\right)^4 \left(\frac{1}{2}\right)(1 + 0.15) + \left(\frac{1}{2}\right)^4 \left(\frac{1}{2}\right)(1 + 0.25) + \left(\frac{1}{2}\right)^4 (0.25) + \left(\frac{1}{2}\right)^4 (0.25) \\
&= \left(\frac{1}{2}\right)^4 \left(\frac{1}{2}\right)(1 + 0.15) + \left(\frac{1}{2}\right)^4 \left(\frac{1}{2}\right)(1 + 0.25) + \left(\frac{1}{2}\right)^4 (0.25) + \left(\frac{1}{2}\right)^4 (0.25) \\
&= 0.10625
\end{aligned}
$$

Pedigrees with related ancestors are so hard to handle using the path analysis. Therefore, the tabular method described in the next session is recommended.

## 11.5.2 Tabular Method

For a pedigree with size $n$ (there are $n$ members in the pedigree of interest), we will calculate the coancestry coefficients between all $n(n - 1)/2$ pairs of relatives. We will then store the coancestry coefficients into a square matrix, called the coancestry matrix,

$$
\Theta = \begin{bmatrix}
\theta_{11} & \theta_{12} & \cdots & \theta_{1n} \\
\theta_{12} & \theta_{22} & \cdots & \theta_{2n} \\
\vdots & \vdots & \ddots & \vdots \\
\theta_{1n} & \theta_{2n} & \cdots & \theta_{nn}
\end{bmatrix}
$$

We can then construct an additive relationship matrix using

$$
A = \begin{bmatrix}
a_{11} & a_{12} & \cdots & a_{1n} \\
a_{12} & a_{22} & \cdots & a_{2n} \\
\vdots & \vdots & \ddots & \vdots \\
a_{1n} & a_{2n} & \cdots & a_{nn}
\end{bmatrix} = 2\Theta = \begin{bmatrix}
2\theta_{11} & 2\theta_{12} & \cdots & 2\theta_{1n} \\
2\theta_{12} & 2\theta_{22} & \cdots & 2\theta_{2n} \\
\vdots & \vdots & \ddots & \vdots \\
2\theta_{1n} & 2\theta_{2n} & \cdots & 2\theta_{nn}
\end{bmatrix}
$$

In the animal breeding literature, matrix $A$ is also called the numerator relationship matrix. The tabular method will calculate the $\Theta$ matrix for you. Even if you are only interested in $\theta_{ij}$, you still have to calculate the entire $\Theta$ matrix, which is an undesirable character of the tabular method. Using the tabular method to calculate the coancestry matrix of a pedigree requires a few rules, which are described below.

(1) Each individual is given a unique identification number (ID) from 1 to $n$ for $n$ individuals in the pedigree. Construct a tabular pedigree (a table) with $n$ rows and three columns. The first column lists the IDs of all members in the pedigree. The second and third columns of the table give the IDs of the sires and the dams, respectively. Individuals in the table are ordered according to their ages, starting from the oldest to the youngest. In fact, the only restriction of the order is that parents cannot appear after their children in the tabular pedigree.

(2) Let $s$ and $d$ be the IDs of the sire and the dam of individual $j$, respectively. The inbreeding coefficient of $j$ equals the coancestry between the two parents, i.e.,

$$f_j = \theta_{sd}$$

where $f_j$ is the inbreeding coefficient of $j$ and $\theta_{sd}$ is the coancestry between $s$ and $d$.

(3) Let $\theta_{jj}$ be the coancestry of individual $j$ with itself and $f_j$ be the inbreeding coefficient of $j$, then

$$\theta_{jj} = \frac{1}{2}\left(1 + f_j\right)$$

(4) If $i$ is not a descendent of $j$, then

$$\theta_{ij} = \frac{1}{2}\left(\theta_{is} + \theta_{id}\right)$$

where $\theta_{is}$ is the coancestry between $i$ and the father of $j$, and $\theta_{id}$ is the coancestry between $i$ and the mother of $j$. Technically, you can interpret the phrase "$i$ is not a descendent of $j$" as "$i$ is older than $j$."

(5) The coancestry between an individual in the pedigree and a missing member is set to zero unless it is specified otherwise.

(6) Consider $m$ founders numbered from 1 to $m$. A founder is defined as an individual whose parents are not identified and not included in the pedigree. The inbreeding coefficients of the founders and the coancestry coefficients between pairs of founders are assumed to be zero by default, unless they are specified as other values by the instructor.

(7) You need to construct an $m \times m$ founder coancestry matrix with values given by the instructor or a diagonal matrix with a constant diagonal element of 1/2. This matrix is the coancestry matrix for the $m$ founders. Starting from this matrix, you expand the matrix one column at a time and one row at a time until the entire coancestry matrix is filled, i.e., becomes an $n \times n$ matrix.

### 11.5.2.1 Example 1

We now use the pedigree shown in Fig. 11.5 to demonstrate the tabular method. The tabular version of the pedigree is shown in Table 11.1. For this particular pedigree, there are six steps to fill the matrix, starting with the two ancestors A and B.

**Step 1**: Constructing a $2 \times 2$ matrix for ancestors A and B (the upper left corner of Table 11.2). The $2 \times 2$ matrix is highlighted with the darkest color. Since neither ancestor is inbred and the two ancestors are not related, we have $\theta_{AA} = \frac{1}{2}(1 + f_A) = \frac{1}{2}(1 + 0) = 0.5$, $\theta_{BB} = \frac{1}{2}(1 + f_B) = \frac{1}{2}(1 + 0) = 0.5$ and $\theta_{AB} = 0$.

**Step 2**: Adding column C and row C to the existing $2 \times 2$ matrix. Since A and B are the parents of C and neither A nor B is a descent of C, we have

**Table 11.1** Tabular form of the pedigree shown in Fig. 11.5

| Child | Sire | Dam |
|-------|------|-----|
| A | * | * |
| B | * | * |
| C | A | B |
| D | A | B |
| P | C | * |
| Q | D | * |
| X | P | Q |

**Table 11.2** Coancestry matrix of the seven members of the pedigree in Fig. 11.5

|   | A | B | C | D | P | Q | X |
|---|-----|-----|-----|-----|-----|-----|-----|
| A | 0.5 | 0 | 0.25 | 0.25 | 0.125 | 0.125 | 0.125 |
| B | 0 | 0.5 | 0.25 | 0.25 | 0.125 | 0.125 | 0.125 |
| C | 0.25 | 0.25 | 0.5 | 0.25 | 0.25 | 0.125 | 0.1875 |
| D | 0.25 | 0.25 | 0.25 | 0.5 | 0.125 | 0.25 | 0.1875 |
| P | 0.125 | 0.125 | 0.25 | 0.125 | 0.5 | 0.0625 | 0.28125 |
| Q | 0.125 | 0.125 | 0.125 | 0.25 | 0.0625 | 0.5 | 0.28125 |
| X | 0.125 | 0.125 | 0.1875 | 0.1875 | 0.28125 | 0.28125 | 0.53125 |

$$\theta_{AC} = \frac{1}{2}(\theta_{AA} + \theta_{AB}) = \frac{1}{2}(0.5 + 0) = 0.25$$

$$\theta_{BC} = \frac{1}{2}(\theta_{AB} + \theta_{BB}) = \frac{1}{2}(0 + 0.5) = 0.25$$

We now need to fill the diagonal element,

$$\theta_{CC} = \frac{1}{2}(1 + f_C) = \frac{1}{2}(1 + \theta_{AB}) = \frac{1}{2}(1 + 0) = 0.5$$

Copy $\theta_{AC}$ and $\theta_{BC}$, and paste them to the places holding $\theta_{CA}$ and $\theta_{CB}$. Remember that a coancestry matrix is symmetric.

**Step 3**: Adding column D and row D to the current $3 \times 3$ matrix. Since A and B are also the parents of D, we have

$$\theta_{AD} = \frac{1}{2}(\theta_{AA} + \theta_{AB}) = \frac{1}{2}(0.5 + 0) = 0.25$$

$$\theta_{BD} = \frac{1}{2}(\theta_{AB} + \theta_{BB}) = \frac{1}{2}(0 + 0.5) = 0.25$$

$$\theta_{CD} = \frac{1}{2}(\theta_{CA} + \theta_{CB}) = \frac{1}{2}(0.25 + 0.25) = 0.25$$

The diagonal element is

$$\theta_{DD} = \frac{1}{2}(1 + f_D) = \frac{1}{2}(1 + \theta_{AB}) = \frac{1}{2}(1 + 0) = 0.5$$

Copy the column vector $[\theta_{AD} \ \theta_{BD} \ \theta_{CD}]^T$ and transpose it into $[\theta_{DA} \ \ \theta_{DB} \ \ \ \theta_{DC}]$, and then paste it into the appropriate places in the matrix. This completes the third step.

**Step 4**: Adding column P and row P to the current $4 \times 4$ matrix. Since C is the parent of P and the other parent is missing, we have

$$\theta_{AP} = \frac{1}{2}(\theta_{AC} + \theta_{A*}) = \frac{1}{2}(0.25 + 0) = 0.125$$

$$\theta_{BP} = \frac{1}{2}(\theta_{BC} + \theta_{B*}) = \frac{1}{2}(0.25 + 0) = 0.125$$

$$\theta_{CP} = \frac{1}{2}(\theta_{CC} + \theta_{C*}) = \frac{1}{2}(0.50 + 0) = 0.250$$

$$\theta_{DP} = \frac{1}{2}(\theta_{DC} + \theta_{D*}) = \frac{1}{2}(0.25 + 0) = 0.125$$

The diagonal element is

$$\theta_{PP} = \frac{1}{2}(1 + f_P) = \frac{1}{2}(1 + 0) = 0.5$$

Copy the column vector $[\theta_{AP} \ \ \theta_{BP} \ \ \theta_{CP} \ \ \theta_{DP}]^T$ and transpose it into $[\theta_{PA} \ \ \theta_{PB} \ \ \theta_{PC} \ \ \theta_{PD}]$, and then paste it into the appropriate places in the matrix. This completes the fourth step.

**Step 5**: Adding column Q and row Q to the current $5 \times 5$ matrix. Since D is the parent of Q and the other parent is missing, we have

$$\theta_{AQ} = \frac{1}{2}\left(\theta_{AD} + \theta_{A*}\right) = \frac{1}{2}(0.25 + 0) = 0.125$$

$$\theta_{BQ} = \frac{1}{2}\left(\theta_{BD} + \theta_{B*}\right) = \frac{1}{2}(0.25 + 0) = 0.125$$

$$\theta_{CQ} = \frac{1}{2}\left(\theta_{CD} + \theta_{C*}\right) = \frac{1}{2}(0.25 + 0) = 0.125$$

$$\theta_{DQ} = \frac{1}{2}\left(\theta_{DD} + \theta_{D*}\right) = \frac{1}{2}(0.50 + 0) = 0.250$$

$$\theta_{PQ} = \frac{1}{2}\left(\theta_{PD} + \theta_{P*}\right) = \frac{1}{2}(0.125 + 0) = 0.0625$$

The diagonal element is

$$\theta_{QQ} = \frac{1}{2}\left(1 + f_Q\right) = \frac{1}{2}(1 + 0) = 0.5$$

Copy the column vector $\begin{bmatrix} \theta_{AQ} & \theta_{BQ} & \theta_{CQ} & \theta_{DQ} & \theta_{PQ} \end{bmatrix}^T$ and transpose it into $\begin{bmatrix} \theta_{QA} & \theta_{QB} & \theta_{QC} & \theta_{QD} & \theta_{QP} \end{bmatrix}$, and then paste it into the appropriate places in the matrix.

**Step 6**: Adding the last column and the last row to the existing 6 × 6 matrix to complete the coancestry matrix. Since P and Q are the parents of X, we have

$$\theta_{AX} = \frac{1}{2}\left(\theta_{AP} + \theta_{AQ}\right) = \frac{1}{2}(0.125 + 0.125) = 0.125$$

$$\theta_{BX} = \frac{1}{2}\left(\theta_{BP} + \theta_{BQ}\right) = \frac{1}{2}(0.125 + 0.125) = 0.125$$

$$\theta_{CX} = \frac{1}{2}\left(\theta_{CP} + \theta_{CQ}\right) = \frac{1}{2}(0.25 + 0.125) = 0.1875$$

$$\theta_{DX} = \frac{1}{2}\left(\theta_{DP} + \theta_{DQ}\right) = \frac{1}{2}(0.125 + 0.25) = 0.1875$$

$$\theta_{PX} = \frac{1}{2}\left(\theta_{PP} + \theta_{PQ}\right) = \frac{1}{2}(0.5 + 0.0625) = 0.28125$$

$$\theta_{QX} = \frac{1}{2}\left(\theta_{QP} + \theta_{QQ}\right) = \frac{1}{2}(0.0625 + 0.5) = 0.28125$$

The diagonal element is

$$\theta_{XX} = \frac{1}{2}\left(1 + f_X\right) = \frac{1}{2}\left(1 + \theta_{PQ}\right) = \frac{1}{2}(1 + 0.0625) = 0.53125$$

Copy the column vector $\begin{bmatrix} \theta_{AX} & \theta_{BX} & \theta_{CX} & \theta_{DX} & \theta_{PX} & \theta_{QX} \end{bmatrix}^T$ and transpose it into $\begin{bmatrix} \theta_{XA} & \theta_{XB} & \theta_{XC} & \theta_{XD} & \theta_{XP} & \theta_{XQ} \end{bmatrix}$, and then paste it into the appropriate places in the matrix to complete the entire coancestry matrix.

### 11.5.2.2 Example 2

For the same pedigree shown in Fig. 11.5 and Table 11.1, let us Assume that $f_A = 0.15, f_B = 0.25$, and $\theta_{AB} = 0.25$, where $A$ and $B$ are the ancestors of the pedigree. We now use the tabular method to calculate the coancestry matrix for all the seven members of the pedigree. The two founders are $A$ and $B$, and thus the 2 × 2 coancestry matrix of the founders is.

|   | A | B |
|---|---|---|
| A | 0.575 | 0.25 |
| B | 0.25 | 0.625 |

where

**Table 11.3** Coancestry matrix for all seven members of the pedigree when the two ancestors are inbred and they are related

|   | A | B | C | D | P | Q | X |
|---|---|---|---|---|---|---|---|
| A | 0.575 | 0.25 | 0.4125 | 0.4125 | 0.20625 | 0.20625 | 0.20625 |
| B | 0.25 | 0.625 | 0.4375 | 0.4375 | 0.21875 | 0.21875 | 0.21875 |
| C | 0.4125 | 0.4375 | 0.625 | 0.425 | 0.3125 | 0.2125 | 0.2625 |
| D | 0.4125 | 0.4375 | 0.425 | 0.625 | 0.2125 | 0.3125 | 0.2625 |
| P | 0.20625 | 0.21875 | 0.3125 | 0.2125 | 0.5 | 0.10625 | 0.303125 |
| Q | 0.20625 | 0.21875 | 0.2125 | 0.3125 | 0.10625 | 0.5 | 0.303125 |
| X | 0.20625 | 0.21875 | 0.2625 | 0.2625 | 0.303125 | 0.303125 | 0.553125 |

$$\theta_{AA} = \frac{1}{2}(1 + f_A) = \frac{1}{2}(1 + 0.15) = 0.575$$

and

$$\theta_{BB} = \frac{1}{2}(1 + f_B) = \frac{1}{2}(1 + 0.25) = 0.625$$

Starting from this the $2 \times 2$ matrix, we expand it one column and one row at a time following exactly the same steps demonstrated earlier to complete the coancestry matrix, which is shown in Table 11.3.

## 11.6   R Package to Calculate a Coancestry Matrix

An R package named "kinship2" is available for calculating the kinship matrix of a pedigree (Sinnwell et al. 2014). The program is very simple and efficient in terms of computational speed for large pedigrees. The only drawback is the inability to handle inbred founders. We now use the Roan Gauntlet (an English bull) pedigree shown in Fig. 11.7 to demonstrate the R code for coancestry matrix calculation. This pedigree is not too complicated and thus we can use both the path analysis and the tabular method to find the inbreeding coefficient of Roan Gauntlet (an English bull), who is labeled as number (11) in the pedigree.

### 11.6.1 Path Analysis

The inbreeding coefficient of number (11) equals the coancestry between number (9) and number (10). Therefore, we only need to calculate $f_{(11)} = \theta_{(9)(10)}$. The paths that connecting (9) and (10) are

$$\text{Path 1} : 10 \leftarrow 6 \leftarrow 3 \leftarrow 1 \rightarrow 4 \rightarrow 8 \rightarrow 9$$
$$\text{Path 2} : 10 \leftarrow 7 \leftarrow 5 \leftarrow 1 \rightarrow 4 \rightarrow 8 \rightarrow 9$$
$$\text{Path 3} : 10 \leftarrow 6 \leftarrow 2 \rightarrow 9$$
$$\text{Path 4} : 10 \leftarrow 7 \leftarrow 2 \rightarrow 9$$

The path values are

$$\text{Path 1} : m_1 = 7; f_1 = 0; \text{value} = \left(\frac{1}{2}\right)^7 (1 + 0) = \left(\frac{1}{2}\right)^7$$

$$\text{Path 2} : m_2 = 7; f_2 = 0; \text{value} = \left(\frac{1}{2}\right)^7 (1 + 0) = \left(\frac{1}{2}\right)^7$$

$$\text{Path 3} : m_3 = 4; f_3 = 0; \text{value} = \left(\frac{1}{2}\right)^4 (1 + 0) = \left(\frac{1}{2}\right)^4$$

$$\text{Path 4} : m_4 = 4; f_4 = 0; \text{value} = \left(\frac{1}{2}\right)^4 (1 + 0) = \left(\frac{1}{2}\right)^4$$

Therefore, the inbreeding coefficient of (11) is

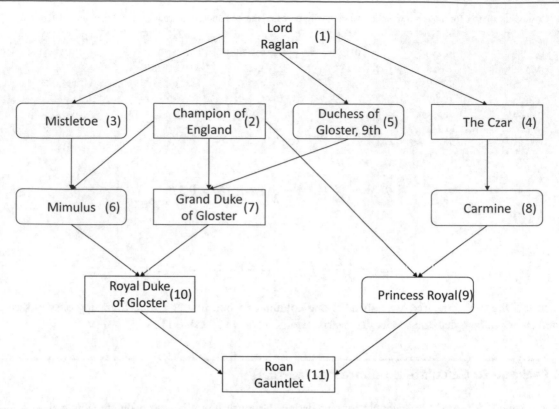

**Fig. 11.7** The Roan Gauntlet (an English bull) pedigree adopted from Lynch and Walsh (Lynch and Walsh 1998)

**Table 11.4** Tabular representation of the Roan Gauntlet pedigree

| Child | Dad | Mom |
|---|---|---|
| 1 | 0 | 0 |
| 2 | 0 | 0 |
| 3 | 1 | 0 |
| 4 | 1 | 0 |
| 5 | 1 | 0 |
| 6 | 2 | 3 |
| 7 | 2 | 5 |
| 8 | 4 | 0 |
| 9 | 2 | 8 |
| 10 | 7 | 6 |
| 11 | 10 | 9 |

$$f_{(11)} = \theta_{(9)(10)} = 2 \times \left(\frac{1}{2}\right)^7 + 2 \times \left(\frac{1}{2}\right)^4 = 0.140625$$

### 11.6.2 Tabular Method

The tabular version of the pedigree is given in Table 11.4. This is the input data required by the R program. You can download the R package "kinship2" (Sinnwell et al. 2014) to calculate the coancestry matrix for any complicated pedigrees. The only input data for the program is the tabular pedigree with 0 or NA indicating a missing parent. Assume that the tabular pedigree is stored in a file named "`data-11-1.txt`". The R code to read the data and calculate the kinship matrix is.

**Table 11.5** Coancestry matrix for the 11 members of the Roan Gauntlet pedigree calculate with kinship2

|    | 1 | 2 | 3 | 4 | 5 | 6 | 7 | 8 | 9 | 10 | 11 |
|----|-----|-----|-----|-----|-----|-----|-----|-----|-----|-----|-----|
| 1  | 0.5 | 0 | 0.25 | 0.25 | 0.25 | 0.125 | 0.125 | 0.125 | 0.0625 | 0.125 | 0.09375 |
| 2  | 0 | 0.5 | 0 | 0 | 0 | 0.25 | 0.25 | 0 | 0.25 | 0.25 | 0.25 |
| 3  | 0.25 | 0 | 0.5 | 0.125 | 0.125 | 0.25 | 0.0625 | 0.0625 | 0.03125 | 0.15625 | 0.09375 |
| 4  | 0.25 | 0 | 0.125 | 0.5 | 0.125 | 0.0625 | 0.0625 | 0.25 | 0.125 | 0.0625 | 0.09375 |
| 5  | 0.25 | 0 | 0.125 | 0.125 | 0.5 | 0.0625 | 0.25 | 0.0625 | 0.03125 | 0.15625 | 0.09375 |
| 6  | 0.125 | 0.25 | 0.25 | 0.0625 | 0.0625 | 0.5 | 0.15625 | 0.03125 | 0.140625 | 0.328125 | 0.234375 |
| 7  | 0.125 | 0.25 | 0.0625 | 0.0625 | 0.25 | 0.15625 | 0.5 | 0.03125 | 0.140625 | 0.328125 | 0.234375 |
| 8  | 0.125 | 0 | 0.0625 | 0.25 | 0.0625 | 0.03125 | 0.03125 | 0.5 | 0.25 | 0.03125 | 0.140625 |
| 9  | 0.0625 | 0.25 | 0.03125 | 0.125 | 0.03125 | 0.140625 | 0.140625 | 0.25 | 0.5 | 0.140625 | 0.320313 |
| 10 | 0.125 | 0.25 | 0.15625 | 0.0625 | 0.15625 | 0.328125 | 0.328125 | 0.03125 | 0.140625 | 0.578125 | 0.359375 |
| 11 | 0.09375 | 0.25 | 0.09375 | 0.09375 | 0.09375 | 0.234375 | 0.234375 | 0.140625 | 0.320313 | 0.359375 | 0.570313 |

```
bull<-read.table(file="data-11-1.txt", header=T)
Theta<-kinship(id=bull$child,dadid=bull$dad,momid=bull$mom)
A<-2*Theta
```

The R data containing the pedigree is called `bull`. Theta ($\Theta$) holds the kinship matrix generated from the `kinship()` function. The additive relationship matrix is $A = 2\Theta$. The kinship matrix generated from `kinship()` is given in Table 11.5.

## 11.7 SAS Program for Calculating a Coancestry Matrix

A SAS procedure called PROC INBREED (SAS Institute Inc. 2009) is particularly designed to calculate coancestry matrices. This program is sufficiently general to handle any complicated pedigrees. In addition, the program allows you to specify the inbreeding coefficients of founders and the coancestry coefficients between founders. Let us calculate the coancestry matrix for the British bull pedigree again using the INBREED procedure in SAS to see whether or not SAS generates the same result as kinship2. The SAS code to read the data and to call PROC INBREED is given below.

```
data bull;
     input child$ dad$ mom$;
cards;
1       .       .
2       .       .
3       1       .
4       1       .
5       1       .
6       2       3
7       2       5
8       4       .
9       2       8
10      7       6
11      10      9
;
proc inbreed data=bull matrix selfdiag;
     var child dad mom;
     ods output InbreedingCoefficient=kinship;
run;
```

**Table 11.6** Coancestry matrix for the 11 members of the Roan Gauntlet pedigree calculate with PROC INBEED

| Inbreeding coefficients | | | | | | | | | | | | | | |
|---|---|---|---|---|---|---|---|---|---|---|---|---|---|
| Child | Dad | Mom | 1 | 2 | 3 | 4 | 5 | 6 | 7 | 8 | 9 | 10 | 11 |
| 1 | | | 0.5000 | . | 0.2500 | 0.2500 | 0.2500 | 0.1250 | 0.1250 | 0.1250 | 0.0625 | 0.1250 | 0.0938 |
| 2 | | | . | 0.5000 | . | . | . | 0.2500 | 0.2500 | . | 0.2500 | 0.2500 | 0.2500 |
| 3 | 1 | | 0.2500 | . | 0.5000 | 0.1250 | 0.1250 | 0.2500 | 0.0625 | 0.0625 | 0.0313 | 0.1563 | 0.0938 |
| 4 | 1 | | 0.2500 | . | 0.1250 | 0.5000 | 0.1250 | 0.0625 | 0.0625 | 0.2500 | 0.1250 | 0.0625 | 0.0938 |
| 5 | 1 | | 0.2500 | . | 0.1250 | 0.1250 | 0.5000 | 0.0625 | 0.2500 | 0.0625 | 0.0313 | 0.1563 | 0.0938 |
| 6 | 2 | 3 | 0.1250 | 0.2500 | 0.2500 | 0.0625 | 0.0625 | 0.5000 | 0.1563 | 0.0313 | 0.1406 | 0.3281 | 0.2344 |
| 7 | 2 | 5 | 0.1250 | 0.2500 | 0.0625 | 0.0625 | 0.2500 | 0.1563 | 0.5000 | 0.0313 | 0.1406 | 0.3281 | 0.2344 |
| 8 | 4 | | 0.1250 | . | 0.0625 | 0.2500 | 0.0625 | 0.0313 | 0.0313 | 0.5000 | 0.2500 | 0.0313 | 0.1406 |
| 9 | 2 | 8 | 0.0625 | 0.2500 | 0.0313 | 0.1250 | 0.0313 | 0.1406 | 0.1406 | 0.2500 | 0.5000 | 0.1406 | 0.3203 |
| 10 | 7 | 6 | 0.1250 | 0.2500 | 0.1563 | 0.0625 | 0.1563 | 0.3281 | 0.3281 | 0.0313 | 0.1406 | 0.5781 | 0.3594 |
| 11 | 10 | 9 | 0.0938 | 0.2500 | 0.0938 | 0.0938 | 0.0938 | 0.2344 | 0.2344 | 0.1406 | 0.3203 | 0.3594 | 0.5703 |

Variables in the input data (tabular pedigree) can be numerical or character but the missing parents cannot be substituted by zero (this is different from the kinship2 program in R). The matrix option in the proc inbreed statement tells the program to deliver the output in a matrix form. The selfdiag option in the proc inbreed statement tells the procedure to report the kinship matrix. Replacing selfdiag by covar will report the additive relationship matrix rather than the kinship matrix. The var statement can hold up to four variables, the first three variables are the child id, the father id, and the mother id. An additional variable, cov, can be placed in the var statement. The kinship matrix generated from the SAS program is given in Table 11.6. It is identical to the matrix generated from the kinship2 package of R.

We now demonstrate the additional features of PROC INBREED for incorporating inbred and genetically related ancestors. For the pedigree in Fig. 11.5, let us assume $f_A = 0.15$, $f_B = 0.25$, and $\theta_{AB} = 0.25$. By default, PROC INBREED assumes that ancestors are non-inbred and are not related to each other. To activate the additional features, we need to create another variable called cov. This variable must be placed in the var statement. Below is the code for pedigree Fig. 11.5 with inbred and related founders.

```
data one;
input child$ dad$ mom$ coeff;
    cov=2*coeff;
datalines;
A       x1      x2      0.15
B       x3      x4      0.25
C       A       B       0.25
D       A       B       0.25
P       C       .       .
Q       D       .       .
X       P       Q       .
;
proc inbreed data=one matrix selfdiag;
    var child dad mom cov;
    ods output InbreedingCoefficient=kinship;
run;

data kinship;
    set kinship;
    drop x1 x2 x3 x4;
    if Individual^="x1" & Individual^="x2" & Individual^="x3" & Individual^="x4";
run;

proc print data=kinship;
run;
```

**Table 11.7** Output of PROC INBREED including four additional members added by users

| Inbreeding coefficients | | | | | | | | | | | | | |
|---|---|---|---|---|---|---|---|---|---|---|---|---|---|
| Child | Dad | Mom | x1 | x2 | A | x3 | x4 | B | C | D | P | Q | X |
| x1 | | | 0.5000 | 0.1500 | 0.3250 | . | . | . | 0.1625 | 0.1625 | 0.0812 | 0.0812 | 0.0812 |
| x2 | | | 0.1500 | 0.5000 | 0.3250 | . | . | . | 0.1625 | 0.1625 | 0.0812 | 0.0812 | 0.0812 |
| A | x1 | x2 | 0.3250 | 0.3250 | 0.5750 | . | . | 0.2500 | 0.4125 | 0.4125 | 0.2062 | 0.2062 | 0.2062 |
| x3 | | | . | . | . | 0.5000 | 0.2500 | 0.3750 | 0.1875 | 0.1875 | 0.0938 | 0.0938 | 0.0938 |
| x4 | | | . | . | . | 0.2500 | 0.5000 | 0.3750 | 0.1875 | 0.1875 | 0.0938 | 0.0938 | 0.0938 |
| B | x3 | x4 | . | . | 0.2500 | 0.3750 | 0.3750 | 0.6250 | 0.4375 | 0.4375 | 0.2188 | 0.2188 | 0.2188 |
| C | A | B | 0.1625 | 0.1625 | 0.4125 | 0.1875 | 0.1875 | 0.4375 | 0.6250 | 0.4250 | 0.3125 | 0.2125 | 0.2625 |
| D | A | B | 0.1625 | 0.1625 | 0.4125 | 0.1875 | 0.1875 | 0.4375 | 0.4250 | 0.6250 | 0.2125 | 0.3125 | 0.2625 |
| P | C | | 0.0812 | 0.0812 | 0.2062 | 0.0938 | 0.0938 | 0.2188 | 0.3125 | 0.2125 | 0.5000 | 0.1062 | 0.3031 |
| Q | D | | 0.0812 | 0.0812 | 0.2062 | 0.0938 | 0.0938 | 0.2188 | 0.2125 | 0.3125 | 0.1062 | 0.5000 | 0.3031 |
| X | P | Q | 0.0812 | 0.0812 | 0.2062 | 0.0938 | 0.0938 | 0.2188 | 0.2625 | 0.2625 | 0.3031 | 0.3031 | 0.5531 |

**Table 11.8** The cleaned kinship matrix for the seven members

| Obs | Individual | Parent1 | Parent2 | A | B | C | D | P | Q | X |
|---|---|---|---|---|---|---|---|---|---|---|
| 1 | A | x1 | x2 | 0.575 | 0.25 | 0.4125 | 0.4125 | 0.2062 | 0.2062 | 0.2062 |
| 2 | B | x3 | x4 | 0.25 | 0.625 | 0.4375 | 0.4375 | 0.2188 | 0.2188 | 0.2188 |
| 3 | C | A | B | 0.4125 | 0.4375 | 0.625 | 0.425 | 0.3125 | 0.2125 | 0.2625 |
| 4 | D | A | B | 0.4125 | 0.4375 | 0.425 | 0.625 | 0.2125 | 0.3125 | 0.2625 |
| 5 | P | C | | 0.2062 | 0.2188 | 0.3125 | 0.2125 | 0.5 | 0.1062 | 0.3031 |
| 6 | Q | D | | 0.2062 | 0.2188 | 0.2125 | 0.3125 | 0.1062 | 0.5 | 0.3031 |
| 7 | X | P | Q | 0.2062 | 0.2188 | 0.2625 | 0.2625 | 0.3031 | 0.3031 | 0.5531 |

The output from `proc inbreed` is an extended kinship matrix including names of the additional parents, x1, x2, x3, and x4 (see Table 11.7). To delete these additional parents from the extended kinship matrix, we created a new SAS data that retrieves the extended kinship matrix and then delete the rows and columns corresponding to the additional parents. The cleaned kinship matrix is shown in Table 11.8. This kinship matrix is exactly the same as the one calculated manually for the same pedigree (Table 11.3) with the same additional assumptions.

# References

Boucher W. Calculation of the inbreeding coefficient. J Math Biol. 1988;26:57–64.

Lynch M, Walsh B. Genetics and analysis of quantitative traits. Sunderland, MA: Sinauer Associates, Inc.; 1998.

SAS Institute Inc. *SAS/STAT: Users' Guide, Version 9.3*. SAS Institute Inc., Cary, NC. 2009.

Sinnwell JP, Therneau TM, Schaid DJ. The kinship2 R package for pedigree data. Hum Hered. 2014;78:91–3.

Sobel E, Lange K. Descent graphs in pedigree analysis: applications to haplotyping, location scores, and marker-sharing statistics. Am J Hum Genet. 1996;58:1323–37.

Wright S. Correlation and causation. J Agric Res. 1921;20:557–85.

The parent-offspring regression method and the sib analysis approach to estimating heritability only apply to homogeneous pedigrees in a population. In other words, the relationship of relatives within a family must be the same across all families. If a population contains many different families and the relationships of relatives within families vary across different families, the approaches we learned so far are not sufficient. This chapter is particularly designed to estimate heritability in a population consisting of complicated pedigree relationships using the additive relationship matrix, i.e., the A matrix. The method is called mixed model analysis (Henderson 1950).

Consider the following population that contains three pedigrees of variable sizes (Fig. 12.1), where the genetic relationships between relatives are different from family to family. For example, $y_1$ and $y_2$ are full-sibs; $y_3$ and $y_4$ are half-sibs; $y_5$ and $y_6$ are full-sibs but both are cousins of $y_7$. The number of siblings per family is also different from family to family. The analysis of variance method we learned before cannot solve such a complicated problem. The problem, however, can be easily solved using the mixed model methodology.

## 12.1 Mixed Model

Let $y_j$ be the phenotypic value of a quantitative trait for $j = 1, \ldots, n$. Each individual phenotype can be expressed as a linear model,

$$y_j = \mu + a_j + e_j$$

where $\mu$ is the grand mean, $a_j$ is the additive genetic effect or breeding value of individual $j$, and $e_j$ is the residual error with an assumed $N(0, \sigma_E^2)$ distribution. The additive genetic effect is assumed to be $N(0, \sigma_A^2)$ distributed, where $\sigma_A^2$ is the additive genetic variance. This model is called the random effect model. However, if the grand mean is replaced by a general term that captures systematic environmental effects and these environmental effects are treated as fixed effects (parameters), then the model is called the mixed model. For $n = 7$ individuals, we have to write seven models, such as

$$y_1 = \mu + a_1 + e_1$$
$$y_2 = \mu + a_2 + e_2$$
$$y_3 = \mu + a_3 + e_3$$
$$y_4 = \mu + a_4 + e_4$$
$$y_5 = \mu + a_5 + e_5$$
$$y_6 = \mu + a_6 + e_6$$
$$y_7 = \mu + a_7 + e_7$$

It will be more convenient to write the above model using a matrix notation,

© Springer Nature Switzerland AG 2022
S. Xu, *Quantitative Genetics*, https://doi.org/10.1007/978-3-030-83940-6_12

$$\begin{bmatrix} y_1 \\ \vdots \\ y_7 \end{bmatrix} = \begin{bmatrix} 1 \\ \vdots \\ 1 \end{bmatrix} \mu + \begin{bmatrix} a_1 \\ \vdots \\ a_7 \end{bmatrix} + \begin{bmatrix} e_1 \\ \vdots \\ e_7 \end{bmatrix}$$

which can be further condensed as

$$y = 1\mu + a + e \tag{12.1}$$

where $y$ is an $n \times 1$ vector of the phenotypic values, 1 is an $n \times 1$ vector of unity, $a$ is an $n \times 1$ vector of the genetic values, and $e$ is an $n \times 1$ vector of the residual errors. In matrix notation, we write $E(y) = 1\mu$ and

$$\text{var}(y) = V = \text{var}(a) + \text{var}(e) = A\sigma_A^2 + I\sigma_E^2$$

where $A$ is an $n \times n$ additive (numerator) relationship matrix, which is twice the coancestry matrix, i.e., $A = 2\Theta$. Recall that the coancestry matrix is calculated based on the pedigree relationships of all individuals in the population. Members from different families are assumed to be uncorrelated. The $A$ matrix in the above example is

$$A = 2\Theta = 2 \times \begin{bmatrix} \frac{1}{2} & \frac{1}{4} & 0 & 0 & 0 & 0 & 0 \\ \frac{1}{4} & \frac{1}{2} & 0 & 0 & 0 & 0 & 0 \\ 0 & 0 & \frac{1}{2} & \frac{1}{8} & 0 & 0 & 0 \\ 0 & 0 & \frac{1}{8} & \frac{1}{2} & 0 & 0 & 0 \\ 0 & 0 & 0 & 0 & \frac{1}{2} & \frac{1}{4} & \frac{1}{32} \\ 0 & 0 & 0 & 0 & \frac{1}{4} & \frac{1}{2} & \frac{1}{32} \\ 0 & 0 & 0 & 0 & \frac{1}{32} & \frac{1}{32} & \frac{1}{2} \end{bmatrix} = \begin{bmatrix} 1 & \frac{1}{2} & 0 & 0 & 0 & 0 & 0 \\ \frac{1}{2} & 1 & 0 & 0 & 0 & 0 & 0 \\ 0 & 0 & 1 & \frac{1}{4} & 0 & 0 & 0 \\ 0 & 0 & \frac{1}{4} & 1 & 0 & 0 & 0 \\ 0 & 0 & 0 & 0 & 1 & \frac{1}{2} & \frac{1}{16} \\ 0 & 0 & 0 & 0 & \frac{1}{2} & 1 & \frac{1}{16} \\ 0 & 0 & 0 & 0 & \frac{1}{16} & \frac{1}{16} & 1 \end{bmatrix}$$

Note that the diagonal elements of $A$ are all one's because no inbreeding has happened in the population. The off-diagonal elements are zero's for unrelated individuals (members from different families) and non-zero's for genetically related individuals. Therefore, the variance matrix of $y$ is

$$\text{var}(y) = V = \begin{bmatrix} 1 & 1/2 & 0 & 0 & 0 & 0 & 0 \\ 1/2 & 1 & 0 & 0 & 0 & 0 & 0 \\ 0 & 0 & 1 & 1/4 & 0 & 0 & 0 \\ 0 & 0 & 1/4 & 1 & 0 & 0 & 0 \\ 0 & 0 & 0 & 0 & 1 & 1/2 & 1/16 \\ 0 & 0 & 0 & 0 & 1/2 & 1 & 1/16 \\ 0 & 0 & 0 & 0 & 1/16 & 1/16 & 1 \end{bmatrix} \sigma_A^2 + \begin{bmatrix} 1 & 0 & 0 & 0 & 0 & 0 & 0 \\ 0 & 1 & 0 & 0 & 0 & 0 & 0 \\ 0 & 0 & 1 & 0 & 0 & 0 & 0 \\ 0 & 0 & 0 & 1 & 0 & 0 & 0 \\ 0 & 0 & 0 & 0 & 1 & 0 & 0 \\ 0 & 0 & 0 & 0 & 0 & 1 & 0 \\ 0 & 0 & 0 & 0 & 0 & 0 & 1 \end{bmatrix} \sigma_E^2$$

Before we write the probability density of the phenotypical values (vector $y$), we first introduce a multivariate normal density in a general form. Let

**Fig. 12.1** A population consists of three pedigrees of variable sizes

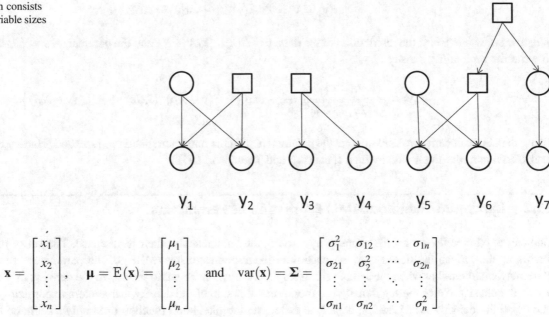

$$x = \begin{bmatrix} x_1 \\ x_2 \\ \vdots \\ x_n \end{bmatrix}, \quad \mu = E(x) = \begin{bmatrix} \mu_1 \\ \mu_2 \\ \vdots \\ \mu_n \end{bmatrix} \quad \text{and} \quad \text{var}(x) = \Sigma = \begin{bmatrix} \sigma_1^2 & \sigma_{12} & \cdots & \sigma_{1n} \\ \sigma_{21} & \sigma_2^2 & \cdots & \sigma_{2n} \\ \vdots & \vdots & \ddots & \vdots \\ \sigma_{n1} & \sigma_{n2} & \cdots & \sigma_n^2 \end{bmatrix}$$

be $n$ variables $x$, the expectation of $x$ and the variance matrix of $x$, respectively. Assume that $x$ follows a multivariate normal distribution. The multivariate normal density is

$$f(x) = \frac{1}{\sqrt{(2\pi)^n \, |\Sigma|}} \exp\left\{ -\frac{1}{2}(x-\mu)^T \Sigma^{-1}(x-\mu) \right\}$$

Traditionally, a vector or a matrix is written in bold face and a single variable is written in plain italic text. However, since matrix is used so often in statistics and in quantitative genetics, we simply italicize all variables, including scalar and matrix, in plain form. Now, let us go to the genetic model for the phenotypic values $y$ from the seven individuals in the population defined in Fig. 12.1. The expectation of $y$ is $E(y) = 1\mu$ and the variance matrix is $\text{var}(y) = V$. The multivariate normal density is written as

$$f(y) = \frac{1}{\sqrt{(2\pi)^n \, |V|}} \exp\left\{ -\frac{1}{2}(y-1\mu)^T V^{-1}(y-1\mu) \right\}$$

Conventionally, for $n$ data points ($n$ is the sample size), a mixed model is written as follows:

$$y = X\beta + Z\gamma + e$$

where $y$ is an $n \times 1$ vector of the response variable, $X$ is an $n \times p$ design matrix for $p$ fixed effects $\beta$, $Z$ is an $n \times q$ design matrix for $q$ random effects $\gamma$, and $e$ is an $n \times 1$ vector of residuals. The fixed effects $\beta$ are parameters (constants) subject to estimation. The random effects $\gamma$ follow a multivariate normal distribution with mean 0 and variance matrix $G$, which represents parameters ($\gamma$ are not parameters). We often write $\gamma \sim N(0, G)$, where $G$ is a $q \times q$ variance matrix. The residuals follow a multivariate normal distribution, denoted by $e \sim N(0, R)$, where $R$ is an $n \times n$ matrix representing the variance matrix of the residuals. To match our random model in Eq. (12.1) to this generic mixed model, we can see that $X = 1$, $\beta = \mu$, $Z = I_n$ (identity matrix), $\gamma = a$, $G = A\sigma_A^2$, and $R = I\sigma_E^2$. The expectation and variance of the typical mixed model are

$$E(y) = X\beta$$

and

$$\text{var}(y) = V = Z\text{var}(\gamma)Z^T + \text{var}(e) = ZGZ^T + R$$

We now know the data $y$, the distribution of the data, $y \sim N(X\beta, ZGZ^T + R)$, and the parameters $\theta = \{\beta, G, R\}$; these allow us to write the probability density of $y$,

$$f(y) = \frac{1}{\sqrt{(2\pi)^n \mid ZGZ^T + R \mid}} \exp\left\{-\frac{1}{2}(y - X\beta)^T (ZGZ^T + R)^{-1}(y - X\beta)\right\}$$

Using the above density, we can estimate the parameters with the maximum likelihood method (Hartley and Rao 1967) or the restricted maximum likelihood method (Patterson and Thompson 1971).

## 12.2  Maximum Likelihood (ML) Estimation of Parameters

One can maximize the probability density by varying the parameters in their legal space. The set of parameter values that maximize the probability density are the maximum likelihood estimates (MLE) of the parameters. Since the density is an extremely small number when the sample size is large, it is not convenient to maximize the density; instead, we maximize the log of the density. Since the log density is a monotonic function of the density, parameters maximizing the log density also maximize the density itself. The log density is called the log likelihood function or simply likelihood function without the word "log." The likelihood function is $L(\theta) = \ln f(y)$ and has the following form,

$$L(\theta) = -\frac{1}{2}\ln \mid ZGZ^T + R \mid -\frac{1}{2}(y - X\beta)^T (ZGZ^T + R)^{-1}(y - X\beta)$$

where a constant, $-\frac{1}{2}n\ln(2\pi)$, has been ignored because it does not contain the parameters and thus is irrelevant to the maximum likelihood estimates of the parameters. The parameters $\theta = \{\beta, G, R\}$ are general forms in linear mixed models. Our mixed model (12.1) is a special case and thus the parameters are $\theta = \{\mu, \sigma_A^2, \sigma_E^2\}$ due to the fact that $G = A\sigma_A^2$ and $R = I\sigma_E^2$, where $A$ is the numerator relationship matrix, not parameters. The actual likelihood function customized into our problem is

$$L(\theta) = -\frac{1}{2}\ln \mid A\sigma_A^2 + I\sigma_E^2 \mid -\frac{1}{2}(y - 1\mu)^T (A\sigma_A^2 + I\sigma_E^2)^{-1}(y - 1\mu) \tag{12.2}$$

The solution is not explicit and thus a numerical optimization method must be resorted. Many computational algorithms are available to numerically solve for the MLE of parameters. Typical algorithms include the Newton-Raphson algorithm, the simplex algorithm and the expectation-maximization (EM) algorithm. The Newton-Raphson iteration equation is

$$\theta^{(t+1)} = \theta^{(t)} - \left[\frac{\partial^2 L(\theta^{(t)})}{\partial\theta\partial\theta^T}\right]^{-1}\left[\frac{\partial L(\theta^{(t)})}{\partial\theta}\right]$$

Let $t = 0$ and $\theta^{(t)} = \theta^{(0)}$ be some initial values of the parameters. The iteration continues until the parameters converge to some constant values. Let $t = t_0$ be the number of iterations that the process converges, i.e.,

$$\left(\theta^{(t_0)} - \theta^{(t_0-1)}\right)^T \left(\theta^{(t_0)} - \theta^{(t_0-1)}\right) < \varepsilon$$

the MLE of the parameters are $\widehat{\theta} = \theta^{(t_0)}$, $\varepsilon$ is a small number such as $\varepsilon = 10^{-8}$. In fact, the MIXED procedure in SAS is very efficient for estimating variance components involved in mixed models, and we do not have to understand the optimization algorithms. More computationally efficient methods will be introduced in Chaps. 19 and 20. The estimated parameters are denoted by $\widehat{\theta} = \{\widehat{\mu}, \widehat{\sigma}_A^2, \widehat{\sigma}_E^2\}$. The asymptotic variance-covariance matrix of the estimated parameters is

$$\text{var}\left(\widehat{\theta}\right) = -\left[\frac{\partial^2 L\left(\widehat{\theta}\right)}{\partial\theta\partial\theta^T}\right]^{-1}$$

where

$$-\left[\frac{\partial^2 L\left(\widehat{\theta}\right)}{\partial\theta\partial\theta^T}\right]^{-1} = -\left[\begin{array}{ccc} \dfrac{\partial^2 L\left(\widehat{\theta}\right)}{\partial\mu\partial\mu} & \dfrac{\partial^2 L\left(\widehat{\theta}\right)}{\partial\mu\partial\sigma_A^2} & \dfrac{\partial^2 L\left(\widehat{\theta}\right)}{\partial\mu\partial\sigma_E^2} \\[2mm] \dfrac{\partial^2 L\left(\widehat{\theta}\right)}{\partial\sigma_A^2\partial\mu} & \dfrac{\partial^2 L\left(\widehat{\theta}\right)}{\partial\sigma_A^2\partial\sigma_A^2} & \dfrac{\partial^2 L\left(\widehat{\theta}\right)}{\partial\sigma_A^2\partial\sigma_E^2} \\[2mm] \dfrac{\partial^2 L\left(\widehat{\theta}\right)}{\partial\sigma_E^2\partial\mu} & \dfrac{\partial^2 L\left(\widehat{\theta}\right)}{\partial\sigma_E^2\partial\sigma_A^2} & \dfrac{\partial^2 L\left(\widehat{\theta}\right)}{\partial\sigma_E^2\partial\sigma_E^2} \end{array}\right]^{-1}$$

is a $3 \times 3$ matrix. It is the variance matrix of the three estimated parameters,

$$\text{var}\begin{bmatrix} \widehat{\mu} \\ \widehat{\sigma}_A^2 \\ \widehat{\sigma}_E^2 \end{bmatrix} = \begin{bmatrix} \text{var}(\widehat{\mu}) & \text{cov}\left(\widehat{\mu},\widehat{\sigma}_A^2\right) & \text{cov}\left(\widehat{\mu},\widehat{\sigma}_E^2\right) \\ \text{cov}\left(\widehat{\mu},\widehat{\sigma}_A^2\right) & \text{var}\left(\widehat{\sigma}_A^2\right) & \text{cov}\left(\widehat{\sigma}_A^2,\widehat{\sigma}_E^2\right) \\ \text{cov}\left(\widehat{\mu},\widehat{\sigma}_E^2\right) & \text{cov}\left(\widehat{\sigma}_A^2,\widehat{\sigma}_E^2\right) & \text{var}\left(\widehat{\sigma}_E^2\right) \end{bmatrix} \tag{12.3}$$

Once the parameters are estimated, $\widehat{\theta} = \{\widehat{\mu},\widehat{\sigma}_A^2,\widehat{\sigma}_E^2\}$, the estimated heritability is obtained by taking the ratio,

$$\widehat{h}^2 = \frac{\widehat{\sigma}_A^2}{\widehat{\sigma}_A^2 + \widehat{\sigma}_E^2}$$

With user defined matrix $A$ as the variance structure of the random effects, the MIXED procedure in SAS will not be able to generate an ANOVA table that allows us to manually calculate the variance of the ratio (estimated heritability) using the delta method introduced in Chap. 10. However, the MIXED procedure does provide an asymptotic variance-covariance matrix for the estimated variance components,

$$\text{var}\begin{bmatrix} \widehat{\sigma}_A^2 \\ \widehat{\sigma}_E^2 \end{bmatrix} = \begin{bmatrix} \text{var}\left(\widehat{\sigma}_A^2\right) & \text{cov}\left(\widehat{\sigma}_A^2,\widehat{\sigma}_E^2\right) \\ \text{cov}\left(\widehat{\sigma}_A^2,\widehat{\sigma}_E^2\right) & \text{var}\left(\widehat{\sigma}_E^2\right) \end{bmatrix}$$

which is the lower right $2 \times 2$ submatrix of the matrix shown in Eq. (12.3). This matrix allows us to find the standard error of $\widehat{h}^2$ even more conveniently than method in Chap. 10. Let $X = \widehat{\sigma}_A^2$ and $Y = \widehat{\sigma}_A^2 + \widehat{\sigma}_E^2$. The estimated heritability is

$$\widehat{h}^2 = \frac{\widehat{\sigma}_A^2}{\widehat{\sigma}_A^2 + \widehat{\sigma}_E^2} = \frac{X}{Y}$$

Therefore, the variance of $\widehat{h}^2$ is approximated by

$$\text{var}\left(\widehat{h}^2\right) = \left(\frac{X}{Y}\right)^2\left[\frac{\text{var}(X)}{X^2} - 2\frac{\text{cov}(X,Y)}{XY} + \frac{\text{var}(Y)}{Y^2}\right]$$

where $\text{var}(X) = \text{var}\left(\widehat{\sigma}_A^2\right)$,

$$\text{var}(Y) = \text{var}(\widehat{\sigma}_A^2 + \widehat{\sigma}_E^2) = \text{var}(\widehat{\sigma}_A^2) + \text{var}(\widehat{\sigma}_E^2) + 2\text{cov}(\widehat{\sigma}_A^2, \widehat{\sigma}_E^2)$$

and

$$\text{cov}(X, Y) = \text{cov}(\widehat{\sigma}_A^2, \widehat{\sigma}_A^2 + \widehat{\sigma}_E^2) = \text{var}(\widehat{\sigma}_A^2) + \text{cov}(\widehat{\sigma}_A^2, \widehat{\sigma}_E^2)$$

## 12.3   Restricted Maximum Likelihood (REML) Estimation of Parameters

The restricted maximum likelihood method for variance estimation was initially proposed by Patterson and Thompson (Patterson and Thompson 1971). The idea was to estimate the variance components without interference from the fixed effects. They proposed a set of error contrasts of $y$ using a transformation matrix $K$ so that $y^* = Ky$, where $K$ is an $(n - p) \times n$ matrix with a property of $KX = 0$. As a result, the error contrasts are

$$y^* = Ky = KX\beta + K(Z\gamma + e) = K(Z\gamma + e)$$

One possible candidate of K is the first $n - p$ rows of matrix $I - X(X^TX)^{-1}X^T$. This matrix has the property of

$$KX = \left[I - X(X^TX)^{-1}X^T\right]X = 0$$

The expectation and variance of the error contrasts are $\text{E}(y^*) = 0$ and

$$\text{var}(y^*) = \text{var}[K(Z\gamma + e)] = K(ZGZ^T + R)K^T$$

respectively. Since the fixed effects have disappeared from the error contrasts, the likelihood function becomes

$$L_R(\theta) = -\frac{1}{2}\ln \mid K(ZGZ^T + R)K^T \mid -\frac{1}{2}yK^T\left[K(ZGZ^T + R)K^T\right]^{-1}Ky$$

where $\theta = \{G, R\}$, which does not include $\beta$. The restricted likelihood function seems to be complicated and hard to handle. Harville (Harville 1974; Harville 1977) showed that the likelihood function can be written in a form similar to the original likelihood function,

$$L_R(\theta) = -\frac{1}{2}\ln \mid V \mid -\frac{1}{2}\ln \mid X^TV^{-1}X \mid -\frac{1}{2}\left(y - X\widehat{\beta}\right)^T V^{-1}\left(y - X\widehat{\beta}\right) \tag{12.4}$$

where $V = ZGZ^T + R$ and

$$\widehat{\beta} = (X^TV^{-1}X)^{-1}X^TV^{-1}y$$

is a function of $V = ZGZ^T + R$ and thus a function of the parameters but is not a parameter itself. The restricted likelihood function differs from the original likelihood function only by an additional term $-\frac{1}{2}\ln \mid X^TV^{-1}X \mid$. A simple proof of Harville's restricted likelihood function (12.4) can be found in Xu (Xu 2019). The same Newton-Raphson iterative equations can be used to find the REML solution of the parameter vector $\theta = \{G, R\}$.

Customizing the restricted likelihood function into our genetic problem, we have

$$L_R(\theta) = -\frac{1}{2}\ln \mid A\sigma_A^2 + I\sigma_E^2 \mid -\frac{1}{2}\ln \mid 1^T(A\sigma_A^2 + I\sigma_E^2)^{-1}1 \mid -\frac{1}{2}(y - 1\widehat{\mu})^T(A\sigma_A^2 + I\sigma_E^2)^{-1}(y - 1\widehat{\mu}) \tag{12.5}$$

where 1 is an $n \times 1$ unity vector (all elements are 1's) and

$$\widehat{\mu} = \left[1^T \left(A\sigma_A^2 + I\sigma_E^2\right)^{-1}1\right]^{-1} \left[1^T \left(A\sigma_A^2 + I\sigma_E^2\right)^{-1}y\right]$$

To show the difference between the maximum likelihood estimate of $\sigma_E^2$ and the restricted maximum likelihood estimate of $\sigma_E^2$, let us define $\lambda = \sigma_A^2/\sigma_E^2$ and rewrite Eq. (12.5) as

$$L_R(\theta) = -\frac{1}{2}\ln|A\lambda + I| - \frac{1}{2}\ln|1^T(A\lambda + I)^{-1}1|$$
$$-\frac{n-1}{2}\ln\left(\sigma_E^2\right) - \frac{1}{2\sigma_E^2}(y - 1\widehat{\mu})^T(A\lambda + I)^{-1}(y - 1\widehat{\mu})$$

Given $\lambda$, the partial derivative of $L_R(\theta)$ with respect to $\sigma_E^2$ is

$$\frac{\partial L_R(\theta)}{\partial \sigma_E^2} = -\frac{n-1}{2\sigma_E^2} + \frac{1}{2\sigma_E^4}(y - 1\widehat{\mu})^T(A\lambda + I)^{-1}(y - 1\widehat{\mu})$$

Set this partial derivative to 0 and solve for $\sigma_E^2$, we get

$$\widehat{\sigma}_E^2 = \frac{1}{n-1}(y - 1\widehat{\mu})^T(A\lambda + I)^{-1}(y - 1\widehat{\mu}) \tag{12.6}$$

However, the likelihood function in Eq. (12.2) is rewritten as

$$L(\theta) = -\frac{1}{2}\ln|A\lambda + I| - \frac{n}{2}\ln\left(\sigma_E^2\right) - \frac{1}{2\sigma_E^2}(y - 1\mu)^T(A\lambda + I)^{-1}(y - 1\mu)$$

Setting

$$\frac{\partial L_R(\theta)}{\partial \sigma_E^2} = -\frac{n}{2\sigma_E^2} + \frac{1}{2\sigma_E^4}(y - 1\widehat{\mu})^T(A\lambda + I)^{-1}(y - 1\widehat{\mu}) = 0$$

and solving for $\sigma_E^2$ leads to

$$\widehat{\sigma}_E^2 = \frac{1}{n}(y - 1\widehat{\mu})^T(A\lambda + I)^{-1}(y - 1\widehat{\mu}) \tag{12.7}$$

Let us put Eq. (12.7) with Eq. (12.6) together,

$$\widehat{\sigma}_E^2 = \begin{cases} (y - 1\widehat{\mu})^T(A\lambda + I)^{-1}(y - 1\widehat{\mu})/n & \text{MLE} \\ (y - 1\widehat{\mu})^T(A\lambda + I)^{-1}(y - 1\widehat{\mu})/(n-1) & \text{REML} \end{cases}$$

We see that the REML estimate of $\sigma_E^2$ has lost one degree of freedom due to removing the fixed effect but the MLE of $\sigma_E^2$ still has $n$ degrees of freedom.

## 12.4  Likelihood Ratio Test

Before we report an estimated heritability, it is necessary to perform a hypothesis test $H_0 : h^2 = 0$. We may calculate the $(1 - \alpha) \times 100\%$ confidence interval (CI) of $\widehat{h}^2$ using $S_{\widehat{h}^2}$. Assume that the sample size is sufficiently large, we can approximate the CI by

$$\widehat{h}^2 \pm z_{1-\alpha/2} S_{\widehat{h}^2}$$

For example, if $\alpha = 0.05$, $z_{1-0.05/2} = z_{0.975} = 1.96$, and thus the CI is $\widehat{h}^2 \pm 1.96 S_{\widehat{h}^2}$. If the interval covers 0, we cannot reject the null hypothesis. An alternative but more appropriate test is the likelihood ratio test for the null hypothesis $H_0 : \sigma_A^2 = 0$. If this hypothesis is accepted, we also accept $H_0 : h^2 = 0$. So, the hypotheses formulated in the two ways are equivalent. Let $\theta_1 = \{\sigma_A^2, \sigma_E^2\}$ and $\theta_0 = \sigma_E^2$ be the parameters under the full model and the reduced model, respectively. The likelihood ratio test statistic is

$$\psi = -2\left[L_R\left(\widehat{\theta}_0\right) - L_R\left(\widehat{\theta}_1\right)\right]$$

where $L_R\left(\widehat{\theta}_0\right)$ is the restricted likelihood value under the null model, and $L_R\left(\widehat{\theta}_1\right)$ is the restricted likelihood value under the full model. If the null hypothesis is true, $\psi$ will follow a mixture of Chi-square 0 and Chi-square 1 distributions with an equal weight, i.e.,

$$\psi \sim \frac{1}{2}\chi_0^2 + \frac{1}{2}\chi_1^2$$

where $\chi_0^2$ is simply a mass at zero. The p-value for such a mixture Chi-square distribution is (Wei and Xu 2016)

$$p = \begin{cases} 1 & \text{if } \psi = 0 \\ \frac{1}{2}\Pr(\chi_1^2 > \psi) & \text{if } \psi > 0 \end{cases}$$

In practice, $\psi$ may be a very small negative number due to floating point errors in the computing system. In that case, the p-value should be calculated as

$$p = \begin{cases} 1 & \text{if } \psi < \varepsilon \\ \frac{1}{2}\Pr(\chi_1^2 > \psi) & \text{if } \psi > \varepsilon \end{cases}$$

where $\varepsilon$ is a small positive number such as $\varepsilon = 10^{-8}$.

## 12.5   Examples

### 12.5.1   Example 1

The phenotypic values and the additive relationship matrix $A$ of the seven members in the population (see Fig. 12.1) are shown in Table 12.1. The data is also stored in "data-12-1.xlsx". The first two columns hold the animal ID and the phenotypic values. The remaining columns hold the additive relationship matrix. The phenotypes and the $A$ matrix should be stored in two

**Table 12.1** The phenotypic values and the additive relationship matrix

| Animal | Y | parm | row | col1 | col2 | col3 | col4 | col5 | col6 | col7 |
|---|---|---|---|---|---|---|---|---|---|---|
| 1 | 53.94636 | 1 | 1 | 1 | 0.5 | 0 | 0 | 0 | 0 | 0 |
| 2 | 52.69279 | 1 | 2 | 0.5 | 1 | 0 | 0 | 0 | 0 | 0 |
| 3 | 52.74754 | 1 | 3 | 0 | 0 | 1 | 0.25 | 0 | 0 | 0 |
| 4 | 48.82705 | 1 | 4 | 0 | 0 | 0.25 | 1 | 0 | 0 | 0 |
| 5 | 35.81209 | 1 | 5 | 0 | 0 | 0 | 0 | 1 | 0.5 | 0.0625 |
| 6 | 46.09668 | 1 | 6 | 0 | 0 | 0 | 0 | 0.5 | 1 | 0.0625 |
| 7 | 47.73397 | 1 | 7 | 0 | 0 | 0 | 0 | 0.0625 | 0.0625 | 1 |

different SAS datasets. The following SAS code reads the whole data, separates the two parts of the data and then calls the MIXED procedure in SAS to estimate variance components.

```
proc import datafile="data-12-1.xlsx" out=one dbms=xlsx replace;
run;
data phe;
    set one;
    keep animal y;
run;
data aa;
    set one;
    drop animal y;
run;
proc mixed data=phe method=ml asycov;
    class animal;
    model y = /solution;
    parms (1) (1) / lowerb=1e-5 1e-5;
    random animal/type=lin(1) ldata=aa solution;
run;
```

The method = ml option in the proc mixed statement calls the maximum likelihood method (default is REML). The asycov option tells the program to display the variance-covariance matrix of the two estimated variance components. Below are the outputs with the information we need, where Table 12.2 shows the estimated variance components and the asymptotic variance-covariance matrix of the estimated variance parameters and Table 12.3 shows the estimated fixed effect (intercept) along with the standard error of the estimated fixed effect. The estimated variance components are.

$$\widehat{\sigma}_A^2 = 25.0585 \quad \text{and} \quad \widehat{\sigma}_E^2 = 6.6558$$

The estimated fixed effect is

$$\widehat{\mu} \pm S_{\widehat{\mu}} = 48.4437 \pm 2.4226$$

The asymptotic variance matrix of the estimated variance components,

**Table 12.2** Output of PROC MIXED for the data in Table 12.1 using the ML method (estimated variance parameters and their asymptotic covariance matrix)

| Covariance parameter estimates | | Asymptotic covariance matrix of estimates | |
|---|---|---|---|
| Cov Parm | Estimate | CovP1 | CovP2 |
| LIN(1) | 25.0585 | 1042.25 | −699.81 |
| Residual | 6.6558 | −699.81 | 660.17 |

**Table 12.3** Output of PROC MIXED for the data in Table 12.1 using the ML method (estimated fixed effect and the standard error of the estimate)

| Solution for fixed effects | | | | | |
|---|---|---|---|---|---|
| Effect | Estimate | Standard error | DF | t Value | Pr > |t| |
| Intercept | 48.4437 | 2.4226 | 6 | 20.00 | <0.0001 |

$$\text{var}\begin{bmatrix} \widehat{\sigma}_A^2 \\ \widehat{\sigma}_E^2 \end{bmatrix} = \begin{bmatrix} \text{var}(\widehat{\sigma}_A^2) & \text{cov}(\widehat{\sigma}_A^2, \widehat{\sigma}_E^2) \\ \text{cov}(\widehat{\sigma}_A^2, \widehat{\sigma}_E^2) & \text{var}(\widehat{\sigma}_E^2) \end{bmatrix} = \begin{bmatrix} 1042.25 & -699.81 \\ -699.81 & 660.17 \end{bmatrix}$$

The estimated heritability is

$$\widehat{h}^2 = \frac{\widehat{\sigma}_A^2}{\widehat{\sigma}_A^2 + \widehat{\sigma}_E^2} = \frac{25.0585}{25.0585 + 6.6558} = 0.7901325$$

Let $X = \widehat{\sigma}_A^2 = 25.0585$ and $Y = \widehat{\sigma}_A^2 + \widehat{\sigma}_E^2 = 31.7143$. The variance of $X$ is

$$\text{var}(X) = \text{var}(\widehat{\sigma}_A^2) = 1042.25$$

The variance of $Y$ is

$$\text{var}(Y) = \text{var}(\widehat{\sigma}_A^2) + \text{var}(\widehat{\sigma}_E^2) + 2\text{cov}(\widehat{\sigma}_A^2, \widehat{\sigma}_E^2) = 1042.25 + 660.17 - 2 \times 699.81 = 302.8$$

The covariance between $X$ and $Y$ is

$$\text{cov}(X, Y) = \text{var}(\widehat{\sigma}_A^2) + \text{cov}(\widehat{\sigma}_A^2, \widehat{\sigma}_E^2) = 1042.25 - 699.81 = 342.44$$

The variance of $\widehat{h}^2$ is

$$\begin{aligned}
\text{var}(\widehat{h}^2) &= \left(\frac{X}{Y}\right)^2 \left[\frac{\text{var}(X)}{X^2} - 2\frac{\text{cov}(X, Y)}{XY} + \frac{\text{var}(Y)}{Y^2}\right] \\
&= \left(\frac{25.0585}{31.7143}\right)^2 \left(\frac{1042.25}{25.0585^2} - 2 \times \frac{342.44}{25.0585 \times 31.7143} + \frac{302.8}{31.7143^2}\right) \\
&= 0.6861673
\end{aligned}$$

The standard error of $\widehat{h}^2$ is

$$S_{\widehat{h}^2} = \sqrt{\text{var}(\widehat{h}^2)} = \sqrt{0.6861673} = 0.8283522$$

The estimated heritability should be reported as

$$\widehat{h}^2 \pm S_{\widehat{h}^2} = 0.7901325 \pm 0.8283522$$

The standard error is so large that the estimated heritability is not reliable, which is expected because the sample size is too small. Under the full model, the likelihood value is $-2L(\widehat{\theta}_1) = 43.7$. We then fitted the null model using the following SAS code,

```
proc mixed data=phe method=ml asycov;
    model y = /solution;
run;
```

Under the null model, the likelihood value is $-2L(\widehat{\theta}_0) = 44.4$. Therefore, the likelihood ratio test statistic is

**Table 12.4** Output of PROC MIXED for the data in Table 12.1 using the REML method (estimated variance parameters and their asymptotic covariance matrix)

| Covariance parameter estimates | | Asymptotic covariance matrix of estimates | |
| --- | --- | --- | --- |
| Cov Parm | Estimate | CovP1 | CovP2 |
| LIN(1) | 36.3216 | 1960.44 | −1170.58 |
| Residual | 1.9231 | −1170.58 | 938.51 |

**Table 12.5** Output of PROC MIXED for the data in Table 12.1 using the REML method (estimated fixed effect and the standard error of the estimate)

| Solution for fixed effects | | | | | |
| --- | --- | --- | --- | --- | --- |
| Effect | Estimate | Standard error | DF | t value | Pr > \|t\| |
| Intercept | 48.4717 | 2.7177 | 6 | 17.84 | <0.0001 |

$$\psi = -2L\left(\widehat{\theta}_0\right) + 2L\left(\widehat{\theta}_1\right) = 44.4 - 43.7 = 0.7$$

The p-value is

$$p = \frac{1}{2}\Pr\left(\chi_1^2 > 0.7\right) = 0.2013918$$

We cannot reject the null hypothesis $H_0 : \sigma_A^2 = 0$ or $H_0 : h^2 = 0$. Let us change the method from ML to REML for the same sample data. The SAS code for the REML method is.

```
proc mixed data=phe method=reml asycov;
    class animal;
    model y = /solution;
    parms (1) (1) / lowerb=1e-5 1e-5;
    random animal/type=lin(1) ldata=aa solution;
run;
```

The output of the REML analysis is given in Table 12.4 (estimated variances and their asymptotic variance-covariance matrix) and Table 12.5 (estimated fixed effect and its standard error). The estimated heritability from the REML analysis is

$$\widehat{h}^2 = \frac{\widehat{\sigma}_A^2}{\widehat{\sigma}_A^2 + \widehat{\sigma}_E^2} = \frac{36.3216}{36.3216 + 1.9231} = 0.9497159$$

This estimated heritability is much higher than the one estimated from the ML method. The sample is too small and thus we cannot say too much about the difference. It is interesting to compare the standard error of the REML with that of the ML. Ignoring the intermediate steps, we simply show the standard error from the REML method,

$$S_{\widehat{h}^2} = \sqrt{\mathrm{var}\left(\widehat{h}^2\right)} = \sqrt{0.6585681} = 0.8115221$$

The estimated heritability should be reported as

$$\widehat{h}^2 \pm S_{\widehat{h}^2} = 0.9497159 \pm 0.8115221$$

The standard error from the REML analysis is slightly smaller than that of the ML analysis (0.8283522). Again, we cannot say anything about the difference because of the small sample size.

**Table 12.6** Partial records of reformatted body weights (gram) of mice measured from 15 full-sib families of laboratory mice (data-12-2.xlsx)

| Animal | Family | Sib | Weight |
|---|---|---|---|
| 1 | 1 | 1 | 19.9 |
| 2 | 1 | 2 | 21.1 |
| 3 | 1 | 3 | 21 |
| 4 | 1 | 4 | 19.2 |
| 5 | 2 | 1 | 18.7 |
| 6 | 2 | 2 | 20.8 |
| 7 | 2 | 3 | 19.6 |
| 8 | 2 | 4 | 21.4 |
| 9 | 3 | 1 | 20.8 |
| 10 | 3 | 2 | 15.5 |
| 11 | 3 | 3 | 11.5 |
| 12 | 3 | 4 | 12.7 |
| 13 | 4 | 1 | 26.1 |
| 14 | 4 | 2 | 25 |
| 15 | 4 | 3 | 22 |
| 16 | 4 | 4 | 36.5 |
| 17 | 5 | 1 | 22.5 |
| 18 | 5 | 2 | 17.8 |
| 19 | 5 | 3 | 23.7 |
| 20 | 5 | 4 | 15.7 |
| . | . | . | . |
| . | . | . | . |
| . | . | . | . |
| . | . | . | . |
| 57 | 15 | 1 | 17.6 |
| 58 | 15 | 2 | 20.5 |
| 59 | 15 | 3 | 33.1 |
| 60 | 15 | 4 | 23.8 |

### 12.5.2 Example 2

For the sib data in Chap. 10 (Table 10.16), we deleted the parental phenotypes and kept only the progeny data for sib analysis to estimate the heritability. The ANOVA in Chap. 10 is called a "family model" in a sense that we treated each family effect as a random variable to estimate the variance between family mean. We now use the "individual model" learned in this chapter by treating each individual effect as a random variable. We provided the additive relationship matrix $A$ for all the 60 individuals in the population and directly estimated $\sigma_A^2 = 2\sigma_B^2$ and $\sigma_E^2 = \sigma_W^2 - \sigma_B^2$ from the MIXED procedure in SAS. Part of the phenotypic data is shown in Table 12.6, where we added an additional column for the animal IDs (data-12-2.xlsx). We then created another data holding the $A$ matrix for the 60 individuals (data-12-3.xlsx). The $A$ matrix is a $60 \times 60$ block diagonal matrix with 15 blocks (one block per family). Members within a family are full-siblings and members between families are not related. Table 12.7 shows the first two blocks (families) of the $60 \times 60$ $A$ matrix. The SAS code to read the data and estimate the variance components is given below.

```
proc import datafile="data-12-2.xlsx" out=phe dbms=xlsx replace;
run;
proc import datafile="data-12-3.xlsx" out=aa dbms=xlsx replace;
run;
proc mixed data=phe method=reml asycov;
    class animal;
    model y = /solution;
    parms (1) (1) / lowerb=1e-5 1e-5;
    random animal/type=lin(1) ldata=aa;
run;
```

**Table 12.7** The first two blocks of the $A$ matrix for the 60 siblings (4 siblings per family) (data-12-3.xlsx)

|      | col1 | col2 | col3 | col4 | col5 | col6 | col7 | col8 |
|------|------|------|------|------|------|------|------|------|
| row1 | 1    | 0.5  | 0.5  | 0.5  | 0    | 0    | 0    | 0    |
| row2 | 0.5  | 1    | 0.5  | 0.5  | 0    | 0    | 0    | 0    |
| row3 | 0.5  | 0.5  | 1    | 0.5  | 0    | 0    | 0    | 0    |
| row4 | 0.5  | 0.5  | 0.5  | 1    | 0    | 0    | 0    | 0    |
| row5 | 0    | 0    | 0    | 0    | 1    | 0.5  | 0.5  | 0.5  |
| row6 | 0    | 0    | 0    | 0    | 0.5  | 1    | 0.5  | 0.5  |
| row7 | 0    | 0    | 0    | 0    | 0.5  | 0.5  | 1    | 0.5  |
| row8 | 0    | 0    | 0    | 0    | 0.5  | 0.5  | 0.5  | 1    |

**Table 12.8** Output of PROC MIXED for the 60 siblings from 15 full-sib families (estimated variance parameters and their asymptotic covariance matrix)

| Covariance parameter estimates | | Asymptotic covariance matrix of estimates | |
|---|---|---|---|
| Cov Parm | Estimate | CovP1 | CovP2 |
| LIN(1)   | 13.3402  | 95.7981  | −60.1028 |
| Residual | 16.7643  | −60.1028 | 60.5609  |

**Table 12.9** Output of PROC MIXED for the 60 siblings from 15 full-sib families (estimated fixed effect and the standard error of the estimate)

| Solution for fixed effects | | | | | |
|---|---|---|---|---|---|
| Effect    | Estimate | Standard error | DF | t Value | Pr > \|t\| |
| Intercept | 20.5067  | 0.9139         | 59 | 22.44   | <0.0001 |

The output is illustrated in Table 12.8 (variance components) and Table 12.9 (fixed effect). The estimated variance components are.

$$\hat{\sigma}_A^2 = 13.3402 \text{ and } \hat{\sigma}_E^2 = 16.7643.$$

The estimated heritability is

$$\hat{h}^2 = \hat{\sigma}_A^2 / \left( \hat{\sigma}_A^2 + \hat{\sigma}_E^2 \right) = 13.3402/(13.3402 + 16.7643) = 0.4431298$$

With the ANOVA method (the family model) in Chap. 10, $\hat{\sigma}_B^2 = 6.67$ and $\hat{\sigma}_W^2 = 23.4344$, and thus the estimated heritability was

$$\hat{h}^2 = 2\hat{\sigma}_B^2 / \left( \hat{\sigma}_B^2 + \hat{\sigma}_W^2 \right) = 2 \times 6.67/(6.67 + 23.4344) = 2 \times 0.2216 = 0.4432$$

Clearly, the estimated heritability from the family model (Chap. 10) and the animal model (Chap. 12) are identical.

## 12.6 Monte Carlo Simulation

Monte Carlo simulation is an important tool to generate data following a particular distribution. Analyzing simulated data will help investigators understand the mechanisms guiding the biological process and the development of diseases and agronomic traits. Under the infinitesimal model of quantitative traits (Fisher 1918), the genetic value of an individual is the cumulative effects of infinite number of loci each with an infinitely small effect. To simulate a quantitative trait, one must simulate many loci each following Mendel's laws of inheritance. The inbreeding and coancestry coefficients allow us to simulate pedigree data directly from the quantitative genetic model without simulating individual loci (Verrier et al. 1991). Assume that a population consists of many pedigrees which are independent from each other. The simulation is conducted one pedigree at a time. The simulation starts with a tabular pedigree and generates the genetic value of each member sequentially from the top to the bottom of the table. Recall that $f_k$ is the inbreeding coefficient of individual $k$ and $\theta_{ij}$ is the coancestry between $i$ and $j$. Let $\sigma_A^2 = V_A$ be the additive genetic variance. If $k$ is an ancestor (neither parent is in the pedigree) or an outside member marrying into the pedigree, $a_k \sim N(0, \sigma_A^2)$. If $k$ is not a founder and his/her parents are $i$ and $j$, then

$$a_k \sim N\left\{\frac{1}{2}\left(a_i + a_j\right), \frac{1}{2}\left[1 - \frac{1}{2}\left(f_i + f_j\right)\right]\sigma_A^2\right\}$$

Verbally, we say that the genetic value of individual $k$ is simulated from a normal distribution with mean being the average of the two parents and variance being the within-family segregation variance (also called Mendelian segregation variance). The mean is

$$M_k = \frac{1}{2}\left(a_i + a_j\right)$$

and the variance is

$$W_k = \frac{1}{2}\left[1 - \frac{1}{2}\left(f_i + f_j\right)\right]\sigma_A^2$$

The Mendelian segregation variance was first derived by Foulley and Chevalet (Foulley and Chevalet 1981) and then by Wang and Xu (2019) using a mixture model. The Mendelian segregation variance we learned before ($W = \frac{1}{2}V_A$) is a special case when neither parent is inbred.

Figure 12.2 shows an extended pedigree with 12 members, where the founders (individuals 1 and 2) are remotely related with a coancestry coefficient of $\theta_{12} = 0.0625$. In addition, one of the two persons who marry into the family (individual 3) is inbred with an inbreeding coefficient of 0.125, i.e., $f_3 = 0.125$. Assume that the additive genetic variance is $\sigma_A^2 = V_A = 5$. Table 12.10 is the tabular version of the pedigree along with the inbreeding coefficients (first four columns), which provides all information needed for the simulation. The mean and the within-family segregation variance used to simulate the breeding value were added to Table 12.10 (columns 5 and 6). The simulated breeding value was added to the table as well (Table 12.10, last column).

We partitioned the 12 members of the pedigree into four categories. Category 1 consists of individuals 1, 2, 3, and 7 who are the founders or married into the family. Their breeding values were simulated from mean 0 and variance $V_A = 5$.

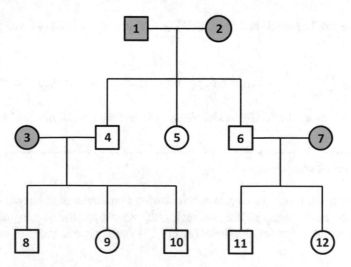

**Fig. 12.2** An extended pedigree for the simulation study

**Table 12.10** Tabular presentation of the pedigree shown in Fig. 12.2

| Individual | Dad | Mom | Inbreeding coefficient | Mean[a] | Variance[b] | Breeding value[c] |
|---|---|---|---|---|---|---|
| 1 | 0 | 0 | 0 | 0 | 5 | −2.312328 |
| 2 | 0 | 0 | 0 | 0 | 5 | 1.612736 |
| 3 | 0 | 0 | 0.1250 | 0 | 5 | −0.876526 |
| 4 | 1 | 2 | 0.0625 | −0.349796 | 2.5 | −2.552231 |
| 5 | 1 | 2 | 0.0625 | −0.349796 | 2.5 | 2.2595757 |
| 6 | 1 | 2 | 0.0625 | −0.349796 | 2.5 | 1.2707087 |
| 7 | 0 | 0 | 0 | 0 | 5 | 3.1076833 |
| 8 | 4 | 3 | 0 | −1.714378 | 2.265625 | −0.133176 |
| 9 | 4 | 3 | 0 | −1.714378 | 2.265625 | −0.945381 |
| 10 | 4 | 3 | 0 | −1.714378 | 2.265625 | −0.857673 |
| 11 | 6 | 7 | 0 | 2.189196 | 2.421875 | 1.962134 |
| 12 | 6 | 7 | 0 | 2.189196 | 2.421875 | −0.664754 |

[a]The mean used to simulate the breeding value
[b]The variance used to simulate the breeding value
[c]The breeding value simulated from a normal distribution with the mean and variance shown in columns 5 and 6, respectively

Category 2 includes individuals 4, 5, and 6, who are full-siblings with individuals 1 and 2 being their parents. Therefore, the mean and variance used to generate their breeding values are

$$\frac{1}{2}(a_1 + a_2) = \frac{1}{2}(1.612736 - 2.312328) = -0.349796$$

and

$$\frac{1}{2}\left[1 - \frac{1}{2}(f_1 + f_2)\right]\sigma_A^2 = \frac{1}{2} \times 5 = 2.5$$

Category 3 consists of individuals 8, 9, and 10, all sharing common parents identified as individuals 3 and 4. Therefore, the mean and variance used to generate the breeding values of these individuals are

$$\frac{1}{2}(a_3 + a_4) = \frac{1}{2}(-0.876526 - 2.552231) = -1.714378$$

and

$$\frac{1}{2}\left[1 - \frac{1}{2}(f_3 + f_4)\right]\sigma_A^2 = \frac{1}{2}\left[1 - \frac{1}{2}(0.125 + 0.0625)\right] \times 5 = 2.265625$$

Finally, category 4 includes individuals 11 and 12, who share common parents labeled as individuals 6 and 7. Therefore, the mean and variance used to simulate these two individuals are

$$\frac{1}{2}(a_6 + a_7) = \frac{1}{2}(1.2707087 + 3.1076833) = 2.189196$$

and

$$\frac{1}{2}\left[1 - \frac{1}{2}(f_6 + f_7)\right]\sigma_A^2 = \frac{1}{2}\left[1 - \frac{1}{2}(0.0625 + 0)\right] \times 5 = 2.421875$$

To simulate a phenotypic value, one simply adds a random error simulated from $N(0, V_E)$ to the breeding value. One can also add a grand mean to the simulated phenotypic value. For example, assume that the grand mean is $\mu = 12$ and the error variance is $V_E = 15$. Let us generate a random environmental error for individual 1 in the pedigree (Table 12.10), say $e_1 \sim N(0, 15)$ and $e_1$ happened to be 4.300362. The simulated phenotypic value for individual 1 should be

$$y_1 = \mu + a_1 + e_1 = 12 - 2.312328 + 4.300362 = 13.98803$$

The Monte Carlo simulation described here is very efficient for large populations consisting of multiple large pedigrees. The simulation does not require the additive relationship matrix (only requires the tabular pedigree). If a pedigree is not very large, say less than a few hundred individuals, an alternative simulation algorithm can be adopted, which needs the additive relationship matrix $A$.

Let us introduce the alternative algorithm of simulation using this 12-member pedigree (Table 12.10) as an example. First, we need to calculate the additive relationship matrix. The SAS code to calculate the numerator relationship matrix is given below.

```
data one;
    input child dad mom coef;
    cov=2*coef;
datalines;
1       .       .       .
2       .       .       .
3       -1      -2      0.125
4       1       2       0.0625
5       1       2       0.0625
6       1       2       0.0625
7       .       .       .
8       4       3       .
9       4       3       .
10      4       3       .
11      6       7       .
12      6       7       .
;

proc inbreed data=one matrix covar;
    var child dad mom cov;
    ods output CovarianceCoefficient=numerator;
run;
data numerator;
    set numerator;
    drop N1 N2;
    if Individual^="-1" & Individual^="-2";
run;
proc print data=numerator;
run;
```

The additive relationship matrix (twice the coancestry matrix) is shown in Table 12.11. Denote the $12 \times 12$ matrix by $A$. Let us perform Cholesky decomposition on $A$ so that $H^T H = A$ where $H$ is an upper triangular matrix. The breeding values for the 12 members (denoted by a $12 \times 1$ vector $a$) can be simulated via

$$a = H^T u \sigma_A$$

where $u \sim N(0, I_{12})$ is a $12 \times 1$ vector containing 12 simulated standardized normal variables. One can verify that

$$\text{var}(a) = H^T \text{var}(u) H \sigma_A^2 = H^T I H \sigma_A^2 = H^T H \sigma_A^2 = A \sigma_A^2$$

**Table 12.11** The numerator (additive) relationship matrix of the 12 members in pedigree Table 12.10

|       | col1   | col2   | col3  | col4   | col5   | col6   | col7 | col8   | col9   | col10  | col11  | col12  |
|-------|--------|--------|-------|--------|--------|--------|------|--------|--------|--------|--------|--------|
| row1  | 1      | 0.125  | 0     | 0.5625 | 0.5625 | 0.5625 | 0    | 0.2813 | 0.2813 | 0.2813 | 0.2813 | 0.2813 |
| row2  | 0.125  | 1      | 0     | 0.5625 | 0.5625 | 0.5625 | 0    | 0.2813 | 0.2813 | 0.2813 | 0.2813 | 0.2813 |
| row3  | 0      | 0      | 1.125 | 0      | 0      | 0      | 0    | 0.5625 | 0.5625 | 0.5625 | 0      | 0      |
| row4  | 0.5625 | 0.5625 | 0     | 1.0625 | 0.5625 | 0.5625 | 0    | 0.5313 | 0.5313 | 0.5313 | 0.2813 | 0.2813 |
| row5  | 0.5625 | 0.5625 | 0     | 0.5625 | 1.0625 | 0.5625 | 0    | 0.2813 | 0.2813 | 0.2813 | 0.2813 | 0.2813 |
| row6  | 0.5625 | 0.5625 | 0     | 0.5625 | 0.5625 | 1.0625 | 0    | 0.2813 | 0.2813 | 0.2813 | 0.5313 | 0.5313 |
| row7  | 0      | 0      | 0     | 0      | 0      | 0      | 1    | 0      | 0      | 0      | 0.5    | 0.5    |
| row8  | 0.2813 | 0.2813 | 0.5625| 0.5313 | 0.2813 | 0.2813 | 0    | 1      | 0.5469 | 0.5469 | 0.1406 | 0.1406 |
| row9  | 0.2813 | 0.2813 | 0.5625| 0.5313 | 0.2813 | 0.2813 | 0    | 0.5469 | 1      | 0.5469 | 0.1406 | 0.1406 |
| row10 | 0.2813 | 0.2813 | 0.5625| 0.5313 | 0.2813 | 0.2813 | 0    | 0.5469 | 0.5469 | 1      | 0.1406 | 0.1406 |
| row11 | 0.2813 | 0.2813 | 0     | 0.2813 | 0.2813 | 0.5313 | 0.5  | 0.1406 | 0.1406 | 0.1406 | 1      | 0.5156 |
| row12 | 0.2813 | 0.2813 | 0     | 0.2813 | 0.2813 | 0.5313 | 0.5  | 0.1406 | 0.1406 | 0.1406 | 0.5156 | 1      |

**Table 12.12** Cholesky decomposition ($H$) of the numerator relationship matrix ($A$)

| Row | Col1 | Col2 | Col3 | Col4 | Col5 | Col6 | Col7 | Col8 | Col9 | Col10 | Col11 | Col12 |
|---|---|---|---|---|---|---|---|---|---|---|---|---|
| 1 | 1 | 0.125 | 0 | 0.5625 | 0.5625 | 0.5625 | 0 | 0.2813 | 0.2813 | 0.2813 | 0.2813 | 0.2813 |
| 2 | 0 | 0.9921567 | 0 | 0.4960784 | 0.4960784 | 0.4960784 | 0 | 0.2480833 | 0.2480833 | 0.2480833 | 0.2480833 | 0.2480833 |
| 3 | 0 | 0 | 1.0606602 | 0 | 0 | 0 | 0 | 0.5303301 | 0.5303301 | 0.5303301 | 0 | 0 |
| 4 | 0 | 0 | 0 | 0.7071068 | 0 | 0 | 0 | 0.3535534 | 0.3535534 | 0.3535534 | 0 | 0 |
| 5 | 0 | 0 | 0 | 0 | 0.7071068 | 0 | 0 | 0 | 0 | 0 | 0 | 0 |
| 6 | 0 | 0 | 0 | 0 | 0 | 0.7071068 | 0 | 0 | 0 | 0 | 0.3535534 | 0.3535534 |
| 7 | 0 | 0 | 0 | 0 | 0 | 0 | 1 | 0 | 0 | 0 | 0.5 | 0.5 |
| 8 | 0 | 0 | 0 | 0 | 0 | 0 | 0 | 0.6731085 | −0.000037 | −0.000037 | −0.000111 | −0.000111 |
| 9 | 0 | 0 | 0 | 0 | 0 | 0 | 0 | 0 | 0.6731085 | −0.000037 | −0.000111 | −0.000111 |
| 10 | 0 | 0 | 0 | 0 | 0 | 0 | 0 | 0 | 0 | 0.6731085 | −0.000111 | −0.000111 |
| 11 | 0 | 0 | 0 | 0 | 0 | 0 | 0 | 0 | 0 | 0 | 0.6959346 | −0.000108 |
| 12 | 0 | 0 | 0 | 0 | 0 | 0 | 0 | 0 | 0 | 0 | 0 | 0.6959346 |

**Table 12.13** A set of simulated standardized normal variables and the corresponding breeding values

| Individual | Standardized normal($u$) | Breeding value ($a$) |
|---|---|---|
| 1 | −0.893451 | −1.997817 |
| 2 | −0.264227 | −0.835923 |
| 3 | −0.306721 | −0.727454 |
| 4 | 0.735829 | −0.253422 |
| 5 | −0.275765 | −1.852892 |
| 6 | 0.8233611 | −0.115022 |
| 7 | −1.667103 | −3.727755 |
| 8 | −0.921723 | −1.877864 |
| 9 | 0.154289 | −0.258264 |
| 10 | −0.410639 | −1.10856 |
| 11 | 0.7437263 | −0.763866 |
| 12 | −1.897187 | −4.873721 |

Because Cholesky decomposition is required, such a method cannot handle very large pedigrees. Table 12.12 shows the upper triangular Cholesky decomposition. Let us generate a vector $u$ from $N(0, I_{12})$ and multiply $H$ by $u$ and $\sigma_A$ to obtain a vector $a$. A simulated $u$ and $a$ are shown in Table 12.13.

# References

Fisher RA. The correlation between relatives on the supposition of Mendelian inheritance. Philos Trans R Soc Edinburgh. 1918;52:399–433.

Foulley JL, Chevalet C. Méthode de prise en compte de la consanguinité dans un modèle simple de simulation de performances. Ann Genet Sel Anim. 1981;13:189–96.

Hartley HO, Rao JNK. Maximum-likelihood estimation for the mixed analysis of variance model. Biometrika. 1967;54:93–108.

Harville DA. Bayesian inference for variance components using only error contrasts. Biometrika. 1974;61:383–5.

Harville DA. Maximum likelihood approaches to variance component estimation and to related problems. J Am Stat Assoc. 1977;72:320–38.

Henderson CR. Estimation of genetic parameters (abstract). Ann Math Stat. 1950;21:309–10.

Patterson HD, Thompson R. Recovery of inter-block information when block sizes are unequal. Biometrika. 1971;58:545–54.

Verrier E, Colleau JJ, Foulley JL. Methods for predicting response to selection in small populations under additive genetic models: a review. Livest Prod Sci. 1991;29:93–114.

Wang M, Xu S. Statistics of Mendelian segregation - a mixture model. J Anim Breed Genet. 2019; https://doi.org/10.1111/jbg.12394.

Wei J, Xu S. A random model approach to QTL mapping in multi-parent advanced generation inter-cross (MAGIC) populations. Genetics. 2016;202:471–86.

Xu S. An alternative derivation of Harville's restricted log likelihood function for variance component estimation. Biom J. 2019;61:157–61.

# Multiple Traits and Genetic Correlation

In any genetic experiment, it is very rare to measure only a single trait. The quality of a new variety of crops or a new breed of animals is often judged by the values of several traits. Although genetic analysis can be performed separately for each trait, the optimal strategy is to analyze all traits jointly, provided that the traits are correlated. This chapter is about multiple trait analysis, where correlation among traits must be considered.

We now present the analysis of two traits as an example of multivariate analysis. Let $X$ and $Y$ be two metric characters measured from the same individual. The phenotypic value of each character can be described by its own linear model,

$$\begin{cases} P_X = A_X + E_X \\ P_Y = A_Y + E_Y \end{cases}$$

The phenotypic covariance can be partitioned into genetic covariance and environmental covariance, as shown below,

$$\underbrace{\text{cov}(P_X, P_Y)}_{\sigma_{P(XY)}} = \text{cov}(A_X + E_X, A_Y + E_Y) = \underbrace{\text{cov}(A_X, A_Y)}_{\sigma_{A(XY)}} + \underbrace{\text{cov}(E_X, E_Y)}_{\sigma_{E(XY)}}$$

where $\sigma_{P(XY)}$ is called the phenotypic covariance, $\sigma_{A(XY)}$ is the (additive) genetic covariance, and $\sigma_{E(XY)}$ is the random environmental covariance. Recall that the environmental covariance between two relatives is often assumed to be zero. However, the environmental covariance between two traits measured from the same individual, $\sigma_{E(XY)}$, should never be assumed zero because $E_X$ and $E_Y$ always happen in the same environment (the same individual). Partitioning of the phenotypic covariance between traits into genetic and environmental covariance is similar to the partitioning of phenotypic variance for an individual trait.

## 13.1 Definition of Genetic Correlation

The phenotypic correlation between $X$ and $Y$ is defined as

$$r_P = r_{P(XY)} = \frac{\text{cov}(P_X, P_Y)}{\sigma_{P(X)}\sigma_{P(Y)}} = \frac{\sigma_{P(XY)}}{\sigma_{P(X)}\sigma_{P(Y)}}$$

The genetic correlation between $X$ and $Y$ is defined by

$$r_A = r_{A(XY)} = \frac{\text{cov}(A_X, A_Y)}{\sigma_{A(X)}\sigma_{A(Y)}} = \frac{\sigma_{A(XY)}}{\sigma_{A(X)}\sigma_{A(Y)}}$$

It is the correlation between the additive genetic effects (breeding values) of two characters expressed in the same individual. The environmental correlation is defined as

© Springer Nature Switzerland AG 2022<br>S. Xu, *Quantitative Genetics*, https://doi.org/10.1007/978-3-030-83940-6_13

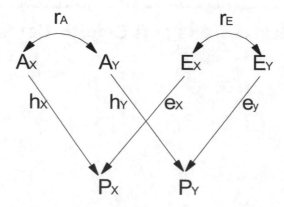

**Fig. 13.1** Path diagram of the relationship of two metric traits measured from the same individual

$$r_E = r_{E(XY)} = \frac{\text{cov}(E_X, E_Y)}{\sigma_{E(X)}\sigma_{E(Y)}} = \frac{\sigma_{E(XY)}}{\sigma_{E(X)}\sigma_{E(Y)}}$$

Let us define $e_X^2$ and $e_Y^2$ as

$$e_X^2 = \frac{\sigma_{E(X)}^2}{\sigma_{P(X)}^2} = 1 - \frac{\sigma_{A(X)}^2}{\sigma_{P(X)}^2} = 1 - h_X^2$$

$$e_Y^2 = \frac{\sigma_{E(Y)}^2}{\sigma_{P(Y)}^2} = 1 - \frac{\sigma_{A(Y)}^2}{\sigma_{P(Y)}^2} = 1 - h_Y^2$$

A partitioning of the phenotypic correlation can be made, as shown below,

$$r_P = r_A h_X h_Y + r_E e_X e_Y \tag{13.1}$$

which can also be seen from the path diagram (Fig. 13.1). There are two paths connecting the phenotypic values of the two traits, one via the genetic reason with a path value $h_X r_A h_Y$ and the other via the environment with a path value $e_X r_E e_Y$. As a result, the phenotypic correlation takes the sum of the two path values. Rules of the path analysis for multiple traits are the same as what we learned in the path analysis for calculating coancestry between two relatives. An alternative expression of Eq. (13.1) is

$$r_P = r_A h_X h_Y + r_E \sqrt{1 - h_X^2}\sqrt{1 - h_Y^2}$$

## 13.2   Causes of Genetic Correlation

There are two causes of the genetic correlation: (a) pleiotropic effect and (b) linkage effect. Pleiotropic effect is a phenomenon that a single gene controls two or more traits. This cause of the genetic correlation is permanent and cannot be decoupled. Linkage effect is due to the fact that genes controlling two or more traits are located in the neighborhood of the same chromosome. When the population is in linkage equilibrium, the genetic correlation caused by linkage will disappear. Therefore, the linkage-effect-caused genetic correlation is temporary. If linkage is perfect, then this cause is of no difference from the pleiotropic effect. In a population with linkage equilibrium, genetic correlation is only caused by pleiotropic effect. Figure 13.2 shows the difference between pleiotropy and linkage.

**Fig. 13.2** Difference between pleiotropy and linkage that cause the genetic correlation between trait $X$ and trait $Y$

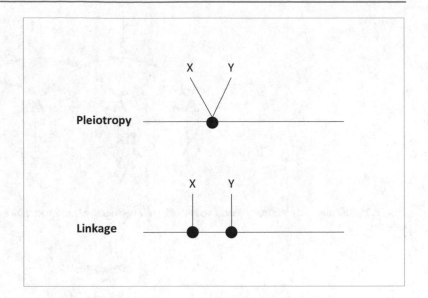

## 13.3   Cross Covariance between Relatives

We now introduce the concept of cross covariance. Cross covariance between two relatives is defined as the covariance between trait $X$ in the first individual and trait $Y$ in the second individual,

$$\text{Individual 1}: \begin{cases} P_{X_1} = A_{X_1} + E_{X_1} \\ P_{Y_1} = A_{Y_1} + \quad E_{Y_1} \end{cases} \quad \text{Individual 2}: \begin{cases} P_{X_2} = A_{X_2} + E_{X_2} \\ P_{Y_2} = A_{Y_2} + E_{Y_2} \end{cases}$$

Assume that the two relatives do not share common environmental effect. The phenotypic covariance between the two relatives is purely caused by their genetic covariance. In this case,

$$\text{cov}(P_{X_1}, P_{Y_2}) = \text{cov}(A_{X_1}, A_{Y_2}) = a_{12}\text{cov}(A_{X_1}, A_{Y_1}) = a_{12}\sigma_{A(XY)}$$

where $a_{12} = 2\theta_{12}$ is the additive relationship between individuals 1 and 2. Similarly,

$$\text{cov}(P_{Y_1}, P_{X_2}) = \text{cov}(A_{Y_1}, A_{X_2}) = a_{12}\text{cov}(A_{Y_1}, A_{X_1}) = a_{12}\sigma_{A(XY)}$$

Therefore, under the additive genetic model, the cross covariance between relatives equals the proportion of the additive genetic covariance between the two traits. This proportion is the additive relationship $a_{12}$.

## 13.4   Estimation of Genetic Correlation

### 13.4.1 Estimate Genetic Correlation Using Parent-Offspring Correlation (Path Analysis)

Genetic correlation can be estimated using path analysis. Let $X$ and $Y$ be the phenotypic values of an offspring for traits $X$ and $Y$. Let $X'$ and $Y'$ be the corresponding trait values measured from a parent. These phenotypic values are connected through the path diagram shown in Fig. 13.3. The genetic correlation, denoted by $r_A$, is not observable, but the cross correlation coefficients can be calculated through data. The following phenotypic correlation coefficients are functions of heritability and genetic correlation,

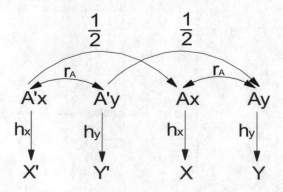

**Fig. 13.3** Path diagram connecting parent and offspring for two traits, where letters with a prime indicate parent and letters without a prime indicate offspring

$$r_{X'X} = \frac{1}{2}h_X^2$$

$$r_{Y'Y} = \frac{1}{2}h_Y^2$$

$$r_{X'Y} = \frac{1}{2}h_X r_A h_Y$$

$$r_{Y'X} = \frac{1}{2}h_X r_A h_Y$$

The last two correlation coefficients are the cross correlation coefficients. Because they have the same expectation, we can combine them together, i.e.,

$$r_{X'Y} + r_{Y'X} = h_X r_A h_Y$$

We now have three unknowns and three equations,

$$\begin{cases} r_{X'X} = \dfrac{1}{2}h_X^2 \\[2mm] r_{Y'Y} = \dfrac{1}{2}h_Y^2 \\[2mm] r_{X'Y} + r_{Y'X} = h_X r_A h_Y \end{cases} \tag{13.2}$$

Let us define

$$\rho_{XY} = \frac{r_{X'Y} + r_{Y'X}}{\sqrt{r_{X'X}\ r_{Y'Y}}}$$

which is observed because the four correlation coefficients are calculated from the data. Substituting these correlations by their path values given in Eq. (13.2), we have

$$\rho_{XY} = \frac{h_X r_A h_Y}{\sqrt{\frac{1}{2}h_X^2 \frac{1}{2}h_Y^2}} = \frac{h_X r_A h_Y}{\frac{1}{2}h_X h_Y} = 2r_A$$

Solving for $r_A$ yields

$$\widehat{r}_A = \frac{1}{2}\rho_{XY} = \frac{r_{X'Y} + r_{Y'X}}{2\sqrt{r_{X'X}\ r_{Y'Y}}}$$

The standard error of the estimated genetic correlation is very difficult to calculate. The denominator involves square root of a product. To find the standard error, we define the ratio by

$$\widehat{\rho}_{XY} = \frac{r_{X'Y} + r_{Y'X}}{\sqrt{r_{X'X} \; r_{Y'Y}}} = \frac{X}{Y}$$

First, the variance of a sample correlation coefficient is (Fisher 1921)

$$\mathrm{var}(r_{XY}) = \frac{1}{n-3}$$

Assume that all the four sample correlations are independent. The variance of the numerator is

$$\mathrm{var}(X) = \mathrm{var}(r_{X'Y} + r_{Y'X}) = \mathrm{var}(r_{X'Y}) + \mathrm{var}(r_{Y'X}) = \frac{2}{n-3}$$

The variance of the denominator is

$$\mathrm{var}(Y) = \mathrm{var}(\sqrt{r_{X'X}r_{Y'Y}}) = \frac{1}{4(n-3)} \left( \frac{r_{Y'Y} + r_{X'X}}{r_{X'X}r_{Y'Y}} \right)$$

The variance of $\widehat{\rho}_{XY}$ is

$$\mathrm{var}(\widehat{\rho}_{XY}) = \left( \frac{X}{Y} \right)^2 \left( \frac{\mathrm{var}(X)}{X^2} + \frac{\mathrm{var}(Y)}{Y^2} \right) = \frac{1}{n-3} \left[ \frac{(r_{X'Y} + r_{Y'X})^2 (r_{Y'Y} + r_{X'X})}{4(r_{X'X} \; r_{Y'Y})^3} + \frac{1}{r_{X'X} \; r_{Y'Y}} \right]$$

Therefore, the standard error of $\widehat{r}_A$ is

$$S_{\widehat{r}_A} = \frac{1}{2} \sqrt{\mathrm{var}(\widehat{\rho}_{XY})} = \frac{1}{2} \sqrt{ \frac{1}{n-3} \left\{ \frac{(r_{X'Y} + r_{Y'X})^2 (r_{Y'Y} + r_{X'X})}{4(r_{X'X} \; r_{Y'Y})^3} + \frac{1}{r_{X'X} \; r_{Y'Y}} \right\} }$$

Let us see an example on how to estimate the genetic correlation using cross correlation coefficients. A total of 35 ewes (female sheep) and their fully grown daughters were measured for two traits: wool length ($X$) and wool yield ($Y$). The data are listed in Table 13.1, also stored in "data-13-1.csv".

```
filename aa "data-13-1.csv";
proc import datafile=aa out=one dbms=csv replace;
run;
ods graphics off;
proc corr data=one;
    var X0 Y0;
    with X1 Y1;
run;
```

The output is shown in Table 13.2 in a $2 \times 2$ table, where $X_0$ and $Y_0$ are the traits from the parents, and $X_1$ and $Y_1$ are the traits from the offspring. Since $X'$ and $Y'$ cannot be used as names of variables in SAS, we replaced them by $X_0$ and $Y_0$. We also renamed traits collected from the offspring by $X_1$ and $Y_1$. There are two numbers within each cell of Table 13.2; the first number is the estimated correlation coefficient, and the second number is the p-value for testing the null hypothesis of $H_0 : r_{XY} = 0$. We can estimate the heritability of each trait and the genetic correlation between the two traits using the above four correlations, as shown below,

**Table 13.1** Wool length ($X$) and wool yield ($Y$) collected from 35 ewes (female sheep) and their 2-year old daughters (盛志廉 and 陈瑶生 1999)

| Family | $X_0$ | $Y_0$ | $X_1$ | $Y_1$ |
|---|---|---|---|---|
| 1 | 14.00 | 4.90 | 12.00 | 5.10 |
| 2 | 8.50 | 5.50 | 9.50 | 4.93 |
| 3 | 8.50 | 5.50 | 10.00 | 5.60 |
| 4 | 8.50 | 5.20 | 9.00 | 6.56 |
| 5 | 9.00 | 5.30 | 8.50 | 4.96 |
| 6 | 8.50 | 3.60 | 11.00 | 6.50 |
| 7 | 9.00 | 4.80 | 8.50 | 4.49 |
| 8 | 9.00 | 5.25 | 10.50 | 5.70 |
| 9 | 11.00 | 4.30 | 9.00 | 6.70 |
| 10 | 11.00 | 5.80 | 11.00 | 4.92 |
| 11 | 11.00 | 4.70 | 11.00 | 4.70 |
| 12 | 11.00 | 4.70 | 10.00 | 5.36 |
| 13 | 9.50 | 4.95 | 8.50 | 5.09 |
| 14 | 8.00 | 4.20 | 8.50 | 5.10 |
| 15 | 8.00 | 4.20 | 9.00 | 5.05 |
| 16 | 8.50 | 7.85 | 8.00 | 7.30 |
| 17 | 10.50 | 4.72 | 10.00 | 5.07 |
| 18 | 8.00 | 3.60 | 9.50 | 4.65 |
| 19 | 9.00 | 5.70 | 9.00 | 6.17 |
| 20 | 8.50 | 5.50 | 9.50 | 5.80 |
| 21 | 12.00 | 5.00 | 9.00 | 4.89 |
| 22 | 7.00 | 3.50 | 8.50 | 4.20 |
| 23 | 9.00 | 5.00 | 9.00 | 4.57 |
| 24 | 9.00 | 3.52 | 9.50 | 4.86 |
| 25 | 10.00 | 5.93 | 9.50 | 5.46 |
| 26 | 9.50 | 5.30 | 9.00 | 5.29 |
| 27 | 9.50 | 5.60 | 10.50 | 3.85 |
| 28 | 9.00 | 4.82 | 8.50 | 3.66 |
| 29 | 9.00 | 4.73 | 11.00 | 5.53 |
| 30 | 8.50 | 4.60 | 11.00 | 5.57 |
| 31 | 10.00 | 5.89 | 10.50 | 6.00 |
| 32 | 10.00 | 7.35 | 10.00 | 5.88 |
| 33 | 9.00 | 4.20 | 8.50 | 5.00 |
| 34 | 8.50 | 5.10 | 9.50 | 5.84 |
| 35 | 10.00 | 8.00 | 9.50 | 5.24 |

Columns $X_0$ and $Y_0$ are trait values collected from the mothers and columns $X_1$ and $Y_1$ are trait values collected from the daughters.

**Table 13.2** Cross generation correlation coefficients for two traits

| Pearson correlation coefficients, N = 35 Prob > |r| under H0: Rho = 0 | | |
|---|---|---|
| | $X_0$ | $Y_0$ |
| $X_1$ | 0.46567 | −0.02727 |
| | 0.0048 | 0.8764 |
| $Y_1$ | −0.01861 | 0.33325 |
| | 0.9155 | 0.0504 |

$$\widehat{h}_X^2 = 2r_{X'X} = 2 \times 0.46566822 = 0.9313$$

$$\widehat{h}_Y^2 = 2r_{Y'Y} = 2 \times 0.33324878 = 0.6664976$$

$$\widehat{r}_A = \frac{r_{X'Y} + r_{Y'X}}{2\sqrt{r_{X'X}\ r_{Y'Y}}} = \frac{(-0.01860521) + (-0.02726918)}{2 \times \sqrt{0.46566822 \times 0.33324878}} = -0.0582261$$

The sample size is $n = 35$ and thus the variance of any correlation coefficient is $1/(n - 3) = 1/(35 - 3) = 0.03125$. The standard error of the estimated genetic correlation is

$$
\begin{aligned}
S_{\widehat{r}_A} &= \frac{1}{2}\sqrt{\frac{1}{n-3}\left\{\frac{(r_{X'Y} + r_{Y'X})^2 (r_{Y'Y} + r_{X'X})}{4(r_{X'X}\ r_{Y'Y})^3} + \frac{1}{r_{X'X}\ r_{Y'Y}}\right\}} \\
&= \frac{1}{2}\sqrt{\frac{1}{35-3}\left\{\frac{(-0.01861 - 0.02727)^2 (0.46567 + 0.33325)}{4(0.33325 \times 0.46567)^3} + \frac{1}{0.33325 \times 0.46567}\right\}} \\
&= 0.2263228
\end{aligned}
$$

Therefore, the genetic correlation should be reported as

$$\widehat{r}_A \pm S_{\widehat{r}_A} = -0.0582 \pm 0.2263$$

## 13.4.2  Estimating Genetic Correlation from Sib Data

Estimating genetic correlation using sib data requires analysis of variances (ANOVA) and analysis of covariance (ANCOVA). Assume that there are $n$ full-sib or half-sib families, each with $m$ siblings. The data structure will look like as given in Table 13.3. The linear models for $X$ and $Y$ are

$$
\begin{aligned}
x_{jk} &= \mu_X + B_{j(X)} + W_{j(X)} \\
y_{jk} &= \mu_Y + B_{j(Y)} + W_{j(Y)}
\end{aligned}
$$

for $j = 1, \ldots, n$ and $k = 1, \ldots, m$. The variances for the two traits are

$$
\begin{aligned}
\mathrm{var}(x_{jk}) &= \mathrm{var}(B_{j(X)}) + \mathrm{var}(W_{j(X)}) = \sigma^2_{B(X)} + \sigma^2_{W(X)} \\
\mathrm{var}(y_{jk}) &= \mathrm{var}(B_{j(Y)}) + \mathrm{var}(W_{j(Y)}) = \sigma^2_{B(Y)} + \sigma^2_{W(Y)}
\end{aligned}
$$

In addition, we have a covariance between $X$ and $Y$,

$$\mathrm{cov}(x_{jk}, y_{jk}) = \mathrm{cov}(B_{j(X)}, B_{j(Y)}) + \mathrm{cov}(W_{j(X)}, W_{j(Y)}) = \sigma_{B(XY)} + \sigma_{W(XY)}$$

Similar to the variance component analysis, the covariance component analysis follows the same expectation and the same relationship between the between-family covariance components and the genetic covariance components. Therefore,

**Table 13.3**  Data structure for sib analysis in estimating genetic correlation between two traits

| Family | Sib 1 | Sib 2 | $\cdots$ | Sib $m$ |
|---|---|---|---|---|
| 1 | $x_{11}, y_{11}$ | $x_{12}, y_{12}$ | $\cdots$ | $x_{1m}, y_{1m}$ |
| 2 | $x_{21}, y_{21}$ | $x_{22}, y_{22}$ | $\cdots$ | $x_{2m}, y_{2m}$ |
| $\vdots$ | $\vdots$ | $\vdots$ | $\cdots$ | $\vdots$ |
| $n$ | $x_{n1}, y_{n1}$ | $x_{n2}, y_{n2}$ | $\cdots$ | $x_{nm}, y_{nm}$ |

**Table 13.4** Analysis of covariance (ANCOVA) table for traits $X$ and $Y$

| Source | Df | SCP | MCP | E(MCP) |
|---|---|---|---|---|
| Between (B) | $df_B = n - 1$ | $SCP_B$ | $MCP_B = SCP_B/df_B$ | $\sigma_{W(XY)} + m\sigma_{B(XY)}$ |
| Within (W) | $df_W = n(m - 1)$ | $SCP_W$ | $MCP_W = SCP_W/df_W$ | $\sigma_{W(XY)}$ |

$$\sigma_{B(XY)} = a_{SIB}\sigma_{A(XY)}$$

where $a_{SIB} = a_{FS} = 0.5$ for full-sibs and $a_{SIB} = a_{HS} = 0.25$ for half-sibs. The within-family covariance component is

$$\sigma_{W(XY)} = (1 - a_{SIB})\sigma_{A(XY)} + \sigma_{E(XY)}$$

The data structure remains the same as the sib analysis in estimating heritability except that we add an additional trait to the table. When you consider trait $X$ only (ignoring trait $Y$), you can perform analysis of variances for that trait and estimate the between-family and within-family variance components. Similarly, we can perform ANOVA for trait $Y$ and ignore trait $X$. When you consider the two traits simultaneously, you can generate another table called ANCOVA (analysis of covariance) table, as demonstrated in Table 13.4.

The covariance components are estimated by equating the expected mean cross products to the observed mean cross products. The solutions are

$$\widehat{\sigma}_{W(XY)} = MCP_W$$
$$\widehat{\sigma}_{B(XY)} = (MCP_B - MCP_W)/m$$

The between-family cross covariance component is a proportion of the genetic covariance, i.e., $\sigma_{B(XY)} = a_{SIB}\sigma_{A(XY)}$. In the meantime, we have already obtained the variance components of individual traits. The genetic correlation is then estimated by

$$\widehat{r}_{A(XY)} = \frac{\widehat{\sigma}_{A(XY)}}{\widehat{\sigma}_{A(X)}\widehat{\sigma}_{A(Y)}} = \frac{\widehat{\sigma}_{B(XY)}/a_{SIB}}{\sqrt{\left(\widehat{\sigma}^2_{B(X)}/a_{SIB}\right)\left(\widehat{\sigma}^2_{B(Y)}/a_{SIB}\right)}} = \frac{\widehat{\sigma}_{B(XY)}}{\widehat{\sigma}_{B(X)}\widehat{\sigma}_{B(Y)}}$$

The computing formulas for the sum of cross products (SCP) are

$$\text{SCP}_B = m\sum_{j=1}^{n}\left(\bar{x}_{j\cdot} - \bar{x}_{\cdot\cdot}\right)\left(\bar{y}_{j\cdot} - \bar{y}_{\cdot\cdot}\right) = m\sum_{j=1}^{n}\bar{x}_{j\cdot}\bar{y}_{j\cdot} - nm\bar{x}_{\cdot\cdot}\bar{y}_{\cdot\cdot}$$

and

$$\text{SCP}_W = \sum_{j=1}^{n}\sum_{k=1}^{m}\left(x_{jk} - \bar{x}_{j\cdot}\right)\left(y_{jk} - \bar{y}_{j\cdot}\right) = \sum_{j=1}^{n}\sum_{k=1}^{m}x_{jk}y_{jk} - m\sum_{j=1}^{n}\bar{x}_{j\cdot}\bar{y}_{j\cdot}$$

where.

$$\bar{x}_{\cdot\cdot} = \frac{1}{nm}\sum_{j=1}^{n}\sum_{k=1}^{m}x_{jk}, \quad \bar{x}_{j\cdot} = \frac{1}{m}\sum_{k=1}^{m}x_{jk}, \quad \bar{y}_{\cdot\cdot} = \frac{1}{nm}\sum_{j=1}^{n}\sum_{k=1}^{m}y_{jk} \quad \text{and} \quad \bar{y}_{j\cdot} = \frac{1}{m}\sum_{k=1}^{m}y_{jk}$$

.

Unbalanced data can also be used to estimate genetic correlation with $m$ substituted by $m_0$ (see Chap. 10 for equation of calculating $m_0$). Genetic correlation can also be estimated via the nested hierarchical mating scheme with either balanced or unbalanced design (see Chap. 10 for equations of calculating the three $k$ values). Of course, the mixed model methodology for general pedigree data analysis can be used to estimate the genetic covariance between two traits, and thus to estimate the genetic correlation. These topics are beyond the scope of this class.

**Table 13.5**  Simulated full-sib family data for two quantitative traits ($X$ and $Y$)

| Family | Sib | X | Y |
|---|---|---|---|
| 1 | 1 | 111.7784 | 168.6859 |
| 1 | 2 | 100.6364 | 149.642 |
| 1 | 3 | 102.8606 | 159.3027 |
| 1 | 4 | 106.7193 | 156.4697 |
| 1 | 5 | 105.9129 | 171.0542 |
| 2 | 1 | 103.8783 | 163.2831 |
| 2 | 2 | 93.05395 | 139.3337 |
| 2 | 3 | 96.93399 | 138.1688 |
| 2 | 4 | 104.087 | 149.7471 |
| 2 | 5 | 111.8017 | 158.8265 |
| 3 | 1 | 102.2341 | 153.0292 |
| 3 | 2 | 78.97513 | 127.063 |
| 3 | 3 | 103.4912 | 153.9898 |
| 3 | 4 | 98.99105 | 143.3288 |
| 3 | 5 | 100.7186 | 145.9164 |
| 4 | 1 | 92.1232 | 141.1175 |
| 4 | 2 | 93.21856 | 152.5122 |
| 4 | 3 | 88.25051 | 135.5489 |
| 4 | 4 | 100.2568 | 144.94 |
| 4 | 5 | 97.87207 | 140.5411 |
| 5 | 1 | 98.67599 | 150.4905 |
| 5 | 2 | 104.2939 | 150.8418 |
| 5 | 3 | 101.5456 | 159.1291 |
| 5 | 4 | 93.15511 | 140.696 |
| 5 | 5 | 105.4987 | 154.1311 |
| ⋮ | ⋮ | ⋮ | ⋮ |
| 100 | 1 | 102.908 | 159.989 |
| 100 | 2 | 106.483 | 159.8893 |
| 100 | 3 | 97.65765 | 148.5592 |
| 100 | 4 | 111.4916 | 160.624 |
| 100 | 5 | 97.95063 | 155.2832 |

**Table 13.6**  Analysis of variances tables for traits $X$ and $Y$, and the analysis of covariance table between traits $X$ and $Y$

| Covariation | Df | SS | MS | E(MS) |
|---|---|---|---|---|
| Between (B) | 99 | 4457.1 | 45.021 | $\sigma^2_{W(X)} + 5\ \sigma^2_{B(X)}$ |
| Within (W) | 400 | 10314.8 | 25.787 | $\sigma^2_{W(X)}$ |
| Between (B) | 99 | 10,952 | 110.626 | $\sigma^2_{W(Y)} + 5\ \sigma^2_{B(Y)}$ |
| Within (W) | 400 | 21,806 | 54.516 | $\sigma^2_{W(Y)}$ |
| Between (B) | 99 | 5176.402 | 52.28689 | $\sigma_{W(XY)} + 5\ \sigma_{B(XY)}$ |
| Within (W) | 400 | 10816.14 | 27.04036 | $\sigma_{W(XY)}$ |

Let us now estimate the genetic correlation coefficient using a simulated data. Table 13.5 shows partial records of a simulated data of two traits with 100 full-sib families, each having five siblings. The data are also stored in "data-13-2.csv". The mean squares and mean cross products are shown in the ANOVA (ANCOVA) table (Table 13.6). The two ANOVA tables allow us to calculate the variance components (see Chap. 10). The ANCOVA table gives us the intermediate result for calculation of the covariance components, as shown below,

$$\sigma_{W(XY)} = MCP_W = 27.04036$$
$$\sigma_{B(XY)} = (MCP_B - MCP_W)/5 = (52.28689 - 27.04035)/5 = 5.049306$$

The estimated between-family and within-family variance and covariance components are given in Table 13.7.

**Table 13.7**  The estimated between-family and within-family variance and covariance components

| | $X$ | | $Y$ |
|---|---|---|---|
| $X$ | $\widehat{\sigma}^2_{B(X)} = 3.846852$ | | $\widehat{\sigma}_{B(XY)} = 5.049306$ |
| $Y$ | $\widehat{\sigma}_{B(XY)} = 5.049306$ | | $\widehat{\sigma}^2_{B(Y)} = 11.222167$ |
| $X$ | $\widehat{\sigma}^2_{W(X)} = 25.78693$ | | $\widehat{\sigma}_{W(XY)} = 27.04036$ |
| $Y$ | $\widehat{\sigma}_{W(XY)} = 27.04036$ | | $\widehat{\sigma}^2_{W(Y)} = 54.51558$ |

The estimated heritability for trait $X$ is

$$\widehat{h}^2_X = \frac{2\widehat{\sigma}^2_{B(X)}}{\widehat{\sigma}^2_{B(X)} + \widehat{\sigma}^2_{W(X)}} = \frac{2 \times 3.846852}{3.846852 + 25.78693} = 0.2596261$$

The estimated heritability for trait $Y$ is

$$\widehat{h}^2_Y = \frac{2\widehat{\sigma}^2_{B(Y)}}{\widehat{\sigma}^2_{B(Y)} + \widehat{\sigma}^2_{W(Y)}} = \frac{2 \times 11.222167}{11.222167 + 54.51558} = 0.3414223$$

The estimated genetic correlation between $X$ and $Y$ is

$$\widehat{r}_{A(XY)} = \frac{\widehat{\sigma}_{B(XY)}}{\sqrt{\widehat{\sigma}^2_{B(X)}\widehat{\sigma}^2_{B(Y)}}} = \frac{5.049306}{\sqrt{3.846852 \times 11.222167}} = 0.7684941$$

There are several different SAS procedures we can use to perform multiple trait analysis. The GLM procedure is preferred for unbalanced data (number of siblings varies across families). The program does not produce the final result we need, but it gives the intermediate result, e.g., the degrees of freedom and the sums of cross products. We must calculate the variance and covariance components manually from the intermediate result. The SAS code for reading data and performing multivariate ANOVA (called MANOVA) is given below.

```
filename aa "data-13-2.csv";
proc import datafile=aa out=one dbms=csv replace;
run;
ods graphics off;
proc glm data=one;
   class family;
   model x y = family;
   random family;
   manova h=family / printh printe;
   ods output HypothesisSSCP=H ErrorSSCP=E;
quit;
```

The output contains many tables, but we only show the ones we need to calculate the variance and covariance components. These intermediate results are summarized in Table 13.8. The sums of cross products (SSCP) listed in Table 13.8 divided by the corresponding degrees of freedom listed in Table 13.6 become the mean squares and the mean cross products. The GLM procedure only provides intermediate results and the final result requires some manual calculation. As a result, the GLM analysis helps investigators understand the model and method for sib analysis.

The MIXED procedure in SAS provides the estimated variance and covariance components automatically without any manual calculation. To use PROC MIXED. The data must be reformatted in a "tall-format," where the response variables must be listed as a single variable, as shown in Table 13.9. The data in the tall-format are stored in "data-13-3.csv".

**Table 13.8** Between-family (H) and within-family (E) sums of squares and sum of cross products produced by PROC GLM for the simulated data

|  | H = Type III SSCP matrix for family | | E = Error SSCP matrix | |
|---|---|---|---|---|
|  | X | Y | X | Y |
| X | 4457.0981239 | 5176.40196 | 10314.772849 | 10816.143924 |
| Y | 5176.40196 | 10952.015478 | 10816.143924 | 21806.233514 |

**Table 13.9** Partial records of the sib data for two traits in a tall-format (3 out of 100 families), where Z is the response variable (including both X and Y)

| Family | Sib | V | Z |
|---|---|---|---|
| 1 | 1 | X | 111.7784294 |
| 1 | 1 | Y | 168.6859347 |
| 1 | 2 | X | 100.6364139 |
| 1 | 2 | Y | 149.6419519 |
| 1 | 3 | X | 102.8605538 |
| 1 | 3 | Y | 159.302705 |
| 1 | 4 | X | 106.7193172 |
| 1 | 4 | Y | 156.4697301 |
| 1 | 5 | X | 105.912939 |
| 1 | 5 | Y | 171.0541689 |
| 2 | 1 | X | 103.8782832 |
| 2 | 1 | Y | 163.2831199 |
| 2 | 2 | X | 93.05394751 |
| 2 | 2 | Y | 139.3336686 |
| 2 | 3 | X | 96.93399023 |
| 2 | 3 | Y | 138.1688416 |
| 2 | 4 | X | 104.0870346 |
| 2 | 4 | Y | 149.7470608 |
| 2 | 5 | X | 111.8017086 |
| 2 | 5 | Y | 158.8264852 |
| 3 | 1 | X | 102.2340963 |
| 3 | 1 | Y | 153.02921 |
| 3 | 2 | X | 78.97513244 |
| 3 | 2 | Y | 127.0630327 |
| 3 | 3 | X | 103.4912068 |
| 3 | 3 | Y | 153.989826 |
| 3 | 4 | X | 98.9910514 |
| 3 | 4 | Y | 143.3287631 |
| 3 | 5 | X | 100.7185757 |
| 3 | 5 | Y | 145.9163895 |

```
filename aa "data-13-2.csv";
proc import datafile=aa out=one dbms=csv replace;
run;
data two;
    set one;
    V='X'; Z=X; output;
    V='Y'; Z=Y; output;
    keep Family Sib Z V;
run;
proc export data=two outfile="data-13-3.csv" dbms=csv replace;
run;
proc mixed data=two method=reml covtest asycov;
    class V Family;
    model Z = V;
```

```
    random V/subject=Family type=UN;
    repeated /subject=Sib type=UN;
run;
```

The **proc import** statement reads the data in the original "wide-format." The **data two** statement converts the data from the "wide-format" into the "tall-format." The **proc mixed** statement analyzes the data. The random statement defines the covariance structure for the random between-family effects. The repeated statement defines the covariance structure of the within-family effects. The output includes several tables but we only need two tables for the purpose of estimating genetic correlation. Table 13.10 shows the estimated variance and covariance components. Table 13.11 gives the asymptotic variance-covariance matrix of the estimated variance and covariance components. The estimated between-family variance and covariance components are $\widehat{\sigma}_{B(X)}^2 = UN(1, 1) = 3.8469$, $\widehat{\sigma}_{B(XY)} = UN(1, 2) = 5.0493$, and $\widehat{\sigma}_{B(Y)}^2 = UN(2, 2) = 11.2222$. The corresponding within-family estimates are $\widehat{\sigma}_{W(X)}^2 = 25.7869$, $\widehat{\sigma}_{W(XY)} = 27.0404$, and $\widehat{\sigma}_{W(Y)}^2 = 54.5156$. These estimates are the same as the estimates from PROC GLM (Tables 13.7). The variance-covariance matrix of the estimated variance components (Table 13.11) allows us to calculate the standard error of the estimated genetic correlation. Note that the diagonal elements of Table 13.11 are the variances of the estimated variance components. Square roots of these variances are the standard errors.

Now let us calculate the standard error of the estimated genetic correlation. The variance-covariance matrix of the estimated between-family variance and covariance components are expressed as

$$
\operatorname{var}\begin{bmatrix} \widehat{\sigma}_{B(X)}^2 \\ \widehat{\sigma}_{B(XY)} \\ \widehat{\sigma}_{B(Y)}^2 \end{bmatrix} = \begin{bmatrix} \operatorname{var}\left(\widehat{\sigma}_{B(X)}^2\right) & \operatorname{cov}\left(\widehat{\sigma}_{B(X)}^2, \widehat{\sigma}_{B(XY)}\right) & \operatorname{cov}\left(\widehat{\sigma}_{B(X)}^2, \widehat{\sigma}_{B(Y)}^2\right) \\ \operatorname{cov}\left(\widehat{\sigma}_{B(XY)}, \widehat{\sigma}_{B(X)}^2\right) & \operatorname{var}\left(\widehat{\sigma}_{B(XY)}\right) & \operatorname{cov}\left(\widehat{\sigma}_{B(XY)}, \widehat{\sigma}_{B(Y)}^2, \right) \\ \operatorname{cov}\left(\widehat{\sigma}_{B(Y)}^2, \widehat{\sigma}_{B(X)}^2\right) & \operatorname{cov}\left(\widehat{\sigma}_{B(Y)}^2, \widehat{\sigma}_{B(XY)}\right) & \operatorname{var}\left(\widehat{\sigma}_{B(Y)}^2\right) \end{bmatrix}
$$

Let us define the estimated genetic correlation as

$$
\widehat{r}_{A(XY)} = \frac{\widehat{\sigma}_{B(XY)}}{\sqrt{\widehat{\sigma}_{B(X)}^2 \widehat{\sigma}_{B(Y)}^2}} = \frac{Z}{\sqrt{WV}} = \frac{5.0493}{\sqrt{3.8469 \times 11.2222}} = 0.7684874
$$

**Table 13.10** Estimated variance and covariance components from PROC MIXED for the sib data

| Covariance parameter estimates | | | | | |
|---|---|---|---|---|---|
| Cov Parm | Subject | Estimate | Standard Error | Z Value | Pr Z |
| UN(1,1) | Family | 3.8469 | 1.3308 | 2.89 | 0.0019 |
| UN(2,1) | Family | 5.0493 | 1.8250 | 2.77 | 0.0057 |
| UN(2,2) | Family | 11.2222 | 3.2379 | 3.47 | 0.0003 |
| UN(1,1) | Sib | 25.7869 | 1.8234 | 14.14 | <0.0001 |
| UN(2,1) | Sib | 27.0404 | 2.3114 | 11.70 | <0.0001 |
| UN(2,2) | Sib | 54.5156 | 3.8548 | 14.14 | <0.0001 |

**Table 13.11** Variance-covariance matrix of the estimated covariance components

| Asymptotic covariance matrix of estimates | | | | | | | |
|---|---|---|---|---|---|---|---|
| Row | Cov Parm | CovP1 | CovP2 | CovP3 | CovP4 | CovP5 | CovP6 |
| 1 | UN(1,1) | 1.7709 | 2.0417 | 2.3555 | −0.6650 | −0.6973 | −0.7312 |
| 2 | UN(2,1) | 2.0417 | 3.3306 | 4.9690 | −0.6973 | −1.0685 | −1.4741 |
| 3 | UN(2,2) | 2.3555 | 4.9690 | 10.4838 | −0.7312 | −1.4741 | −2.9719 |
| 4 | UN(1,1) | −0.6650 | −0.6973 | −0.7312 | 3.3248 | 3.4864 | 3.6559 |
| 5 | UN(2,1) | −0.6973 | −1.0685 | −1.4741 | 3.4864 | 5.3424 | 7.3706 |
| 6 | UN(2,2) | −0.7312 | −1.4741 | −2.9719 | 3.6559 | 7.3706 | 14.8597 |

where $W = \hat{\sigma}_{B(X)}^2 = 3.8469$, $Z = \hat{\sigma}_{B(XY)} = 5.0493$, and $V = \hat{\sigma}_{B(Y)}^2 = 11.2222$. Let us define the partial derivatives of the estimated genetic correlation with respect to the three variance and covariance components as

$$
\nabla = \begin{bmatrix} \dfrac{\partial \hat{r}_{A(XY)}}{\partial W} \\[2ex] \dfrac{\partial \hat{r}_{A(XY)}}{\partial Z} \\[2ex] \dfrac{\partial \hat{r}_{A(XY)}}{\partial V} \end{bmatrix} = \begin{bmatrix} -\dfrac{Z}{2\sqrt{W^3 V}} \\[2ex] \dfrac{1}{\sqrt{WV}} \\[2ex] -\dfrac{Z}{2\sqrt{V^3 W}} \end{bmatrix} = \begin{bmatrix} -0.09988398 \\[1ex] 0.15219681 \\[1ex] -0.03423960 \end{bmatrix}
$$

and

$$
\Sigma = \begin{bmatrix} \mathrm{var}(W) & \mathrm{cov}(W,Z) & \mathrm{cov}(W,V) \\ \mathrm{cov}(Z,W) & \mathrm{var}(Z) & \mathrm{cov}(Z,V) \\ \mathrm{cov}(V,W) & \mathrm{cov}(V,Z) & \mathrm{var}(V) \end{bmatrix} = \begin{bmatrix} 1.7709 & 2.0417 & 2.3555 \\ 2.0417 & 3.3306 & 4.9690 \\ 2.3555 & 4.9690 & 10.4838 \end{bmatrix}
$$

The variance of $\hat{r}_{A(XY)}$ is

$$
\mathrm{var}\left(\hat{r}_{A(XY)}\right) = \nabla \Sigma \nabla^T = 0.009355331
$$

The standard error is

$$
S_{\hat{r}_{A(XY)}} = \sqrt{\mathrm{var}\left(\hat{r}_{A(XY)}\right)} = \sqrt{0.009355331} = 0.09672296
$$

So, the estimated genetic correlation should be reported as

$$
\hat{r}_{A(XY)} \pm S_{\hat{r}_{A(XY)}} = 0.7685 \pm 0.0967
$$

The standard error is so small compared with the estimated genetic correlation because of the large sample $(n = 500)$.

### 13.4.3 Estimating Genetic Correlation Using a Nested Mating Design

A nested design is represented by a mating system with sibs nested within a dam family and multiple dam families nested within a sire family and the entire population may consist of several sire families. We have learned such a nested design for estimating heritability in Chap. 10. The same mating design can also be used to estimating genetic correlation. Assume that there are $s$ sires and each sire is mated with $d$ dams and each dam has $n$ progeny. The ANOVA tables for traits $X$ and $Y$ are the same as what we learned in Chap. 10. The ANCOVA is shown in Table 13.12. The covariance components are estimated by equating the expected mean cross products to the observed mean cross products. The solutions are

**Table 13.12** ANCOVA table for the nested mating design

| Source | Df | SCP | MCP | E(MCP) |
|---|---|---|---|---|
| Sire | $df_S = s - 1$ | $SCP_S$ | $MCP_S = SCP_S/df_S$ | $\sigma_{W(XY)} + n\sigma_{D(XY)} + dn\sigma_{S(XY)}$ |
| Dam | $df_D = s(d - 1)$ | $SCP_D$ | $MCP_D = SCP_D/df_D$ | $\sigma_{W(XY)} + n\sigma_{D(XY)}$ |
| Sib | $df_W = sd(n - 1)$ | $SCP_W$ | $MCP_W = SCP_W/df_W$ | $\sigma_{W(XY)}$ |

$$\widehat{\sigma}_{W(XY)} = MCP_W$$

$$\widehat{\sigma}_{D(XY)} = (MCP_D - MCP_W)/n$$

$$\widehat{\sigma}_{S(XY)} = (MCP_S - MCP_D)/(dn)$$

From the estimated variance and covariance components, we can estimate the genetic correlation in three different ways,

$$\widehat{r}_{A(XY)} = \frac{\sigma_{S(XY)}}{\sqrt{\sigma_{S(X)}^2 \sigma_{S(Y)}^2}}$$

$$\widehat{r}_{A(XY)} = \frac{\sigma_{D(XY)}}{\sqrt{\sigma_{D(X)}^2 \sigma_{D(Y)}^2}}$$

$$\widehat{r}_{A(XY)} = \frac{1}{df_S + df_D}\left(df_S \frac{\sigma_{S(XY)}}{\sqrt{\sigma_{S(X)}^2 \sigma_{S(Y)}^2}} + df_D \frac{\sigma_{D(XY)}}{\sqrt{\sigma_{D(X)}^2 \sigma_{D(Y)}^2}}\right)$$

The third way is an average of the correlation from the sire component and the correlation from the dam component weighted by their degrees of freedom. If the data are unbalanced, we should replace the $n$ in the dam's expected mean cross products by $k_1$, the $n$ in the sire's expected mean cross products by $k_2$ and the $dn$ in the sire's expected mean cross products by $k_3$. These $k$ values are given in Chap. 10. Let us look at an example of a nested design with $s = 10$ sire families, each sire family consists of $d = 10$ dam families and each dam family consists of $n = 5$ siblings. Partial records of the data are shown in Table 13.13. Again, we will use two SAS procedures to estimate the variance and covariance components. The GLM procedure reports the intermediate results and the MIXED procedure reports the estimated variance and covariance components. The MIXED procedure requires the "tall-format" for the data, as shown in Table 13.14, where the phenotypic values of the two traits are listed in one column named $Z$ with trait names identified by an additional character variable named $V$.

The SAS code to read the data and call the GLM procedure is shown below.

**Table 13.13** Partial records of two traits from 5 out of 100 dam (full-sib) families (data-13-4.csv)

| Sire | Dam | Sib | X | Y |
|------|-----|-----|-----|-----|
| 1 | 1 | 1 | 113.7784 | 170.6859 |
| 1 | 1 | 2 | 102.6364 | 151.642 |
| 1 | 1 | 3 | 104.8606 | 161.3027 |
| 1 | 1 | 4 | 108.7193 | 158.4697 |
| 1 | 1 | 5 | 107.9129 | 173.0542 |
| 1 | 2 | 1 | 105.8783 | 165.2831 |
| 1 | 2 | 2 | 95.05395 | 141.3337 |
| 1 | 2 | 3 | 98.93399 | 140.1688 |
| 1 | 2 | 4 | 106.087 | 151.7471 |
| 1 | 2 | 5 | 113.8017 | 160.8265 |
| 1 | 3 | 1 | 104.2341 | 155.0292 |
| 1 | 3 | 2 | 80.97513 | 129.063 |
| 1 | 3 | 3 | 105.4912 | 155.9898 |
| 1 | 3 | 4 | 100.9911 | 145.3288 |
| 1 | 3 | 5 | 102.7186 | 147.9164 |
| 1 | 4 | 1 | 94.1232 | 143.1175 |
| 1 | 4 | 2 | 95.21856 | 154.5122 |
| 1 | 4 | 3 | 90.25051 | 137.5489 |
| 1 | 4 | 4 | 102.2568 | 146.94 |
| 1 | 4 | 5 | 99.87207 | 142.5411 |
| 1 | 5 | 1 | 100.676 | 152.4905 |
| 1 | 5 | 2 | 106.2939 | 152.8418 |
| 1 | 5 | 3 | 103.5456 | 161.1291 |
| 1 | 5 | 4 | 95.15511 | 142.696 |

**Table 13.14** Partial records of two traits from 3 out of 100 dam (full-sib) families with the tall-format (data-13-5.xlsx)

| Sire | Dam | Sib | V | Z |
|------|-----|-----|---|---|
| 1 | 1 | 1 | X | 113.7784294 |
| 1 | 1 | 1 | Y | 170.6859347 |
| 1 | 1 | 2 | X | 102.6364139 |
| 1 | 1 | 2 | Y | 151.6419519 |
| 1 | 1 | 3 | X | 104.8605538 |
| 1 | 1 | 3 | Y | 161.302705 |
| 1 | 1 | 4 | X | 108.7193172 |
| 1 | 1 | 4 | Y | 158.4697301 |
| 1 | 1 | 5 | X | 107.912939 |
| 1 | 1 | 5 | Y | 173.0541689 |
| 1 | 2 | 1 | X | 105.8782832 |
| 1 | 2 | 1 | Y | 165.2831199 |
| 1 | 2 | 2 | X | 95.05394751 |
| 1 | 2 | 2 | Y | 141.3336686 |
| 1 | 2 | 3 | X | 98.93399023 |
| 1 | 2 | 3 | Y | 140.1688416 |
| 1 | 2 | 4 | X | 106.0870346 |
| 1 | 2 | 4 | Y | 151.7470608 |
| 1 | 2 | 5 | X | 113.8017086 |
| 1 | 2 | 5 | Y | 160.8264852 |
| 1 | 3 | 1 | X | 104.2340963 |
| 1 | 3 | 1 | Y | 155.02921 |
| 1 | 3 | 2 | X | 80.97513244 |
| 1 | 3 | 2 | Y | 129.0630327 |
| 1 | 3 | 3 | X | 105.4912068 |
| 1 | 3 | 3 | Y | 155.989826 |
| 1 | 3 | 4 | X | 100.9910514 |
| 1 | 3 | 4 | Y | 145.3287631 |
| 1 | 3 | 5 | X | 102.7185757 |
| 1 | 3 | 5 | Y | 147.9163895 |

```
filename aa "data-13-4.csv";
proc import datafile=aa out=one dbms=csv replace;
run;
ods graphics off;
proc glm data=one;
   class sire dam sib;
   model x y =sire dam(sire);
   random sire dam(sire);
   manova h=sire / printh printe;
   manova h=dam(sire)/printh;
quit;
```

The output includes several ANOVA tables. Here, we only show Table 13.15 and Table 13.16 for trait X. The corresponding tables for trait Y and for the sums of cross products and mean of cross products are not listed in the text. The only statistics we need from these two tables are the degrees of freedom, $df_S = 10 - 1 = 9$ and $df_D = s (d - 1) = 10 \times (10 - 1) = 90$ and $df_W = sd(n - 1) = 10 \times 10 \times (5 - 1) = 400$. The output also includes the expected mean squares table (Table 13.17), which allows users to calculate the estimated variance and covariance components. The GLM procedure delivers three additional tables showing the sum of squares and sum of cross products (SSCP), one is the between sire SSCP, one is the between dam within sire SSCP and one is the residual (error) SSCP, all of which are given in Table 13.18.

**Table 13.15**  ANOVA table for the overall model for variable X produced by PROC GLM

| Source | DF | Sum of squares | Mean square | F value | Pr > F |
|---|---|---|---|---|---|
| Model | 99 | 21957.84539 | 221.79642 | 8.60 | <0.0001 |
| Error | 400 | 10314.77284 | 25.78693 | | |
| Corrected total | 499 | 32272.61823 | | | |

**Table 13.16**  ANOVA table for sire and dam effects for variable X produced by PROC GLM

| Source | DF | Type III SS | Mean square | F value | Pr > F |
|---|---|---|---|---|---|
| Sire | 9 | 17861.22528 | 1984.58059 | 76.96 | <0.0001 |
| Dam(Sire) | 90 | 4096.62010 | 45.51800 | 1.77 | 0.0001 |

**Table 13.17**  Expected mean squares and mean cross products

| Source | Type III expected mean square |
|---|---|
| Sire | Var(Error) + 5 Var(Dam(Sire)) + 50 Var(Sire) |
| Dam(Sire) | Var(Error) + 5 Var(Dam(Sire)) |

**Table 13.18**  The SSCP tables delivered by the GLM procedure

| | H = Type III SSCP matrix for sire | | H = Type III SSCP matrix for dam (Sire) | | E = Error SSCP matrix | |
|---|---|---|---|---|---|---|
| | X | Y | X | Y | X | Y |
| X | 17861.22 | 19400.94 | 4096.62 | 4573.10 | 10314.77 | 10816.14 |
| Y | 19400.94 | 21816.69 | 4573.10 | 9229.85 | 10816.14 | 21806.23 |

**Table 13.19**  The MSCP tables obtained by dividing the SSCP tables by the degrees of freedom

| | H = Type III MSCP Matrix for Sire | | H = Type III MSCP Matrix for Dam(Sire) | | E = Error MSCP Matrix | |
|---|---|---|---|---|---|---|
| | X | Y | X | Y | X | Y |
| X | 1984.581 | 2155.660 | 45.5180 | 50.8122 | 25.7869 | 27.0403 |
| Y | 2155.660 | 2424.077 | 50.8122 | 102.5540 | 27.0403 | 54.5155 |

Dividing the above three matrices by their corresponding degrees of freedom results in three mean squares and mean cross products (MSCP) (Table 13.19). Finally, we obtain the variance and covariance component matrices for the sire term

$$
\begin{bmatrix} \widehat{\sigma}^2_{S(X)} & \widehat{\sigma}_{S(XY)} \\ \widehat{\sigma}_{S(XY)} & \widehat{\sigma}^2_{S(X)} \end{bmatrix} = \begin{bmatrix} (1984.581 - 45.518)/50 & (2155.66 - 50.81229)/50 \\ (2155.66 - 50.81229)/50 & (2424.077 - 102.554)/50 \end{bmatrix}
$$

which equals

$$
\begin{bmatrix} \widehat{\sigma}^2_{S(X)} & \widehat{\sigma}_{S(XY)} \\ \widehat{\sigma}_{S(XY)} & \widehat{\sigma}^2_{S(X)} \end{bmatrix} = \begin{bmatrix} 38.78125 & 42.09695 \\ 42.09695 & 46.43047 \end{bmatrix} \tag{13.3}
$$

The variance and covariance component matrices for the dam term is

$$
\begin{bmatrix} \widehat{\sigma}^2_{D(X)} & \widehat{\sigma}_{D(XY)} \\ \widehat{\sigma}_{D(XY)} & \widehat{\sigma}^2_{D(X)} \end{bmatrix} = \begin{bmatrix} (45.518 - 25.78693)/5 & (50.81229 - 27.04036)/5 \\ (50.81229 - 27.04036)/5 & (102.554 - 54.51558)/5 \end{bmatrix}
$$

which equals

$$\begin{bmatrix} \widehat{\sigma}^2_{D(X)} & \widehat{\sigma}_{D(XY)} \\ \widehat{\sigma}_{D(XY)} & \widehat{\sigma}^2_{D(X)} \end{bmatrix} = \begin{bmatrix} 3.946214 & 4.754385 \\ 4.754385 & 9.607683 \end{bmatrix} \tag{13.4}$$

The variance and covariance component matrices for the siblings within dam is

$$\begin{bmatrix} \widehat{\sigma}^2_{W(X)} & \widehat{\sigma}_{W(XY)} \\ \widehat{\sigma}_{W(XY)} & \widehat{\sigma}^2_{W(X)} \end{bmatrix} = \begin{bmatrix} 25.78693 & 27.04036 \\ 27.04036 & 54.51558 \end{bmatrix} \tag{13.5}$$

The three estimated genetic correlations are

$$\widehat{r}_{A(XY)} = \frac{\sigma_{S(XY)}}{\sqrt{\sigma^2_{S(X)}\sigma^2_{S(Y)}}} = \frac{42.09695}{\sqrt{38.78125 \times 46.43047}} = 0.992061$$

$$\widehat{r}_{A(XY)} = \frac{\sigma_{D(XY)}}{\sqrt{\sigma^2_{D(X)}\sigma^2_{D(Y)}}} = \frac{4.754385}{\sqrt{3.946214 \times 9.607683}} = 0.772138$$

$$\widehat{r}_{A(XY)} = \frac{1}{9+90}(9 \times 0.992061 + 90 \times 0.772138) = 0.792131$$

The same data can be analyzed with the MIXED procedure in SAS. The SAS code to convert the data from the wide-format (Table 13.13) to the tall-format (Table 13.14) and to estimate the variance and covariance components is

```
data two;
    set one;
    V='X'; Z=X; output;
    V='Y'; Z=Y; output;
    keep Sire Dam Sib Z V;
run;
proc export data=two outfile="data-13-5.csv" dbms=csv replace;
run;
proc import out=two datafile="data-13-5.csv" dbms=csv replace;
run;
proc mixed data=two method=reml covtest asycov;
    class V Sire Dam;
    model Z = V;
    random V/subject=Sire type=UN;
    random V/subject=Dam(Sire) type=UN;
    repeated /subject=Sib type=UN;
run;
```

The output includes two tables, one being the estimated variance and covariance components (Table 13.20) and the other being the asymptotic variance-covariance matrix of the estimated variance and covariance components (Table 13.21). The estimated parameters from the MIXED procedure (Table 13.20) are exactly the same as the estimates from the GLM procedures, see Eqs. (13.3), (13.4), and (13.5).

Genetic correlation can also be estimated from the pedigree data. If the response variables are arranged in a single column sorted by individuals, the model is

$$y = X\beta + \xi + e$$

where $\xi$ is a vector of breeding values for all $n$ individuals for all $m$ traits. Let $A$ be the $n \times n$ numerator relationship matrix and $I$ be an $n \times n$ identity matrix. Denote the genetic variance-covariance matrix and the residual variance-covariance matrix for the $m$ traits as

**Table 13.20** The estimated variance and covariance components from the MIXED procedure in SAS

| Covariance parameter estimates | | | | | |
|---|---|---|---|---|---|
| Cov Parm | Subject | Estimate | Standard error | Z value | Pr Z |
| UN(1,1) | Sire | 38.7813 | 18.7113 | 2.07 | 0.0191 |
| UN(2,1) | Sire | 42.0970 | 20.5030 | 2.05 | 0.0401 |
| UN(2,2) | Sire | 46.4305 | 22.8565 | 2.03 | 0.0211 |
| UN(1,1) | Dam(Sire) | 3.9462 | 1.4052 | 2.81 | 0.0025 |
| UN(2,1) | Dam(Sire) | 4.7544 | 1.8536 | 2.56 | 0.0103 |
| UN(2,2) | Dam(Sire) | 9.6077 | 3.1533 | 3.05 | 0.0012 |
| UN(1,1) | Sib | 25.7869 | 1.8234 | 14.14 | <0.0001 |
| UN(2,1) | Sib | 27.0404 | 2.3114 | 11.70 | <0.0001 |
| UN(2,2) | Sib | 54.5156 | 3.8548 | 14.14 | <0.0001 |

**Table 13.21** Asymptotic variance-covariance matrix for the estimates from the MIXED procedure in SAS

| Asymptotic covariance matrix of estimates | | | | | | | | | | |
|---|---|---|---|---|---|---|---|---|---|---|
| Row | Cov Parm | CovP1 | CovP2 | CovP3 | CovP4 | CovP5 | CovP6 | CovP7 | CovP8 | CovP9 |
| 1 | UN(1,1) | 350.11 | 380.29 | 413.08 | −0.1842 | −0.2056 | −0.2295 | -276E-13 | -396E-13 | -572E-13 |
| 2 | UN(2,1) | 380.29 | 420.37 | 464.53 | −0.2056 | −0.3222 | −0.4632 | -304E-13 | -437E-13 | -631E-13 |
| 3 | UN(2,2) | 413.08 | 464.53 | 522.42 | −0.2295 | −0.4632 | −0.9349 | -335E-13 | -481E-13 | -696E-13 |
| 4 | UN(1,1) | −0.1842 | −0.2056 | −0.2295 | 1.9747 | 2.1953 | 2.4412 | −0.6650 | −0.6973 | −0.7312 |
| 5 | UN(2,1) | −0.2056 | −0.3222 | −0.4632 | 2.1953 | 3.4359 | 4.9268 | −0.6973 | −1.0685 | −1.4741 |
| 6 | UN(2,2) | −0.2295 | −0.4632 | −0.9349 | 2.4412 | 4.9268 | 9.9431 | −0.7312 | −1.4741 | −2.9719 |
| 7 | UN(1,1) | -276E-13 | -304E-13 | -335E-13 | −0.6650 | −0.6973 | −0.7312 | 3.3248 | 3.4864 | 3.6559 |
| 8 | UN(2,1) | -396E-13 | -437E-13 | -481E-13 | −0.6973 | −1.0685 | −1.4741 | 3.4864 | 5.3424 | 7.3706 |
| 9 | UN(2,2) | -572E-13 | -631E-13 | -696E-13 | −0.7312 | −1.4741 | −2.9719 | 3.6559 | 7.3706 | 14.8597 |

$$
\Pi = \begin{bmatrix} \Pi_{11} & \Pi_{12} & \cdots & \Pi_{1m} \\ \Pi_{21} & \Pi_{22} & \cdots & \Pi_{2m} \\ \vdots & \vdots & \ddots & \vdots \\ \Pi_{m1} & \Pi_{m2} & \cdots & \Pi_{mm} \end{bmatrix} \text{ and } \Sigma = \begin{bmatrix} \Sigma_{11} & \Sigma_{12} & \cdots & \Sigma_{1m} \\ \Sigma_{21} & \Sigma_{22} & \cdots & \Sigma_{2m} \\ \vdots & \vdots & \ddots & \vdots \\ \Sigma_{m1} & \Sigma_{m2} & \cdots & \Sigma_{mm} \end{bmatrix}
$$

.

The expectation of $y$ is $\mathrm{E}(y) = X\beta$ and the variance matrix of $y$ is

$$
\mathrm{var}(y) = \mathrm{var}(\xi) + \mathrm{var}(e) = A \otimes \Pi + I \otimes \Sigma
$$

where $\otimes$ represents Kronecker product of two matrices. The ML or REML method can be used to estimate parameters $\theta = \{\Pi, \Sigma\}$. Unfortunately, the MIXED procedure does not have an option to specify this structure $(A \otimes \Pi)$ for random effects. One may have to write an R program to estimate genetic correlation with pedigree data.

# References

Fisher RA. On the 'probable error' of a coefficient of correlation deduced from a small sample. Metron. 1921;1:3–32.
盛志廉, 陈瑶生. 数量遗传学. 科学出版社, 北京. 1999.

# Concept and Theory of Selection

## 14.1 Evolutionary Forces

Evolutionary forces are factors that can change the genetic properties of a population. There are four major evolutionary forces: selection, mutation, migration, and genetic drift. Selection is considered to be the most important one. Selection is defined as a phenomenon that parents who contribute to the next generation are not a random sample but selected based on some criteria (Falconer and Mackay 1996). Selection can be classified into artificial selection and natural selection. If the selection criteria are determined by human, the selection is called artificial selection. In most cases, the selection criteria are determined by nature, and such a selection is called natural selection. Domesticated animals and agricultural crops around us are mainly the results of long-term artificial selection. All wild animals and wild plants, looking the way as they look and behaving the way as they behave, are results of natural selection.

The genetic properties of a population are determined by its gene frequencies and genotype frequencies. However, gene and genotypic frequencies are not always observable. What we can observe are the population means and variances of traits that reflect gene and genotype frequencies of the population.

## 14.2 Change in Gene Frequency and Genotype Frequencies

Here, we use a deleterious recessive gene as an example. Assume that allele $A_1$ is completely dominant over allele $A_2$ so that $A_1A_1$ and $A_1A_2$ are indistinguishable in terms of selection. Table 14.1 shows the genotype frequencies, selection coefficients, and fitness of the three genotypes. The fitness ($w$) is defined as $w = 1 - s$ where $s$ is the selection coefficient. The gametic contribution of each genotype is the initial frequency multiplied by the fitness. After selection, the genotype frequencies will change, i.e., the genotype frequencies in the selected parents will be different from the frequencies of the initial unselected population. Table 14.2 shows the new genotype frequencies of the selected parental population. Define the initial frequencies of $A_1$ and $A_2$ by $p_0$ and $q_0$, respectively, then the new allelic frequencies for $A_1$ and $A_2$ are

$$p_1 = P_1 + \frac{1}{2}H_1 = \left[\frac{p_0^2}{1 - sq_0^2}\right] + \frac{1}{2}\left[\frac{2p_0q_0}{1 - sq_0^2}\right] = \frac{p_0}{1 - sq_0^2}$$

$$q_1 = Q_1 + \frac{1}{2}H_1 = \left[\frac{(1-s)q_0^2}{1 - sq_0^2}\right] + \frac{1}{2}\left[\frac{2p_0q_0}{1 - sq_0^2}\right] = \frac{q_0 - sq_0^2}{1 - sq_0^2}$$

The change of gene frequency in one generation of selection is

$$\Delta q = q_1 - q_0 = \frac{-sq_0^2(1 - q_0)}{1 - sq_0^2} \leq 0$$

S. Xu, *Quantitative Genetics*, https://doi.org/10.1007/978-3-030-83940-6_14

**Table 14.1** Change in gene frequency and genotypic frequencies

| Genotype | $A_1A_1$ | $A_1A_2$ | $A_2A_2$ | Total |
|---|---|---|---|---|
| Initial freq | $p^2$ | $2pq$ | $q^2$ | 1 |
| Selection coefficient | 0 | 0 | $s$ | |
| Fitness | 1 | 1 | $1-s$ | |
| Gametic contribution | $p^2$ | $2pq$ | $(1-s)q^2$ | $1-sq^2$ |

**Table 14.2** Genotypic frequencies of the selected parents

| Genotype | $A_1A_1$ | $A_1A_2$ | $A_2A_2$ | Total |
|---|---|---|---|---|
| Frequency in selected Parents | $\frac{p^2}{1-sq^2}$ | $\frac{2pq}{1-sq^2}$ | $\frac{(1-s)q^2}{1-sq^2}$ | 1 |

**Table 14.3** Genotypic frequencies in the next generation

| Genotype | Frequency |
|---|---|
| $A_1A_1$ | $(1-q_1)^2$ |
| $A_1A_2$ | $2q_1(1-q_1)$ |
| $A_2A_2$ | $q_1^2$ |

So, $q_1 \leq q_0$, meaning that the frequency of $A_2$ has been decreased due to selection against this allele. If the selected parents are randomly mated, the genotype frequencies in the offspring will be formed based on the Hardy-Weinberg law, as given in Table 14.3. Therefore, after one generation of selection, both the gene and genotype frequencies have changed. As a result, the population means and genetic variances of traits controlled by this gene have also changed. The genetic effect of selection is the change in gene and genotype frequencies.

## 14.3  Artificial Selection

Artificial selection is also called truncation selection. The fitness function is represented by a clear threshold in the scale of the phenotypic value of a trait. In contrast, natural selection rarely happens this way because nature often selects individuals based on a smooth fitness function (Walsh and Lynch 2018).

### 14.3.1  Directional Selection

Directional truncation selection is the most important type of selection in animal and plant breeding. An individual is selected if its phenotypic value is greater than a threshold $T$. The fitness function is defined as

$$w(y) = \begin{cases} 1 & \text{if} \quad y \geq T \\ 0 & \text{if} \quad y < T \end{cases},$$

where $y \sim N\left(\mu, \sigma_P^2\right)$ is the phenotypic value of a trait and it follows a normal distribution with mean $\mu$ and variance $\sigma_P^2$. The fitness function is not smooth. The probability density of the phenotypic value is

$$f(y) = \frac{1}{\sqrt{2\pi}\sigma_P} \exp\left\{-\frac{1}{2}\left(\frac{y-\mu}{\sigma_P}\right)^2\right\}.$$

The selection favors either higher values of a trait or lower values of the trait but not both. This explains why the selection is called directional selection. The expectation of the fitness, i.e., the average fitness of the population, is

$$\overline{W} = \int_{-\infty}^{\infty} w(y)f(y)dy = \int_{T}^{\infty} \frac{1}{\sqrt{2\pi}\sigma_P} \exp\left\{-\frac{1}{2}\left(\frac{y-\mu}{\sigma_P}\right)^2\right\}dy = 1 - \Phi\left(\frac{T-\mu}{\sigma_P}\right). \tag{14.1}$$

Let the truncation point in the standardized scale be

$$t = \frac{T-\mu}{\sigma_P}.$$

The average fitness is denoted by

$$\overline{W} = 1 - \Phi(t) = p,$$

which is the proportion selected. The expectation of the selected population is

$$\begin{aligned}\mu_S &= \frac{1}{\overline{W}} \int_{-\infty}^{\infty} yw(y)f(y)dy \\ &= \frac{1}{\overline{W}} \int_{T}^{\infty} yf(y)dy \\ &= \mu + \frac{\sigma_P}{p\sqrt{2\pi}} \exp\left\{-\frac{1}{2}\left(\frac{T-\mu}{\sigma_P}\right)^2\right\} \\ &= \mu + \frac{\sigma_P}{p}\phi(t)\end{aligned} \tag{14.2}$$

where

$$\phi(t) = \frac{1}{\sqrt{2\pi}} \exp\left\{-\frac{1}{2}t^2\right\}$$

is the height of the standardized normal density at the truncation point $t$. *Selection differential*, the difference between the mean of the selected population and the mean of the whole population before selection takes place, is defined as

$$S = \mu_S - \mu = \frac{\sigma_P}{p}\phi(t). \tag{14.3}$$

Selection differential is also the covariance between the fitness and the phenotypic value (Walsh and Lynch 2018). This can be shown by rearranging Eq. (14.2)

$$S = \mu_S - \mu = \frac{1}{\overline{W}} \int_{-\infty}^{\infty} yw(y)f(y)dy - \mu \tag{14.4}$$

The expectation of the fitness $w$ is

$$E(w) = \int_{-\infty}^{\infty} w(y)f(y)dy = \overline{W}$$

The expectation of the trait $y$ is

$$E(y) = \int_{-\infty}^{\infty} yf(y)dy = \mu$$

The expectation of the product of $w$ and $y$ is

$$E(wy) = \int_{-\infty}^{\infty} yw(y)f(y)dy$$

The covariance between $w$ and $y$ is

$$\text{cov}(w, y) = E(wy) - E(w)E(y) = \int_{-\infty}^{\infty} yw(y)f(y)dy - \mu\overline{W} \tag{14.5}$$

Dividing both sides of Eq. (14.5) by $\overline{W}$ leads to

$$\frac{1}{\overline{W}}\text{cov}(w, y) = \frac{1}{\overline{W}} \int_{-\infty}^{\infty} yw(y)f(y)dy - \mu,$$

which is identical to the selection differential shown in Eq. (14.4). Therefore,

$$S = \mu_S - \mu = \frac{1}{\overline{W}}\text{cov}(w, y)$$

So, selection differential is proportional (not exactly equal) to the covariance between the fitness and the trait value. Selection differential represents the change of population mean within the current generation. *Response to selection* is defined as the difference between the mean of the progeny from the selected parents (mean of the next generation) and the mean of the parental generation (before selection). Denote the response to selection by $R$, the relationship between $R$ and $S$ is given by

$$R = h^2 S,$$

where $h^2$ is the heritability. This means that only the heritable part of the selection differential can be passed (transformed) into the gain in the next generation. The response equation is also called the "breeders' equation" because it is important in plant and animal breeding. The breeders' equation is also useful in phenotypic evolution (Lande 1976). Selection differentials cannot be compared for different traits. A large value of selection differential does not mean a strong selection. The strength of a selection is measured by *selection intensity*, denoted by $i$, which is a standardized selection differential. It is defined as the selection differential divided by the phenotypic standard deviation, i.e.,

$$i = \frac{S}{\sigma_P}, \tag{14.6}$$

where $\sigma_P$ is the standard deviation of the phenotypic value of the trait under selection, i.e., $\sigma_P = \sqrt{\sigma_P^2} = \sqrt{V_P}$, where $V_P$ is the conventional notation for phenotypic variance. We now express the selection response as a function of the selection intensity,

$$R = h^2 S = h^2 i \sigma_P = h i \sigma_A$$

So, the breeders' equation has many different forms. The proportion selected, $p$, is defined as the average fitness (14.1). It is also the ratio of the number of the selected individuals to the total number of the individuals measured under the truncation selection scheme. Theoretically, $p$ is also defined by

$$p = 1 - \int_{-\infty}^{T} f(y)dy = 1 - \int_{-\infty}^{t} f(x)dx = 1 - \Phi(t),$$

where $x = (y - \mu)/\sigma_P$ is the standardized normal variable and $t = (T - \mu)/\sigma_P$ is the standardized truncation point in the $N(0, 1)$ scale. If the selected character is normally distributed, there is a simple relationship between $i$ and $p$. Let $z = \phi(t)$ be the density of the standardized normal variable at the standardized truncation point. Substituting Eq. (14.3) into Eq. (14.6) yields

$$i = \frac{S}{\sigma_P} = \frac{\mu_S - \mu}{\sigma_P} = \frac{\sigma_P \phi(t)}{\sigma_P p} = \frac{\phi(t)}{p} = \frac{z}{p}$$

Figure 14.1 shows the geometric relationship between parameters in the phenotypic distribution. The upper panel (b) of Fig. 14.1 shows the normal distribution of the phenotypic value in its original scale, $y \sim N(\mu, \sigma_P^2)$. The lower panel (b) shows

**Fig. 14.1** Normal distribution (**a**) and standardized normal distribution (**b**)

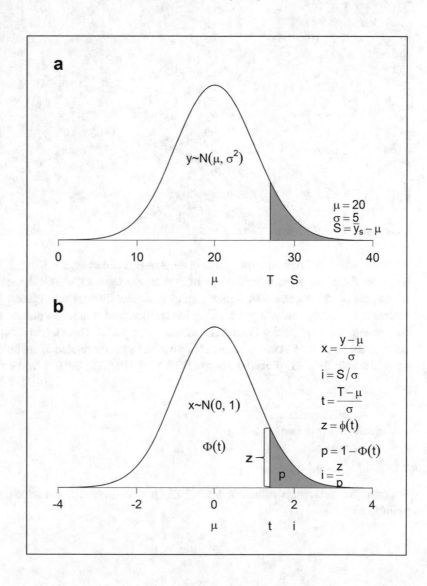

**Table 14.4** The *tip* relationship among the standardized truncation point ($t$), the selection intensity ($i$), and the proportion selected ($p$)

| $p$ | $t$ | $i$ | $p$ | $t$ | $i$ | $p$ | $t$ | $i$ |
|------|--------|--------|------|---------|--------|------|---------|--------|
| 0.01 | 2.3263 | 2.6652 | 0.34 | 0.4125 | 1.0777 | 0.67 | −0.4399 | 0.5405 |
| 0.02 | 2.0537 | 2.4209 | 0.35 | 0.3853 | 1.0583 | 0.68 | −0.4677 | 0.5259 |
| 0.03 | 1.8808 | 2.2681 | 0.36 | 0.3585 | 1.0392 | 0.69 | −0.4959 | 0.5113 |
| 0.04 | 1.7507 | 2.1543 | 0.37 | 0.3319 | 1.0205 | 0.70 | −0.5244 | 0.4967 |
| 0.05 | 1.6449 | 2.0627 | 0.38 | 0.3055 | 1.0020 | 0.71 | −0.5534 | 0.4821 |
| 0.06 | 1.5548 | 1.9854 | 0.39 | 0.2793 | 0.9838 | 0.72 | −0.5828 | 0.4675 |
| 0.07 | 1.4758 | 1.9181 | 0.40 | 0.2533 | 0.9659 | 0.73 | −0.6128 | 0.4529 |
| 0.08 | 1.4051 | 1.8583 | 0.41 | 0.2275 | 0.9482 | 0.74 | −0.6433 | 0.4383 |
| 0.09 | 1.3408 | 1.8043 | 0.42 | 0.2019 | 0.9307 | 0.75 | −0.6745 | 0.4237 |
| 0.10 | 1.2816 | 1.7550 | 0.43 | 0.1764 | 0.9135 | 0.76 | −0.7063 | 0.4090 |
| 0.11 | 1.2265 | 1.7094 | 0.44 | 0.1510 | 0.8964 | 0.77 | −0.7388 | 0.3943 |
| 0.12 | 1.1750 | 1.6670 | 0.45 | 0.1257 | 0.8796 | 0.78 | −0.7722 | 0.3796 |
| 0.13 | 1.1264 | 1.6273 | 0.46 | 0.1004 | 0.8629 | 0.79 | −0.8064 | 0.3648 |
| 0.14 | 1.0803 | 1.5898 | 0.47 | 0.0753 | 0.8464 | 0.80 | −0.8416 | 0.3500 |
| 0.15 | 1.0364 | 1.5544 | 0.48 | 0.0502 | 0.8301 | 0.81 | −0.8779 | 0.3350 |
| 0.16 | 0.9945 | 1.5207 | 0.49 | 0.0251 | 0.8139 | 0.82 | −0.9154 | 0.3200 |
| 0.17 | 0.9542 | 1.4886 | 0.50 | 0.0000 | 0.7979 | 0.83 | −0.9542 | 0.3049 |
| 0.18 | 0.9154 | 1.4578 | 0.51 | −0.0251 | 0.7820 | 0.84 | −0.9945 | 0.2897 |
| 0.19 | 0.8779 | 1.4282 | 0.52 | −0.0502 | 0.7662 | 0.85 | −1.0364 | 0.2743 |
| 0.20 | 0.8416 | 1.3998 | 0.53 | −0.0753 | 0.7506 | 0.86 | −1.0803 | 0.2588 |
| 0.21 | 0.8064 | 1.3724 | 0.54 | −0.1004 | 0.7351 | 0.87 | −1.1264 | 0.2432 |
| 0.22 | 0.7722 | 1.3459 | 0.55 | −0.1257 | 0.7196 | 0.88 | −1.1750 | 0.2273 |
| 0.23 | 0.7388 | 1.3202 | 0.56 | −0.1510 | 0.7043 | 0.89 | −1.2265 | 0.2113 |
| 0.24 | 0.7063 | 1.2953 | 0.57 | −0.1764 | 0.6891 | 0.90 | −1.2816 | 0.1950 |
| 0.25 | 0.6745 | 1.2711 | 0.58 | −0.2019 | 0.6740 | 0.91 | −1.3408 | 0.1785 |
| 0.26 | 0.6433 | 1.2476 | 0.59 | −0.2275 | 0.6589 | 0.92 | −1.4051 | 0.1616 |
| 0.27 | 0.6128 | 1.2246 | 0.60 | −0.2533 | 0.6439 | 0.93 | −1.4758 | 0.1444 |
| 0.28 | 0.5828 | 1.2022 | 0.61 | −0.2793 | 0.6290 | 0.94 | −1.5548 | 0.1267 |
| 0.29 | 0.5534 | 1.1804 | 0.62 | −0.3055 | 0.6141 | 0.95 | −1.6449 | 0.1086 |
| 0.30 | 0.5244 | 1.1590 | 0.63 | −0.3319 | 0.5993 | 0.96 | −1.7507 | 0.0898 |
| 0.31 | 0.4959 | 1.1380 | 0.64 | −0.3585 | 0.5846 | 0.97 | −1.8808 | 0.0701 |
| 0.32 | 0.4677 | 1.1175 | 0.65 | −0.3853 | 0.5698 | 0.98 | −2.0537 | 0.0494 |
| 0.33 | 0.4399 | 1.0974 | 0.66 | −0.4125 | 0.5552 | 0.99 | −2.3263 | 0.0269 |

the distribution of the phenotypic value in the standardized scale, $x \sim N(0, 1)$. From panel (b) of this figure, we see that $z$ is the height of the curve at the truncation point, $p$ is the shaded area, $\Phi(t)$ is the standardized cumulative distribution function (the white area of the normal distribution), and $i$ is the conditional expectation (mean) of distribution in the shaded area. The selection intensity is always greater than the standardized truncation point, $i > t$. Table 14.4 shows the numeric relationship between the three quantities in artificial selection, $t$, $i$, and $p$. Their relationship is also called the *tip* relationship. The *tip* table is not as useful as it was before because they can be easily calculated using built-in functions in many computer languages. For example, with the IML procedure or the DATA step in SAS, the functional relationships can be called as follows:

```
p = 1 - cdf('normal',t);
t = quantile('normal',1-p);
z = pdf('normal',t);
i = z/p;
```

For example, if the proportion selected is 10% ($p = 0.10$), what is the selection intensity? First, we need to find the truncation point $t$ using

$$t = \Phi^{-1}(1 - p) = \text{quantile}('\text{normal}', 1 - 0.10) = 1.28155.$$

We then calculate the density (height) of the normal distribution at the truncation point,

$$z = \phi(t) = \text{pdf}('\text{normal}', 1.28155) = 0.17549.$$

Finally, we divide the height by the proportion selected to get the selection intensity,

$$i = z/p = 0.17549/0.10 = 1.75498.$$

When males and females have different selection differentials, their average should be used. Let $S_m$ and $S_f$ be the selection differentials for males and females, respectively, the corresponding selection intensities are.

$$i_m = \frac{S_m}{\sigma_p} \text{ and } i_f = \frac{S_f}{\sigma_p}$$

The overall selection differential and selection intensity are

$$S = \frac{1}{2}\left(S_m + S_f\right)$$

and

$$i = \frac{1}{2}\left(i_m + i_f\right).$$

Therefore, the selection response under different selection intensities for males and females is

$$R = h^2 S = \frac{1}{2}h^2 S_m + \frac{1}{2}h^2 S_f = \frac{1}{2}h^2 i_m \sigma_p + \frac{1}{2}h^2 i_f \sigma_p.$$

Note that the overall selection response must take the simple average of males and females because males and females contribute equally to the next generation. We now show an example on how to calculate the selection response. Assume that $\sigma_P = 5$, $\mu = 25$, and $p = 0.20$, we want to find the truncation point, selection intensity, selection differential, and response to selection. From Table 14.4, we get $t = 0.84$ and $i = 1.40$ when $p = 0.20$. Note that $t$ is the truncation point in the standardized scale. What is the truncation point in the real scale $T$? Because $t = (T - \mu)/\sigma_p$, we have

$$T = t\sigma_P + \mu = 0.842 \times 5 + 25 = 29.21.$$

The selection differential is $S = i\sigma_P = 1.40 \times 5.0 = 7.0$. Assume that both the males and the females have the same selection intensity and the heritability of the trait is $h^2 = 0.50$. The selection response is

$$R = h^2 S = 0.5 \times 7.0 = 3.5.$$

The population mean of the offspring (next generation) will be $\mu + R = 25.0 + 3.5 = 28.5$. In addition to SAS, R is a much moreconvenient computer language to calculate the response to selection.

```
(1) Given p  =  0.2, find t
p<-0.2
t<-qnorm(1-p)
t
[1] 0.8416212
```

```
(2) Given t = 0.84, find p
t<-0.84
p<-1-pnorm(t)
p
[1] 0.2004542


(3) Given t = 0.84, find z
t<-0.84
z<-dnorm(t)
z
[1] 0.2803438


(4) Given p = 0.2, find i
p<-0.2
t<-qnorm(1-p)
z<-dnorm(t)
i<-z/p
i
[1] 1.39981


(5) Given t = 0.84, find i
t<-0.84
z<-dnorm(t)
i<-z/p
i
[1] 1.401719
```

The selection response we discussed so far is the genetic improvement for one generation (assuming non-overlapping generations), where parents only reproduce in one generation interval (Walsh and Lynch 2018). Domesticated animals, perennial plants, and many species in nature live multiple years and can have progeny over more than 1 year. In such cases, the response should be expressed in terms of response per year. To express the breeders' equation in terms of rate of response, we first need to compute the generation intervals (the average age of parents when their progenies are born) for both sexes. Let $L_m$ and $L_f$ be the generation intervals for male and female parents, respectively. The yearly genetic progress is (Walsh and Lynch 2018)

$$R_{year} = \left( \frac{i_m + i_f}{L_m + L_f} \right) h^2 \sigma_P = \left( \frac{i_m + i_f}{L_m + L_f} \right) h \sigma_A$$

Marker-assisted selection (Lande and Thompson 1990) is a way to shorten the generation interval and thus improve genetic gain per unit time.

### 14.3.2 Stabilizing Selection

Stabilizing selection can happen in breeding programs to improve the uniformity of agricultural products. The fitness function is

$$w(y) = \begin{cases} 1 & \text{if } T_1 \leq y \leq T_2 \\ 0 & \text{otherwise} \end{cases}$$

**Fig. 14.2** Six different selection modes, where the panels on the left show artificial selections and the panels on the right show natural selections. Shaded areas show the proportions selected. The $f(y)$ function is the density of the trait (selection criterion) and the $w(y)$ function is the fitness function

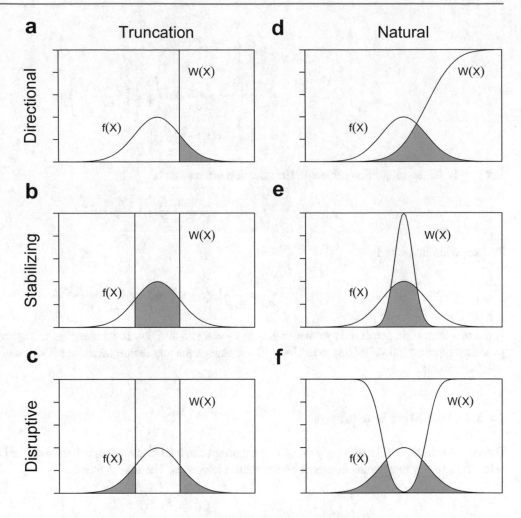

Figure 14.2b illustrates the truncated stabilizing selection. The average fitness of the population (proportion selected) is

$$\overline{W} = \int_{-\infty}^{\infty} w(y)f(y)dy$$

$$= \int_{T_1}^{T_2} \frac{1}{\sqrt{2\pi}\sigma_P} \exp\left\{-\frac{1}{2}\left(\frac{y-\mu}{\sigma_P}\right)^2\right\}dy$$

$$= \Phi\left(\frac{T_2-\mu}{\sigma_P}\right) - \Phi\left(\frac{T_1-\mu}{\sigma_P}\right)$$

Therefore, the proportion selected is

$$p = \Phi(t_2) - \Phi(t_1),$$

where $t_1$ and $t_2$ are the two truncation points in the standardized scale. The average value of the selected parents is obtained via

$$\mu_S = \frac{1}{\overline{W}} \int\limits_{-\infty}^{\infty} y w(y) f(y) dy$$

$$= \frac{1}{\overline{W}} \int\limits_{T_1}^{T_2} y \frac{1}{\sqrt{2\pi}\sigma_P} \exp\left\{-\frac{1}{2}\left(\frac{y-\mu}{\sigma_P}\right)^2\right\} dy$$

$$= \mu + \frac{\sigma_P}{\overline{W}}[\phi(t_1) - \phi(t_2)]$$

Let $p = \overline{W}$ be the proportion selected. The selection differential is

$$S = \mu_S - \mu = \frac{\sigma_P}{p}[\phi(t_1) - \phi(t_2)]$$

The selection intensity is

$$i = \frac{S}{\sigma_P} = \frac{\sigma_P}{\sigma_P p}[\phi(t_1) - \phi(t_2)] = \frac{\phi(t_1) - \phi(t_2)}{p}$$

If the two truncation points are symmetrical, then $\mu_S = \mu$ and thus $S = 0$; no selection response will occur. This means that a perfectly symmetrical stabilizing selection will not change the population mean, but it will reduce the genetic variance if the trait is heritable.

### 14.3.3  Disruptive Selection

Disruptive selection will never happen in breeding programs, but it can happen in nature and in laboratorial experiments. It is a selection strategy that avoids competition for natural resources. The fitness function is

$$w(y) = \begin{cases} 0 & \text{if } T_1 \leq y \leq T_2 \\ 1 & \text{otherwise} \end{cases},$$

which is just the complementary or opposite of the stabilizing selection (see Fig. 14.2c). The average fitness of the population is,

$$\overline{W} = 1 - \int\limits_{-\infty}^{\infty} w(y) f(y) dy$$

$$= 1 - \int\limits_{T_1}^{T_2} \frac{1}{\sqrt{2\pi}\sigma_P} \exp\left\{-\frac{1}{2}\left(\frac{y-\mu}{\sigma_P}\right)^2\right\} dy$$

$$= 1 - \Phi\left(\frac{T_2 - \mu}{\sigma_P}\right) + \Phi\left(\frac{T_1 - \mu}{\sigma_P}\right)$$

Therefore, the proportion selected is

$$p = 1 - \Phi(t_2) + \Phi(t_1)$$

The average value of the selected parents is obtained via

$$\mu_S = \frac{1}{\overline{W}} \int\limits_{-\infty}^{\infty} y w(y) f(y) dy$$

$$= \frac{1}{\overline{W}} \int\limits_{-\infty}^{T_1} \frac{y}{\sqrt{2\pi}\sigma_P} \exp\left\{ -\frac{1}{2}\left(\frac{y-\mu}{\sigma_P}\right)^2 \right\} dy + \frac{1}{\overline{W}} \int\limits_{T_2}^{\infty} \frac{y}{\sqrt{2\pi}\sigma_P} \exp\left\{ -\frac{1}{2}\left(\frac{y-\mu}{\sigma_P}\right)^2 \right\} dy$$

$$= \mu + \frac{\sigma_P}{\overline{W}}[\phi(t_2) - \phi(t_1)]$$

The selection differential is

$$S = \mu_S - \mu = \frac{\sigma_P}{p}[\phi(t_2) - \phi(t_1)]$$

The selection intensity is

$$i = \frac{S}{\sigma_P} = \frac{\sigma_P}{\sigma_P p}[\phi(t_2) - \phi(t_1)] = \frac{\phi(t_2) - \phi(t_1)}{p}$$

If the two truncation points are symmetrical (balanced disruptive selection), then no selection response will be expected. However, balanced disruptive selection will increase the genetic variance if the trait is heritable.

## 14.4 Natural Selection

Natural selection often assigns a fitness to an individual by a probability between 0 and 1 and the fitness is a smooth function of the phenotypic value. Figure 14.2d, e, f illustrates three types of natural selection, called directional natural selection, stabilizing natural selection, and disruptive natural selection.

### 14.4.1 Directional Selection

The fitness is a smooth function of the phenotypic value of a trait, which can be defined as

$$w(y) = \int\limits_{-\infty}^{y} \frac{1}{\sqrt{2\pi}\omega} \exp\left\{ -\frac{1}{2}\left(\frac{y-\theta}{\omega}\right)^2 \right\} dx,$$

where $\theta$ is a selection parameter representing the direction that natural selection aims to. Figure 14.2d illustrates this type of selection. If $\theta > \mu$, the population will move in the upper direction; otherwise, it will move in the lower direction. Parameter $\omega$ represents the smoothness of the fitness function. If $\omega \to 0$, then $w(y) \to 1$ for $y > \theta$ and $w(y) \to 0$ for $y \leq \theta$. So, the artificial directional selection is a special case of natural directional selection when $\omega = 0$. The average fitness of the population is the proportion selected (Walsh and Lynch 2018),

$$\overline{W} = \int\limits_{-\infty}^{\infty} w(y) f(y) dy = \cdots = 1 - \Phi\left(\frac{\theta - \mu}{\sqrt{\sigma_P^2 + \omega^2}}\right)$$

Comparing this with its counterpart in truncation selection, we can see that

$$t = \frac{\theta - \mu}{\sqrt{\sigma_P^2 + \omega^2}}$$

So, directional selection in nature is equivalent to directional truncation selection with a truncation point being a function of population parameters. Natural selection is weaker than truncation selection when $\omega > 0$. The mean of the selected individuals is

$$\mu_S = \frac{1}{\overline{W}} \int_{-\infty}^{\infty} y w(y) f(y) dy$$

$$= \mu + \frac{\sigma_P}{\sqrt{\omega^2 + \sigma_P^2}} \frac{1}{\overline{W}} \frac{\sigma_P}{\sqrt{2\pi}} \exp\left\{-\frac{1}{2}t^2\right\}$$

$$= \mu + \sqrt{\frac{\sigma_P^2}{\omega^2 + \sigma_P^2}} \frac{\sigma_P}{\overline{W}} \phi(t)$$

where

$$\phi(t) = \frac{1}{\sqrt{2\pi}} \exp\left\{-\frac{1}{2}t^2\right\}$$

Therefore, the selection differential is

$$S = \mu_S - \mu = \sqrt{\frac{\sigma_P^2}{\omega^2 + \sigma_P^2}} \frac{\sigma_P}{\overline{W}} \phi(t)$$

When the population is already at the optimal fitness level, $\theta = \mu$, the selection differential will be zero. The selection intensity is

$$i = \frac{S}{\sigma_P} = \frac{1}{\sigma_P} \sqrt{\frac{\sigma_P^2}{\omega^2 + \sigma_P^2}} \frac{\sigma_P}{\overline{W}} \phi(t) = \frac{1}{p} \phi(t) \sqrt{\frac{\sigma_P^2}{\omega^2 + \sigma_P^2}}$$

where $p = \overline{W}$ is the proportion selected.

## 14.4.2 Stabilizing Selection

Stabilizing selection may be the most common mode of selection in nature. The fitness function proposed by Haldane (1954) is

$$w(y) = \exp\left\{-\frac{1}{2}\left(\frac{y - \theta}{\omega}\right)^2\right\},$$

where individuals with the trait value equal to the optimal parameter $\theta$ will have the maximum fitness and deviation from this optimal value will reduce the fitness. The strength of the selection is represented by the smoothness parameter $\omega$. The average fitness of the population is,

$$\overline{W} = \int_{-\infty}^{\infty} w(y) f(y) dy = \cdots = \sqrt{\frac{\omega^2}{\omega^2 + \sigma_P^2}} \exp\left\{-\frac{1}{2}\left(\frac{\theta - \mu}{\sqrt{\sigma_P^2 + \omega^2}}\right)^2\right\}$$

The average value of the selected parents is obtained via

$$\mu_S = \frac{1}{\overline{W}} \int\limits_{-\infty}^{\infty} y w(y) f(y) dy = \cdots = \mu + \frac{\sigma_P^2}{\omega^2 + \sigma_P^2} (\theta - \mu)$$

The selection differential is

$$S = \mu_S - \mu = \frac{\sigma_P^2}{\omega^2 + \sigma_P^2} (\theta - \mu) \tag{14.7}$$

If $\theta = \mu$, the selection differential will be zero and thus there will be no response to such a stabilizing selection. However, the population variance will be reduced.

### 14.4.3  Disruptive Selection

Disruptive selection may happen in nature to avoid competition for limited resources. The fitness function proposed by Haldane (1954) is

$$w(y) = 1 - \exp\left\{ -\frac{1}{2} \left( \frac{y - \theta}{\omega} \right)^2 \right\},$$

where individuals with the trait value deviating away from parameter $\theta$ will have greater fitness than individuals having mediocre trait values. The strength of the selection is represented by the smoothness parameter $\omega$. The average fitness of the population is the proportion selected,

$$\overline{W} = \int\limits_{-\infty}^{\infty} w(y) f(y) dy = \cdots = 1 - \sqrt{\frac{\omega^2}{\omega^2 + \sigma_P^2}} \exp\left\{ -\frac{1}{2} \left( \frac{\theta - \mu}{\sqrt{\sigma_P^2 + \omega^2}} \right)^2 \right\}$$

The average value of the selected parents is obtained via

$$\mu_S = \frac{1}{\overline{W}} \int\limits_{-\infty}^{\infty} y w(y) f(y) dy = \cdots = \mu + \frac{\sigma_P^2}{\omega^2 + \sigma_P^2} (\mu - \theta)$$

The selection differential is

$$S = \mu_S - \mu = \frac{\sigma_P^2}{\omega^2 + \sigma_P^2} (\mu - \theta) \tag{14.8}$$

If $\theta = \mu$, there will be no selection response, but the population variance will be increased. Comparing eqs. (14.8) with (14.7), we see that stabilizing selection and disruptive selection only differs by the sign of the selection differential.

## 14.5   Change of Genetic Variance After Selection

### 14.5.1  Single Trait

Not only does selection change the population means of traits affected by selection, but also changes the population variances for the affected traits. We now use the directional truncation selection as an example to show the changes in phenotypic and genetic variances. The effect of selection on the genetic variance was first discovered by Bulmer (1971) and thus it is called the

Bulmer effect (Walsh and Lynch 2018). To clarify the notation, let us denote the phenotypic and genetic variances in the parental generation by $\sigma_P^2(0)$ and $\sigma_A^2(0)$, respectively. The corresponding variances in the progeny (next generation) are denoted by $\sigma_P^2(1)$ and $\sigma_A^2(1)$. Without selection, the variances will remain the same across generations (Hardy-Weinberg equilibrium). Let $P^*$ and $A^*$ be the changed phenotypic value and breeding value in the parental generation. Let us look at the change of the phenotypic variance in the current generation. Assume that both the male and female parents have the same selection intensity. Under the normal assumption of the phenotypic value, the variance of the phenotype ($P^*$) in the selected parents is

$$
\begin{aligned}
\sigma_{P^*}^2(0) &= \frac{1}{\overline{W}} \int_{-\infty}^{\infty} w(y)(y - \mu)^2 f(y) dy \\
&= \frac{1}{\overline{W}} \int_{T}^{\infty} (y - \mu)^2 f(y) dy \\
&= [1 - i(i - t)]\sigma_P^2(0) \\
&= (1 - \kappa)\sigma_P^2(0)
\end{aligned}
$$

where $t$ is the standardized truncation point, $i$ is the selection intensity and $\kappa = i(i - t)$ is the proportion of reduction in the phenotypic variance due to directional truncation selection. The $\kappa$ values in other types of selection will be different. In most types of selection, $\kappa$ is positive and thus selection will reduce the variance. In a balanced disruptive selection, however, $\kappa$ will be negative, which causes an increase in variance. Bulmer (1971) defined $d = -\kappa$ and claimed that $d$ is caused by linkage disequilibrium under the infinitesimal model. Detail of the truncated normal distribution can be found in the truncated normal distribution theory (Cohen 1991). The genetic variance in the current population after selection is (Robertson 1977)

$$
\sigma_{A^*}^2(0) = \left(1 - h^2\kappa\right)\sigma_A^2(0),
$$

where $h^2$ is the heritability of the trait. In fact, replacing $h^2$ by $r_{AP}^2$ is more appropriate because this parameter is really the squared correlation between the phenotype (subject to selection) and the breeding value. The phenotypic variance in the next generation is (Bulmer 1971)

$$
\sigma_P^2(1) = \left(1 - \frac{1}{2}h^4\kappa\right)\sigma_P^2(0)
$$

This change in phenotypic variance is entirely due to the change in genetic variance because selection does not change the environmental variance. The genetic variance of the next generation after selection is (Walsh and Lynch 2018)

$$
\sigma_A^2(1) = \left(1 - \frac{1}{2}h^2\kappa\right)\sigma_A^2(0) \tag{14.9}
$$

## 14.5.2  Multiple Traits

The next step is to look at the change in the covariance between two traits across generations. Let $X$, $Y$, and $Z$ be three correlated traits and the selection is on trait $Z$. The change of variance for $Z$ follows what we learned previously for a single trait. Here, we want to examine the changes of variances for traits $X$ and $Y$. More importantly, we are particularly interested in the change of covariance between $X$ and $Y$ due to the selection on $Z$. The phenotypic covariance after the selection in the current generation is

$$\sigma^*_{P(XY)}(0) = \left(1 - \frac{r_{P(XZ)}r_{P(YZ)}}{r_{P(XY)}}\kappa\right)\sigma_{P(XY)}(0), \tag{14.10}$$

where $r_{P(XY)}$ is the phenotypic correlation between $X$ and $Y$. Proof of this equation will be deferred to the end of this chapter. The corresponding change in the genetic covariance of the current generation is

$$\sigma^*_{A(XY)}(0) = \left(1 - h_Z^2\frac{r_{A(XZ)}r_{A(YZ)}}{r_{A(XY)}}\kappa\right)\sigma_{A(XY)}(0),$$

where $h_Z^2$ is the heritability of trait $Z$ and $r_{A(XY)}$ is the genetic correlation between the two traits ($X$ and $Y$). Now let us see how the changes are translated into the changes in the next generation. The phenotypic covariance in the next generation is

$$\sigma_{P(XY)}(1) = \left(1 - \frac{1}{2}h_X^2 h_Y^2\frac{r_{P(XZ)}r_{P(YZ)}}{r_{P(XY)}}\kappa\right)\sigma_{P(XY)}(0) \tag{14.11}$$

The corresponding change in genetic covariance in the next generation is

$$\sigma_{A(XY)}(1) = \left(1 - \frac{1}{2}h_Z^2\frac{r_{A(XZ)}r_{A(YZ)}}{r_{A(XY)}}\kappa\right)\sigma_{A(XY)}(0) \tag{14.12}$$

The change of variance is a special case of the change of covariance. For example, the change in variance for trait $X$ is simply a special case of the covariance when $Y$ is replaced by $X$. Let us see what the variance version of Eq. (14.12) is. Replacing $Y$ by $X$ leads to

$$\sigma_{A(XX)}(1) = \left(1 - \frac{1}{2}h_Z^2\frac{r_{A(XZ)}r_{A(XZ)}}{r_{A(XX)}}\kappa\right)\sigma_{A(XX)}(0),$$

which is

$$\sigma^2_{A(X)}(1) = \left(1 - \frac{1}{2}h_Z^2 r^2_{A(XZ)}\kappa\right)\sigma^2_{A(X)}(0)$$

Similarly, the variance version of the covariance for trait $Y$ is obtained by replacing $X$ by $Y$ and Eq. (14.12) becomes

$$\sigma_{A(YY)}(1) = \left(1 - \frac{1}{2}h_Z^2\frac{r_{A(YZ)}r_{A(YZ)}}{r_{A(YY)}}\kappa\right)\sigma_{A(YY)}(0),$$

which is

$$\sigma^2_{A(Y)}(1) = \left(1 - \frac{1}{2}h_Z^2 r^2_{A(YZ)}\kappa\right)\sigma^2_{A(Y)}(0)$$

The general formula also allows us to see the change in variance when one of the two traits is subject to selection. For example, if trait $X$ is subject to selection, what is the change in covariance between $X$ and $Y$? This time, we need to replace $Z$ by $X$. This version of Eq. (14.12) is

$$\sigma_{A(XY)}(1) = \left(1 - \frac{1}{2}h_X^2\frac{r_{A(XX)}r_{A(XY)}}{r_{A(XY)}}\kappa\right)\sigma_{A(XY)}(0),$$

which is

$$\sigma_{A(XY)}(1) = \left(1 - \frac{1}{2}h_X^2\kappa\right)\sigma_{A(XY)}(0).$$

Finally, what is the formula for the change of variance when the trait itself is selected? You simply let $X = Y = Z$ and modify Eq. (14.12) into

$$\sigma_{A(ZZ)}(1) = \left(1 - \frac{1}{2}h_Z^2\frac{r_{A(ZZ)}r_{A(ZZ)}}{r_{A(ZZ)}}\kappa\right)\sigma_{A(ZZ)}(0),,$$

which is

$$\sigma_{A(Z)}^2(1) = \left(1 - \frac{1}{2}h_Z^2\kappa\right)\sigma_{A(Z)}^2(0)$$

This is exactly the same as the single trait version of the change in variance shown in Eq. (14.9).

### 14.5.3  The $\kappa$ Values in Other Types of Selection

Walsh and Lynch (2018) provided the $\kappa$ values for directional truncation selection, symmetrically truncated stabilizing selection and symmetrically truncated disruptive selection. We now provide general formulas for the three types of truncation selection.

#### 14.5.3.1  Directional Truncated Selection

The $\kappa$ value for this type of selection is usually expressed as $\kappa = i(i - t)$. However, the original formula is

$$\kappa = \frac{1}{1 - \Phi(t)}\phi(t)\left(\frac{1}{1 - \Phi(t)}\phi(t) - t\right),$$

where $p = 1 - \Phi(t)$ is the proportion selected and $i = [1 - \Phi(t)]^{-1}\phi(t)$ is the selection intensity.

#### 14.5.3.2  Stabilizing Truncation Selection

Let $t_1 = (T_1 - \mu)/\sigma_P$ and $t_2 = (T_2 - \mu)/\sigma_P$ be the two truncation points in the standardized scale so that $\sigma_P t_1 + \mu \leq y \leq \sigma_P t_2 + \mu$. The relative reduction of variance in the doubly truncated sample is (Johnson et al. 1994)

$$\kappa = \frac{t_2\phi(t_2) - t_1\phi(t_1)}{\Phi(t_2) - \Phi(t_1)} + \left(\frac{\phi(t_2) - \phi(t_1)}{\Phi(t_2) - \Phi(t_1)}\right)^2$$

When the selection is symmetrical (balanced), i.e., $t_1 = -t_2$, so that the proportion selected is

$$p = \Phi(t_2) - \Phi(t_1) = \Phi(t_2) - \Phi(-t_2) = \Phi(t_2) - [1 - \Phi(t_2)] = 2\Phi(t_2) - 1.$$

The truncation point can then be expressed as a quantile of the standardized normal distribution

$$t_2 = \Phi^{-1}[(1 + p)/2] = t_{(1+p)/2}.$$

Therefore,

$$\kappa = \frac{t_2\phi(t_2) + t_2\phi(t_2)}{p} + \left(\frac{\phi(t_2) - \phi(t_2)}{p}\right)^2 = \frac{2t_2\phi(t_2)}{p} = \frac{2t_{(1+p)/2}\phi(t_{(1+p)/2})}{p},$$

which is the same as the one given by Walsh and Lynch (2018).

### 14.5.3.3 Disruptive Truncation Selection

The selection is made of two parts, one being the upper directional selection and one being the lower directional selection. The expectation and variance for the lower directional selection part are

$$\mu_1 = \mu - \frac{\sigma_P}{p_1}\phi(t_1)$$

and

$$\sigma_1^2 = \sigma_P^2\left\{1 - \frac{\phi(t_1)}{p_1}\left(\frac{\phi(t_1)}{p_1} + t_1\right)\right\},$$

respectively, where $p_1 = \Phi(t_1)$ is the proportion selected for the lower part. For the upper directional selection, the expectation and variance are

$$\mu_2 = \mu + \frac{\sigma_P}{p_2}\phi(t_2)$$

and

$$\sigma_2^2 = \sigma_P^2\left\{1 - \frac{\phi(t_2)}{p_2}\left(\frac{\phi(t_2)}{p_2} - t_2\right)\right\},$$

respectively, where $p_2 = 1 - \Phi(t_2)$ is the proportion selected in the upper selection part. The variance of the disruptive selection is obtained by merging the variances of the two parts using the total variance theory

$$\text{var}(y) = \text{E}\left[\text{var}(y|d)\right] + \text{var}[\text{E}(y|d)],$$

where $d = 1$ indicates the lower part and $d = 2$ indicates the upper part. Let us look at the expectation of the conditional variance

$$\Theta_1 = \text{E}\left[\text{var}(y|d)\right] = \frac{1}{p_1 + p_2}\left(p_1\sigma_1^2 + p_2\sigma_2^2\right)$$

The variance of the conditional mean is

$$\Theta_2 = \text{var}[\text{E}(y|d)] = \frac{1}{p_1 + p_2}\left(p_1\mu_1^2 + p_2\mu_2^2\right) - \left[\frac{1}{p_1 + p_2}\left(p_1\mu_1 + p_2\mu_2\right)\right]^2$$

The $\kappa$ value is defined as

$$\kappa = 1 - \frac{\Theta_1 + \Theta_2}{\sigma_P^2} \tag{14.13}$$

The final expression is explicit but too complicated to write. However, if the selection is symmetrical (balanced), a simple expression is available (Walsh and Lynch 2018). In this special case, $p_1 = p_2 = p/2$, $\sigma_1^2 = \sigma_2^2$ and thus

$$\Theta_1 = \frac{1}{p_1 + p_2}\left(p_1\sigma_1^2 + p_2\sigma_2^2\right) = \sigma_P^2\left\{1 - \frac{\phi(t_2)}{p_2}\left(\frac{\phi(t_2)}{p_2} - t_2\right)\right\}$$

and

$$\Theta_2 = \frac{1}{p_1 + p_2}\left(p_1\mu_1^2 + p_2\mu_2^2\right) - \left[\frac{1}{p_1 + p_2}\left(p_1\mu_1 + p_2\mu_2\right)\right]^2$$

$$= \frac{1}{2}\left(\mu_1^2 + \mu_2^2\right) - \mu^2$$

$$= \frac{1}{2}\left\{2\mu^2 + 2\left[\frac{\sigma_P}{p_2}\phi(t_2)\right]^2\right\} - \mu^2$$

$$= \sigma_P^2\left[\frac{\phi(t_2)}{p_2}\right]^2$$

Adding the two terms together, we have

$$\Theta_1 + \Theta_2 = \sigma^2\left\{1 - \frac{\phi(t_2)}{p_2}\left(\frac{\phi(t_2)}{p_2} - t_2\right)\right\} + \sigma^2\left[\frac{\phi(t_2)}{p_2}\right]^2 = \sigma^2\left[1 + \frac{\phi(t_2)}{p_2}t_2\right] \tag{14.14}$$

Substituting (14.14) into (14.13) yields

$$\kappa = 1 - \frac{\Theta_1 + \Theta_2}{\sigma_P^2} = 1 - \frac{1}{\sigma_P^2}\left\{\sigma_P^2\left[1 + \frac{\phi(t_2)}{p_2}t_2\right]\right\} = \frac{\phi(t_2)}{p_2}t_2 \tag{14.15}$$

Note that given the proportion selected $p$, the truncation point is

$$t_2 = \Phi^{-1}(1 - p/2) = t_{(1-p/2)}$$

Substitution of $t_2$ in Eq. (14.15) by $t_{(1 - p/2)}$ results in

$$\kappa = \frac{\phi\left(t_{(1-p/2)}\right)}{p/2}t_{(1-p/2)} = 2t_{(1-p/2)}\frac{\phi\left(t_{(1-p/2)}\right)}{p}$$

which is identical to the formula given by Walsh and Lynch (2018). The $\kappa$ values for the three natural selection counterparts are not available. Theoretically, they can be derived, but it is beyond the scope of this course.

### 14.5.4  Derivation of Change in Covariance

#### 14.5.4.1  Change in Covariance Within-Generation

Equation (14.10) is a general formula for within-generation change in covariance, from which results for any special cases can be easily derived. We now remove the $P$ (phenotype) from $r_{P(XY)}$ to indicate the generality of Eq. (14.10). Let $Z$ be a variable subject to selection in any selection mode. Define $X$ and $Y$ as two variables correlated to Z, and they are also correlated each other. The two linear models are

$$\begin{aligned} X &= a_X + b_{XZ}Z + e_X \\ Y &= a_Y + b_{YZ}Z + e_Y \end{aligned} \tag{14.16}$$

The covariance between $X$ and $Y$, $\operatorname{cov}(X, Y) = \sigma_{XY}$, is.

$$\sigma_{XY} = b_{XZ}\operatorname{var}(Z)b_{YZ} + \operatorname{cov}(e_X, e_Y) = b_{XZ}b_{YZ}\sigma_Z^2 + \operatorname{cov}(e_X, e_Y) \tag{14.17}$$

Let

$$\mathrm{cov}(e_X, e_Y) = \sigma_{XY} - b_{XZ}b_{YZ}\sigma_Z^2 = \left(1 - b_{XZ}b_{YZ}\sigma_Z^2/\sigma_{XY}\right)\sigma_{XY},$$

then Eq. (14.17) is rewritten as

$$\sigma_{XY} = b_{XZ}b_{YZ}\sigma_Z^2 + \left(1 - b_{XZ}b_{YZ}\sigma_Z^2/\sigma_{XY}\right)\sigma_{XY}.$$

Selection for $Z$ leads to

$$\sigma_{Z^*}^2 = (1 - \kappa)\sigma_Z^2.$$

Therefore, the covariance between $X$ and $Y$ becomes

$$
\begin{aligned}
\sigma_{XY}^* &= b_{XZ}b_{YZ}\sigma_{Z^*}^2 + \left(1 - b_{XZ}b_{YZ}\sigma_Z^2/\sigma_{XY}\right)\sigma_{XY} \\
&= b_{XZ}b_{YZ}(1 - \kappa)\sigma_Z^2 + \left(1 - b_{XZ}b_{YZ}\sigma_Z^2/\sigma_{XY}\right)\sigma_{XY} \\
&= \left[1 + (1 - \kappa)b_{XZ}b_{YZ}\sigma_Z^2/\sigma_{XY} - b_{XZ}b_{YZ}\sigma_Z^2/\sigma_{XY}\right]\sigma_{XY} \\
&= \left\{1 + [(1 - \kappa) - 1]b_{XZ}b_{YZ}\sigma_Z^2/\sigma_{XY}\right\}\sigma_{XY} \\
&= \left(1 - \kappa b_{XZ}b_{YZ}\sigma_Z^2/\sigma_{XY}\right)\sigma_{XY}
\end{aligned}
$$

Let us simplify the above equation by replacing the regression by correlation

$$\frac{b_{XZ}b_{YZ}\sigma_Z^2}{\sigma_{XY}} = \frac{\sigma_{XZ}\sigma_{YZ}\sigma_X\sigma_Y\sigma_Z^2}{\sigma_Z^2\sigma_Z^2\sigma_X\sigma_Y\sigma_{XY}} = \frac{\sigma_{XZ}\sigma_{YZ}\sigma_X\sigma_Y}{\sigma_X\sigma_Z\sigma_Y\sigma_Z\sigma_{XY}} = \frac{r_{XZ}r_{YZ}}{r_{XY}}$$

Therefore, we have

$$\sigma_{XY}^* = \left(1 - \frac{r_{XZ}r_{YZ}}{r_{XY}}\kappa\right)\sigma_{XY} \tag{14.18}$$

This formula was first presented in genetics literature by Alan Robertson (Robertson 1977), perhaps based on his imagination. He did not cite any source of the equation and verbally described how he ended up with this formula. He said that the covariance was partitioned into one part with $\sigma_Z^2$ subject to change and one part in residual. The change in the covariance was only due to the change in the part subject to selection, not the residual. The genetic covariance within-generation after selection was

$$\sigma_{A(XY)}^* = \left(1 - \frac{r_{A_XP_Z}r_{A_YP_Z}}{r_{A(XY)}}\kappa\right)\sigma_{A(XY)} = \left(1 - h_Z^2\frac{r_{A(XZ)}r_{A(YZ)}}{r_{A(XY)}}\kappa\right)\sigma_{A(XY)}$$

The change in covariance within-generation was purely statistical and has nothing to do with genetics.

When I was a PhD student at Purdue University (1987–1989), Professor Thomas Kuczek suggested me to use the linear combinations resented in Eq. (14.16) to derive the general formula (14.18). Under Professor Kuczek's direction, I was able to derive the general formula. I was excited to tell Professor Truman Martin but only learned from him that Alan Robertson already derived the general formula a decade ago.

### 14.5.4.2 Change in Covariance Between-Generation

A simple form of the between-generation change in covariance requires some assumption of genetics: (1) the selected parents are not inbred, (2) they are not genetically related, and (3) both sexes are selected equally, i.e., male and female parents have the same selection intensity. When these assumptions are violated, average inbreeding coefficients of the male parents and average inbreeding coefficient of the female parents must be given. The average coancestry between the male and female parents is also needed. In this chapter, we only introduce the simple form of the covariance change.

The derivation of this formula depends on $r_{P_X A_X} = h_X$ and $r_{P_X A_Y} = h_X r_{A(XY)}$. Let $P_X$, $P_X^m$, and $P_X^f$ be the phenotypic values of trait $X$ for the progeny, the male parent, and the female parent, respectively. The phenotypic values corresponding to trait $Y$ are $P_Y$, $P_Y^m$, and $P_Y^f$, respectively. The two traits from the offspring are modeled by

$$P_X = \mu_X + \frac{1}{2}h_X^2 P_X^m + \frac{1}{2}h_X^2 P_X^f + e_X$$

$$P_Y = \mu_Y + \frac{1}{2}h_Y^2 P_Y^m + \frac{1}{2}h_Y^2 P_Y^f + e_Y,$$

where $e_X$ and $e_Y$ are residuals (including Mendelian segregation effects and environmental effects). Without selection, the covariance between $P_X$ and $P_Y$ is.

$$\sigma_{P(XY)}(1) = \frac{1}{4}h_X^2 h_Y^2 \mathrm{cov}\left(P_X^m, P_Y^m\right) + \frac{1}{4}h_X^2 h_Y^2 \mathrm{cov}\left(P_X^f, P_Y^f\right) + \mathrm{cov}(e_X, e_Y)$$

$$= \frac{1}{2}h_X^2 h_Y^2 \sigma_{P(XY)}(0) + \mathrm{cov}(e_X, e_Y) \tag{14.19}$$

where

$$\mathrm{cov}(e_X, e_Y) = \sigma_{P(XY)}(0) - \frac{1}{2}h_X^2 h_Y^2 \sigma_{P(XY)}(0) \tag{14.20}$$

Substituting (14.20) into (14.19) leads to

$$\sigma_{P(XY)}(1) = \frac{1}{2}h_X^2 h_Y^2 \sigma_{P(XY)}(0) + \left(1 - \frac{1}{2}h_X^2 h_Y^2\right)\sigma_{P(XY)}(0)$$

Selection will only change the covariance in the first term and thus

$$\sigma_{P(XY)}(1) = \frac{1}{2}h_X^2 h_Y^2 \sigma_{P(XY)}^*(0) + \left(1 - \frac{1}{2}h_X^2 h_Y^2\right)\sigma_{P(XY)}(0)$$

$$= \frac{1}{2}h_X^2 h_Y^2 \left(1 - \frac{r_{P(XZ)} r_{P(YZ)}}{r_{P(XY)}}\kappa\right)\sigma_{XY}(0) + \left(1 - \frac{1}{2}h_X^2 h_Y^2\right)\sigma_{P(XY)}(0)$$

$$= \left(1 - \frac{1}{2}h_X^2 h_Y^2 \frac{r_{P(XZ)} r_{P(YZ)}}{r_{P(XY)}}\kappa\right)\sigma_{P(XY)}(0)$$

This equation is identical to Eq. (14.11). The genetic models for the next generations are

$$A_X = \frac{1}{2}A_X^m + \frac{1}{2}A_X^f + \omega_X$$

$$A_Y = \frac{1}{2}A_Y^m + \frac{1}{2}A_Y^f + \omega_Y$$

where $\mathrm{cov}(\omega_X, \omega_Y) = \frac{1}{2}\sigma_{A(XY)}$ is the covariance due to Mendelian segregation. The covariance between $A_X$ and $A_Y$ is.

$$\sigma_{A(XY)}(1) = \frac{1}{4}\mathrm{cov}\left(A_X^m, A_Y^m\right) + \frac{1}{4}\mathrm{cov}\left(A_X^f, A_Y^f\right) + \mathrm{cov}(\omega_X, \omega_Y)$$

$$= \frac{1}{2}\sigma_{A(XY)}(0) + \frac{1}{2}\sigma_{A(XY)}(0)$$

After selection,

$$\sigma_{A(XY)}(1) = \frac{1}{2}\sigma_{A(XY)}^*(0) + \frac{1}{2}\sigma_{A(XY)}(0)$$

$$= \frac{1}{2}\left(1 - h_Z^2 \frac{r_{A(XZ)}r_{A(YZ)}}{r_{A(XY)}}\kappa\right)\sigma_{A(XY)}(0) + \frac{1}{2}\sigma_{A(XY)}(0)$$

$$= \left(1 - \frac{1}{2}h_Z^2 \frac{r_{A(XZ)}r_{A(YZ)}}{r_{A(XY)}}\kappa\right)\sigma_{A(XY)}(0)$$

which is identical to Eq. (14.12).

## References

Bulmer MG. The effect of selection on genetic variability. Am Nat. 1971;105:201–11.

Cohen AC. Truncated and censored samples. New York: Marcel Dekker; 1991.

Falconer DS, Mackay TFC. Introduction to quantitative genetics. Harlow: Addison Wesley Longman; 1996.

Haldane JB. The measurement of natural selection. In: The 9th international congress of genetics, 1954; Vol 1, pp. 480–487.

Johnson NL, Kotz S, Balakrishnan N. Continuous univariate distributions. New York: Wiley; 1994.

Lande R. Natural selection and random genetic drift in phenotypic evolution. Evolution. 1976;30:314–34.

Lande R, Thompson R. Efficiency of marker-assisted selection in the improvement of quantitative traits. Genetics. 1990;124:743–56.

Robertson A. The effect of selection on the estimation of genetic parameters. Z Tierzuecht Zuechtungsbiol. 1977;94:131–5.

Walsh B, Lynch M. Evolution and selection of quantitative traits. Oxford: Oxford University Press; 2018.

# Methods of Artificial Selection

# 15

We have learned the concept and theory of artificial selection. This chapter will deal with particular methods of artificial selection. Different methods of selection often have different efficiencies. Therefore, methods of selection should be chosen based on the characteristics of traits and characteristics of populations.

## 15.1 Objective of Selection

Before we do any artificial selection, we must know what we want and what we are trying to accomplish. Therefore, we must understand the objective of selection. The objective of selection is a variable to be improved or to be predicted. In plant and animal breeding, the objective is the "breeding value" of a candidate to be selected as a parent. So, breeding value is our objective of selection. We are still targeting a single trait, and therefore the objective of selection is the breeding value of a single trait. When a selection program involves multiple traits, the objective of selection will be redefined as a collection of multiple traits.

## 15.2 Criteria of Selection

Normally, the objective of selection, e.g., the breeding value, is not observable and thus cannot be directly selected. We must select some variables that (a) can be observed and (b) are correlated with the objective. Such a variable is called the selection criterion. For example, the phenotypic value is one of the criteria we can use to select animals and plants. Phenotypic values of other traits can also be used as the selection criteria if these traits are highly correlated with the target trait. There are several methods of selection with the breeding value of a single trait as the sole target. Different methods are defined based on different selection criteria. Figure 15.1 shows all sources of information that can be used as criteria of selection. In all methods of selection, the selection objective is the breeding value of the candidate, denoted by $A$ in the figure. The $P$ directly connected to $A$ is the phenotypic value of this candidate. All other $P$'s are phenotypic values of the candidate's relatives.

## 15.3 Methods of Selection

### 15.3.1 Individual Selection

The selection criterion is solely the phenotypic value of the candidate, i.e., $P$ in Fig. 15.1. Individual selection is also called mass selection. The selection method used by our ancestors is the individual selection method. You select what you can directly see. All domestic animals and plants were selected from their wild relatives by our ancestors through mass selection. Mass selection can be very effective for highly heritable traits. For example, plant height, grain size, and human height are highly heritable and selection on these traits is very effective in terms of improvement of these traits.

© Springer Nature Switzerland AG 2022
S. Xu, *Quantitative Genetics*, https://doi.org/10.1007/978-3-030-83940-6_15

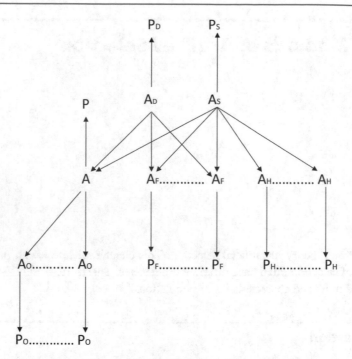

**Fig. 15.1** Different sources of information available for selection. The objective of the selection is the genetic value ($A$) of the candidate. All variables with $P$ are observable and thus can be used as selection criteria

### 15.3.2 Family Selection

The selection criterion is the mean of the family to which the candidate belongs. Note that all candidates in the same family have the same criterion. Therefore, the whole family will be selected or rejected as a unit. Families can be full-sib families ($P_F$) or half-sib families ($P_H$) (Fig. 15.1). Family selection can be effective for traits with low heritability and family members do not share much common environment. When the family size is large the family mean represents the genetic value of the family because errors between individuals within the family tend to be cancelled out each other by taking the average. Grain yield in agricultural crops and stress resistance traits in animals often have low heritability. Fertility-related traits in animals also have low heritability and thus family selection should be considered.

### 15.3.3 Sib Selection

The selection criterion is the mean of full-sibs ($P_F$) or half-sibs ($P_H$) of the candidate. Sib selection differs from family selection in that the candidate does not contribute to the estimate of the mean of siblings. When the family size is large, this method is not much different from the method of family selection. It can be efficient when the heritability of the trait is low and there is little common environmental variance. In addition, for sex-limited traits (traits only expressed in one sex), using sibling data may be the only way to select individuals who do not have measurements of these sex-limited traits. Typical examples include milk yield in dairy cattle where bulls do not have milk yield trait but yields of their sisters are certainly useful as criteria of selection.

### 15.3.4 Progeny Testing

The selection criterion is the mean of progenies ($P_O$) of the candidate (see Fig. 15.1). This method is also similar to family selection. It can be used when the heritability of the trait is low. Progeny testing is the most effective method in terms of genetic gain because the criterion of selection is close enough to the objective of selection. The mean value of an infinite number of progeny of a candidate (expressed as deviation from the population mean) is half the breeding value of the candidate. Progeny testing is very important in dairy cattle industry for selecting bulls. The problem with this method is the long generation interval compared with other methods. We have to wait until the candidate has reproduced and has measurement of traits from progenies of the candidate. Although the response to selection per generation can be high but the yearly progress may be low.

### 15.3.5 Within-Family Selection

The selection criterion for within-family selection is the deviation of the candidate phenotype from the mean of the family to which the candidate belongs. Similar to family selection, this method can be used when the heritability of the trait is low. However, it is only used when family members share common environmental effects, i.e., the common environmental variance is high. Common environmental variance includes variation of maternal effect, cultural effect, and so on where family members share very similar environments. In the swine industry, weaning weights of piglets is a trait often affected by the maternal effect due to the fact that their mother provides the same milk and the same maternal care to every piglet. The ELC college admission policy in California actually adopts the within-family selection strategy. The University of California's "Eligible in the Local Context" (ELC) policy of college admission states that if you are a state resident who has met the minimum requirements of UC college admissions and are not admitted to any campus to which you apply, you will be offered a spot at another campus if space is available, provided that you rank in the top 9% of your graduating class at any high school in California. If a high school is located in a poor school district and the school is not well equipped with good facility and high quality teachers, the average student scores may be low compared with other schools, not because the students are bad but because of the bad school from which they receive the education. The scores of the top students from this unprivileged school may be lower than the average score of students in other privileged schools, but they are admitted to colleges with higher priority than the students with higher scores but educated in privileged high schools.

### 15.3.6 Pedigree Selection

The selection criterion is determined by the phenotypic values of parents, grandparents, and so on. This method of selection can be used in the early stages of development, i.e., an individual can be selected before it is fully grown, or even before it is born. Pedigree selection can shorten generation interval and thus may improve annual genetic gain.

### 15.3.7 Combined Selection

The selection criterion is determined based on all available information, including candidate's own phenotype, family information, pedigrees, and so on. All information is used but information from each source is properly weighted in significance. The method is also called index selection, which will be discussed later. Index selection is the most efficient method of selection. Figure 15.1 illustrates all pieces of information that can be used for selection.

The above seven methods of selection are used in classical breeding programs. With genomic data, a new method emerged as an effective selection method, called genomic selection (Meuwissen et al. 2001), which will be introduced towards the end of this course.

## 15.4 Evaluation of a Selection Method

In individual selection, the objective of selection is $A$ (the breeding value of a candidate) and the selection criterion is $P$ (the phenotypic value of the candidate). As we have learned before, the selection response ($R$) is

$$R = h^2 S = ih\sigma_A$$

where $h = \sigma_A/\sigma_P$ is the square root of heritability, which can be shown to be the correlation coefficient between $A$ and $P$, both from the same individual,

$$r_{AP} = \frac{\text{cov}(A, P)}{\sigma_A \sigma_P} = \frac{\text{cov}(A, A + R)}{\sigma_A \sigma_P} = \frac{\text{cov}(A, A)}{\sigma_A \sigma_P} = \frac{\sigma_A^2}{\sigma_A \sigma_P} = \frac{\sigma_A}{\sigma_P} = h$$

Therefore, the selection response can be rewritten as

$$R = ir_{AP}\sigma_A$$

In general, the selection response is expressed as

$$R_X = i_X r_{AX} \sigma_A \tag{15.1}$$

where the subscript $X$ indicates the method of selection, $i_X$ is the selection intensity for a particular selection method, $A$ is the breeding value of the candidate (the objective of the selection), and $\sigma_A$ is the standard deviation of the breeding value (standard deviation of the selection objective). The selection criterion $X$ can be anything that is observable and also correlated with $A$. Eq. (15.1) is the generalized breeders' equation and it applies to all methods of selection. We now present the derivation of the generalized selection equation. Recall that the regression coefficient of $b_{AX}$ means the amount of change in $A$ per unit change in $X$. Therefore, we can write

$$\underbrace{\overline{A}_S - \overline{A}}_{R} = b_{AX} \underbrace{\left(\overline{X}_S - \overline{X}\right)}_{S}$$

The definition of the regression coefficient is

$$b_{AX} = \frac{\text{cov}(A, X)}{\text{var}(X)} = \frac{\text{cov}(A, X)}{\sigma_A \sigma_X} \frac{\sigma_A}{\sigma_X} = r_{AX} \frac{\sigma_A}{\sigma_X}$$

Therefore,

$$\underbrace{\overline{A}_S - \overline{A}}_{R_X} = b_{AX} \underbrace{\left(\overline{X}_S - \overline{X}\right)}_{S_X} = r_{AX} \frac{\left(\overline{X}_S - \overline{X}\right)}{\sigma_X} \sigma_A = r_{AX} i_X \sigma_A$$

## 15.5   Examples

### 15.5.1 Family Selection

The objective of family selection is still the breeding value of the candidate. The selection criterion is the mean of $n$ members of the family where the candidate belongs to. Note that the $n$ members of the family include the candidate itself. The response to selection given in the standard textbook (Falconer and Mackay 1996) is

$$R = i\sigma_P h^2 \frac{1 + (n-1)a_{SIB}}{\sqrt{n\left[1 + (n-1)a_{SIB}h^2\right]}} \tag{15.2}$$

where $a_{FS} = 1/2$ for full-sib family and $a_{HS} = 1/4$ for half-sib family, $n$ is the number of sibs in the family. Note that the subscript SIB means either FS or HS. Eq. (15.2) consists of a basic part of mass selection ($i\sigma_P h^2$) and an additional factor. All methods of selection can be compared with the mass selection.

Recall that $R = i_X r_{AX} \sigma_A$, where $X$ is the mean phenotypic value of $n$ family members, i.e.,

$$X = \frac{1}{n} \sum_{j=1}^{n} P_j$$

where $P_j$ is the phenotypic value of the $j$th sibling in the family. The correlation coefficient between the breeding value of the candidate and the family mean is

$$r_{AX} = \frac{\mathrm{cov}(A, X)}{\sigma_A \sigma_X}$$

where

$$
\begin{aligned}
\sigma_X^2 &= \frac{1}{n^2} \mathrm{var}\left( \sum_{j=1}^{n} P_j \right) \\
&= \frac{1}{n^2} \left[ \sum_{j=1}^{n} \mathrm{var}(P_j) + 2 \sum_{j'>j}^{n} \mathrm{cov}(P_j, P_{j'}) \right] \\
&= \frac{1}{n^2} \left[ n\sigma_P^2 + 2 \frac{n(n-1)}{2} \mathrm{cov}(\mathrm{SIB}) \right] \\
&= \frac{1}{n^2} \left[ n\sigma_P^2 + n(n-1) a_{\mathrm{SIB}} h^2 \sigma_P^2 \right] \\
&= \frac{1 + (n-1) a_{\mathrm{SIB}} h^2}{n} \sigma_P^2
\end{aligned}
$$

Note that the derivation involves

$$\mathrm{cov}(SIB) = a_{SIB} \sigma_A^2 = a_{SIB} h^2 \sigma_P^2$$

Let us deal with the covariance,

$$
\begin{aligned}
\mathrm{cov}(A, X) &= \frac{1}{n} \mathrm{cov}\left( A, \sum_{j=1}^{n} P_j \right) \\
&= \frac{1}{n} \sum_{j=1}^{n} \mathrm{cov}(A, P_j) \\
&= \frac{1}{n} \left( \sum_{j=1}^{n-1} a_{SIB} \sigma_A^2 + \sigma_A^2 \right)
\end{aligned}
$$

Therefore

$$R = i_X \sigma_A r_{AX} = i\sigma_A \frac{\frac{1}{n}[1 + (n-1)a_{SIB}]\sigma_A^2}{\sigma_A \sqrt{\frac{1}{n}[1 + (n-1)a_{SIB}h^2]\sigma_P^2}} = i\sigma_P h^2 \frac{1 + (n-1)a_{SIB}}{\sqrt{n[1 + (n-1)a_{SIB}h^2]}}$$

## 15.5.2  Within-Family Selection

The objective of within-family selection is again the breeding value of the candidate. The selection criterion is the deviation of the phenotypic value of the candidate from the mean of $n$ members of the family to which the candidate belongs. Note that the $n$ members include the candidate itself. The response to selection is given in the classical textbook of quantitative genetics (Falconer and Mackay 1996)

$$R = i\sigma_P h^2 (1 - a_{SIB}) \sqrt{\frac{n-1}{n(1 - \rho_{SIB})}}$$

where $\rho_{SIB} = a_{SIB} h^2 + \sigma_C^2 / \sigma_P^2$ and $\sigma_C^2$ is the common environmental variance. Remember that within-family selection is effective when $\sigma_C^2 > 0$. Recall that $R = i_X r_{AX} \sigma_A$, where $X$ is the deviation of the phenotypic value of the candidate (individual $j$) from the mean phenotypic value of $n$ siblings, i.e.,

$$X = P_j - \frac{1}{n} \sum_{j'=1}^{n} P_{j'}$$

where $P_j$ is the phenotypic value of the candidate and $P_{j'}$ is the phenotypic value of member $j'$ of the family for $j' \neq j$. The correlation coefficient between the breeding value of the candidate and the within-family deviation is

$$r_{AX} = \frac{\text{cov}(A, X)}{\sigma_A \sigma_X}$$

The variance of $X$ is

$$\sigma_X^2 = \text{var}\left( P_j - \frac{1}{n} \sum_{j'=1}^{n} P_{j'} \right) = \text{var}(P_j) + \text{var}\left( \frac{1}{n} \sum_{j'=1}^{n} P_{j'} \right) - 2\text{cov}\left( P_j, \frac{1}{n} \sum_{j'=1}^{n} P_{j'} \right)$$

where

$$\text{var}(P_j) = \sigma_P^2, \quad \text{var}\left( \frac{1}{n} \sum_{j'=1}^{n} P_{j'} \right) = \frac{1 + (n-1)\rho_{SIB}}{n} \sigma_P^2$$

and

$$\text{cov}\left( P_j, \frac{1}{n} \sum_{j'=1}^{n} P_{j'} \right) = \frac{1}{n} \left[ \sum_{j' \neq j}^{n} \text{cov}(P_j, P_{j'}) + \text{cov}(P_j, P_j) \right]$$

$$= \frac{1}{n} \left[ \sum_{j' \neq j}^{n} \rho_{SIB} \sigma_P^2 + \text{var}(P_j) \right]$$

$$= \frac{1}{n} \left[ (n-1)\rho_{SIB} \sigma_P^2 + \sigma_P^2 \right]$$

$$= \frac{1}{n} \left[ (n-1)\rho_{SIB} + 1 \right] \sigma_P^2$$

Therefore, the variance of $X$ is

$$\sigma_X^2 = \sigma_P^2 + \frac{1 + (n-1)\rho_{SIB}}{n}\sigma_P^2 - \frac{2}{n}[(n-1)\rho_{SIB} + 1]\sigma_P^2$$

$$= \frac{n + 1 + (n-1)\rho_{SIB} - 2(n-1)\rho_{SIB} - 2}{n}\sigma_P^2$$

$$= \frac{(n-1) - (n-1)\rho_{SIB}}{n}\sigma_P^2$$

$$= \frac{(n-1)(1 - \rho_{SIB})}{n}\sigma_P^2$$

The covariance between $A$ and $X$ is

$$\text{cov}(A, X) = \text{cov}\left(A, P_j - \frac{1}{n}\sum_{j'=1}^{n} P_{j'}\right)$$

$$= \text{cov}(A, P_j) - \text{cov}\left(A, \frac{1}{n}\sum_{j=1}^{n} P_j\right)$$

$$= \text{cov}(A, A + R) - \text{cov}\left(A, \frac{1}{n}\sum_{j=1}^{n} P_j\right)$$

$$= \text{var}(A) - \text{cov}\left(A, \frac{1}{n}\sum_{j=1}^{n} P_j\right)$$

$$= \sigma_A^2 - \frac{1}{n}[(n-1)a_{SIB} + 1]\sigma_A^2$$

$$= \frac{n - (n-1)a_{SIB} - 1}{n}\sigma_A^2$$

$$= \frac{(n-1)(1 - a_{SIB})}{n}\sigma_A^2$$

Therefore,

$$R = i_X r_{AX}\sigma_A = i\sigma_A \frac{\frac{(n-1)(1-a_{SIB})}{n}\sigma_A^2}{\sigma_A\sqrt{\frac{(n-1)(1-\rho_{SIB})}{n}\sigma_P^2}} = i\sigma_P h^2(1 - a_{SIB})\sqrt{\frac{n-1}{n(1 - \rho_{SIB})}}$$

### 15.5.3  Sib Selection

Sib selection is much the same as family selection except that the mean of the siblings does not include the candidate. Let $n$ be the family size and $n - 1$ be the number of siblings of the candidate. Let the candidate be indexed as the last individual, i.e., the $n^{\text{th}}$ member of the family. The mean of the siblings is

$$X = \frac{1}{n-1}\sum_{j=1}^{n-1} P_j$$

The variance of $X$ is

$$\text{var}(X) = \left(\frac{1}{n-1}\right)^2 \left\{ \sum_{j=1}^{n-1} \text{var}(P_j) + 2 \sum_{j=1}^{n-2} \sum_{j'=j+1}^{n-1} \text{cov}(P_j, P_{j'}) \right\}$$

$$= \left(\frac{1}{n-1}\right)^2 \left\{ \sum_{j=1}^{n-1} \sigma_P^2 + 2 \sum_{j=1}^{n-2} \sum_{j'=j+1}^{n-1} a_{SIB} \sigma_A^2 \right\}$$

$$= \left(\frac{1}{n-1}\right)^2 \left\{ (n-1)\sigma_P^2 + (n-1)(n-2)a_{SIB}\sigma_A^2 \right\}$$

$$= \left(\frac{1}{n-1}\right)^2 \left\{ (n-1) + (n-1)(n-2)a_{SIB}h^2 \right\} \sigma_P^2$$

The covariance between $A$ and $X$ is

$$\text{cov}(A, X) = \frac{1}{n-1} \sum_{j=1}^{n-1} \text{cov}(A, P_j) = \frac{1}{n-1} \sum_{j=1}^{n-1} a_{SIB}\sigma_A^2 = \frac{1}{n-1} a_{SIB}h^2\sigma_P^2$$

The correlation between $A$ and $X$ is

$$r_{AX} = \frac{\text{cov}(A, X)}{\sqrt{\text{var}(X)\text{var}(A)}} = \frac{a_{SIB}h^2\sigma_P^2}{\sqrt{\left[(n-1) + (n-1)(n-2)a_{SIB}h^2\right]\sigma_P^2\sigma_A^2}}$$

After a few steps of simplification, we have

$$r_{AX} = \frac{a_{SIB}h}{\sqrt{(n-1)\left[1 + (n-2)a_{SIB}h^2\right]}}$$

The selection response is

$$R = i\sigma_A r_{AX} = i\sigma_P h^2 \frac{a_{SIB}}{\sqrt{(n-1)\left[1 + (n-2)a_{SIB}h^2\right]}}$$

### 15.5.4  Pedigree Selection

The selection criterion is $X = (P_m + P_f)/2$, the middle parent value. The correlation between $A$ and $X$ is

$$\text{cov}(A, X) = \frac{1}{2}\text{cov}(A, P_m) + \frac{1}{2}\text{cov}(A, P_f) = \frac{1}{4}\sigma_A^2 + \frac{1}{4}\sigma_A^2 = \frac{1}{2}\sigma_A^2$$

The variance of $X$ is

$$\text{var}(X) = \frac{1}{4}\text{var}(P_m) + \frac{1}{4}\text{var}(P_f) = \frac{1}{2}\sigma_P^2$$

The covariance between $A$ and $X$ is

$$r_{AX} = \frac{\frac{1}{2}\sigma_A^2}{\sigma_A\sqrt{\frac{1}{2}\sigma_P^2}} = \frac{\frac{1}{2}}{\sqrt{\frac{1}{2}}}\frac{\sigma_A}{\sigma_P} = \sqrt{\frac{1}{2}}h$$

The selection response is

$$R = i\sigma_A r_{AX} = i\sigma_P h^2\left(\sqrt{\frac{1}{2}}h\right)$$

So, pedigree selection is much less efficient than mass selection, but candidates can be selected early before they reach the adulthood.

### 15.5.5 Combined Selection

Combined selection takes advantage of all available information. Let $X_1 = P$ be the phenotypic value of the candidate, $X_2 = (P_m + P_f)/2$ be the mean of the two parents and $X_3 = (P_{F_1} + P_{F_2})/2$ be the mean of two full-siblings of the candidate. The selection criterion is $X = X_1 + X_2 + X_3$. The simple sum of the three pieces of information is not an optimal index. Let us find the variance of $X$,

$$\text{var}(X) = \text{var}(X_1) + \text{var}(X_2) + \text{var}(X_3) + 2\text{cov}(X_1, X_2) + 2\text{cov}(X_1, X_3) + 2\text{cov}(X_2, X_3)$$

where $\text{var}(X_1) = \text{var}(P) = \sigma_P^2$, $\text{var}(X_2) = \frac{1}{4}\text{var}(P_m + P_f) = \frac{1}{2}\sigma_P^2$, and.

$$\text{var}(X_3) = \frac{1}{4}\left[\text{var}(P_{F_1}) + \text{var}(P_{F_2}) + 2\text{cov}(P_{F_1}, P_{F_2})\right] = \frac{1}{4}\left[\sigma_P^2 + \sigma_P^2 + 2 \times \frac{1}{2}\sigma_A^2\right] = \frac{1}{4}(2 + h^2)\sigma_P^2$$

The covariances between the three sources of information are

$$\text{cov}(X_1, X_2) = \frac{1}{2}\text{cov}(P, P_m) + \frac{1}{2}\text{cov}(P, P_f) = \frac{1}{2}\sigma_A^2$$

$$\text{cov}(X_1, X_3) = \frac{1}{2}\text{cov}(P, P_{F_1}) + \frac{1}{2}\text{cov}(P, P_{F_2}) = \frac{1}{2}\sigma_A^2$$

and

$$\text{cov}(X_2, X_3) = \frac{1}{4}\left[\text{cov}(P_m, P_{F_1}) + \text{cov}(P_m, P_{F_2}) + \text{cov}(P_f, P_{F_1}) + \text{cov}(P_f, P_{F_2})\right]$$
$$= \frac{1}{4}\left[\frac{1}{2}\sigma_A^2 + \frac{1}{2}\sigma_A^2 + \frac{1}{2}\sigma_A^2 + \frac{1}{2}\sigma_A^2\right]$$
$$= \frac{1}{2}\sigma_A^2$$

Therefore,

$$\text{var}(X) = \sigma_P^2 + \frac{1}{2}\sigma_P^2 + \frac{1}{4}(2 + h^2)\sigma_P^2 + 2 \times \frac{1}{2}\sigma_A^2 + 2 \times \frac{1}{2}\sigma_A^2 + 2 \times \frac{1}{2}\sigma_A^2$$
$$= \sigma_P^2 + \frac{1}{2}\sigma_P^2 + \frac{1}{4}(2 + h^2)\sigma_P^2 + 3h^2\sigma_P^2$$
$$= \left(2 + \frac{1}{4}h^2 + 3h^2\right)\sigma_P^2$$

The covariance between $A$ and $X$ is

$$\mathrm{cov}(A, X) = \mathrm{cov}(A, P) + \frac{1}{2}\mathrm{cov}(A, P_m) + \frac{1}{2}\mathrm{cov}\left(A, P_f\right) + \frac{1}{2}\mathrm{cov}(A, P_{F_1}) + \frac{1}{2}\mathrm{cov}(A, P_{F_2})$$

$$= \sigma_A^2 + \frac{1}{4}\sigma_A^2 + \frac{1}{4}\sigma_A^2 + \frac{1}{4}\sigma_A^2 + \frac{1}{4}\sigma_A^2$$

$$= 2\sigma_A^2$$

Therefore, the correlation is

$$r_{AX} = \frac{\mathrm{cov}(A, X)}{\sqrt{\mathrm{var}(A)\mathrm{var}(X)}} = \frac{2h}{\sqrt{2 + \frac{1}{4}h^2 + 3h^2}}$$

The response to selection is

$$R = i\sigma_A r_{AX} = i\sigma_P h^2 \frac{2}{\sqrt{2 + \frac{1}{4}h^2 + 3h^2}}$$

Interestingly, if $h^2 \to 0$, the combined selection ($\sqrt{2} = 1.4132$) is more efficient than the mass selection. However, if $h^2 \to 1$, the combined selection ($2/\sqrt{5.25} = 0.8729$) is less efficient than the mass selection. The reason for the lower efficiency of the combined selection is due to the fact that the three pieces of information are highly correlated and are not combined in an optimal way. We will learn index selection where pieces of information are combined in an optimal way in Chap. 16.

## References

Falconer DS, Mackay TFC. Introduction to quantitative genetics. Harlow: Addison Wesley Longman; 1996.
Meuwissen THE, Hayes BJ, Goddard ME. Prediction of total genetic value using genome-wide dense marker maps. Genetics. 2001;157:1819.

# Selection Index and the Best Linear Unbiased Prediction

In Chap. 15, we learned all methods of selection commonly used in breeding programs. In the end, we introduced a method called combined selection, where all pieces of information are used. If different pieces of information are simply added together, we may not achieve the maximum genetic gain. Therefore, it is important that different pieces of information are weighted in an optimal way. Such a weighted sum is called selection index. Selection index was first developed by Smith and Hazel for improving multiple traits. However, selection index for a single trait adopts the same idea as a multiple trait selection index simply by replacing different traits by different types of relatives.

## 16.1 Selection Index

Our purpose here is to extract as much information as we can to predict the breeding value of a candidate. Information that can be used includes: (1) individual's own phenotypic value, (2) phenotypic values of relatives, and (3) correlated characters. Consider that we use $m$ sources of information to predict the breeding value of a candidate. The breeding value is always denoted by $A$. The information comes from $m$ sources, described by an $m \times 1$ vector,

$$X = [X_1 \ X_2 \ \cdots \ X_m]^{\mathrm{T}}$$

The selection criterion is a linear combination of these different sources of information, called the selection index.

$$I = b_1 X_1 + b_2 X_2 + \cdots + b_m X_m = b^{\mathrm{T}} X$$

where

$$b = [b_1 \ \ b_2 \ \ \cdots \ \ b_m]^{\mathrm{T}}$$

are called the weights of the selection index. The objective of the selection is $A$ and the selection criterion is $I$, thus the response to selection is

$$R_I = i_I r_{IA} \sigma_A$$

where $r_{IA}$ is the correlation coefficient between $A$ and $I$, $i_I$ is the selection intensity, and $\sigma_A$ is the genetic standard deviation. The purpose here is to find the optimal set of $b$ such that $r_{IA}$ is maximum. If $r_{IA}$ is maximum, then $R_I$ will also be maximum.

### 16.1.1 Derivation of the Index Weights

We will derive $b$ using $I = b_1 X_1 + b_2 X_2$ as an example. Let us define

$$V_X = \text{var}(X) = \begin{bmatrix} \text{var}(X_1) & \text{cov}(X_1, X_2) \\ \text{cov}(X_2, X_1) & \text{var}(X_2) \end{bmatrix}$$

as the variance matrix of $X$ and

$$C_{XA} = \text{cov}(X, A) = \begin{bmatrix} \text{cov}(X_1, A) \\ \text{cov}(X_2, A) \end{bmatrix}$$

as the covariance matrix between $A$ and $X$. The variance of the selection index is

$$\sigma_I^2 = \text{var}(I) = \text{var}(b^T X) = b^T \text{var}(X) b = b^T V_X b$$

and

$$\text{cov}(I, A) = \text{cov}(b^T X, A) - b^T \text{cov}(X, A) = b^T C_{XA}$$

Therefore,

$$r_{IA}^2 = \frac{\text{cov}^2(I, A)}{\sigma_I^2 \sigma_A^2} = \frac{b^T C_{XA} C_{AX} b}{(b^T V_X b)\ \sigma_A^2}$$

where $C_{AX} = C_{XA}^T$. The weights of the index are found by maximizing the squared correlation coefficient between $A$ and $I$. However, it is much easier to maximize the log of the squared correlation, just like the maximum likelihood method for estimating parameters where we often maximize the log likelihood. If the weights maximize the log of the squared correlation, they also maximize the squared correlation coefficient. Let us define

$$Q = \ln\left(r_{IA}^2\right) = 2 \ln\left(b^T C_{XA}\right) - \ln\left(b^T V_X b\right) - \ln\left(\sigma_A^2\right)$$

To maximize $Q$, we need to take the partial derivatives of $Q$ with respect to $b$ and set the derivatives to 0, which are

$$\frac{\partial Q}{\partial b} = \begin{bmatrix} \partial Q / \partial b_1 \\ \partial Q / \partial b_2 \end{bmatrix} = 2 \frac{C_{XA}}{b^T C_{XA}} - \frac{2 V_X b}{b^T V_X b} = 0$$

and they are further reduced to

$$C_{XA} - \left(\frac{b^T C_{XA}}{b^T V_X b}\right) V_X b = 0$$

Let us make a restriction (constraint),

$$c = \frac{b^T C_{XA}}{b^T V_X b} = 1$$

so that the solution of $b$ is

$$b = V_X^{-1} C_{XA}$$

Note that $c$ is simply a constant multiplied to every element of $b$. As a result, ignorance of $c$ will not change the relative rank of selection indices for individual candidates.

## 16.1.2  Evaluation of Index Selection

Response to index selection is calculated via $R_I = i_I r_{IA} \sigma_A$, where

$$r_{IA} = \frac{\text{cov}(I, A)}{\sigma_I \sigma_A}$$

is the correlation coefficient between $A$ and $I$. Furthermore,

$$\text{cov}(I, A) = b^{\text{T}} \text{cov}(X, A) = b^{\text{T}} C_{XA} = \left(V_X^{-1} C_{XA}\right)^{\text{T}} C_{XA} = C_{AX} V_X^{-1} C_{XA}$$

and

$$\sigma_I^2 = b^{\text{T}} \text{var}(X) b = b^{\text{T}} V_X b = \left(V_X^{-1} C_{XA}\right)^{\text{T}} V_X \left(V_X^{-1} C_{XA}\right) = C_{AX} V_X^{-1} C_{XA}$$

Therefore,

$$\text{cov}(I, A) = \sigma_I^2 = C_{AX} V_X^{-1} C_{XA}$$

and

$$r_{IA} = \frac{\text{cov}(I, A)}{\sigma_I \sigma_A} = \frac{\sigma_I^2}{\sigma_I \sigma_A} = \frac{\sigma_I}{\sigma_A}$$

Finally, we get

$$R_I = i_I r_{IA} \sigma_A = i_I \, \sigma_I$$

## 16.1.3  Comparison of Index Selection with a Simple Combined Selection

Towards the end of Chap. 15, we evaluated the accuracy of a combined selection by simply adding different sources of information together, where we had three pieces of information, $X_1 = P$ is the phenotypic value of the candidate, $X_2 = (P_m + P_f)/2$ is the mean of the two parents, and $X_3 = (P_{F_1} + P_{F_2})/2$ is the mean of two full-siblings of the candidate. The selection criterion of the simple combined selection is $X = X_1 + X_2 + X_3$. The simple sum of the three pieces of information is not an optimal index. With the index selection, the selection criterion is

$$I = b_1 X_1 + b_2 X_2 + b_3 X_3$$

We now compare the accuracy of the optimal index selection with the simple combined selection. The variance matrix of the three pieces of information is

$$V_X = \begin{bmatrix} \text{var}(X_1) & \text{cov}(X_1, X_2) & \text{cov}(X_1, X_3) \\ \text{cov}(X_1, X_2) & \text{var}(X_2) & \text{cov}(X_2, X_3) \\ \text{cov}(X_1, X_3) & \text{cov}(X_2, X_3) & \text{var}(X_3) \end{bmatrix} = \begin{bmatrix} \sigma_P^2 & \frac{1}{2}\sigma_A^2 & \frac{1}{2}\sigma_A^2 \\ \frac{1}{2}\sigma_A^2 & \frac{1}{2}\sigma_P^2 & \frac{1}{2}\sigma_A^2 \\ \frac{1}{2}\sigma_A^2 & \frac{1}{2}\sigma_A^2 & \frac{1}{4}\left(2 + h^2\right)\sigma_P^2 \end{bmatrix}$$

The covariance between $A$ and the three pieces of information is

$$C_{XA} = \begin{bmatrix} \mathrm{cov}(X_1, A) \\ \mathrm{cov}(X_2, A) \\ \mathrm{cov}(X_3, A) \end{bmatrix} = \begin{bmatrix} \sigma_A^2 \\ \frac{1}{2}\sigma_A^2 \\ \frac{1}{2}\sigma_A^2 \end{bmatrix}$$

The weights of the selection index are

$$\begin{bmatrix} b_1 \\ b_2 \\ b_3 \end{bmatrix} = \begin{bmatrix} \sigma_P^2 & \frac{1}{2}\sigma_A^2 & \frac{1}{2}\sigma_A^2 \\ \frac{1}{2}\sigma_A^2 & \frac{1}{2}\sigma_P^2 & \frac{1}{2}\sigma_A^2 \\ \frac{1}{2}\sigma_A^2 & \frac{1}{2}\sigma_A^2 & \frac{1}{4}(2+h^2)\sigma_P^2 \end{bmatrix}^{-1} \begin{bmatrix} \sigma_A^2 \\ \frac{1}{2}\sigma_A^2 \\ \frac{1}{2}\sigma_A^2 \end{bmatrix} = \begin{bmatrix} 1 & \frac{1}{2}h^2 & \frac{1}{2}h^2 \\ \frac{1}{2}h^2 & \frac{1}{2} & \frac{1}{2}h^2 \\ \frac{1}{2}h^2 & \frac{1}{2}h^2 & \frac{1}{4}(2+h^2) \end{bmatrix}^{-1} \begin{bmatrix} h^2 \\ \frac{1}{2}h^2 \\ \frac{1}{2}h^2 \end{bmatrix}$$

An explicit solution is available but is too complicated to show the detail. Therefore, we will numerically evaluate the selection accuracy. The variance of the selection index is

$$\sigma_I^2 = b^T V_X b = \sigma_P^2 [b_1 \quad b_2 \quad b_3] \begin{bmatrix} 1 & \frac{1}{2}h^2 & \frac{1}{2}h^2 \\ \frac{1}{2}h^2 & \frac{1}{2} & \frac{1}{2}h^2 \\ \frac{1}{2}h^2 & \frac{1}{2}h^2 & \frac{1}{4}(2+h^2) \end{bmatrix} \begin{bmatrix} b_1 \\ b_2 \\ b_3 \end{bmatrix}$$

Let us define $U_X$ as

$$U_X = \begin{bmatrix} 1 & \frac{1}{2}h^2 & \frac{1}{2}h^2 \\ \frac{1}{2}h^2 & \frac{1}{2} & \frac{1}{2}h^2 \\ \frac{1}{2}h^2 & \frac{1}{2}h^2 & \frac{1}{4}(2+h^2) \end{bmatrix}$$

so that $V_X = \sigma_P^2 U_X$ and $\sigma_I^2 = b^T V_X b = b^T U_X b \sigma_P^2$. Note that the covariance between $I$ and $A$ is

$$\mathrm{cov}(I, A) = b^T C_{XA} = \left(b_1 + \frac{1}{2}b_2 + \frac{1}{2}b_3\right)h^2 \sigma_P^2$$

Therefore, the accuracy of the index selection is

$$r_{IA} = \frac{\left(b_1 + \frac{1}{2}b_2 + \frac{1}{2}b_3\right)h^2 \sigma_P^2}{\sqrt{h^2 \sigma_P^2 \sigma_P^2 b^T U_X b}} = \frac{\left(b_1 + \frac{1}{2}b_2 + \frac{1}{2}b_3\right)h}{\sqrt{b^T U_X b}}$$

Recall that the accuracy of the simple combined selection is

$$r_{AX} = \frac{2h}{\sqrt{2 + \frac{1}{4}h^2 + 3h^2}}$$

The accuracy of mass selection is $r_{AP} = h$. We can now compare the accuracies of the three methods under any given value of the heritability. For example, when $h^2 = 0.45$, the three accuracies are $r_{AP} = 0.6708$, $r_{AX} = 0.7210$, and $r_{IA} = 0.7426$. Clearly, index selection is more efficient than the simple combined selection, which is in turn better than the mass selection. The weights of the selection index are

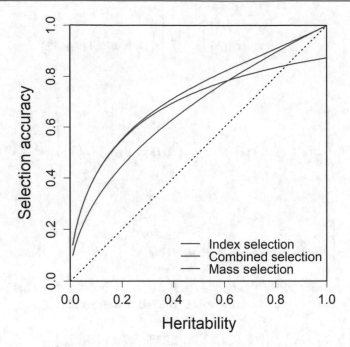

**Fig. 16.1** Comparison of the accuracies of three selection methods (index selection, combined selection, and mass selection)

$$\begin{bmatrix} b_1 \\ b_2 \\ b_3 \end{bmatrix} = \begin{bmatrix} 0.3669399 \\ 0.2159215 \\ 0.1532346 \end{bmatrix}$$

It states that the phenotypic value of the candidate contains more information than the mid-parent value which in turn gives more information than the average of the two siblings. In Chap. 15, we found that the simple combined selection is not always better than the mass selection in some range of $h^2$. We now evaluate the accuracies of the three methods in the whole range of the heritability. The results are illustrated in Fig. 16.1. We can see that when $h^2 < 0.615$, the combined selection is more effective than the mass selection, but the effectiveness is reversed when $h^2 > 0.615$. However, the accuracy of the index selection is always higher than both the mass selection and the combined selection in the whole range of heritability. Therefore, index selection is the optimal method of selection.

### 16.1.4 Index Selection Combining Candidate Phenotype with the Family Mean

The index is a linear combination of individual phenotype and the family mean. The selection index is

$$I = b_1 X_1 + b_2 X_2$$

where $X_1$ is the phenotypic value of the candidate, $X_2$ is the mean of $n$ members of the family (e.g., full-sib family) including the phenotypic value of the candidate. You may prove that

$$V_X = \begin{bmatrix} \mathrm{var}(X_1) & \mathrm{cov}(X_1, X_2) \\ \mathrm{cov}(X_1, X_2) & \mathrm{var}(X_2) \end{bmatrix} = V_P \begin{bmatrix} 1 & \frac{1}{n}\left[1 + (n-1)a_{FS}h^2\right] \\ \frac{1}{n}\left[1 + (n-1)a_{FS}h^2\right] & \frac{1}{n}\left[1 + (n-1)a_{FS}h^2\right] \end{bmatrix}$$

The covariance matrix is

$$C_{XA} = \begin{bmatrix} \text{cov}(X_1, A) \\ \text{cov}(X_2, A) \end{bmatrix} = V_P \begin{bmatrix} h^2 \\ \frac{1}{n}[1 + (n-1)a_{FS}]h^2 \end{bmatrix}$$

Therefore,

$$b = V_X^{-1} C_{XA} = \begin{bmatrix} 1 & \frac{1}{n}\left[1 + (n-1)a_{FS}h^2\right] \\ \frac{1}{n}\left[1 + (n-1)a_{FS}h^2\right] & \frac{1}{n}\left[1 + (n-1)a_{FS}h^2\right] \end{bmatrix}^{-1} \begin{bmatrix} h^2 \\ \frac{1}{n}[1 + (n-1)a_{FS}]h^2 \end{bmatrix}$$

which can be simplified into

$$b = \begin{bmatrix} n & 1 + (n-1)a_{FS}h^2 \\ 1 + (n-1)a_{FS}h^2 & 1 + (n-1)a_{FS}h^2 \end{bmatrix}^{-1} \begin{bmatrix} nh^2 \\ h^2 + (n-1)a_{FS}h^2 \end{bmatrix}$$

where $a_{FS} = 1/2$ is the additive relationship between full-sibs. If the family is defined as a half-sib family, then $a_{FS}$ should be replaced by $a_{HS} = 1/4$. Let $n = 5$, $h^2 = 0.25$, and $a_{FS} = 1/2$. The index weights are

$$\begin{bmatrix} b_1 \\ b_2 \end{bmatrix} = \begin{bmatrix} 5 & 1.5 \\ 1.5 & 1.5 \end{bmatrix}^{-1} \begin{bmatrix} 1.25 \\ 0.75 \end{bmatrix} = \begin{bmatrix} 0.1428571 \\ 0.3571429 \end{bmatrix}$$

The family mean captures more information than the phenotypic value of the candidate and thus $X_2$ carries a much heavier weight than $X_1$. Let $X_{j1}$ be the phenotypic value of candidate $j$ and $X_{j2}$ be the mean of the family where $j$ belongs to. The index score for candidate $j$ is

$$I_j = X_{j1}b_1 + X_{j2}b_2$$

Table 16.1 gives the two $X$ values (expressed as deviations from the population means) and the index scores ($I$) for five hypothetic individuals. If one candidate needs to be selected from the five, candidate number 4 is certainly the best choice because it has the highest index score. In index selection, the individual pieces of information, $X_1, X_2$, and so on, are expressed as deviations from the population mean. For example, if $Y_{j1} = 150$ and $Y_{j2} = 100$ are the actual values of the two traits for individual $j$ and the population means of the two traits are $\mu_1 = 120$ and $\mu_2 = 110$, then $X_{j1} = Y_{j1} - \mu_1$ and $X_{j2} = Y_{j2} - \mu_2$. The index should be

$$I_j = b_1(Y_{j1} - \mu_1) + b_2(Y_{j2} - \mu_2) = b_1 X_{j1} + b_2 X_{j2}$$

The index selection introduced here has a serious limitation. All candidates should have the same sources of information and use the same set of index weights. In reality, different candidates may have different types of relatives available. In order to use index selection, we may have to use different weights for different candidates because their relatives are different. Such a selection index is the best linear unbiased prediction (BLUP), which will be introduced in the next section of the chapter.

**Table 16.1** Selection index scores for five hypothetical candidates

| Candidate | $X_1$ | $X_2$ | I | Rank |
|---|---|---|---|---|
| 1 | 10.25 | 14.51 | 6.646429 | 2 |
| 2 | 12.05 | 10.36 | 5.421429 | 4 |
| 3 | 11.33 | 12.08 | 5.932857 | 3 |
| 4 | 15.43 | 13.75 | 7.115000 | 1 |
| 5 | 9.38 | 11.32 | 5.382857 | 5 |

## 16.2 Best Linear Unbiased Prediction (BLUP)

### 16.2.1 Relationship Between Selection Index and BLUP

The BLUP is a generalized selection index procedure to predict the breeding values of candidates for selection. It is always associated with the mixed model methodology. The method incorporates all available information, including candidate's own phenotypic value (if available) and the phenotypic values of all relatives of the candidate. The method requires the additive genetic relationship matrix ($A = 2\Theta$) of all individuals with phenotypic records and the additive genetic relationship matrix between the recorded individuals and the candidate. It also requires genetic parameters (already estimated prior to the prediction). In fact, the BLUP is performed simultaneously for all candidates of interest. Figure 16.2 gives the pedigree of all 14 members in a sample population we learned in Chap. 12. Not only can we predict the breeding values for individuals in class $X$, but also predict the breeding values for individuals already having phenotypic records. For example, the predicted breeding value for $y_1$ using an index selection should be better than using its phenotypic value alone because the index selection will incorporate the record of $y_2$ (her sister). Breeding values of individuals 1 and 2 can be predicted using indexes

$$I_1 = a_1 y_1 + a_2 y_2$$
$$I_2 = a_2 y_1 + a_1 y_2$$

where $a_1$ is the weight for the phenotypic value of the candidate and $a_2$ is the weight for the sibling trait. Similar argument also applies to individuals 3 and 4

$$I_3 = b_1 y_3 + b_2 y_4$$
$$I_4 = b_2 y_3 + b_1 y_4$$

For the last three individuals, the indexes are

$$I_5 = c_1 y_5 + c_2 y_6 + c_3 y_7$$
$$I_6 = c_2 y_6 + c_1 y_5 + c_3 y_7$$
$$I_7 = d_1 y_6 + d_2 y_5 + d_3 y_7$$

One can build a general selection index using all seven records for individual $j$, as shown below

$$I_j = b_{j1} y_1 + b_{j2} y_2 + b_{j3} y_3 + b_{j4} y_4 + b_{j5} y_5 + b_{j6} y_6 + b_{j7} y_7$$

**Fig. 16.2** A hypothetical population consisting of three independent families, where individuals labeled $X$ do not have phenotypic records, but individuals labeled $Y$ do

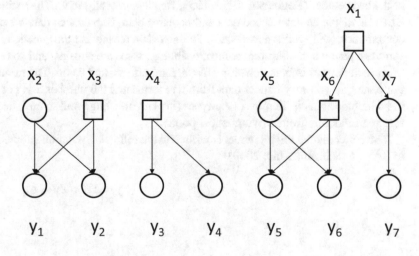

This time, the indexes of all individuals use the same set of phenotypic records, but the weights of the indexes vary across individuals. Instead of calling it an index, we call is the BLUP and denote it by

$$\widehat{y}_j = b_{j1}y_1 + b_{j2}y_2 + b_{j3}y_3 + b_{j4}y_4 + b_{j5}y_5 + b_{j6}y_6 + b_{j7}y_7$$

In matrix notation, we write $\widehat{y}_j = b_j^T y$, where $y$ is a $7 \times 1$ vector for the phenotypic values of all records and $b_j$ is a $7 \times 1$ vector of the weights for the $j$th individual. With this generalized index selection, we can also predict the breeding values of an ancestor,

$$\widehat{x}_j = b_{j1}y_1 + b_{j2}y_2 + b_{j3}y_3 + b_{j4}y_4 + b_{j5}y_5 + b_{j6}y_6 + b_{j7}y_7$$

where the weights also vary from one ancestor to another. Now, let us define

$$B = \begin{bmatrix} b_{11} & b_{12} & \cdots & b_{17} \\ b_{21} & b_{22} & \cdots & b_{27} \\ \vdots & \vdots & \ddots & \vdots \\ b_{71} & b_{72} & \cdots & b_{77} \end{bmatrix}$$

as the weights for all candidates from all relatives. The predicted breeding values for all the seven members can be written in matrix notation as

$$\widehat{y} = By \tag{16.1}$$

If the $B$ matrix is calculated using a proper optimization algorithm, the predicted $y$ is called the BLUP, which is the focus of this chapter.

## 16.2.2  Theory of BLUP

In mixed model analysis, model effects include fixed effects and random effects. The fixed effects often have fewer levels than the random effects, and they are often set by the experimenters (not randomly selected from a population of levels). Their effects are treated as parameters and subject to estimation. The estimated fixed effects are called the best linear unbiased estimates (BLUE). In contrast, random effects often have many levels, and they represent a random sample from an "infinite" number of levels. As random effects, they follow a distribution (normal distribution is almost always assumed). The variance among the levels of random effects is the parameter of interest. The random effect themselves are not parameters and their estimates are called the BLUP, which was first developed by a famous animal breeder Charles Roy Henderson in early 1950s of the last century (Henderson 1950, 1953; Henderson et al. 1959). The main purpose of developing the mixed model and the BLUP is to handle unbalanced data and pedigree data. Unbalanced data means that the number of observations per treatment combination (cell) varies across cells. Pedigree data means that the genetic relationship among animals is heterogeneous, i.e., some are parent and offspring, some are siblings, some are cousins, and so on. In animal and plant breeding, the fixed effects often include year effects, location effects, farm effects, and so on. The number of levels for each fixed effect is often small. For example, you may collect data from five farms and the number 5 is not a large number. In contrast, the random effects often include animals, families, genotypes (in plant breeding), and so on. The number of levels can be huge, and they are often selected randomly from a hypothetical pool.

The conventional mixed model is written in the following notation adopted from the user manual of the MIXED procedure in SAS (SAS Institute Inc. 2009).

$$y = X\beta + Z\gamma + \varepsilon$$

where $y$ is the response variable, $X$ is the design matrix for the fixed effects, $\beta$ is a vector of the fixed effects, $Z$ is the design matrix for the random effects, $\gamma$ is a vector of the random effects, $\varepsilon$ is a vector of the residuals. The random effects are assumed to be normally distributed, $\gamma \sim N(0, G)$, where $G$ is the variance matrix for the random effects. The residuals are also assumed to be normally distributed, $\varepsilon \sim N(0, R)$, where $R$ is the residual variance matrix. The residuals may be contributed by environmental errors and all other effects not fitted in the model if they affect the response variable. The parameters are denoted by $\theta = \{\beta, G, R\}$. You can see that $\gamma$ are excluded from $\theta$ because they are not parameters. Note that "$\gamma$" is treated as a plural word because it contains multiple elements. Estimation of parameters has been studied in Chap. 12. The BLUE and BLUP of the model effects are obtained assuming that $G$ and $R$ are known true values. If the variance matrices are estimated from the data, the estimated fixed effects and the predicted random effects are no longer the BLUE and the BLUP; rather, they are called empirical Bayes estimates. However, in the mixed model literature, we almost always call them BLUE and BLUP, although they are not, from the theoretical point of view. Assume that $G$ and $R$ are known. The BLUE and BLUP are obtained simultaneously from Henderson's mixed model equation (MME) (Henderson 1950), which is

$$\begin{bmatrix} X^T R^{-1} X & X^T R^{-1} Z \\ Z^T R^{-1} X & Z^T R^{-1} Z + G^{-1} \end{bmatrix} \begin{bmatrix} \beta \\ \gamma \end{bmatrix} = \begin{bmatrix} X^T R^{-1} y \\ Z^T R^{-1} y \end{bmatrix}$$

Therefore,

$$\begin{bmatrix} \widehat{\beta} \\ \widehat{\gamma} \end{bmatrix} = \begin{bmatrix} X^T R^{-1} X & X^T R^{-1} Z \\ Z^T R^{-1} X & Z^T R^{-1} Z + G^{-1} \end{bmatrix}^{-1} \begin{bmatrix} X^T R^{-1} y \\ Z^T R^{-1} y \end{bmatrix}$$

The variance-covariance matrix of the BLUE and BLUP is

$$\mathrm{var} \begin{bmatrix} \widehat{\beta} \\ \widehat{\gamma} \end{bmatrix} = \begin{bmatrix} X^T R^{-1} X & X^T R^{-1} Z \\ Z^T R^{-1} X & Z^T R^{-1} Z + G^{-1} \end{bmatrix}^{-1} = \begin{bmatrix} C_{\beta\beta} & C_{\beta\gamma} \\ C_{\gamma\beta} & C_{\gamma\gamma} \end{bmatrix}$$

This means that $\mathrm{var}(\widehat{\gamma}) = C_{\gamma\gamma}$ and square roots of the diagonal elements of $C_{\gamma\gamma}$ are the standard errors of the prediction.

There are several different ways to derive Henderson's MME. The one presented by Robinson (1991) is adopted from Henderson (1973) and is very intuitive and easy to follow. The joint distribution of $y$ and $\gamma$ is assumed to be normal with expectation

$$\mathrm{E} \begin{bmatrix} y \\ \gamma \end{bmatrix} = \begin{bmatrix} X\beta + Z\gamma \\ 0 \end{bmatrix}$$

and variance

$$\mathrm{var} \begin{bmatrix} y \\ \gamma \end{bmatrix} = \begin{bmatrix} R & 0 \\ 0 & G \end{bmatrix}$$

The expectation and variance of $y$ are the conditional expectation $\mathrm{E}(y|\gamma) = X\beta + Z\gamma$ and conditional variance $\mathrm{var}(y|\gamma) = R$ given $\gamma$. The log joint density is

$$L(\beta, \gamma) = -\frac{1}{2} \ln \det \begin{bmatrix} R & 0 \\ 0 & G \end{bmatrix} - \frac{1}{2} [y - X\beta - Z\gamma \quad \gamma] \begin{bmatrix} R & 0 \\ 0 & G \end{bmatrix}^{-1} \begin{bmatrix} y - X\beta - Z\gamma \\ y - X\beta - Z\gamma \end{bmatrix}$$

The first term is irrelevant to $\beta$ and $\gamma$, and thus ignored when maximizing the log density. So, the quantity to be maximized is

$$Q(\beta,\gamma) = -\frac{1}{2}\left[(y - X\beta - Z\gamma)^T \quad \gamma^T\right]\begin{bmatrix} R & 0 \\ 0 & G \end{bmatrix}^{-1}\begin{bmatrix} y - X\beta - Z\gamma \\ \gamma \end{bmatrix}$$

$$= -\frac{1}{2}(y - X\beta - Z\gamma)^T R^{-1}(y - X\beta - Z\gamma) - \frac{1}{2}\gamma^T G^{-1}\gamma$$

The partial derivatives of $Q(\beta, \gamma)$ with respect to $\beta$ and $\gamma$ are

$$\frac{\partial}{\partial \beta}Q(\beta,\gamma) = X^T R^{-1}(y - X\beta - Z\gamma) = X^T R^{-1}y - X^T R^{-1}X\beta - X^T R^{-1}Z\gamma$$

and

$$\frac{\partial}{\partial \gamma}Q(\beta,\gamma) = Z^T R^{-1}(y - X\beta - Z\gamma) - G^{-1}\gamma = Z^T R^{-1}y - Z^T R^{-1}X\beta - Z^T R^{-1}Z\gamma - G^{-1}\gamma$$

Setting the partial derivatives to zero, we have

$$X^T R^{-1}y = X^T R^{-1}X\beta + X^T R^{-1}Z\gamma$$

$$Z^T R^{-1}y = Z^T R^{-1}X\beta + \left(Z^T R^{-1}Z + G^{-1}\right)\gamma$$

Rearranging the above two equations leads to the Henderson's MME

$$\begin{bmatrix} X^T R^{-1}X & X^T R^{-1}Z \\ Z^T R^{-1}X & Z^T R^{-1}Z + G^{-1} \end{bmatrix}\begin{bmatrix} \beta \\ \gamma \end{bmatrix} = \begin{bmatrix} X^T R^{-1}y \\ Z^T R^{-1}y \end{bmatrix}$$

Therefore,

$$\begin{bmatrix} \widehat{\beta} \\ \widehat{\gamma} \end{bmatrix} = \begin{bmatrix} C_{\beta\beta} & C_{\beta\gamma} \\ C_{\gamma\beta} & C_{\gamma\gamma} \end{bmatrix}\begin{bmatrix} X^T R^{-1}y \\ Z^T R^{-1}y \end{bmatrix} = \begin{bmatrix} C_{\beta\beta}X^T R^{-1}y + C_{\beta\gamma}Z^T R^{-1}y \\ C_{\gamma\beta}X^T R^{-1}y + C_{\gamma\gamma}Z^T R^{-1}y \end{bmatrix}$$

The predicted random effects are

$$\widehat{\gamma} = \left(C_{\gamma\beta}X^T + C_{\gamma\gamma}Z^T\right)R^{-1}y = By \tag{16.2}$$

which is the same as the selection index given in Eq. (16.1), where

$$B = \left(C_{\gamma\beta}X^T + C_{\gamma\gamma}Z^T\right)R^{-1}$$

This concludes the derivation of the BLUP and the equivalence between selection index and BLUP. An alternative expression of Eq. (16.2) is the conditional expectation of the random effects conditional on the phenotypic values,

$$\widehat{\gamma} = \mathrm{E}(\gamma|y) = Z^T GV^{-1}\left(y - X\widehat{\beta}\right) = Z^T G\left(ZGZ^T + R\right)^{-1}\left(y - X\widehat{\beta}\right)$$

where

$$\widehat{\beta} = \left(X^T V^{-1}X\right)^{-1}X^T V^{-1}y$$

The BLUP of the random effects can be written as

$$\widehat{\gamma} = Z^T G \left( Z G Z^T + R \right)^{-1} \left[ I - X \left( X^T V^{-1} X \right)^{-1} X^T V^{-1} \right] y$$

In terms of selection index, $\widehat{\gamma} = By$, the weights matrix of the index is

$$B = Z^T G \left( Z G Z^T + R \right)^{-1} \left[ I - X \left( X^T V^{-1} X \right)^{-1} X^T V^{-1} \right]$$

## 16.3 Examples and SAS Programs

### 16.3.1 Example 1

We now use the full pedigree (Fig. 16.2) and the phenotypic values of the 14 individuals to demonstrate the BLUP method for predicting breeding values of all members in the population. Missing values are filled with periods (.), which are not zeros. Table 16.2 shows that among the 14 members, only the last 7 members have phenotypic values. Therefore, the parameters can only be estimated from the 7 members with phenotypic records. The model for the 7 members is

$$y = X\beta + Z\gamma + \varepsilon$$

where $X$ is a $7 \times 1$ unity vector (all elements are 1's), $\beta$ is the population mean, $Z$ is an $7 \times 7$ identity matrix, $\gamma$ is a $7 \times 1$ vector of genetic values of the 7 members, and $\varepsilon$ is the residual vector. The random effects are $\gamma \sim N(0, G)$ distributed where $G = A_{22}\sigma_A^2$ and $\sigma_A^2$ is the genetic variance. Matrix $A_{22}$ is the additive relationship matrix of the seven members (last generation). The residuals are $N(0, R)$ distributed where $R = I\sigma^2$ and $\sigma^2$ is the residual variance. The population covers three generations. The seven ancestors do not have phenotypes. Even if generation number may affect the trait, we do not have enough information to fit generation numbers as fixed effects in the model.

First, we use PROC INBREED to calculate the additive relationship matrix for all the 14 members (but only use the relationship matrix for individuals in the last generation). The SAS code to read the data and calculate the numerator relationship matrix is given below.

Table 16.2 Tabular pedigree and phenotypic values of 14 members in the pedigree

| Child | Dad | Mom | Generation | y |
| --- | --- | --- | --- | --- |
| x1 | . | . | 1 | . |
| x2 | . | . | 2 | . |
| x3 | . | . | 2 | . |
| x4 | . | . | 2 | . |
| x5 | . | . | 2 | . |
| x6 | x1 | . | 2 | . |
| x7 | x1 | . | 2 | . |
| y1 | x3 | x2 | 3 | 53.94636 |
| y2 | x3 | x2 | 3 | 52.69279 |
| y3 | x4 | . | 3 | 52.74754 |
| y4 | x4 | . | 3 | 48.82705 |
| y5 | x6 | x5 | 3 | 35.81209 |
| y6 | x6 | x5 | 3 | 46.09668 |
| y7 | . | x7 | 3 | 47.73397 |

```
data ped;
    input child$ dad$ mom$ gen$ y;
datalines;
x1       .       .       1       .
x2       .       .       2       .
x3       .       .       2       .
x4       .       .       2       .
x5       .       .       2       .
x6       x1      .       2       .
x7       x1      .       2       .
y1       x3      x2      3       53.94636
y2       x3      x2      3       52.69279
y3       x4      .       3       52.74754
y4       x4      .       3       48.82705
y5       x6      x5      3       35.81209
y6       x6      x5      3       46.09668
y7       .       x7      3       47.73397
;


proc inbreed data=ped matrix covar;
    var child dad mom;
run;
```

The output is a $14 \times 14$ numerator relationship matrix (Table 16.3) because of the `covar` option in the PROC INBREED statement. Let $A$ be the $14 \times 14$ numerator relationship matrix, it is partitioned into $2 \times 2$ blocks, each block being a $7 \times 7$ matrix.

$$A = \begin{bmatrix} A_{11} & A_{12} \\ A_{21} & A_{22} \end{bmatrix}$$

The lower right $7 \times 7$ block ($A_{22}$) corresponds to the last 7 members, and this matrix is the one we need to estimate the parameters. Once the parameters are estimated, we can predict the genetic values of the 7 members and also the 7 ancestors who do not have phenotypic records.

Let $n = n_1 + n_2$ be the total sample size, where $n_2 = 7$ be the number of individuals (class $y$) with phenotype records and $n_1 = 7$ be the number of individuals (class $x$) without phenotype records. Let $y_1$ be an $n_1 \times 1$ vector of the missing values, and $\mathbf{y}_2$ be an $n_2 \times 1$ vector of observed phenotypic values. Let $y = \begin{bmatrix} y_1^T & y_2^T \end{bmatrix}^T$ be an $n \times 1$ vector of phenotypic values for the entire sample. Let $\theta = \{ \beta, \sigma_A^2, \sigma^2 \}$ be the known parameter values or estimated parameter values prior to the BLUP prediction.

**Table 16.3**  The numerator relationship matrix for the 14 members in the population

| Child | Dad | Mom | x1 | x2 | x3 | x4 | x5 | x6 | x7 | y1 | y2 | y3 | y4 | y5 | y6 | y7 |
|-------|-----|-----|----|----|----|----|----|----|----|----|----|----|----|----|----|----|
| x1 |    |    | 1 | . | . | . | . | 0.5 | 0.5 | . | . | . | . | 0.25 | 0.25 | 0.25 |
| x2 |    |    | . | 1 | . | . | . | . | . | 0.5 | 0.5 | . | . | . | . | . |
| x3 |    |    | . | . | 1 | . | . | . | . | 0.5 | 0.5 | . | . | . | . | . |
| x4 |    |    | . | . | . | 1 | . | . | . | . | . | 0.5 | 0.5 | . | . | . |
| x5 |    |    | . | . | . | . | 1 | . | . | . | . | . | . | 0.5 | 0.5 | . |
| x6 | x1 |    | 0.5 | . | . | . | . | 1 | 0.25 | . | . | . | . | 0.5 | 0.5 | 0.125 |
| x7 | x1 |    | 0.5 | . | . | . | . | 0.25 | 1 | . | . | . | . | 0.125 | 0.125 | 0.5 |
| y1 | x3 | x2 | . | 0.5 | 0.5 | . | . | . | . | 1 | 0.5 | . | . | . | . | . |
| y2 | x3 | x2 | . | 0.5 | 0.5 | . | . | . | . | 0.5 | 1 | . | . | . | . | . |
| y3 | x4 |    | . | . | . | 0.5 | . | . | . | . | . | 1 | 0.25 | . | . | . |
| y4 | x4 |    | . | . | . | 0.5 | . | . | . | . | . | 0.25 | 1 | . | . | . |
| y5 | x6 | x5 | 0.25 | . | . | . | 0.5 | 0.5 | 0.125 | . | . | . | . | 1 | 0.5 | 0.0625 |
| y6 | x6 | x5 | 0.25 | . | . | . | 0.5 | 0.5 | 0.125 | . | . | . | . | 0.5 | 1 | 0.0625 |
| y7 |    | x7 | 0.25 | . | . | . | . | 0.125 | 0.5 | . | . | . | . | 0.0625 | 0.0625 | 1 |

Note that $\beta$ is the intercept (defined as $\mu$ before). Let $X = \begin{bmatrix} X_1^T & X_2^T \end{bmatrix}^T$ be a unity vector of dimension $n \times 1$ with the same partitioning corresponding to that of $y$. Define the phenotypic variance matrix by $V = A\sigma_A^2 + I\sigma^2$. Corresponding to the partitioning of matrix $A$, matrix $V$ is partitioned into

$$V = \begin{bmatrix} V_{11} & V_{12} \\ V_{21} & V_{22} \end{bmatrix}$$

The predicted phenotypic values of individuals in class $x$ given individuals in class $y$ are the conditional means of $y_1$ given $y_2$,

$$\widehat{y}_1 = \sigma_A^2 A_{12} V_{22}^{-1} (y_2 - X_2\beta) + X_1\beta$$

which is the expected best linear unbiased prediction (BLUP) of the genetic values for individuals without phenotypic records. Although we have already observed vector $y_2$, we can still predict it using

$$\widehat{y}_2 = \sigma_A^2 A_{22} V_{22}^{-1} (y_2 - X_2\beta) + X_2\beta$$

Vertical concatenation of $\widehat{y}_1$ and $\widehat{y}_2$ leads to the following expression

$$\widehat{y} = \begin{bmatrix} \widehat{y}_1 \\ \widehat{y}_2 \end{bmatrix} = \sigma_A^2 \begin{bmatrix} A_{12} \\ A_{22} \end{bmatrix} V_{22}^{-1} (y_2 - X_2\beta) + X\beta$$

If required, we can provide the prediction errors, which take the square roots of the diagonal elements of the following variance matrix:

$$\mathrm{var}(\widehat{y}) = \sigma_A^2 A - \sigma_A^2 \begin{bmatrix} A_{12} \\ A_{22} \end{bmatrix} V_{22}^{-1} \begin{bmatrix} A_{21} & A_{22} \end{bmatrix} \sigma_A^2$$

When building the selection index, the population mean (fixed effect) is supposed to be removed from the original records. However, the BLUP of the phenotypic values $\widehat{y}$ is not the selection index yet. We need to remove the mean or the fixed effect from the predicted $y$ to get the actual selection index,

$$\widehat{I} = \begin{bmatrix} \widehat{I}_1 \\ \widehat{I}_2 \end{bmatrix} = \sigma_A^2 \begin{bmatrix} A_{12} \\ A_{22} \end{bmatrix} V_{22}^{-1} (y_2 - X_2\beta)$$

The MIXED procedure in SAS allows missing values to be filled with period (.) for the response variable $y$. Therefore, SAS can take the whole population (14 members) as an input file. Internally, PROC MIXED uses the conditional expectation and conditional variance approach to predict individuals without records conditional on individuals with records. The following code reads the data and performs mixed model analysis to estimate the fixed effects and the variance components,

```
filename aa "aa.csv";
filename bb "phe.csv";
proc import datafile=aa out=one dbms=csv replace;
run;
proc import datafile=bb out=two dbms=csv replace;
run;

proc mixed data=two method=reml noprofile;
    class child;
    model y = /solution outp=yhat;
    random child /type=lin(1) ldata=one solution;
run;
proc print data=yhat;
run;
```

**Table 16.4**  Estimated parameters by PROC MIXED from the sample data with 14 members

| Parameter | Estimate | Standard error |
|---|---|---|
| $\beta$ | 48.4717 | 2.7177 |
| $\sigma_A^2$ | 36.3215 | 44.27686 |
| $\sigma^2$ | 1.9231 | 30.63495 |

**Table 16.5**  Predicted breeding values for the 14 members of the sample data

| Solution for random effects | | | | | |
|---|---|---|---|---|---|
| Child | Estimate | Std err pred | DF | t value | Pr > \|t\| |
| x1 | −2.4494 | 5.7959 | 0 | −0.42 | . |
| x2 | 3.1217 | 5.2621 | 0 | 0.59 | . |
| x3 | 3.1217 | 5.2621 | 0 | 0.59 | . |
| x4 | 1.7772 | 5.1705 | 0 | 0.34 | . |
| x5 | −4.8356 | 5.2256 | 0 | −0.93 | . |
| x6 | −4.8514 | 5.3093 | 0 | −0.91 | . |
| x7 | 1.2722 | 5.4635 | 0 | −0.23 | . |
| y1 | 5.2493 | 2.9478 | 0 | 1.78 | . |
| y2 | 4.1158 | 2.9478 | 0 | 1.40 | . |
| y3 | 4.0525 | 2.9358 | 0 | 1.38 | . |
| y4 | 0.3905 | 2.9358 | 0 | 0.13 | . |
| y5 | −11.9112 | 2.9523 | 0 | −4.03 | . |
| y6 | −2.6114 | 2.9523 | 0 | −0.88 | . |
| y7 | −0.7310 | 2.9226 | 0 | −0.25 | . |

Table 16.3 was directly copied from the SAS output. The estimated parameters are listed in Table 16.4. You can tell that the standard errors for the two estimated variance components are huge due to the extremely small sample size. Nobody would use 7 records to estimate genetic parameters, other than using the sample for demonstration purpose. The estimated heritability is

$$\widehat{h}^2 = \frac{\widehat{\sigma}_A^2}{\widehat{\sigma}_A^2 + \widehat{\sigma}^2} = \frac{36.315}{36.315 + 1.9231} = 0.9497$$

The `solution` option in the `random` statement allows the program to display the BLUP of the random effects (see Table 16.5). The `outp = yhat` option in the `model` displays the predicted phenotypic values $(\widehat{y} = X\widehat{\beta} + Z\widehat{\gamma})$, which are listed in Table 16.6. Even though the sample size is very small, we can still see that the prediction errors for the individuals without phenotype records are much larger than the ones with phenotype records (the fourth column of Table 16.6).

We used the data to estimate the parameters and conduct prediction within the same MIXED procedure. If the parameters are already known or estimated from other much larger dataset, the MIXED procedure allows users to provide known parameter values, escape the estimation step and directly go to the prediction step. Here is the SAS code to take user provided variance parameters.

```
proc mixed data=two method=reml noprofile;
    class child;
    model y = /solution outp=yhat;
    parms (36.3125) (1.9231) /noiter;
    random child /type=lin(1) ldata=one solution;
run;
```

**Table 16.6**  Predicted phenotypic values by PROC MIXED for the sample data

| Child | y | Pred | StdErrPred | DF | Alpha | Lower | Upper | Resid |
|-------|---|------|------------|----|-------|-------|-------|-------|
| x1 | . | 46.0223 | 5.7656 | 0 | 0.05 | . | . | . |
| x2 | . | 51.5934 | 5.05604 | 0 | 0.05 | . | . | . |
| x3 | . | 51.5934 | 5.05604 | 0 | 0.05 | . | . | . |
| x4 | . | 50.2489 | 4.77311 | 0 | 0.05 | . | . | . |
| x5 | . | 43.6361 | 5.06975 | 0 | 0.05 | . | . | . |
| x6 | . | 43.6203 | 4.99643 | 0 | 0.05 | . | . | . |
| x7 | . | 47.1995 | 5.33919 | 0 | 0.05 | . | . | . |
| y1 | 53.9464 | 53.7211 | 1.34419 | 0 | 0.05 | . | . | 0.2253 |
| y2 | 52.6928 | 52.5875 | 1.34419 | 0 | 0.05 | . | . | 0.10527 |
| y3 | 52.7475 | 52.5242 | 1.35382 | 0 | 0.05 | . | . | 0.22336 |
| y4 | 48.8271 | 48.8622 | 1.35382 | 0 | 0.05 | . | . | −0.0352 |
| y5 | 35.8121 | 36.5605 | 1.34379 | 0 | 0.05 | . | . | −0.7484 |
| y6 | 46.0967 | 45.8603 | 1.34379 | 0 | 0.05 | . | . | 0.23637 |
| y7 | 47.734 | 47.7407 | 1.35717 | 0 | 0.05 | . | . | −0.0067 |

```
proc mixed data=two method=reml noprofile;
    class child;
    model y = /solution outp=yhat;
    parms (36.3125) (1.9231) /hold=1,2;
    random child /type=lin(1) ldata=one solution;
run;
```

The `parms` statement gives users an opportunity to provide the parameter values and tells the program not to iterate (the first block of codes) or to hold the parameter values (the second block of codes). The predicted values are the same as those presented in Table 16.5 and Table 16.5 because the user provided parameters are the same as the estimated from the sample data. If we provide slightly different parameters, we can see the differences in the predicted random effects.

```
proc mixed data=two method=reml noprofile;
    class child;
    model y = /solution outp=yhat;
    parms (30) (5) /noiter;
    random child /type=lin(1) ldata=one solution;
run;
```

The predicted random effects with the user provided parameters (Table 16.7) are quite different from the predicted values from the parameters internally estimated from the sample data.

### 16.3.2  Example 2

This is the British bull pedigree with hypothetical phenotypic values. The tabular pedigree and the phenotypes are given in Table 16.8. In this example, we will fit a fixed effect (the sex effect). We will estimate the parameters and predict the phenotypic values in the same program. We already used PROC INBREED to calculate the numerator relationship matrix (Table 16.9). Using the two datasets (Tables 16.8 and 16.9) as input data. The SAS code to read the data and perform prediction are given as follows:

```
filename aa "data-16-3.csv";
filename phe "data-16-4.csv";
proc import datafile=aa out=aa dbms=csv replace;
proc import datafile=phe out=phe dbms=csv replace;
run;
proc mixed data=phe asycov;
```

**Table 16.7**  Predicted random effects using a different set of variance parameters

| Solution for random effects | | | | | | |
|---|---|---|---|---|---|---|
| Child | | Estimate | Std Err Pred | DF | t value | Pr > \|t\| |
| Child | x1 | −2.2821 | 5.2835 | 0 | −0.43 | . |
| Child | x2 | 2.9183 | 4.8347 | 0 | 0.60 | . |
| Child | x3 | 2.9183 | 4.8347 | 0 | 0.60 | . |
| Child | x4 | 1.6458 | 4.7643 | 0 | 0.35 | . |
| Child | x5 | −4.4957 | 4.8071 | 0 | −0.94 | . |
| Child | x6 | −4.5128 | 4.8745 | 0 | −0.93 | . |
| Child | x7 | −1.1924 | 5.0150 | 0 | −0.24 | . |
| Child | y1 | 4.8476 | 3.0770 | 0 | 1.58 | . |
| Child | y2 | 3.9074 | 3.0770 | 0 | 1.27 | . |
| Child | y3 | 3.6611 | 3.0638 | 0 | 1.19 | . |
| Child | y4 | 0.4534 | 3.0638 | 0 | 0.15 | . |
| Child | y5 | −10.6088 | 3.0864 | 0 | −3.44 | . |
| Child | y6 | −2.8953 | 3.0864 | 0 | −0.94 | . |
| Child | y7 | −0.6989 | 3.0405 | 0 | −0.23 | . |

**Table 16.8**  The British bull pedigree and simulated phenotypic values

| Name | Sire | Dam | Sex | y |
|---|---|---|---|---|
| 1 | 0 | 0 | 1 | 104.148 |
| 2 | 0 | 0 | 1 | 97.9 |
| 3 | 1 | 0 | 0 | 101.711 |
| 4 | 1 | 0 | 1 | 106.457 |
| 5 | 1 | 0 | 0 | 104.551 |
| 6 | 2 | 3 | 0 | 102.861 |
| 7 | 2 | 5 | 1 | 101.75 |
| 8 | 4 | 0 | 0 | 98.888 |
| 9 | 2 | 8 | 0 | 97.463 |
| 10 | 7 | 6 | 1 | 99.067 |
| 11 | 10 | 9 | 1 | 102.02 |

**Table 16.9**  the numerator relationship matrix for the 11 members of the British bull pedigree

| Parm | Row | Col1 | Col2 | Col3 | Col4 | Col5 | Col6 | Col7 | Col8 | Col9 | Col10 | Col11 |
|---|---|---|---|---|---|---|---|---|---|---|---|---|
| 1 | 1 | 1 | 0 | 0.5 | 0.5 | 0.5 | 0.25 | 0.25 | 0.25 | 0.125 | 0.25 | 0.1875 |
| 1 | 2 | 0 | 1 | 0 | 0 | 0 | 0.5 | 0.5 | 0 | 0.5 | 0.5 | 0.5 |
| 1 | 3 | 0.5 | 0 | 1 | 0.25 | 0.25 | 0.5 | 0.125 | 0.125 | 0.0625 | 0.3125 | 0.1875 |
| 1 | 4 | 0.5 | 0 | 0.25 | 1 | 0.25 | 0.125 | 0.125 | 0.5 | 0.25 | 0.125 | 0.1875 |
| 1 | 5 | 0.5 | 0 | 0.25 | 0.25 | 1 | 0.125 | 0.5 | 0.125 | 0.0625 | 0.3125 | 0.1875 |
| 1 | 6 | 0.25 | 0.5 | 0.5 | 0.125 | 0.125 | 1 | 0.3125 | 0.0625 | 0.28125 | 0.65625 | 0.46875 |
| 1 | 7 | 0.25 | 0.5 | 0.125 | 0.125 | 0.5 | 0.3125 | 1 | 0.0625 | 0.28125 | 0.65625 | 0.46875 |
| 1 | 8 | 0.25 | 0 | 0.125 | 0.5 | 0.125 | 0.0625 | 0.0625 | 1 | 0.5 | 0.0625 | 0.28125 |
| 1 | 9 | 0.125 | 0.5 | 0.0625 | 0.25 | 0.0625 | 0.28125 | 0.28125 | 0.5 | 1 | 0.28125 | 0.640626 |
| 1 | 10 | 0.25 | 0.5 | 0.3125 | 0.125 | 0.3125 | 0.65625 | 0.65625 | 0.0625 | 0.28125 | 1.15625 | 0.71875 |
| 1 | 11 | 0.1875 | 0.5 | 0.1875 | 0.1875 | 0.1875 | 0.46875 | 0.46875 | 0.28125 | 0.640626 | 0.71875 | 1.140626 |

```
   class name;
   model y = sex / solution outp=yhat;
   random name / type=lin(1) ldata=aa solution;
run;


proc print data=yhat;
run;
```

**Table 16.10** Estimated parameters for the British bull data by PROC MIXED of SAS

| Parameter | Estimate | Standard error |
|---|---|---|
| Intercept | 101.00 | 1.7191 |
| Sex | 1.0229 | 1.6556 |
| Genetic variance | 6.1283 | 7.009329 |
| Residual variance | 4.0021 | 4.079473 |

**Table 16.11** Predicted random effects for the 11 members of the British bull pedigree

| Solution for random effects | | | | | | |
|---|---|---|---|---|---|---|
| Name | | Estimate | Std Err Pred | DF | t Value | Pr > \|t\| |
| Name | 1 | 2.0124 | 1.9925 | 0 | 1.01 | . |
| Name | 2 | −2.5329 | 1.8689 | 0 | −1.36 | . |
| Name | 3 | 1.1397 | 1.9249 | 0 | 0.59 | . |
| Name | 4 | 2.3811 | 1.9728 | 0 | 1.21 | . |
| Name | 5 | 2.2508 | 1.8896 | 0 | 1.19 | . |
| Name | 6 | 0.1336 | 1.9422 | 0 | 0.07 | . |
| Name | 7 | −0.4771 | 2.0227 | 0 | −0.24 | . |
| Name | 8 | −0.7816 | 1.9102 | 0 | −0.41 | . |
| Name | 9 | −2.2541 | 1.9376 | 0 | −1.16 | . |
| Name | 10 | −1.1608 | 2.0926 | 0 | −0.55 | . |
| Name | 11 | −1.0010 | 2.0830 | 0 | −0.48 | . |

**Table 16.12** Predicted phenotypic effects for the 11 members of the British bull pedigree

| Obs | Sex | y | Pred | StdErrPred | DF | Alpha | Lower | Upper | Resid |
|---|---|---|---|---|---|---|---|---|---|
| 1 | 1 | 104.148 | 104.032 | 1.51677 | 0 | 0.05 | . | . | 0.11558 |
| 2 | 1 | 97.9 | 99.487 | 1.48631 | 0 | 0.05 | . | . | −1.58711 |
| 3 | 0 | 101.711 | 102.137 | 1.52354 | 0 | 0.05 | . | . | −0.42588 |
| 4 | 1 | 106.457 | 104.401 | 1.55006 | 0 | 0.05 | . | . | 2.05582 |
| 5 | 0 | 104.551 | 103.248 | 1.59363 | 0 | 0.05 | . | . | 1.30311 |
| 6 | 0 | 102.861 | 101.131 | 1.49468 | 0 | 0.05 | . | . | 1.73025 |
| 7 | 1 | 101.75 | 101.543 | 1.41431 | 0 | 0.05 | . | . | 0.20704 |
| 8 | 0 | 98.888 | 100.215 | 1.53200 | 0 | 0.05 | . | . | −1.32749 |
| 9 | 0 | 97.463 | 98.743 | 1.49535 | 0 | 0.05 | . | . | −1.28000 |
| 10 | 1 | 99.067 | 100.859 | 1.41297 | 0 | 0.05 | . | . | −1.79226 |
| 11 | 1 | 102.02 | 101.019 | 1.46496 | 0 | 0.05 | . | . | 1.00093 |

The estimated parameters are shown in Table 16.10. The predicted random effects and predicted phenotypic values are listed in Tables 16.11 and 16.12. The standard errors for the estimated variance components are very large due to the small sample size. The estimated heritability is

$$\widehat{h}^2 = \frac{\widehat{\sigma}_A^2}{\widehat{\sigma}_A^2 + \widehat{\sigma}^2} = \frac{6.1283}{6.1283 + 4.0021} = 0.6049416$$

Again, the sample size is very small and do not take this estimate seriously. The predicted phenotypic value is obtained from

$$\widehat{y}_j = \widehat{\beta}_1 + X_{j2}\widehat{\beta}_2 + \widehat{\gamma}_j$$

where $\widehat{\beta}_1 = 101.00$, $\widehat{\beta}_2 = 1.0229$, $X_{j2} = 0$ for female, and $X_{j2} = 1$ for male. For example,

$$\widehat{y}_2 = 101.00 + 1.0229 - 2.5329 = 99.49$$

The last column shows the estimated residual, which is

$$\widehat{\varepsilon}_2 = y_2 - \widehat{y}_2 = 97.9 - 99.49 = -1.5871$$

The predicted phenotypic value for the third individual is

$$\widehat{y}_3 = 101.00 + 0 + 1.1397 = 102.1397$$

The estimated residual is

$$\widehat{\varepsilon}_3 = y_3 - \widehat{y}_3 = 101.711 - 102.1397 = -0.4287$$

Since BLUP is identical to selection index and it is the optimal method for selection, there is no need to compare this method with any other sub-optimal methods. The selection response is the average predicted breeding value of all selected parents, assuming that both sexes are selected identically, i.e., with the same selection intensity. If male and female parents have different selection intensities, the response takes the average selection intensity of the two sexes.

# References

Henderson CR. Estimation of genetic parameters (abstract). Ann Math Stat. 1950;21:309–10.

Henderson CR. Estimation of variance and covariance components. Biometrics. 1953;9:226–52.

Henderson CR. Sire evaluation and genetic trends. In: Animal breeding and genetics symposium in honor of Dr Jay L Lush, 1973. pp. 10–41. American Society of Animal Science, American Dairy Science Association, Poultry Science Association, Champaign

Henderson CR, Kempthorne O, Searle SR, von Krosigk CM. The estimation of environmental and genetic trends from records subject to culling. Biometrics. 1959;15:192–218.

Robinson GK. That BLUP is a good thing: the estimation of random effects. Statistical Science. 1991;6:15–32.

SAS Institute Inc. SAS/STAT: users' guide, version 9.3. Cary, NC: SAS Institute Inc.; 2009.

The selection methods we learned so far are all focused on a single trait. The economic values of animals and plants, however, often depend on the quality of several traits. It is very rare for breeders to select only one trait and ignore all other traits. Selection must be considered on multiple traits simultaneously.

## 17.1 Common Methods of Multiple Trait Selection

### 17.1.1 Tandem Selection

Tandem selection is a method of selection where only one trait is selected at any given time. Selection takes place for each character singly in successive generations. For example, if two traits are of interest, we may select the first trait for a few generations until the trait value reaches a desired level and then select the second trait until it reaches a desired level for the second trait. This method is the simplest multiple trait selection method, but it is not optimal. If the two traits are genetically correlated, the first trait will be changed when the second trait is selected. Assume that your purpose of selection is to increase the values for both traits, but the genetic correlation is negative between the two traits. When one trait is increased, the other trait will be decreased. Tandem selection will never be effective in this situation.

### 17.1.2 Independent Culling Level Selection

We set up a minimum criterion (threshold) for each trait. A candidate can only be selected if all traits pass the minimum thresholds. The selection takes place for all the characters at the same time but in an independent manner. We reject all individuals that fail to come up to the standard for any character, regardless of their values for other characters. The advantage of independent culling level selection is that if traits are expressed in different stages, this method will allow breeders to select in several stages, referred to as multistage selection. Multistage selection is a cost-effective selection method in large animals and trees. For example, individuals can be selected in early stages in life if traits are expressed sequentially in time. Individuals failing to meet the selection criteria in early stages will be eliminated immediately without occupying the space, facilities, and so on.

### 17.1.3 Index Selection

Selection index for multiple traits was first invented by Smith and Hazel in the late 1930s and early 1940s of the last century (Smith 1936; Hazel and Lush 1942; Hazel 1943) and thus it is also called the Smith-Hazel selection index. The selection criterion is an index, which is a linear combination of the phenotypic values of multiple traits. The difference between independent culling level selection and index selection for two traits is illustrated in Fig. 17.1, where $X_1$ and $X_2$ are the phenotypic values of two traits and $I = 0.75X_1 + 0.25X_2$ is the selection index. The minimum culling levels are $X_1 > 23$ and $X_2 > 53$ for the independent culling level selection. The minimum threshold for the selection index is $I > 34$. The index

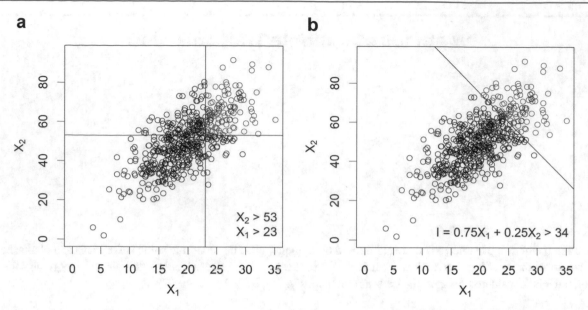

**Fig. 17.1** Comparison of index selection with independent culling level selection. The left panel (**a**) shows the independent culling level selection and the right panel (**b**) shows the index selection. Individuals tagged with red are selected for breeding

selection gives individuals with extremely high values of some traits an opportunity to be selected. This method is the most efficient method for multiple trait selection and will be the focus of this chapter.

### 17.1.4 Multistage Index Selection

At any given stage, the selection criterion is an index combining all traits available up to that stage. This is a combination of independent culling level selection and index selection. If traits are measured in multiple stages, this method is the most efficient method if the cost of trait measurement and animal maintenance is taken into account.

### 17.1.5 Other Types of Selection

Selection for extreme values is a method opposite to the independent culling level selection. Let $t_1$ and $t_2$ be the cutoff points for trait $X_1$ and $X_2$. The selection criterion is "$X_1 > t_1$ or $X_2 > t_2$" while the independent culling level selection uses a criterion of "$X_1 > t_1$ and $X_2 > t_2$." This method of selection is rarely used in practice.

Another selection method is called indirect selection. Let $Y$ be the primary trait of interest, but it is hard to measure and the heritability of $Y$ is low. Let $X$ be a trait of no economic value, but it is easy to measure and the trait has a high heritability. In addition, $X$ is highly correlated to $Y$. Instead of selecting $Y$ to improve $Y$ (direct selection), we may consider selecting $X$ to improve $Y$. Such a selection method is called indirect selection, while $Y$ is called the primary trait and $X$ is called the secondary traits. The selection response from the direct selection is

$$R_{\text{Direct}} = i_Y h_Y \sigma_{A(Y)}$$

The selection response from the indirect selection is

$$R_{\text{Indirect}} = i_X r_{A(XY)} h_X \sigma_{A(Y)}$$

The relative efficiency of indirect selection to direct selection is

$$\rho = \frac{R_{\text{Indirect}}}{R_{\text{Direct}}} = \frac{i_X r_{A(XY)} h_X \sigma_{A(Y)}}{i_Y h_Y \sigma_{A(Y)}} = \frac{i_X r_{A(XY)} h_X}{i_Y h_Y}$$

Indirect selection will be beneficial if $\rho > 1$. If the two selection methods have the same selection intensity, the efficiency becomes

$$\rho = \frac{r_{A(XY)} h_X}{h_Y}$$

Therefore, the secondary trait must be highly heritable and highly correlated to the primary trait, i.e., $r_{A(XY)} h_X > h_Y$, in order to consider indirect selection.

## 17.2 Index Selection

### 17.2.1 Variance and Covariance Matrices

Define $X = [X_1 \ X_2 \ \cdots \ X_m]^T$ as a vector of phenotypic values for $m$ traits measured from the same individual, $A = [A_1 \ A_2 \ \cdots \ A_m]^T$ as a vector of breeding values of the $m$ traits, and $R = [R_1 \ R_2 \ \cdots \ R_m]^T$ as a vector of the environmental error effects. The multivariate model is

$$X = A + R$$

Here, we assume that $X$ is already expressed as a deviation from the population mean, we have $\text{E}(X) = \text{E}(A) + \text{E}(R) = 0$. The variance covariance matrix of $X$ is

$$\text{var}(X) = \text{var}(A) + \text{var}(R) \tag{17.1}$$

Let $P = \text{var}(X)$, $G = \text{var}(A)$, and $E = \text{var}(R)$ be the corresponding variance-covariance matrices for the three components. Eq. (17.1) now can be rewritten as

$$P = G + E$$

where

$$P = \begin{bmatrix} \text{var}(X_1) & \text{cov}(X_1, X_2) & \cdots & \text{cov}(X_1, X_m) \\ \text{cov}(X_1, X_2) & \text{var}(X_2) & \cdots & \text{cov}(X_2, X_m) \\ \vdots & \vdots & \ddots & \vdots \\ \text{cov}(X_1, X_m) & \text{cov}(X_2, X_m) & \cdots & \text{var}(X_m) \end{bmatrix}$$

is the phenotypic variance-covariance matrix of the $m$ traits,

$$G = \begin{bmatrix} \text{var}(A_1) & \text{cov}(A_1, A_2) & \cdots & \text{cov}(A_1, A_m) \\ \text{cov}(A_1 A_2) & \text{var}(A_2) & \cdots & \text{cov}(A_2, A_m) \\ \vdots & \vdots & \ddots & \vdots \\ \text{cov}(A_1, A_m) & \text{cov}(A_2, A_m) & \cdots & \text{var}(A_m) \end{bmatrix}$$

is the genetic variance-covariance matrix for the $m$ characters, and

$$E = \begin{bmatrix} \text{var}(R_1) & \text{cov}(R_1, R_2) & \cdots & \text{cov}(R_1, R_m) \\ \text{cov}(R_1, R_2) & \text{var}(R_2) & \cdots & \text{cov}(R_2, R_m) \\ \cdots & \cdots & \cdots & \cdots \\ \text{cov}(R_1, R_m) & \text{cov}(R_2, R_m) & \cdots & \text{var}(R_m) \end{bmatrix}$$

is the environmental variance-covariance matrix.

Let $\sigma^2_{P(i)}$ and $\sigma^2_{A(i)}$ be the phenotypic and genetic variances for the $i$th trait, and $\sigma_{P(ij)}$ and $\sigma_{A(ij)}$ be the corresponding covariance between trait $i$ and trait $j$. Let $r_{P(ij)}$ and $r_{A(ij)}$ be the phenotypic and genetic correlation coefficients between traits $i$ and $j$. Denote the heritability of trait $i$ by $h_i^2$. One can obtain the $P$ and $G$ matrices using

$$P_{ii} = \sigma^2_{P(i)}$$
$$P_{ij} = r_{P(ij)}\sigma_{P(i)}\sigma_{P(j)}$$

and

$$G_{ii} = \sigma^2_{A(i)} = h_i^2 \sigma^2_{P(i)}$$
$$G_{ij} = r_{A(ij)}\sigma_{A(i)}\sigma_{A(j)} = r_{A(ij)}h_i h_j \sigma_{P(i)}\sigma_{P(j)}$$

The elements of matrix $E$ can be obtained similarly, as shown below

$$E_{ii} = \sigma^2_{E(i)} = \left(1 - h_i^2\right)\sigma^2_{P(i)}$$

and

$$E_{ij} = \sigma_{E(ij)} = r_{E(ij)}\sigma_{E(i)}\sigma_{E(j)} = \left(r_{P(ij)} - r_{A(ij)}h_i h_j\right)\sigma_{P(i)}\sigma_{P(j)}$$

Recall that

$$r_{P(ij)} = r_{A(ij)}h_i h_j + r_{E(ij)}e_i e_j = r_{A(ij)}h_i h_j + r_{E(ij)}\sqrt{\left(1 - h_i^2\right)\left(1 - h_j^2\right)}$$

Therefore,

$$r_{E(ij)} = \frac{r_{P(ij)} - r_{A(ij)}h_i h_j}{\sqrt{\left(1 - h_i^2\right)\left(1 - h_j^2\right)}}$$

In other words, the three variance-covariance matrices ($P$, $G$, and $E$) can be constructed based on heritability and phenotypic variance of each trait, and the genetic and phenotypic correlation between each pair of traits. For example, let $\sigma^2_{P(1)} = 10$, $\sigma^2_{P(2)} = 15$, and $r_{P(12)} = 0.5$. The phenotypic variance matrix, $P$, is

$$\begin{bmatrix} P_{11} & P_{12} \\ P_{21} & P_{22} \end{bmatrix} = \begin{bmatrix} \sigma^2_{P(1)} & r_{P(12)}\sqrt{\sigma^2_{P(1)}\sigma^2_{P(2)}} \\ r_{P(12)}\sqrt{\sigma^2_{P(1)}\sigma^2_{P(2)}} & \sigma^2_{P(2)} \end{bmatrix} = \begin{bmatrix} 10 & 6.123724 \\ 6.123724 & 15 \end{bmatrix}$$

Let $h_1^2 = 0.67$, $h_2^2 = 0.35$, and $r_{A(12)} = 0.45$. From the above information, we can get $\sigma^2_{A(1)} = h_1^2 \sigma^2_{P(1)} = 6.5$, $\sigma^2_{A(2)} = h_2^2 \sigma^2_{P(2)} = 5.25$, and $\sigma_{A(12)} = r_{A(12)}\sqrt{\sigma^2_{A(1)}\sigma^2_{A(2)}} = 2.7788$. The genetic variance matrix, $G$, is

$$\begin{bmatrix} G_{11} & G_{12} \\ G_{21} & G_{22} \end{bmatrix} = \begin{bmatrix} \sigma_{A(1)}^2 & \sigma_{A(12)} \\ \sigma_{A(21)} & \sigma_{A(2)}^2 \end{bmatrix} = \begin{bmatrix} 6.5 & 2.7788 \\ 2.7788 & 5.25 \end{bmatrix}$$

We can also define the covariance matrix between the phenotypic values and the genetic values using

$$\text{cov}(X, A) = \text{cov}(A + R, A) = \text{cov}(A, A) = \text{var}(A) = G \tag{17.2}$$

So, the covariance matrix between the phenotypic values and the genetic values is the same as the genetic variance matrix. Note that Eq. (17.2) holds only if the traits defined in $X$ are the same set of traits defined in $A$. If they are not the same, $G$ is called the genetic covariance matrix. The selection index to be introduced here assumes that the number of traits in $X$ and the number of traits in $A$ are the same. However, extension to the general case is very easy. In the extreme case, the traits in $X$ may not be the same set of traits in $A$. In general, let $n$ be the number of traits in $X$ and $m$ be the number of traits in $A$, the $G$ matrix is defined as an $n \times m$ genetic covariance matrix.

### 17.2.2 Selection Index (Smith-Hazel Index)

In index selection, the selection criterion is an index,

$$I = b_1 X_1 + b_2 X_2 + \cdots + b_m X_m = b^{\mathsf{T}} X$$

which is a linear combination of the phenotypic values. These $X$s are the phenotypic values of the traits measured from the same individual and are expressed as deviations from the population means. They are not different sources of information of the same trait from different relatives in the selection index described in Chap. 16.

The objective of index selection is a linear combination of breeding values for all characters from the same candidate, defined as

$$H = w_1 A_1 + w_2 A_2 + \cdots + w_m A_m = w^{\mathsf{T}} A$$

where

$$w = \begin{bmatrix} w_1 & w_2 & \cdots & w_m \end{bmatrix}^{\mathsf{T}}$$

are called economic weights, which are determined by breeders (not the job of quantitative geneticists). However, the weights are often expressed in units of the genetic standard deviations. For example, if we want to assign 50% weight on the first trait, 25% weight on the second trait, and another 25% weight on the third trait in a three-trait selection index, the actual economic weights are $w_1 = 0.5/\sigma_{A(1)}$, $w_2 = 0.25/\sigma_{A(2)}$, and $w_3 = 0.25/\sigma_{A(3)}$. Therefore,

$$H = w_1 A_1 + w_2 A_2 + w_3 A_3 = \frac{0.5}{\sigma_{A(1)}} A_1 + \frac{0.25}{\sigma_{A(2)}} A_2 + \frac{0.25}{\sigma_{A(3)}} A_3$$

Since the breeding values are also expressed as deviations from the means, the $H$ appears as the weighted sum of standardized breeding values. The selection objective, $H$, is also called the "aggregate breeding value". The index weights, $b$, are found by maximizing the correlation between $I$ and $H$, denoted by $r_{IH}$. The set of index weights that maximize this correlation is

$$b = [\text{var}(X)]^{-1} [\text{cov}(X, H)]$$

where $\text{var}(X) = P$ and $\text{cov}(X, H) = \text{cov}(X, w^{\mathsf{T}} A) = \text{cov}(X, A)w = \text{var}(A)w = Gw$.

Therefore, the index weights are

$$b = P^{-1}Gw$$

Derivation of $b$ is presented at the end of this section. Once we have the index weights, we can calculate the index score for each candidate using $I = b^{\mathrm{T}}X$. Individuals are ranked based on their index scores. The index scores are treated as observed "phenotypic values." Only the best individuals are selected for breeding. This selection method is called index selection.

### 17.2.3 Response to Index Selection

#### 17.2.3.1 Gain in Aggregate Breeding Value

The selection response of the aggregate breeding value is denoted by $\Delta H$, which can be calculated using the breeders' equation,

$$\Delta H = i_I r_{IH} \sigma_H$$

where

$$\sigma_H^2 = \mathrm{var}(H) = \mathrm{var}(w^T A) = w^T \mathrm{var}(A)w = w^T Gw$$

The variance of the selection index is

$$\sigma_I^2 = \mathrm{var}(I) = \mathrm{var}(b^T X) = b^T \mathrm{var}(X)b = b^T Pb$$

The covariance between $I$ and $H$ is

$$\mathrm{cov}(I, H) = \mathrm{cov}(b^T X, w^T A) = b^T \mathrm{cov}(X, A)w = b^T Gw$$

Recall that $b = P^{-1}Gw$, which allows us to further express the covariance between $I$ and $H$ as

$$\mathrm{cov}(I, H) = b^T Gw = b^T PP^{-1}Gw = b^T Pb$$

Therefore,

$$r_{IH} = \frac{\mathrm{cov}(I, H)}{\sigma_H \sigma_I} = \frac{b^T Pb}{\sigma_H \sigma_I} = \frac{\sigma_I^2}{\sigma_H \sigma_I} = \frac{\sigma_I}{\sigma_H}$$

Finally, the overall gain in the aggregate breeding value is

$$\Delta H = i_I \frac{\sigma_I}{\sigma_H} \sigma_H = i_I \sigma_I \qquad (17.3)$$

#### 17.2.3.2 Gains of Individual Traits

The aggregate breeding value consists of $m$ components,

$$\Delta H = w_1 \Delta A_1 + w_2 \Delta A_2 + \cdots + w_m \Delta A_m$$

where $\Delta A_i$ is the genetic change (response) of the $i$th character. Define

$$\Delta A = \begin{bmatrix} \Delta A_1 & \Delta A_2 & \cdots & \Delta A_m \end{bmatrix}^T$$

which is calculated as

$$\Delta A = b_{AI}(\bar{I}_s - \bar{I}) = b_{AI}\Delta I = \frac{\mathrm{cov}(A,I)}{\mathrm{var}(I)}\Delta I = \frac{G^T b}{\sigma_I}\frac{\Delta I}{\sigma_I} = \frac{i_I}{\sigma_I}G^T b$$

Note that $i_I = \Delta I/\sigma_I$ is the selection intensity (a scalar not a matrix) and

$$\mathrm{cov}(A,I) = \mathrm{cov}(A, b^T X) = \mathrm{cov}(A,X)b = G^T b$$

The detailed expression of $\Delta A$ is

$$\Delta A = \begin{bmatrix} \Delta A_1 \\ \Delta A_2 \\ \vdots \\ \Delta A_m \end{bmatrix} = \frac{i_I}{\sigma_I}\begin{bmatrix} G_{11} & G_{12} & \cdots & G_{1m} \\ G_{21} & G_{22} & \cdots & G_{2m} \\ \vdots & \vdots & \ddots & \vdots \\ G_{m1} & G_{m1} & \cdots & G_{mm} \end{bmatrix}\begin{bmatrix} b_1 \\ b_2 \\ \vdots \\ b_m \end{bmatrix}$$

Selection index can be treated as a "synthetic trait." As a trait, a selection index also has a heritability, which is defined as (Walsh and Lynch 2018)

$$h_I^2 = \frac{\sigma_{I(A)}^2}{\sigma_{I(P)}^2} = \frac{b^T G b}{b^T P b}$$

The reason for proposing a heritability for selection index is to generalize the breeders' equation to index selection,

$$\Delta H = h_I^2 \Delta I = i_I h_I^2 \sigma_I$$

However, this contradicts to the actual definition of the overall gain shown in Eq. (17.3). Therefore, I do not think $h_I^2$ is a useful parameter.

## 17.2.4 Derivation of the Smith-Hazel Selection Index

Given $\mathrm{cov}(I, H) = b^T G w$, $\mathrm{var}(I) = b^T P b$ and $\mathrm{var}(H) = w^T G w$. The squared correlation between $I$ and $H$ is

$$r_{IH}^2 = \frac{\mathrm{cov}^2(I,H)}{\mathrm{var}(I)\mathrm{var}(H)} = \frac{(b^T G w)^2}{(b^T P b)(w^T G w)}$$

The index weights are found by maximizing

$$Q(b) = \ln(r_{IH}^2) = 2\ln(b^T G w) - \ln(b^T P b) - \ln(w^T G w)$$

The partial derivative of $Q(b)$ with respect to $b$ is

$$\frac{\partial}{\partial b}Q(b) = \frac{2Gw}{(b^T G w)} - \frac{2Pb}{(b^T P b)} = 0$$

which is rewritten as

$$\frac{Gw}{(b^T G w)} - \frac{Pb}{(b^T P b)} = 0$$

Simplifying it leads to

$$Pb = \frac{(b^T Pb)}{(b^T Gw)} Gw$$

The solution for $b$ is

$$b = \frac{(b^T Pb)}{(b^T Gw)} P^{-1} Gw$$

Let the ratio equal to unity, i.e.,

$$(b^T Pb) = (b^T Gw)$$

The final form of the index weights is

$$b = P^{-1} Gw$$

Why do we set $(b^T Pb)/(b^T Gw) = 1$? Can we set $(b^T Pb)/(b^T Gw) = 2$ or $(b^T Pb)/(b^T Gw) = \rho$, where $\rho$ is any constant? Let us choose $\rho$ as the constant, we have

$$b^{(\rho)} = \rho P^{-1} Gw = \rho b$$

The index from $b^{(\rho)}$ will be

$$I^{(\rho)} = X^T b^{(\rho)} = \rho X^T b = \rho I$$

Although the index value has changed, the ranking based on the two indices remains the same. Therefore, the result of selection will be the same and choosing $\rho = 1$ is just a convenient way to simplify the construction of the selection index. For example, if $b = \begin{bmatrix} 1.5 & 0.8 \end{bmatrix}^T$ and $\rho = 2.0$, the two indices are

$$I = 1.5X_1 + 0.8X_2$$

and

$$I^{(\rho)} = 2 \times 1.5X_1 + 2 \times 0.8X_2 = 3X_1 + 1.6X_2$$

For two candidates labeled by $i$ and $j$, let $X_{i1} = 5.2$ and $X_{i1} = 2.5$ be the trait values for individual $i$, and $X_{j1} = 4.5$ and $X_{j1} = 3.1$ be the trait values for individual $j$. The indices for the two individuals are

$$I_i = 1.5X_{i1} + 0.8X_{i2} = 1.5 \times 5.2 + 0.8 \times 2.5 = 9.80$$
$$I_j = 1.5X_{j1} + 0.8X_{j2} = 1.5 \times 4.5 + 0.8 \times 3.1 = 9.23$$

and

$$I_i^{(\rho)} = 3.0X_{i1} + 1.6X_{i2} = 3.0 \times 5.2 + 1.6 \times 2.5 = 19.60$$
$$I_j^{(\rho)} = 3.0X_{j1} + 1.6X_{j2} = 3.0 \times 4.5 + 1.6 \times 3.1 = 18.46$$

The ranking for $I$ is $I_i > I_j$ and the ranking for $I^{(\rho)}$ remains $I_i^{(\rho)} > I_j^{(\rho)}$. As long as the ranking remains the same, the result of selection will be the same.

## 17.2.5 An Example of Index Selection

The example is adopted from Falconer and Mackay (Falconer and Mackay 1996). The purpose of the selection is to improve body weight ($A_1$) and tail length ($A_2$) using an index combining both the observed body weight ($X_1$) and the tail length ($X_2$). Let the economic weights for the two traits be $w = \begin{bmatrix} w_1 & w_2 \end{bmatrix}^T = \begin{bmatrix} 0.40 & -1.89 \end{bmatrix}^T$. The selection objective is $H = w_1 A_1 + w_2 A_2$ but the selection criterion is $I = b_1 X_1 + b_2 X_2$. Assume that the following phenotypic and genetic variance-covariance matrices are estimated from a large experiment,

$$P = \begin{bmatrix} 6.37 & 0.601 \\ 0.601 & 0.28 \end{bmatrix}$$

and

$$G = \begin{bmatrix} 2.2932 & 0.1557 \\ 0.1557 & 0.1232 \end{bmatrix}$$

The index weights are

$$b = P^{-1} Gw = \begin{bmatrix} 6.37 & 0.601 \\ 0.601 & 0.28 \end{bmatrix}^{-1} \begin{bmatrix} 2.2932 & 0.1557 \\ 0.1557 & 0.1232 \end{bmatrix} \begin{bmatrix} 0.40 \\ -1.89 \end{bmatrix} = \begin{bmatrix} 0.1947 \\ -1.027 \end{bmatrix}$$

Therefore,

$$\begin{bmatrix} b_1 \\ b_2 \end{bmatrix} = \begin{bmatrix} 0.1947 \\ -1.027 \end{bmatrix}$$

and the selection index is

$$I = b^T X = b_1 X_1 + b_2 X_2 = 0.1947 X_1 - 1.027 X_2$$

The variance of the index is

$$\sigma_I^2 = b^T Pb = \begin{bmatrix} 0.1947 & -1.027 \end{bmatrix} \begin{bmatrix} 6.37 & 0.601 \\ 0.601 & 0.28 \end{bmatrix} \begin{bmatrix} 0.1947 \\ -1.027 \end{bmatrix} = 0.2965$$

Assume that the proportion selected is $p = 0.3755$ so that the selection intensity is $i_I = 1.01$, the selection response for the aggregate breeding value is

$$\Delta H = i_I \sigma_I = 1.01 \times \sqrt{0.2965} = 0.5501$$

The selection responses of the two component traits in the index are

$$\Delta A = \frac{i_I}{\sigma_I} G^T b = \frac{1.01}{\sqrt{0.2965}} \times \begin{bmatrix} 2.2932 & 0.1557 \\ 0.1557 & 0.1232 \end{bmatrix} \begin{bmatrix} 0.1947 \\ -1.027 \end{bmatrix} = \begin{bmatrix} 0.5316 \\ -0.1785 \end{bmatrix}$$

This means that

$$\begin{bmatrix} \Delta A_1 \\ \Delta A_2 \end{bmatrix} = \begin{bmatrix} 0.5316 \\ -0.1785 \end{bmatrix}$$

One can verify that

$$\Delta H = w_1 \Delta A_1 + w_2 \Delta A_2 = 0.40 \times 0.5316 + (-1.89) \times (-0.1785) = 0.5501$$

Theoretically, we can show $\Delta H = w^T \Delta A$ based on the following derivation:

$$\Delta H = w^T \Delta A = \frac{i_I}{\sigma_I} w^T G^T b = \frac{i_I}{\sigma_I} w^T G^T \left( P^{-1} P \right) b = \frac{i_I}{\sigma_I} b^T P b = i_I \sigma_I$$

The genetic variance of the index is

$$\sigma_{I(A)}^2 = b^T G b = 0.154631$$

The heritability of the index is

$$h_I^2 = \frac{b^T G b}{b^T P b} = \frac{0.15463}{0.2964951} = 0.5215295$$

The PROC IML code to perform the above calculations is given below.

```
proc iml;
    q=0.3755; *proportion selected (p);
    t=probit(1-q);
    z=pdf('normal',t,0,1);
    i=z/q;
    P={6.37 0.601, 0.601 0.28};
    G={2.2932 0.1557, 0.1557 0.1232};
    w={0.4, -1.89};
    b=P**-1*G*w;
    sI=sqrt(b'*P*b);
    dH=i*si;
    dA=(i/si)*G*b;
    print (b||dA);
    print (si||dH||w'*dA);
    print(i*si);
    print(b'*P*b);
    print i;
    print(b'*G*b);
    print(b'*G*b/(b'*P*b));
quit;
```

## 17.3  Restricted Selection Index

Recall that the aggregate breeding value $H = w^T A$ is a weighted sum of the breeding values of individuals. If a trait, say trait $k$, has no economic value, we may set $w_k = 0$. If you do not want trait $k$ to change, setting its economic value to zero will not give you what you want. The trait will change due to correlations with other traits included in the index. In order to keep a trait constant, Kempthorne and Nordskog (Kempthorne and Nordskog 1959) proposed a restricted selection index. The weights of the index are found like the Smith-Hazel index so that the correlation between the index and the aggregated breeding value is maximized, but subject to some constraints for the restricted index. The constraints are defined as $\text{cov}(I, A_k) = G_k b = 0$ if we do not want to change trait $k$, where $G_k$ is the $k$th row of matrix $G$. You can define a matrix $C$ so that $CG = G_k$ and the restriction can be described as $CGb = 0$. For example, if there are three traits in both $H$ (aggregate breeding value) and $I$ (selection index) and we do not want the first two traits to change, the $C$ matrix should be

$$C = \begin{bmatrix} 1 & 0 & 0 \\ 0 & 1 & 0 \end{bmatrix}$$

You can see that

$$CGb = \begin{bmatrix} 1 & 0 & 0 \\ 0 & 1 & 0 \end{bmatrix} \begin{bmatrix} G_{11} & G_{12} & G_{13} \\ G_{21} & G_{22} & G_{23} \\ G_{31} & G_{32} & G_{33} \end{bmatrix} \begin{bmatrix} b_1 \\ b_2 \\ b_3 \end{bmatrix} = \begin{bmatrix} G_{11} & G_{12} & G_{13} \\ G_{21} & G_{22} & G_{23} \end{bmatrix} \begin{bmatrix} b_1 \\ b_2 \\ b_3 \end{bmatrix} = \begin{bmatrix} 0 \\ 0 \end{bmatrix}$$

is indeed identical to

$$\begin{bmatrix} \mathrm{cov}(A_1, I) \\ \mathrm{cov}(A_2, I) \end{bmatrix} = \begin{bmatrix} G_{11}b_1 + G_{12}b_2 + G_{13}b_3 \\ G_{21}b_1 + G_{22}b_2 + G_{23}b_3 \end{bmatrix} = \begin{bmatrix} 0 \\ 0 \end{bmatrix}$$

Recall that

$$r_{HI}^2 = \frac{\mathrm{cov}^2(I, H)}{\sigma_I^2 \sigma_H^2} = \frac{(b^T Gw)^2}{(w^T Gw)(b^T Pb)}$$

The objective function for maximization is the log of $r_{IH}^2$ plus the constraints, which is

$$Q(b) = 2\ln(b^T Gw) - \ln(w^T Gw) - \ln(b^T Pb) - 2b^T GC^T \lambda$$

where $\lambda = [\lambda_1 \quad \lambda_2]^T$ are Lagrange multipliers for constraints $CGb = 0$. The partial derivative of the objective function with respect to $b$ is

$$\frac{\partial Q(b)}{\partial b} = \frac{2Gw}{(w^T Gb)} - \frac{2Pb}{(b^T Pb)} - 2GC^T \lambda = 0$$

Rearranging this equation, we get

$$\frac{Pb}{(b^T Pb)} = \frac{Gw}{(w^T Gb)} - GC^T \lambda$$

Further manipulation of this equation yields

$$Pb = \frac{(b^T Pb)}{(w^T Gb)} Gw - (b^T Pb)GC^T \lambda$$

and thus

$$b = \frac{(b^T Pb)}{(w^T Gb)} P^{-1} Gw - (b^T Pb)P^{-1}GC^T \lambda \tag{17.4}$$

Multiplying both sides of Eq. (17.4) by $CG$, we have

$$CGb = \frac{(b^T Pb)}{(w^T Gb)} CGP^{-1} Gw - (b^T Pb)CGP^{-1}GC^T \lambda = 0$$

due to the fact that $CGb = 0$. Rearrangement of the above equation leads to

$$\frac{(b^T Pb)}{(w^T Gb)} CGP^{-1}Gw - (b^T Pb)CGP^{-1}GC^T\lambda = 0$$

Solving for the Lagrange multipliers, we get

$$\lambda = \left[(b^T Pb)CGP^{-1}GC^T\right]^{-1}\frac{(b^T Pb)}{(w^T Gb)}CGP^{-1}Gw \qquad (17.5)$$

Substituting Eq. (71.5) into Eq. (17.4), we obtain

$$\begin{aligned}
b &= \frac{(b^T Pb)}{(w^T Gb)}P^{-1}Gw - (b^T Pb)P^{-1}GC^T\left[(b^T Pb)CGP^{-1}GC^T\right]^{-1}\frac{(b^T Pb)}{(w^T Gb)}CGP^{-1}Gw \\
&= \frac{(b^T Pb)}{(w^T Gb)}\left[P^{-1}Gw - P^{-1}GC^T(CGP^{-1}GC^T)^{-1}CGP^{-1}Gw\right] \\
&= \frac{(b^T Pb)}{(w^T Gb)}\left[I - P^{-1}GC^T(CGP^{-1}GC^T)^{-1}CG\right]P^{-1}Gw
\end{aligned}$$

Setting

$$\frac{b^T Pb}{w^T Gb} = 1$$

results in the final form of the weights for the restricted index of Kempthorne and Nordskog (Kempthorne and Nordskog 1959),

$$b = \left[I - P^{-1}GC^T(CGP^{-1}GC^T)^{-1}CG\right]P^{-1}Gw \qquad (17.6)$$

In the classical Smith-Hazel index, $C = 0$, and the weights of the restricted index become $b = P^{-1}Gw$. So, the restricted selection index is a general one with the Smith-Hazel index being a special case. The derivation presented here is slightly different from Kempthorne and Nordskog (Kempthorne and Nordskog 1959) who explicitly put a restriction, $b^T Pb = 1$, with an extra Lagrange multiplier for this constraint.

The example in 2.5 is used to demonstrate the restricted index selection. Everything else stays the same as the usual index selection except that the constraint is $\text{cov}(I, A_2) = 0$, which is represented by $C = [\,0 \quad 1\,]$. The Smith-Hazel index weights are

$$b = P^{-1}Gw = \begin{bmatrix} 0.1947086 \\ -1.0271 \end{bmatrix}$$

The adjustment factor is

$$\left[I - P^{-1}GC^T(CGP^{-1}GC^T)^{-1}CG\right] = \begin{bmatrix} 1.0589511 & 0.046646 \\ -1.338301 & -0.058951 \end{bmatrix}$$

The restricted selection index weights are

$$b_r = \begin{bmatrix} 1.0589511 & 0.046646 \\ -1.338301 & -0.058951 \end{bmatrix}\begin{bmatrix} 0.1947086 \\ -1.0271 \end{bmatrix} = \begin{bmatrix} 0.1582768 \\ -0.20003 \end{bmatrix}$$

The gains of individual traits are

$$\begin{bmatrix} \Delta A_1 \\ \Delta A_2 \end{bmatrix} = \begin{bmatrix} 0.9201391 \\ 0 \end{bmatrix}$$

The overall gain in the aggregate breeding value is $\Delta H = 0.3680556$, which is smaller than the gain from the Smith-Hazel index selection ($\Delta H = 0.5501$). There is a price to pay for the restriction, only $0.3680556/0.5501 = 66.91\%$ of that under no restriction.

Tallis's restriction selection index (Tallis 1962) is a generalization of Kempthorne and Nodskog (Kempthorne and Nordskog 1959) restriction index where the constraints are defined as $CGb = d$ where $d$ is a vector with any values desired by the investigator. If $d = 0$, the selection is the usual restricted selection index. For some traits, the investigator may want the genetic progresses to be some fixed values (predetermined values). The derivation starts by multiplying both sides of Eq. (17.4) by $CG$ so that

$$CGb = \rho CGP^{-1}Gw - (b^T Pb)CGP^{-1}GC^T\lambda = d$$

where

$$\rho = \frac{b^T Pb}{w^T Gb}$$

We then solve for $\lambda$,

$$\lambda = [(b^T Pb)CGP^{-1}GC^T]^{-1}(\rho CGP^{-1}Gw - d)$$

Substituting the $\lambda$ back into Eq. (17.4), we get

$$\begin{aligned} b &= \rho P^{-1}Gw - (b^T Pb)P^{-1}GC^T[(b^T Pb)CGP^{-1}GC^T]^{-1}(\rho CGP^{-1}Gw - d) \\ &= \rho P^{-1}Gw - P^{-1}GC^T(CGP^{-1}GC^T)^{-1}(\rho CGP^{-1}Gw - d) \\ &= \rho \left[ I - P^{-1}GC^T(CGP^{-1}GC^T)^{-1}CG \right]P^{-1}Gw + P^{-1}GC^T(CGP^{-1}GC^T)^{-1}d \end{aligned}$$

Set $\rho = 1$ to give the final form of the weights,

$$b = \left[ I - P^{-1}GC^T(CGP^{-1}GC^T)^{-1}CG \right]P^{-1}Gw + P^{-1}GC^T(CGP^{-1}GC^T)^{-1}d$$

which is the same as that given by Walsh and Lynch (Walsh and Lynch 2018). This time, $\rho$ affects the weights of the first part equally, but the second part does not contain $\rho$. This restricted selection index may not give what you need.

## 17.4  Desired Gain Selection Index

Tallis' restricted selection index is a kind of desired gain selection index because vector $d$ contains gains desired by investigators. However, if desired gains are of primary concerns for traits, the aggregated breeding value involved in Tallis's index is redundant. Yamada et al. (Yamada et al. 1975) proposed a robust desired gain selection index that does not require the aggregate breeding value and thus does not require economic weights. The justification is that if people want to improve traits at some desired levels, the aggregated breeding value is irrelevant. Let $n$ be the number of traits in the index and $m$ be the number of traits to be improved with desired (predetermined) gains. Denote the phenotypic variance matrix of traits in the index by $\text{var}(X) = P_{n \times n}$ and the covariance between $X$ and the breeding values of the traits to be improved by $\text{cov}(X, A) = G_{n \times m}$. In the Smith-Hazel index selection, the genetic gains are $(i/\sigma_I)G^T b = \Delta A$. If we replace the Smith-Hazel gains by the desired gains $\Delta A = d$, then.

$$(i/\sigma_I)G^T b = d$$

Given $d$, we can solve $b$ retrospectively so that

$$b = (\sigma_I/i)(G^T)^{-1}d$$

The inverse of $G$ does not exist unless the traits in the index and the traits to be improved are the same set of traits. Yamada et al. (Yamada et al. 1975) and Itoh and Yamanda (Itoh and Yamanda 1986) replaced the non-existing inverse of $G^T$ by a generalized inverse to give the final form of the weights of the desired gain index

$$b = (\sigma_I/i)P^{-1}G(G^TP^{-1}G)^{-1}d \tag{17.7}$$

The constant $c = \sigma_I/i$ is proportional to all elements of $b$ and thus is irrelevant to the selection and thus can be set to 1 for the moment. Itoh and Yamanda (Itoh and Yamanda 1986) provided an elegant derivation of Eq. (17.7) by minimizing $\sigma_I^2 = b^TPb$ subject to $G^Tb = d$. The actual objective function to be minimized is

$$Q(b,\lambda) = \frac{1}{2}b^TPb + \lambda^T(G^Tb - d)$$

where $\lambda^T = [\lambda_1 \quad \lambda_2 \quad \cdots \quad \lambda_m]$ is a vector of Lagrange multipliers. The partial derivatives of $Q(b, \lambda)$ with respect to $b$ and $\lambda$ are

$$\frac{\partial}{\partial b}Q(b,\lambda) = Pb + G\lambda$$

$$\frac{\partial}{\partial \lambda}Q(b,\lambda) = G^Tb - d$$

Setting these partial derivatives to zero yields the following equation system,

$$\begin{bmatrix} P & G \\ G^T & 0 \end{bmatrix}\begin{bmatrix} b \\ \lambda \end{bmatrix} = \begin{bmatrix} 0 \\ d \end{bmatrix}$$

The solution for $b$ is

$$b = P^{-1}G(G^TP^{-1}G)^{-1}d \tag{17.8}$$

which is proportional to Eq. (17.7) by a constant $c = \sigma_I/i$ that is already set to 1.

The same example of Itoh and Yamanda (Itoh and Yamanda 1986) is used here for illustration. The index includes three traits but only the first two traits are to be improved with desired gains. The phenotypic variance matrix, the genetic covariance matrix, and the desired gains are.

$$P = \begin{bmatrix} 100 & 30 & 15 \\ 30 & 36 & 18 \\ 15 & 18 & 25 \end{bmatrix}, G = \begin{bmatrix} 25 & 9 \\ 9 & 12.96 \\ 6 & 8.64 \end{bmatrix} \text{ and } d = \begin{bmatrix} 1 \\ 2 \end{bmatrix}$$

The index weights calculated from Eq. (17.8) are

$$\begin{bmatrix} b_1 \\ b_2 \\ b_3 \end{bmatrix} = \begin{bmatrix} -0.020741 \\ 0.1330913 \\ 0.0534495 \end{bmatrix}$$

The standard deviation of the desired gain index is

$$\sigma_I = \sqrt{b^T P b} = 0.8996261$$

The expected gains are

$$\Delta A = (i/\sigma_I) G^T b = \begin{bmatrix} \Delta A_1 \\ \Delta A_2 \end{bmatrix} = (i/\sigma_I) \begin{bmatrix} 1 \\ 2 \end{bmatrix} = c^{-1} \begin{bmatrix} 1 \\ 2 \end{bmatrix}$$

The gain of the second trait is indeed twice the gain of the first trait. The gains of the two traits are proportional to the desired gains. The actual gains also depend on the selection intensity. For example, if the selection intensity is $i = 0.8996261$, then $c = \sigma_I/i = 1$ and the gains are just $\Delta A_1 = 1$ and $\Delta A_2 = 2$. If the selection intensity is $i = 1.4$, corresponding to $p = 0.20$, we have $c^{-1} = i/\sigma_I = 1.4/0.8996 = 1.556247$. The actual gains are

$$\Delta A = \begin{bmatrix} \Delta A_1 \\ \Delta A_2 \end{bmatrix} = 1.556247 \times \begin{bmatrix} 1 \\ 2 \end{bmatrix} = \begin{bmatrix} 1.556247 \\ 3.112494 \end{bmatrix}$$

The proportions of the gains for the two traits remain 1:2.

## 17.5 Multistage Index Selection

### 17.5.1 Concept of Multistage Index Selection

Multistage index selection was first proposed by Cunningham (Cunningham 1975). The idea was to take advantage of the cost saving of independent culling level selection and the high gain of index selection. Assume that there are $m$ traits and measured in $m$ different stages (one trait is measured at any given stage). In the first stage, only $X_1$ is measured and selection can be performed based on the phenotype of this trait. The selection criterion can be a rescaled $X_1$ denoted by $I_1 = b_{11}X_1$. Note that the individual ranking will not change whether you select $X_1$ or $b_{11}X_1$, where $b_{11}$ is a scalar. In the second stage, $X_2$ is available and selection in the second stage can be performed based an index combining both traits, $I_2 = b_{21}X_1 + b_{22}X_2$. In the $m$th stage (the last stage) where all traits are available, the selection criterion is an index combining all traits,

$$I_m = b_{m1}X_1 + b_{m2}X_2 + \cdots + b_{mm}X_m$$

This multistage selection scheme can be summarized as

$$I_1 = b_{11}X_1$$
$$I_2 = b_{21}X_1 + b_{22}X_2$$
$$\cdots\cdots\cdots\cdots\cdots\cdots\cdots\cdots\cdots$$
$$I_m = b_{m1}X_1 + b_{m2}X_2 + \cdots + b_{mm}X_m$$

The objective of the index selection remains the same as

$$H = w^T A = \sum_{k=1}^{m} w_k A_k$$

but the criterion of selection varies across different stages. This selection should be more efficient than the traditional independent culling level selection because each stage uses all information available up to that stage. The weights of selection index at each stage are calculated based on the method we learned before. Cunningham (Cunningham 1975) realized that selection in a later state only uses an already selected population. The selected population will have different phenotypic variance matrix and different genetic variance matrix from the base population (prior to selection) due to the Bulmer effect (Bulmer 1971). Therefore, an adjustment of these variance matrices should be conducted to take into account the effects of selection prior to the current stage.

## 17.5.2  Cunningham's Weights of Multistage Selection Indices

Cunningham (Cunningham 1975) multistage selection can deal with a general situation where several traits may be available and be selected simultaneously in any given stage. For example, among the $m$ traits, $m_1 \geq 1$ traits are available simultaneously in stage one and $m_2 \geq 1$ may be available in stage two. If $m_1 + m_2 = m$, we will only need to select individuals in two stages. The selection scheme appears to be

$$I_1 = b_{11}X_1 + \cdots + b_{1m_1}X_{m_1}$$
$$I_2 = b_{21}X_1 + \cdots + b_{2m_1}X_{m_1} + b_{2(m_1+1)}X_{m_1+1} + \cdots + b_{2m}X_m$$

We now derive the selection indices using a three-stage selection as an example. Let $X = \begin{bmatrix} X_1 & X_2 & X_3 & X_4 \end{bmatrix}^T$ be the phenotypic values of four traits. Assume that the first two traits are measured in the first stage and one additional trait is available in each of the remaining stages. The selection scheme can be described by

$$I_1 = b_{11}X_1 + b_{12}X_2$$
$$I_2 = b_{21}X_1 + b_{22}X_2 + b_{23}X_3$$
$$I_3 = b_{31}X_1 + b_{32}X_2 + b_{32}X_3 + b_{34}X_4$$

In matrix notation, let $\mathbf{b}_1 = \begin{bmatrix} b_{11} & b_{12} \end{bmatrix}^T$ and $\mathbf{X}_1 = \begin{bmatrix} X_1 & X_2 \end{bmatrix}^T$, which are written in bold face to reflect the first stage. Define $\mathbf{b}_2 = \begin{bmatrix} b_{21} & b_{22} & b_{23} \end{bmatrix}^T$ and $\mathbf{X}_2 = \begin{bmatrix} X_1 & X_2 & X_3 \end{bmatrix}^T$ for the second stage. Furthermore, define $\mathbf{b}_3 = \begin{bmatrix} b_{31} & b_{32} & b_{33} & b_{34} \end{bmatrix}^T$ and $\mathbf{X}_3 = \begin{bmatrix} X_1 & X_2 & X_3 & X_4 \end{bmatrix}^T$ for the weights and trait values in the last stage. The $k$th stage selection index is $I_k = \mathbf{b}_k^T \mathbf{X}_k$ for $k = 1, 2, 3$. Corresponding to the stages, we define the following matrices. For stage one, the matrices are.

$$\mathbf{P}_{11} = \begin{bmatrix} P_{11} & P_{12} \\ P_{21} & P_{22} \end{bmatrix} \text{ and } \mathbf{G}_1 = \begin{bmatrix} G_{11} & G_{12} & G_{13} & G_{14} \\ G_{21} & G_{22} & G_{23} & G_{24} \end{bmatrix}$$

In the second stage, the phenotypic and genetic variance matrices are

$$\mathbf{P}_{22} = \begin{bmatrix} P_{11} & P_{12} & P_{13} \\ P_{21} & P_{22} & P_{23} \\ P_{31} & P_{32} & P_{33} \end{bmatrix} \text{ and } \mathbf{G}_2 = \begin{bmatrix} G_{11} & G_{12} & G_{13} & G_{14} \\ G_{21} & G_{22} & G_{23} & G_{24} \\ G_{31} & G_{32} & G_{33} & G_{34} \end{bmatrix}$$

In the last stage, the corresponding matrices are

$$\mathbf{P}_{33} = \begin{bmatrix} P_{11} & P_{12} & P_{13} & P_{14} \\ P_{21} & P_{22} & P_{23} & P_{24} \\ P_{31} & P_{32} & P_{33} & P_{34} \\ P_{41} & P_{42} & P_{43} & P_{44} \end{bmatrix} \text{ and } \mathbf{G}_3 = \begin{bmatrix} G_{11} & G_{12} & G_{13} & G_{14} \\ G_{21} & G_{22} & G_{23} & G_{24} \\ G_{31} & G_{32} & G_{33} & G_{34} \\ G_{41} & G_{42} & G_{43} & G_{44} \end{bmatrix}$$

The selection index weights for the $k$th stage are

$$\mathbf{b}_k = \mathbf{P}_{kk}^{-1} \mathbf{G}_k \mathbf{w}$$

for $k = 1, 2, 3$, where $\mathbf{w} = \begin{bmatrix} w_1 & \cdots & w_m \end{bmatrix}^T$ is the economic weights. This set of index weights are derived by maximizing the correlation between $I_k$ and $H$ for all stages.

### 17.5.3 Xu-Muir's Weights of Multistage Selection Indices

Cunningham's multistage selection indices are correlated among different stages. Therefore, the changes of phenotypic and genetic variance matrices must be taken into account to validate the multistage index selection. Just taking into account the changes in variances is not enough because the distribution of the traits after selection will not be normal. Xu and Muir (Xu and Muir 1992) proposed an alternative method of multistage selection that shares the characteristics of the partial least squares method. In the first stage, the weights are derived by maximizing

$$r_{I_1 H} = \frac{\text{cov}(H, I_1)}{\sigma_H \sigma_{I_1}}$$

where $\sigma_{I_1}^2 = \mathbf{b}_1^T \mathbf{P}_{11} \mathbf{b}_1$. The optimal weights are

$$\mathbf{b}_1 = \mathbf{P}_{11}^{-1} \mathbf{G}_1 \mathbf{w}$$

As a result,

$$\text{cov}(I_1, H) = \mathbf{b}_1^T \mathbf{G}_1 \mathbf{w} = \mathbf{b}_1^T \mathbf{P}_{11} \mathbf{P}_{11}^{-1} \mathbf{G}_1 \mathbf{w} = \mathbf{b}_1^T \mathbf{P}_{11} \mathbf{b}_1 = \sigma_{I_1}^2$$

and thus

$$r_{I_1 H} = \frac{\text{cov}(H, I_1)}{\sigma_H \sigma_{I_1}} = \frac{\sigma_{I_1}}{\sigma_H}$$

The weights of the second-stage index are derived by maximizing $r_{I_2 H}$ subject to

$$\text{cov}(I_2, I_1) = \mathbf{b}_2^T \mathbf{P}_{21} \mathbf{b}_1 = \mathbf{0}$$

The solution for $\mathbf{b}_2$ will make sure that $I_2$ and $I_1$ are independent. As a result, selection in the first stage will not affect selection in the second stage. The weights in the second stage are

$$\mathbf{b}_2 = \left[ \mathbf{I} - \mathbf{P}_{22}^{-1} \mathbf{R}_{21} \left( \mathbf{R}_{12} \mathbf{P}_{22}^{-1} \mathbf{R}_{21} \right)^{-1} \mathbf{R}_{12} \right] \mathbf{P}_{22}^{-1} \mathbf{G}_2 \mathbf{w}$$

where $\mathbf{R}_{21} = \mathbf{P}_{21} \mathbf{b}_1$ and $\mathbf{R}_{12} = \mathbf{R}_{21}^T$. You can see that the weights are much like the weights in the restricted selection index given in Eq. (17.6). In the third stage, the weights are found by maximizing $r_{I_3 H}$ subject to

$$\text{cov}(I_3, I_2) = \mathbf{b}_3^T \mathbf{P}_{32} \mathbf{b}_2 = \mathbf{0}$$
$$\text{cov}(I_3, I_1) = \mathbf{b}_3^T \mathbf{P}_{31} \mathbf{b}_1 = \mathbf{0}$$

The solutions are

$$\mathbf{b}_3 = \left[ \mathbf{I} - \mathbf{P}_{33}^{-1} \mathbf{R}_{32} \left( \mathbf{R}_{23} \mathbf{P}_{33}^{-1} \mathbf{R}_{32} \right)^{-1} \mathbf{R}_{23} \right] \mathbf{P}_{33}^{-1} \mathbf{G}_3 \mathbf{w}$$

where $\mathbf{R}_{32} = \mathbf{P}_{32} \mathbf{B}_2$, $\mathbf{B}_2 = [\mathbf{b}_1 \| \mathbf{b}_2]$ and $\mathbf{R}_{23} = \mathbf{R}_{32}^T$. Note that $[\mathbf{b}_1 \| \mathbf{b}_2]$ represents horizontal concatenation of the two vectors. The lengths of the two vectors are different, but the shorter one should be extended to the same length as the longer one by filling the void spaces by zeroes. For example,

$$\mathbf{B}_2 = \left[\mathbf{b}_1 \middle\| \mathbf{b}_2\right] = \begin{bmatrix} b_{11} & b_{21} \\ b_{12} & b_{22} \\ 0 & b_{23} \end{bmatrix}$$

In general, the index weights for the $k$th stage is

$$\mathbf{b}_k = \left[\mathbf{I} - \mathbf{P}_{kk}^{-1}\mathbf{R}_{k(k-1)}\left(\mathbf{R}_{(k-1)k}\mathbf{P}_{kk}^{-1}\mathbf{R}_{k(k-1)}\right)^{-1}\mathbf{R}_{(k-1)k}\right]\mathbf{P}_{kk}^{-1}\mathbf{G}_k\mathbf{w}$$

where

$$\mathbf{R}_{k(k-1)} = \mathbf{P}_{k(k-1)}\mathbf{B}_{k-1}$$

The independent nature of the multistage indices provides an easy way to calculate the gain of each stage. For the $k$th stage selection, the response of the aggregated breeding values is

$$\Delta H_k = i_k\sigma_{I_k}$$

and the overall gain in all stages takes the sum of all stage-specific gains,

$$\Delta H = \sum_{k=1}^{s} i_k\sigma_{I_k}$$

where $s = 3$ is the total number of stages. The gains for the $m$ component traits are

$$\Delta A = \sum_{k=1}^{s} \frac{i_k}{\sigma_{I_k}}\mathbf{G}_k^{\mathrm{T}}\mathbf{b}_k$$

The nature of independence between indices of different stages allows the gains to be calculated for later stages without adjustment for the changes of variance and covariance due to selection in earlier stages. It also allows breeders to develop optimal proportions of selection for all stages by minimizing the total cost (Xu and Muir 1992).

### 17.5.4 An Example for Multistage Index Selection

Let $X = \begin{bmatrix} X_1 & X_2 & X_3 & X_4 \end{bmatrix}^{\mathrm{T}}$ be four quantitative traits and $A = \begin{bmatrix} A_1 & A_2 & A_3 & A_4 \end{bmatrix}^{\mathrm{T}}$ be the breeding values of the four traits. The phenotypic variance matrix is

$$P = \begin{bmatrix} 137.178 & -90.957 & 0.136 & 0.564 \\ -90.957 & 201.558 & 1.103 & -1.231 \\ 0.136 & 1.103 & 0.202 & 0.104 \\ 0.564 & -1.231 & 0.104 & 2.874 \end{bmatrix}$$

and the genetic variance matrix is

$$G = \begin{bmatrix} 14.634 & -18.356 & -0.109 & 1.233 \\ -18.356 & 32.029 & 0.103 & -2.574 \\ -0.109 & 0.103 & 0.089 & 0.023 \\ 1.233 & -2.574 & 0.023 & 1.225 \end{bmatrix}$$

Assume that the economic weights of the four traits are

$$\mathbf{w} = \begin{bmatrix} w_1 & w_2 & w_3 & w_4 \end{bmatrix}^{\mathrm{T}} = \begin{bmatrix} -3.555 & 19.536 & -113.746 & 48.307 \end{bmatrix}^{\mathrm{T}}$$

The multistage selection scheme is described as follows:

$$I_1 = b_{11}X_1 + b_{12}X_2$$
$$I_2 = b_{21}X_1 + b_{22}X_2 + b_{23}X_3$$
$$I_3 = b_{31}X_1 + b_{32}X_2 + b_{32}X_3 + b_{34}X_4$$

### 17.5.4.1 Cunningham's Weights of Multistage Selection Indices

Stage 1: The phenotypic variance and genetic covariance matrices are.

$$\mathbf{P}_{11} = \begin{bmatrix} 137.178 & -90.957 \\ -90.957 & 201.558 \end{bmatrix} \text{ and } \mathbf{G}_1 = \begin{bmatrix} 14.634 & -18.356 & -0.109 & 1.233 \\ -18.356 & 32.029 & 0.103 & -2.574 \end{bmatrix}$$

.

The weights of the index in the first stage are calculated as $\mathbf{b}_1 = \mathbf{P}_{11}^{-1}\mathbf{G}_1\mathbf{w}$, which are

$$\mathbf{b}_1 = \begin{bmatrix} b_{11} \\ b_{12} \end{bmatrix} = \begin{bmatrix} -0.9180003 \\ 2.3388678 \end{bmatrix}$$

Therefore, the index in the first stage is

$$I_1 = -0.9180003X_1 + 2.3388678X_2$$

Stage 2: The corresponding matrices in stage two are.

$$\mathbf{P}_{22} = \begin{bmatrix} 137.178 & -90.957 & 0.136 \\ -90.957 & 201.558 & 1.103 \\ 0.136 & 1.103 & 0.202 \end{bmatrix} \text{ and } \mathbf{G}_2 = \begin{bmatrix} 14.634 & -18.356 & -0.109 & 1.233 \\ -18.356 & 32.029 & 0.103 & -2.574 \\ -1.09 & 0.103 & 0.089 & 0.023 \end{bmatrix}$$

The index weights, $\mathbf{b}_2 = \mathbf{P}_{22}^{-1}\mathbf{G}_2\mathbf{w}$, are

$$\mathbf{b}_2 = \begin{bmatrix} b_{21} \\ b_{22} \\ b_{23} \end{bmatrix} = \begin{bmatrix} -0.6063094 \\ 2.7382556 \\ -47.2795438 \end{bmatrix}$$

So, the index of the second stage is

$$I_2 = -0.6063094X_1 + 2.7382556X_2 - 47.2795438X_3$$

Stage 3: The $P$ and $G$ matrices in the last stage are

$$\mathbf{P}_{33} = \begin{bmatrix} 137.178 & -90.957 & 0.136 & 0.564 \\ -90.957 & 201.558 & 1.103 & -1.231 \\ 0.136 & 1.103 & 0.202 & 0.104 \\ 0.564 & -1.231 & 0.104 & 2.874 \end{bmatrix}$$

and

$$\mathbf{G}_3 = \begin{bmatrix} 14.634 & -18.356 & -0.109 & 1.233 \\ -18.356 & 32.029 & 0.103 & -2.574 \\ -0.109 & 0.103 & 0.089 & 0.023 \\ 1.233 & -2.574 & 0.023 & 1.225 \end{bmatrix}$$

The weights of the index, $\mathbf{b}_3 = \mathbf{P}_{33}^{-1}\mathbf{G}_3\mathbf{w}$, are

$$\mathbf{b}_3 = \begin{bmatrix} b_{31} \\ b_{32} \\ b_{33} \\ b_{34} \end{bmatrix} = \begin{bmatrix} -0.5923588 \\ 2.7793332 \\ -49.4459478 \\ 3.7539200 \end{bmatrix}$$

Therefore, the index in the last stage is

$$I_3 = -0.5923588X_1 + 2.7793332X_2 - 49.4459478X_3 + 3.7539200X_4$$

The Cunningham's multistage selection scheme is summarized as

$$I_1 = -0.9180003X_1 + 2.3388678X_2$$
$$I_2 = -0.6063094X_1 + 2.7382556X_2 - 47.2795438X_3$$
$$I_3 = -0.5923588X_1 + 2.7793332X_2 - 49.4459478X_3 + 3.7539200X_4$$

### 17.5.4.2 Xu and Muir's Weights of Multistage Selection Indices

Stage 1: The results are identical to the Cunningham's selection index in the first stage. The variance matrices are.

$$\mathbf{P}_{11} = \begin{bmatrix} 137.178 & -90.957 \\ -90.957 & 201.558 \end{bmatrix} \text{ and } \mathbf{G}_1 = \begin{bmatrix} 14.634 & -18.356 & -0.109 & 1.233 \\ -18.356 & 32.029 & 0.103 & -2.574 \end{bmatrix}$$

The index weights, $\mathbf{b}_1 = \mathbf{P}_{11}^{-1}\mathbf{G}_1\mathbf{w}$, are

$$\mathbf{b}_1 = \begin{bmatrix} b_{11} \\ b_{12} \end{bmatrix} = \begin{bmatrix} -0.9180003 \\ 2.3388678 \end{bmatrix}$$

Stage 2: The variance and covariance matrices are.

$$\mathbf{P}_{22} = \begin{bmatrix} 137.178 & -90.957 & 0.136 \\ -90.957 & 201.558 & 1.103 \\ 0.136 & 1.103 & 0.202 \end{bmatrix} \text{ and } \mathbf{G}_2 = \begin{bmatrix} 14.634 & -18.356 & -0.109 & 1.233 \\ -18.356 & 32.029 & 0.103 & -2.574 \\ -0.109 & 0.103 & 0.089 & 0.023 \end{bmatrix}$$

The $\mathbf{R}$ matrix is

$$\mathbf{R}_{21} = \mathbf{P}_{21}\mathbf{b}_1 = \begin{bmatrix} 137.178 & -90.957 \\ -90.957 & 201.558 \\ 0.136 & 1.103 \end{bmatrix} \begin{bmatrix} -0.9180003 \\ 2.3388678 \end{bmatrix} = \begin{bmatrix} -338.665841 \\ 554.916068 \\ 2.45492300 \end{bmatrix}$$

The index weights, $\mathbf{b}_2 = \left[\mathbf{I} - \mathbf{P}_{22}^{-1}\mathbf{R}_{21}\left(\mathbf{R}_{12}\mathbf{P}_{22}^{-1}\mathbf{R}_{21}\right)^{-1}\mathbf{R}_{12}\right]\mathbf{P}_{22}^{-1}\mathbf{G}_2\mathbf{w}$, are

$$\mathbf{b}_2 = \begin{bmatrix} b_{21} \\ b_{22} \\ b_{23} \end{bmatrix} = \begin{bmatrix} 0.3116909 \\ 0.3993878 \\ -47.27954 \end{bmatrix}$$

Stage 3: The variance and covariance matrices are

$$\mathbf{P}_{33} = \begin{bmatrix} 137.178 & -90.957 & 0.136 & 0.564 \\ -90.957 & 201.558 & 1.103 & -1.231 \\ 0.136 & 1.103 & 0.202 & 0.104 \\ 0.564 & -1.231 & 0.104 & 2.874 \end{bmatrix}$$

and

$$\mathbf{G}_3 = \begin{bmatrix} 14.634 & -18.356 & -0.109 & 1.233 \\ -18.356 & 32.029 & 0.103 & -2.574 \\ -0.109 & 0.103 & 0.089 & 0.023 \\ 1.233 & -2.574 & 0.023 & 1.225 \end{bmatrix}$$

The $\mathbf{R}$ matrix is

$$\mathbf{R}_{32} = \mathbf{P}_{32}\mathbf{B}_2 = \begin{bmatrix} 137.178 & -90.957 & 0.136 \\ -90.957 & 201.558 & 1.103 \\ 0.136 & 1.103 & 0.202 \\ 0.564 & -1.231 & 0.104 \end{bmatrix} \begin{bmatrix} -0.9180003 & 0.3116909 \\ 2.3388678 & 0.3993878 \\ 0 & -47.27954 \end{bmatrix}$$

$$= \begin{bmatrix} -338.665841 & 0 \\ 554.916068 & 0 \\ 2.454923 & -9.067553 \\ -3.396898 & -5.232925 \end{bmatrix}$$

The index weights, $\mathbf{b}_3 = \left[ \mathbf{I} - \mathbf{P}_{33}^{-1}\mathbf{R}_{32}\left(\mathbf{R}_{23}\mathbf{P}_{33}^{-1}\mathbf{R}_{32}\right)^{-1}\mathbf{R}_{23} \right] \mathbf{P}_{33}^{-1}\mathbf{G}_3\mathbf{w}$, are

$$\mathbf{b}_3 = \begin{bmatrix} b_{31} \\ b_{32} \\ b_{33} \\ b_{34} \end{bmatrix} = \begin{bmatrix} 0.0139506 \\ 0.0410776 \\ -2.166404 \\ 3.75392 \end{bmatrix}$$

The multistage selection scheme is summarized as

$$I_1 = -0.9180003X_1 + 2.3388678X_2$$
$$I_2 = 0.3116909X_1 + 0.3993878X_2 - 47.27954X_3$$
$$I_3 = 0.0139506X_1 + 0.0410776X_2 - 2.166404X_3 + 3.75392X_4$$

Assume that the proportion selected in each stage is 0.5, i.e., the overall proportion selected is $0.5 \times 0.5 \times 0.5 = 0.125$, we can calculate the gains of the four traits in each stage. Define the following matrix of weights,

$$\mathbf{B} = \mathbf{b}_1 \| \mathbf{b}_2 \| \mathbf{b}_3 = \begin{bmatrix} -0.9180003 & 0.3116909 & 0.0139506 \\ 2.3388678 & 0.3993878 & 0.0410776 \\ 0 & -47.27954 & -2.166404 \\ 0 & 0 & 3.75392 \end{bmatrix}$$

The selection intensity corresponding to $p = 0.5$ is $i_1 = i_2 = i_3 = 0.7978846$. The standard deviations of the three indices are

$$\sigma_{I_1} = \sqrt{\mathbf{b}_1^T \mathbf{P}_{11} \mathbf{b}_1} = 40.109483$$

$$\sigma_{I_2} = \sqrt{\mathbf{b}_2^T \mathbf{P}_{22} \mathbf{b}_2} = 20.705308$$

$$\sigma_{I_3} = \sqrt{\mathbf{b}_3^T \mathbf{P}_{33} \mathbf{b}_3} = 6.284433$$

The ratios of the selection intensity to the standard deviation are

$$i_1/\sigma_{I_1} = 0.7978846/40.109483 = 0.0198927$$
$$i_2/\sigma_{I_2} = 0.7978846/20.705308 = 0.0385353$$
$$i_3/\sigma_{I_3} = 0.7978846/6.284433 = 0.1269621$$

Stage 1: $\Delta H_1 = i_1 \sigma_{I_1} = 32.002738$ and

$$\Delta A = \frac{i_1}{\sigma_{I_1}} \mathbf{G}_1^T \mathbf{b}_1 = \begin{bmatrix} -1.121275 \\ 1.825399 \\ 0.0067827 \\ -0.142275 \end{bmatrix}$$

Stage 2: $\Delta H_2 = i_2 \sigma_{I_2} = 16.520446$ and

$$\Delta A = \frac{i_2}{\sigma_{I_2}} \mathbf{G}_2^T \mathbf{b}_2 = \begin{bmatrix} 0.0918524 \\ 0.0848084 \\ -0.161876 \\ -0.06671 \end{bmatrix}$$

Stage 3: $\Delta H_3 = i_3 \sigma_{I_3} = 5.014252$ and

$$\Delta A = \frac{i_3}{\sigma_{I_3}} \mathbf{G}_3^T \mathbf{b}_3 = \begin{bmatrix} 0.5478227 \\ -1.120584 \\ -0.013174 \\ 0.5662752 \end{bmatrix}$$

The overall gain in aggregate breeding value after all three stages of selection is

$$\Delta H = \Delta H_1 + \Delta H_2 + \Delta H_3 = 32.002738 + 16.520446 + 5.014252 = 53.537435$$

Without the orthogonality of the indices between stages, we cannot calculate the gains as easily as we did. However, the price to pay for the orthogonality is the reduced gain compared with the Cunningham's multistage index selection.

## 17.6  Selection Gradients

In natural selection, the selection agents may not act on a single "index,"; rather, they may simultaneously act on all $m$ traits. Therefore, the "selection differential" consists of a vector of selection differentials for all the $m$ traits. Let us denote the vector of selection differentials by

$$\mathbf{S} = \begin{bmatrix} S_1 & S_2 & \cdots & S_m \end{bmatrix}^{\mathrm{T}}$$

Lande (Lande 1976) showed that $\mathbf{S}$ is a vector of the covariance between the fitness and the phenotypic values of the $m$ traits. Let $w$ be the fitness and $\mathbf{X} = \begin{bmatrix} X_1 & X_2 & \cdots & X_m \end{bmatrix}^{\mathrm{T}}$ be an $m \times 1$ vector for the phenotypic values, the vector of selection differentials is

$$\mathbf{S} = \mathrm{cov}(w, \mathbf{X}) = \begin{bmatrix} \mathrm{cov}(w, X_1) & \mathrm{cov}(w, X_2) & \cdots & \mathrm{cov}(w, X_m) \end{bmatrix}^{\mathrm{T}}$$

Let us denote the vector of selection responses for the $m$ traits by

$$\Delta \mathbf{A} = \begin{bmatrix} \Delta A_1 & \Delta A_2 & \cdots & \Delta A_m \end{bmatrix}^{\mathrm{T}}$$

or

$$\mathbf{R} = \begin{bmatrix} R_1 & R_2 & \cdots & R_m \end{bmatrix}^{\mathrm{T}}$$

to be consistent with the R notation for selection response. Lande (Lande 1976) also showed that

$$\mathbf{R} = \mathbf{G}\mathbf{P}^{-1}\mathbf{S}$$

This is actually the multivariate version of the breeder's equation. Imagine that for a single trait, $\mathbf{G} = V_A$ is the additive genetic variance and $\mathbf{P} = V_P$ is the phenotypic variance, then $\mathbf{G}\mathbf{P}^{-1} = V_A/V_P = h^2$ is the narrow sense heritability. Therefore, the multivariate breeders' equation becomes $R = h^2 S$, equivalent to the single variable breeder's equation. As a result, Lande (Lande 1976) called $\mathbf{G}\mathbf{P}^{-1}$ the "heritability matrix" for multiple traits. Lande also defined $\boldsymbol{\beta} = \mathbf{P}^{-1}\mathbf{S}$ as a vector of *selection gradients* to represent the strength of natural selection, which leads to another version of the multivariate breeders' equation,

$$\mathbf{R} = \mathbf{G}\boldsymbol{\beta}$$

This area of research is a hybrid between quantitative genetics and evolution and has been called the "phenotypic evolution."

## References

Bulmer MG. The effect of selection on genetic variability. Am Nat. 1971;105:201–11.
Cunningham EP. Multi-stage index selection. Theoret Appl Genet. 1975;46:55–61.
Falconer DS, Mackay TFC. Introduction to quantitative genetics. Harlow, Essex: Addison Wesley Longman; 1996.
Hazel LN. The genetic basis for constructing selection indexes. Genetics. 1943;28:476–90.
Hazel LN, Lush JL. The efficiency of three methods of selection. J Hered. 1942;33:393–9.
Itoh Y, Yamanda Y. Re-examination of selection index for desired gains. Genet Sel Evol. 1986;18:499–504.
Kempthorne O, Nordskog AW. Restricted selection indices. Biometrics. 1959;15:10–9.
Lande R. Natural selection and random genetic drift in phenotypic evolution. Evolution. 1976;30:314–34.
Smith FH. A discriminate function for plant selection. Ann Eugen. 1936;7:240–50.
Tallis GM. A selection index for optimal genotype. Biometrics. 1962;18:120–2.
Walsh B, Lynch M. Evolution and selection of quantitative traits. Oxford: Oxford University Press; 2018.
Xu S, Muir WM. Selection index updating. Theor Appl Genet. 1992;83:451–8.
Yamada Y, Yokouchi K, Nishida A. Selection index when genetic gains of individual traits are of primary concern. Japanese J Genet. 1975;50:33–41.

# Mapping Quantitative Trait Loci

The most commonly used population for quantitative trait locus (QTL) mapping is the $F_2$ population derived from the cross of two inbred lines. The classical method for QTL mapping is the interval mapping (IM) procedure (Lander and Botstein 1989). However, the most commonly used method is the composite interval mapping (CIM) method (Zeng 1993; Jansen and Stam 1994; Zeng 1994). This chapter will introduce the IM and CIM methods for QTL mapping in an $F_2$ population. Towards the end of the chapter, we will introduce a QTL mapping method using linear mixed models.

Interval mapping was originally developed by Lander and Botstein (1989) and further modified by numerous authors. Interval mapping has revolutionized genetic mapping because we can really pinpoint the exact location of a QTL. In this chapter, we will introduce several statistical methods of interval mapping based on an $F_2$ design. The method of interval mapping for a BC design is straightforward and thus will not be discussed in this chapter. The maximum likelihood (ML) method of interval mapping is the optimal method for interval mapping. The least squares (LS) method (Haley and Knott 1992) is a simplified approximation of the Lander and Botstein method. The iteratively reweighted least squares (IRWLS) method (Xu 1998b) is a further improved method over the least squares method. Feenstra et al. (Feenstra et al. 2006) developed an estimating equation (EE) method for QTL mapping, which is an extension of the IRWLS with improved performance. Han and Xu (2008) developed a Fisher scoring algorithm (FISHER) for QTL mapping. Both the EE and FISHER algorithms maximize the same likelihood function and thus they generate identical result. In this chapter, we introduce the methods based on their simplicities rather than their chronological orders of development. Therefore, the methods will be introduced in the following order: LS, IRWLS, FISHER, and ML.

## 18.1 Linkage Disequilibrium

QTL mapping is based on an important phenomenon of population genetics, linkage disequilibrium (LD), which describes the statistical relationship between two loci in the genome of an organism. Physical linkage is a primary cause for LD between two loci. Of course, two loci from different chromosomes may also show LD due to natural selection and other evolutionary forces. In major gene detection (Chap. 8), we assume that the QTL overlaps with a fully informative marker, i.e., the genotype of the QTL is observed. In QTL mapping, we do not know the genotypes of a QTL, but only know genotypes of markers. If the QTL is located nearby a marker, say marker A, the estimated marker (A) effect, $a_A$, is reduced to

$$a_A = (1 - 2r_{AQ})a_Q \tag{18.1}$$

where $a_Q$ is the effect of the QTL, $r_{AQ}$ is the recombination fraction between marker A and the QTL, and $(1 - 2r_{AQ}) = \rho_{Z_A Z_Q}$ is the correlation coefficient between the genotype indicators of the marker and the genotype indicator of the QTL (see Fig. 18.1 for the relationship of the QTL and the two flanking markers). The genotype indicators are defined as

**Fig. 18.1** Linkage relationship of a QTL with two flanking markers (A and B), where $r_{AQ}$ and $r_{QB}$ are the recombination fractions between the QTL and markers A and B, respectively, and $r_{AB}$ is the recombination fraction between markers A and B. The three rows (a, b, and c) represent three different putative positions of the QTL

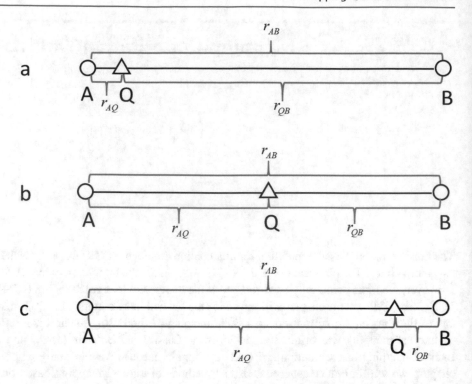

**Table 18.1** Joint distribution of the genotypic indicator variables between the QTL and marker A. The last row and last columns are the marginal probabilities

|  | $Q_1Q_1(1)$ | $Q_1Q_2(0)$ | $Q_2Q_2(-1)$ |  |
|---|---|---|---|---|
| $A_1A_1(1)$ | $\frac{1}{4}(1-r_{AQ})^2$ | $\frac{1}{2}(1-r_{AQ})r_{AQ}$ | $\frac{1}{4}r_{AQ}^2$ | 1/4 |
| $A_1A_2(0)$ | $\frac{1}{2}(1-r_{AQ})r_{AQ}$ | $\frac{1}{2}(1-r_{AQ})^2+\frac{1}{2}r_{AQ}^2$ | $\frac{1}{2}(1-r_{AQ})r_{AQ}$ | 1/2 |
| $A_2A_2(-1)$ | $\frac{1}{4}r_{AQ}^2$ | $\frac{1}{2}(1-r_{AQ})r_{AQ}$ | $\frac{1}{4}(1-r_{AQ})^2$ | 1/4 |
|  | 1/4 | 1/4 | 1/4 |  |

$$Z_A = \begin{cases} +1 & \text{for } A_1A_1 \text{ with probability } 1/4 \\ 0 & \text{for } A_1A_2 \text{ with probability } 1/2 \\ -1 & \text{for } A_2A_2 \text{ with probability } 1/4 \end{cases}$$

and

$$Z_Q = \begin{cases} +1 & \text{for } Q_1Q_1 \text{ with probability } 1/4 \\ 0 & \text{for } Q_1Q_2 \text{ with probability } 1/2 \\ -1 & \text{for } Q_2Q_2 \text{ with probability } 1/4 \end{cases}$$

Imagine that if the QTL is far away from the markers, $r_{AQ} \to 1/2$ and thus $1 - 2r_{AQ} \to 0$. As a result, $a_A \to 0$ and we cannot detect this QTL. Therefore, the success of a QTL mapping project depends on a sufficiently high density of marker map. Table 18.1 shows the joint distribution of the genotype indicators between marker A and the QTL, from which we can find the expectations, the variances, and the covariance between $Z_A$ and $Z_Q$.

From the marginal distribution, we calculate variances of $Z_A$ and $Z_Q$,

$$\text{var}(Z_A) = \text{E}\left(Z_A^2\right) - \text{E}^2(Z_A)$$
$$= \frac{1}{4} \times (1)^2 + \frac{1}{2} \times (0)^2 + \frac{1}{4} \times (-1)^2 - \left[\frac{1}{4} \times 1 + \frac{1}{2} \times 0 + \frac{1}{4} \times (-1)\right]^2$$
$$= \frac{1}{2}$$

and

$$\text{var}(Z_Q) = \text{E}\left(Z_Q^2\right) - \text{E}^2(Z_Q)$$
$$= \frac{1}{4} \times (1)^2 + \frac{1}{2} \times (0)^2 + \frac{1}{4} \times (-1)^2 - \left[\frac{1}{4} \times 1 + \frac{1}{2} \times 0 + \frac{1}{4} \times (-1)\right]^2$$
$$= \frac{1}{2}$$

From the joint distribution, we calculate the covariance between $Z_A$ and $Z_Q$,

$$\text{cov}(Z_A, Z_Q) = \text{E}(Z_A Z_Q) - \text{E}(Z_A)\text{E}(Z_Q)$$
$$= \frac{1}{4}(1 - r_{AQ})^2(1)(1) + \frac{1}{4}r_{AQ}^2(1)(-1) + \frac{1}{4}r_{AQ}^2(-1)(1) + \frac{1}{4}(1 - r_{AQ})^2(-1)(-1)$$
$$= \frac{1}{2}(1 - r_{AQ})^2 - \frac{1}{2}r_{AQ}^2$$
$$= \frac{1}{2}(1 - 2r_{AQ})$$

Therefore, the correlation between $Z_A$ and $Z_Q$ is

$$\rho_{Z_A Z_Q} = \frac{\text{cov}(Z_A, Z_Q)}{\sqrt{\text{var}(Z_A)\text{var}(Z_{AQ})}} = \frac{\frac{1}{2}(1 - 2r_{AQ})}{\sqrt{\frac{1}{2} \times \frac{1}{2}}} = 1 - 2r_{AQ}$$

This explains why Eq. (18.1) takes that form.

Genomic distance between two loci can be measured by the recombination fraction between the two loci. However, recombination fraction is not additive in a sense that $r_{AB} \neq r_{AQ} + r_{QB}$. An additive distance is the number of cross-overs between two loci. Let $d_{AB} = d_{AQ} + d_{QB}$ be the number of crossovers between loci A and B, where $d_{AQ}$ and $d_{QB}$ are the numbers of crossovers between A and Q and between Q and B, respectively. Distances measured by the number of crossovers are additive. Therefore, when the test statistic of a genome location is plotted against the location, the position of genomic location should be measured by the additive distance. The unit of additive distance is Morgan (M) or centiMorgan (cM). A Morgan is defined as the length of segment of the genome having, on average, one crossover per meiosis. One hundredth of a Morgan is a cM. The relationship between the distance measured in Morgan and the distance measured in recombination fraction is

$$r_{AB} = \frac{1}{2}[1 - \exp(-2d_{AB})] \tag{18.2}$$

The reverse relationship is

$$d_{AB} = -\frac{1}{2}\ln(1 - 2r_{AB}) \tag{18.3}$$

Equations (18.2) and (18.3) are the famous Haldane mapping function (Haldane 1919). For example, if $d_{AB} = 0.75$, the recombination coefficient is

$$r_{AB} = \frac{1}{2}[1 - \exp(-2 \times 0.75)] = 0.3884349$$

If $r_{AB} = 0.25$, then the corresponding additive distance is

$$d_{AB} = -\frac{1}{2}\ln(1 - 2 \times 0.25) = 0.3465736$$

## 18.2   Interval Mapping

Interval mapping is an extension of the individual marker analysis, i.e., major gene detection presented in Chap. 8, so that two markers are analyzed at a time. In major gene detection, we cannot estimate the exact position of a QTL. With interval mapping, we use two markers to define an interval, within which a putative QTL position is proposed (see Fig. 18.1 for the QTL interval bracketed by markers A and B). The genotype of the putative QTL is not observable but can be inferred with a certain probability using the three-point or multipoint method (Jiang and Zeng 1997). Once the genotype of the QTL is inferred, we can estimate and test the QTL effect at that particular position. We divide the interval into many putative positions of QTL with one or two cM apart and investigate every putative position within the interval. Once we have searched the current interval, we move on to the next interval and so on until all intervals have been searched. The putative QTL position (not necessarily at a marker) that has the maximum test score is the estimated QTL position, provided that the test statistic at that point passes a predetermined threshold.

We now show how to calculate the conditional probability of the QTL taking one of the three genotypes given the genotypes of the two flanking markers. The conditional probability is denoted by

$$\Pr(Q = k | A = i, B = j) = \frac{\Pr(A = i, B = j | Q = k)\Pr(Q = k)}{\sum_{k'=1}^{3}\Pr(A = i, B = j | Q = k')\Pr(Q = k')}$$
$$= \frac{\Pr(A = i | Q = k)\Pr(B = j | Q = k)\Pr(Q = k)}{\sum_{k'=1}^{3}\Pr(A = i | Q = k')\Pr(B = j | Q = k')\Pr(Q = k')} \tag{18.4}$$

The conditional probability eventually is expressed as a function of the marginal probability and two locus transition probabilities. The two locus transition probability matrix is given in Table 18.2. Each element of the table is a conditional probability of marker A taking one of the three genotypes given the genotype of the QTL. The transition probability matrix from the QTL to marker B is obtained from Table 18.2 by replacing A by B. Let us calculate

$$\Pr(Q = 1 | A = 2, B = 3) = \frac{\Pr(A = 2 | Q = 1)\Pr(B = 3 | Q = 1)\Pr(Q = 1)}{\sum_{k'=1}^{3}\Pr(A = 2 | Q = k')\Pr(B = 3 | Q = k')\Pr(Q = k')} \tag{18.5}$$

where $\Pr(Q = 1) = \Pr(Q = 3) = \Pr(Q = 2)/2 = 1/4$ are the marginal probabilities and the conditional probabilities are

**Table 18.2**  Transition probability matrix from QTL to marker A

|              | $A_1A_1$ (1)      | $A_1A_2$ (0)               | $A_2A_2$ (−1)      |
|--------------|-------------------|----------------------------|--------------------|
| $Q_1Q_1$ (1) | $(1 - r_{AQ})^2$  | $2(1 - r_{AQ})r_{AQ}$      | $r_{AQ}^2$         |
| $Q_1Q_2$ (0) | $(1 - r_{AQ})r_{AQ}$ | $(1 - r_{AQ})^2 + r_{AQ}^2$ | $(1 - r_{AQ})r_{AQ}$ |
| $Q_2Q_2$ (−1)| $r_{AQ}^2$        | $2(1 - r_{AQ})r_{AQ}$      | $(1 - r_{AQ})^2$   |

$$\Pr(A = 2|Q = 1) = 2r_{AQ}(1 - r_{AQ})$$
$$\Pr(A = 2|Q = 2) = r_{AQ}^2 + (1 - r_{AQ})^2$$
$$\Pr(A = 2|Q = 3) = 2r_{AQ}(1 - r_{AQ})$$

and

$$\Pr(B = 3|Q = 1) = r_{QB}^2$$
$$\Pr(B = 3|Q = 2) = r_{QB}(1 - r_{QB})$$
$$\Pr(B = 3|Q = 3) = (1 - r_{QB})^2$$

The numerator of Eq. (18.5) is

$$\Pr(A = 2|Q = 1)\Pr(B = 3|Q = 1)\Pr(Q = 1) = \frac{1}{2}r_{AQ}(1 - r_{AQ})r_{QB}^2$$

The denominator of Eq. (18.5) is the sum of the following three terms,

$$\Pr(A = 2|Q = 1)\Pr(B = 3|Q = 1)\Pr(Q = 1) = \frac{1}{2}r_{AQ}(1 - r_{AQ})r_{QB}^2$$

$$\Pr(A = 2|Q = 2)\Pr(B = 3|Q = 2)\Pr(Q = 2) = \frac{1}{2}\left[(1 - r_{AQ})^2 + r_{AQ}^2\right](1 - r_{QB})r_{QB}$$

$$\Pr(A = 2|Q = 3)\Pr(B = 3|Q = 3)\Pr(Q = 3) = \frac{1}{2}(1 - r_{AQ})r_{AQ}(1 - r_{QB})^2$$

Note that

$$r_{AB} = r_{AQ}(1 - r_{QB}) + r_{QB}(1 - r_{AQ})$$

and

$$1 - r_{AB} = r_{AQ}r_{QB} + (1 - r_{QB})(1 - r_{AQ})$$

After simplification, the sum of the three terms in the denominator is

$$\sum_{k'=1}^{3} \Pr(A = 2|Q = k')\Pr(B = 3|Q = k')\Pr(Q = k') = \frac{1}{2}r_{AB}(1 - r_{AB})$$

which is $\Pr(A = 2, B = 3)$, the joint distribution between markers A and B. Finally, the conditional probability is

$$\Pr(Q = 1|A = 2, B = 3) = \frac{r_{AQ}(1 - r_{AQ})r_{QB}^2}{r_{AB}(1 - r_{AB})} \tag{18.6}$$

The other two conditional probabilities of QTL given $A = 2$ and $B = 3$ are found similarly and they are

$$\Pr(Q = 2|A = 2, B = 3) = \frac{\left[(1 - r_{AQ})^2 + r_{AQ}^2\right](1 - r_{QB})r_{QB}}{r_{AB}(1 - r_{AB})} \tag{18.7}$$

and

**Fig. 18.2** The LOD test statistics for markers (**a**), virtual markers (**b**), and every virtual marker of a hypothetic chromosome

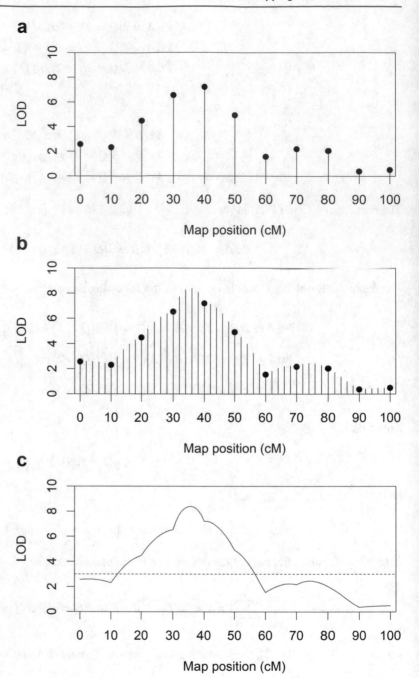

$$\Pr(Q = 3 | A = 2, B = 3) = \frac{(1 - r_{AQ}) r_{AQ} (1 - r_{QB})^2}{r_{AB}(1 - r_{AB})} \qquad (18.8)$$

Of course, these conditional probabilities are easily calculated via computer programs. Once the three conditional probabilities of QTL are calculated, they are used to estimate the effect and test the significance of a putative position between the two flanking markers. Figure 18.2 demonstrates the process of interval mapping, where the top panel (**a**) shows the test statistics of all markers, the panel in the middle (**b**) shows tests of all markers along with the inserted virtual markers and the panel at the bottom (**c**) shows the test statistic profile (a smooth curve connecting all the test statistics of the chromosome).

Two statistical methods are often used to perform interval mapping. One is the mixture model maximum likelihood method (Lander and Botstein 1989) and the other is the least squares method (Haley and Knott 1992). Both methods will be introduced in this chapter. Although interval mapping is rarely used after the composite interval mapping became available, it is the basis from which more advance technologies were developed. To maintain the continuity of the topics, we will present the interval mapping in detail before studying the composite interval mapping and other state-of-the-art methods of QTL mapping.

## 18.2.1 Least Squares Method (the Haley-Knott Method)

The LS method was introduced by Haley and Knott (1992) aiming to improving the computational speed. The statistical model for the phenotypic value of the $j$th individual is

$$y_j = X_j\beta + Z_j\gamma + \varepsilon_j \tag{18.9}$$

where $\beta$ is a $p \times 1$ vector for some model effects (nuisance parameters) that are irrelevant to QTL, $X_j$ is a $1 \times p$ known design matrix for the nuisance parameters, $\gamma = [a \ d]^T$ is a $2 \times 1$ vector for QTL effects of a putative locus ($a$ for the additive effect and $d$ for the dominance effect), $Z_j$ is a $1 \times 2$ vector for the genotype indicator variables defined as

$$Z_j = \begin{cases} H_1 & \text{for} & A_1A_1 \\ H_2 & \text{for} & A_1A_2 \\ H_3 & \text{for} & A_2A_2 \end{cases}$$

where $H_k$ for $k = 1, 2, 3$ is the $k$th row of matrix

$$H = \begin{bmatrix} +1 & 0 \\ 0 & 1 \\ -1 & 0 \end{bmatrix}$$

For the additive model, the $H$ matrix only has the first column. The residual error $\varepsilon_j$ is assumed to be distributed normally with zero mean and variance $\sigma^2$. Although normal distribution for $\varepsilon_j$ is not a required assumption for the LS method, it is required for the maximum likelihood method. Normality of the error term is also required for the statistical test in all parametric methods. It is important to include non-QTL effects $\beta$ in the model to reduce the residual error variance. For example, location and year effects are common in replicated experiments. These effects are not related to QTL but will contribute to the residual error if not included in the model. If there is no such a non-QTL effect to consider in a nice designed experiment, $\beta$ will be a single parameter (intercept) and $X_j$ will be just a unity across all $j = 1, \cdots, n$.

With interval mapping, the QTL genotype is never known unless the putative QTL position overlaps with a fully informative marker. Therefore, Haley and Knott (Haley and Knott 1992) suggested replacing the unknown $Z_j$ by the expectation of $Z_j$ conditional on flanking marker genotypes. Let $p_j(+1)$, $p_j(0)$, and $p_j(-1)$ be the conditional probabilities for the three genotypes of the QTL given genotypes of the flanking markers. Eq. (18.4) is a general formula to calculate the three conditional probabilities. The genotype index is $k = 1$ for $Q_1Q_1$, $k = 2$ for $Q_1Q_2$, and $k = 3$ for $Q_2Q_2$. The special notation for the conditional probability was adopted from Zeng (Zeng 1994). For the $k$th genotype, the conditional probability is denoted by $p_j(2 - k)$. In general, we have

$$p_j(2 - k) = \begin{cases} p_j(-1) & \text{for} \ k = 1 \\ p_j(0) & \text{for} \ k = 2 \\ p_j(+1) & \text{for} \ k = 3 \end{cases}$$

The LS model of Haley and Knott (Haley and Knott 1992) is

$$y_j = X_j\beta + U_j\gamma + e_j$$

where

$$U_j = \mathrm{E}(Z_j) = p_j(+1)H_1 + p_j(0)H_2 + p_j(-1)H_3$$

is the conditional expectation of $Z_j$. The residual error $e_j$ (different from $\varepsilon_j$) remains normal with mean zero and variance $\sigma^2$ although this assumption has been violated. The least squares estimates of $\beta$ and $\gamma$ are.

$$\begin{bmatrix} \widehat{\beta} \\ \widehat{\gamma} \end{bmatrix} = \begin{bmatrix} \sum_{j=1}^{n} X_j^T X_j & \sum_{j=1}^{n} X_j^T U_j \\ \sum_{j=1}^{n} U_j^T X_j & \sum_{j=1}^{n} U_j^T U_j \end{bmatrix}^{-1} \begin{bmatrix} \sum_{j=1}^{n} X_j^T y_j \\ \sum_{j=1}^{n} U_j^T y_j \end{bmatrix}$$

and the estimated residual error variance is

$$\widehat{\sigma}^2 = \frac{1}{n-p-2} \sum_{j=1}^{n} \left( y_j - X_j\widehat{\beta} - U_j\widehat{\gamma} \right)^2$$

The variance-covariance matrix of the estimated parameters is

$$\mathrm{var}\begin{bmatrix} \widehat{\beta} \\ \widehat{\gamma} \end{bmatrix} = \begin{bmatrix} \sum_{j=1}^{n} X_j^T X_j & \sum_{j=1}^{n} X_j^T U_j \\ \sum_{j=1}^{n} U_j^T X_j & \sum_{j=1}^{n} U_j^T U_j \end{bmatrix}^{-1} \widehat{\sigma}^2 = \begin{bmatrix} C_{XX} & C_{XU} \\ C_{UX} & C_{UU} \end{bmatrix} \widehat{\sigma}^2 \tag{18.10}$$

which is a $(p+2) \times (p+2)$ matrix. The variance matrix of the estimated QTL effects is

$$\mathrm{var}(\widehat{\gamma}) = C_{UU}\widehat{\sigma}^2 = \begin{bmatrix} \mathrm{var}(\widehat{a}) & \mathrm{cov}\left(\widehat{a},\widehat{d}\right) \\ \mathrm{cov}\left(\widehat{a},\widehat{d}\right) & \mathrm{var}\left(\widehat{d}\right) \end{bmatrix} \tag{18.11}$$

which is the $2 \times 2$ lower right diagonal bock of matrix (18.10). Note that $C_{UU}$ corresponds to the lower right block of the inverse matrix, not the inverse of $\sum_{j=1}^{n} U_j^T U_j$. The standard errors of the estimated additive and dominance effects are the square roots of the diagonal elements of matrix (18.11).

We can use the F-test to test the hypothesis of $H_0 : \gamma = 0$. The F-test statistic is

$$F = \frac{1}{2}\widehat{\gamma}^T (C_{UU}^{-1}/\widehat{\sigma}^2)\widehat{\gamma} = \frac{1}{2}\begin{bmatrix} \widehat{a} & \widehat{d} \end{bmatrix} \begin{bmatrix} \mathrm{var}(\widehat{a}) & \mathrm{cov}\left(\widehat{a},\widehat{d}\right) \\ \mathrm{cov}\left(\widehat{a},\widehat{d}\right) & \mathrm{var}\left(\widehat{d}\right) \end{bmatrix}^{-1} \begin{bmatrix} \widehat{a} \\ \widehat{d} \end{bmatrix}$$

Under the null model, $F$ follows an F distribution with $df_1 = 2$ and $df_2 = n - p - 2$.

The likelihood ratio test statistic can also be applied if we assume that $e_j \sim N(0, \sigma^2)$ for all $j = 1, \cdots, n$. The log likelihood function for the full model is

$$L_1 = -\frac{n}{2}\ln\left(\widehat{\sigma}^2\right) - \frac{1}{2\widehat{\sigma}^2}\sum_{j=1}^{n}\left(y - X_j\widehat{\beta} - U_j\widehat{\gamma}\right)^2 \approx -\frac{n}{2}\left[\ln\left(\widehat{\sigma}^2\right) + 1\right]$$

The log likelihood function under the reduced model $H_0 : \gamma = 0$ is.

$$L_0 = -\frac{n}{2} \ln \left( \tilde{\sigma}^2 \right) - \frac{1}{2\tilde{\sigma}^2} \sum_{j=1}^{n} \left( y - X_j \tilde{\beta} \right)^2 \approx -\frac{n}{2} \left[ \ln \left( \tilde{\sigma}^2 \right) + 1 \right]$$

where

$$\tilde{\beta} = \left[ \sum_{j=1}^{n} X_j^T X_j \right]^{-1} \left[ \sum_{j=1}^{n} X_j^T y_j \right]$$

and

$$\tilde{\sigma}^2 = \frac{1}{n-p} \sum_{j=1}^{n} \left( y_j - X_j \tilde{\beta} \right)^2$$

The likelihood ratio test statistic is

$$\lambda = -2(L_0 - L_1)$$

Under the null hypothesis, $\lambda$ follows a Chi-square distribution with two degrees of freedom. For the additive model (excluding the dominance effect), the likelihood ratio test follows a Chi-square distribution with one degree of freedom.

## 18.2.2 Iteratively Reweighted Least Squares (IRWLS)

The LS method is flawed because the residual variance is heterogeneous after replacing $Z_j$ by its conditional expectation $U_j$. The conditional variance of $Z_j$ given marker information varies from one individual to another and it will contribute to the residual variance. Xu (1995, 1998b, 1998a) modified the exact linear model (18.9) by

$$y_j = X_j \beta + U_j \gamma + (Z_j - U_j)\gamma + \varepsilon_j$$

which differs from the Haley and Knott's (Haley and Knott 1992) model by a term $(Z_j - U_j)\gamma$. Since $Z_j$ is not observable, this additional term is merged into the residual error if it is ignored. Let

$$e_j = (Z_j - U_j)\gamma + \varepsilon_j$$

be the new residual error. The Haley and Knott's model can be rewritten as

$$y_j = X_j \beta + U_j \gamma + e_j$$

Although we assume $\varepsilon_j \sim N(0, \sigma^2)$, this does not validate the normal and i.i.d. assumption of $e_j$. The expectation for $e_j$ is

$$E(e_j) = [E(Z_j) - U_j]\gamma + E(\varepsilon_j) = 0$$

The variance of $e_j$ is

$$\text{var}(e_j) = \sigma_j^2 = \gamma^T \text{var}(Z_j)\gamma + \sigma^2 = \left( \frac{1}{\sigma^2} \gamma^T \Sigma_j \gamma + 1 \right) \sigma^2$$

where $\Sigma_j = \text{var}(Z_j)$, which is defined as a conditional variance-covariance matrix given flanking marker information. The explicit form of $\Sigma_j$ is

$$\Sigma_j = \mathrm{E}\left(Z_j^T Z_j\right) - \mathrm{E}\left(Z_j^T\right)\mathrm{E}(Z_j)$$

where

$$
\begin{aligned}
\mathrm{E}\left(Z_j^T Z_j\right) &= p_j(+1)H_1^T H_1 + p_j(0)H_2^T H_2 + p_j(-1)H_3^T H_3 \\
&= \begin{bmatrix} p_j(+1) + p_j(-1) & 0 \\ 0 & p_j(0) \end{bmatrix}
\end{aligned}
$$

is a $2 \times 2$ matrix and

$$
\begin{aligned}
\mathrm{E}(Z_j) &= p_j(+1)H_1 + p_j(0)H_2 + p_j(-1)H_3 \\
&= \begin{bmatrix} p_j(+1) - p_j(-1) & p_j(0) \end{bmatrix}
\end{aligned}
$$

is a $1 \times 2$ vector. Therefore,

$$
\Sigma_j = \begin{bmatrix} \left[p_j(+1) + p_j(-1)\right] - \left[p_j(+1) - p_j(-1)\right]^2 & -p_j(0)\left[p_j(+1) - p_j(-1)\right] \\ -p_j(0)\left[p_j(+1) - p_j(-1)\right] & p_j(0)\left[1 - p_j(0)\right] \end{bmatrix}
$$

Let

$$
\sigma_j^2 = \left(\frac{1}{\sigma^2}\gamma^T \Sigma_j \gamma + 1\right)\sigma^2 = \frac{1}{W_j}\sigma^2
$$

where

$$
W_j = \left(\frac{1}{\sigma^2}\gamma^T \Sigma_j \gamma + 1\right)^{-1}
$$

is the weight variable for the $j$th individual. The weighted least squares estimates of the parameters are

$$
\begin{bmatrix} \widehat{\beta} \\ \widehat{\gamma} \end{bmatrix} = \begin{bmatrix} \sum_{j=1}^{n} X_j^T W_j X_j & \sum_{j=1}^{n} X_j^T W_j U_j \\ \sum_{j=1}^{n} U_j^T W_j X_j & \sum_{j=1}^{n} U_j^T W_j U_j \end{bmatrix}^{-1} \begin{bmatrix} \sum_{j=1}^{n} X_j^T W_j y_j \\ \sum_{j=1}^{n} U_j^T W_j y_j \end{bmatrix} \tag{18.12}
$$

and

$$
\widehat{\sigma}^2 = \frac{1}{n-p-2}\sum_{j=1}^{n} W_j\left(y_j - X_j\widehat{\beta} - U_j\widehat{\gamma}\right)^2 \tag{18.13}
$$

Since $W_j$ is a function of $\gamma$ and $\sigma^2$, iterations are required. The iteration process is demonstrated below,

1. Initialize $\gamma$ and $\sigma^2$.
2. Update $\beta$ and $\gamma$ using Eq. (18.12).
3. Update $\sigma^2$ using Eq. (18.13).
4. Repeat Step 2 to Step 3 until a certain criterion of convergence is satisfied.

The iteration process is very fast, usually taking less than 5 iterations to converge. Since the weight is not a constant (it is a function of the parameters), repeatedly updating the weight is required. Therefore, the weighted least squares method is also called iteratively reweighted least squares (IRWLS). The few cycles of iterations make the results of IRWLS very close to that of the maximum likelihood method (to be introduce later). A nice property of IRWLS is that the variance-covariance matrix of the estimated parameters is automatically given as a by-product of the iteration process. This matrix is

$$
\operatorname{var}\begin{bmatrix} \widehat{\beta} \\ \widehat{\gamma} \end{bmatrix} = \begin{bmatrix} \sum_{j=1}^{n} X_j^T W_j X_j & \sum_{j=1}^{n} X_j^T W_j U_j \\ \sum_{j=1}^{n} U_j^T W_j X_j & \sum_{j=1}^{n} U_j^T W_j U_j \end{bmatrix}^{-1} \widehat{\sigma}^2
$$

As a result, the F-test statistic can be used for significance test. Like the least squares method, a likelihood ratio test statistic can also be established for significance test. The $L_0$ under the null model is the same as that described in the section of least squares method. The $L_1$ under the alternative model is

$$
L_1 = -\frac{n}{2} \ln\left(\widehat{\sigma}^2\right) + \frac{1}{2} \sum_{j=1}^{n} \ln\left(W_j\right) - \frac{1}{2\widehat{\sigma}^2} \sum_{j=1}^{n} W_j \left(y - X_j\widehat{\beta} - U_j\widehat{\gamma}\right)^2
$$

$$
\approx -\frac{n}{2} \left[\ln\left(\widehat{\sigma}^2\right) + 1\right] + \frac{1}{2} \sum_{j=1}^{n} \ln\left(W_j\right)
$$

### 18.2.3 Maximum Likelihood Method

The maximum likelihood method (Lander and Botstein 1989) is the optimal one compared to all other interval mapping procedures described so far. The linear model for the phenotypic value of $y_j$ is

$$
y_j = X_j\beta + Z_j\gamma + \varepsilon_j
$$

where $\varepsilon_j \sim N(0, \sigma^2)$ is assumed. The genotype indicator variable $Z_j$ is a missing value because we cannot observe the genotype of a putative QTL. Rather than replacing $Z_j$ by $U_j$ as done in the least squares and the weighted least squares methods, the maximum likelihood method takes into consideration the mixture distribution of $y_j$. When the genotype of the putative QTL is observed, the probability density of $y_j$ is.

$$
f_k(y_j) = \Pr(y_j | Z_j = H_k) = \frac{1}{\sqrt{2\pi}\sigma} \exp\left[-\frac{1}{2\sigma^2}\left(y_j - X_j\beta - H_k\gamma\right)^2\right]
$$

When flanking marker information is used, the conditional probability that $Z_j = H_k$ is

$$
p_j(2 - k) = \Pr(Z_j = H_k), \forall k = 1, 2, 3
$$

for the three genotypes, $Q_1Q_1$, $Q_1Q_2$, and $Q_2Q_2$. These probabilities are different from the Mendelian segregation ratio (0.25, 0.5, 0.25). They are the conditional probabilities given marker information and thus vary from one individual to another because different individuals may have different marker genotypes. Using the conditional probabilities as the weights, we get the mixture distribution

$$
f(y_j) = \sum_{k=1}^{3} p_j(2 - k) f_k(y_j)
$$

where

$$p_j(2-k) = \begin{cases} p_j(+1) & \text{for } k=1 \\ p_j(0) & \text{for } k=2 \\ p_j(-1) & \text{for } k=3 \end{cases}$$

is a special notation for the conditional probability and should not be interpreted as $p_j \times (2-k)$. The log likelihood function is

$$L(\theta) = \sum_{j=1}^{n} \ln f(y_j) = \sum_{j=1}^{n} \ln \sum_{k=1}^{3} p_j(2-k) f_k(y_j)$$

The parameters are held in a parameter vector $\theta = \{\beta, \gamma, \sigma^2\}$. The maximum likelihood estimates of the parameters are obtained from any convenient optimization algorithms. However, for the mixture model with known mixing proportions, $p_j(2-k)$, for every observation, the EM algorithm (Dempster et al. 1977) is the most convenient method. Assume that the genotypes of all individuals are observed, the maximum likelihood estimates of parameters would be

$$\begin{bmatrix} \beta \\ \gamma \end{bmatrix} = \begin{bmatrix} \sum_{j=1}^{n} X_j^T X_j & \sum_{j=1}^{n} X_j^T Z_j \\ \sum_{j=1}^{n} Z_j^T X_j & \sum_{j=1}^{n} Z_j^T Z_j \end{bmatrix}^{-1} \begin{bmatrix} \sum_{j=1}^{n} X_j^T y_j \\ \sum_{j=1}^{n} Z_j^T y_j \end{bmatrix}$$

and

$$\sigma^2 = \frac{1}{n} \sum_{j=1}^{n} \left(y_j - X_j\beta - Z_j\gamma\right)^2$$

The EM algorithm takes advantage of the above explicit solutions of the parameters by substituting all entities containing the missing value $Z_j$ by their posterior expectations,

$$\begin{bmatrix} \beta \\ \gamma \end{bmatrix} = \begin{bmatrix} \sum_{j=1}^{n} X_j^T X_j & \sum_{j=1}^{n} X_j^T \mathrm{E}(Z_j) \\ \sum_{j=1}^{n} \mathrm{E}\left(Z_j^T\right) X_j & \sum_{j=1}^{n} \mathrm{E}\left(Z_j^T Z_j\right) \end{bmatrix}^{-1} \begin{bmatrix} \sum_{j=1}^{n} X_j^T y_j \\ \sum_{j=1}^{n} \mathrm{E}\left(Z_j^T\right) y_j \end{bmatrix} \tag{18.14}$$

and

$$\sigma^2 = \frac{1}{n} \sum_{j=1}^{n} \mathrm{E}\left[\left(y_j - X_j\beta - Z_j\gamma\right)^2\right] \tag{18.15}$$

For the residual error variance, the denominator is $n$, not $n-3$, as we normally use. Also, the posterior expectation is taken with respect to the whole square term, not $\left(y_j - X_j\beta - \mathrm{E}(Z_j)\gamma\right)^2$. The posterior expectations are taken using the posterior probabilities of QTL genotypes, which are defined as

$$p_j^*(2-k) = \frac{p_j(2-k) f_k(y_j)}{\sum_{k'=1}^{3} p_j(2-k') f_{k'}(y_j)}, \forall k = 1, 2, 3 \tag{18.16}$$

The posterior expectations are

$$E(Z_j) = \sum_{k=1}^{3} p_j^*(2 - k)H_k$$

$$E\left(Z_j^T Z_j\right) = \sum_{k=1}^{3} p_j^*(2 - k)H_k^T H_k \tag{18.17}$$

$$E\left[\left(y_j - X_j\beta - Z_j\gamma\right)^2\right] = \sum_{k=1}^{3} p_j^*(2 - k)\left(y_j - X_j\beta - H_k\gamma\right)^2$$

Since $f_k(y_j)$ is a function of parameters and thus $p_j^*(2 - k)$ is also a function of the parameters. However, the parameters are unknown and they are the very quantities we want to find out. Therefore, iterations are required. Here is the iteration process,

Step (0): Initialize $\theta = \theta^{(t)}$.
Step (1): Calculate the posterior expectations using Eqs. (18.16) and (18.17).
Step (2): Update parameters using Eqs. (18.14) and (18.15).
Step (3): Increment $t$ by 1 and repeat Step (1) to Step (2) until a certain criterion of convergence is satisfied.

There is no simple way to calculate the variance-covariance matrix for the estimated parameters under the EM algorithm, although advanced methods are available, such as the Louis' (Louis 1982) method. Therefore, we cannot perform the F test easily. The likelihood ratio test may be the only choice for hypothesis test with the EM algorithm. The log likelihood function used to construct the likelihood ratio test statistic is $L(\theta)$. The likelihood ratio test statistic is

$$\lambda = -2(L_0 - L_1)$$

where $L_1 = L\left(\widehat{\theta}\right)$ is the observed log likelihood function evaluated at $\widehat{\theta} = \left\{\widehat{\beta}, \widehat{\gamma}, \widehat{\sigma}^2\right\}$ and $L_0 = L\left(\widetilde{\theta}\right)$ is the log likelihood function evaluated at $\widetilde{\theta} = \left\{\widetilde{\beta}, 0, \widetilde{\sigma}^2\right\}$ under the restricted model. The estimated parameter $\widetilde{\theta}$ under the restricted model and $L_0$ are the same as those given in the section of the least squares method.

## 18.3  Composite Interval Mapping

A basic assumption of the interval mapping is that a chromosome carries at most one QTL. If there are more than one QTL, the estimated effect of a QTL will be biased due to LD with other QTLs. In other words, QTL on the same chromosome cannot be separated effectively with the interval mapping. In addition, QTLs on other chromosomes will be absorbed by the residual error. Quantitative traits, by definition, are controlled by multiple QTL. Theoretically, a multiple regression model should be used for mapping multiple QTLs (Zeng 1993). Since the number and positions of QTLs are not known, the model dimension is also not known. A model selection approach, e.g., step-wise regression, may be adopted. Model selection, however, has its own issues such as inconsistency and unstable result. Zeng (1993) studied the theoretical basis of multiple regression in terms of QTL mapping. Based on the theory, Zeng (1994) proposed a composite interval mapping technique. Let $n$ be the sample size and $m$ be the total number of markers. The composite interval mapping model is

$$y_j = b_0 + x_j^* b^* + \sum_{k \neq i, i+1}^{m} x_{jk} b_k + e_j$$

where $y_j$ is the phenotypic value of the trait of interest measured from the $j$th individual for $j = 1, \cdots, n$, $b_0$ is the intercept, $b^*$ is the effect of a candidate QTL in an interval bracketed by markers $i$ and $i + 1$, $x_j^*$ is the genotype indicator variable of the candidate QTL (missing value), $b_k$ is the effect of marker $k$, $x_{jk}$ is the genotype indicator variable of marker $k$ for $k = 1, \cdots, m$, and $e_j$ is the residual with an assumed $N(0, \sigma^2)$ distribution. Note that $x_{jk}$ is called a co-factor or covariate. These co-factors do not include the two flanking markers that defines the putative interval. If the marker density is very high, including all markers in the list of co-factors will cause multiple collinearity and model overfitting. It is recommended to select representative markers evenly across the genome to serve as co-factors. Zeng (Zeng 1994) pointed out three properties of his new method. (1) Effect of the putative QTL (partial regression coefficient) is only affected by QTLs within the interval bracketed by markers $i$ and $i + 1$, not affected by QTLs in other intervals. This statement may not be exactly true. QTLs between markers $i - 1$ and $i$ will be absorbed by the candidate QTL, so will QTLs between markers $i + 1$ and $i + 2$, because markers $i$ and $i + 1$ are not included in the list of co-factors; rather, they are used to calculate the distribution of the candidate QTL genotype. (2) Effects of QTLs on other chromosomes will be captured by the co-factors, leading to a reduced error variance and thus an increased statistical power. (3) Conditional on linked markers, the multiple regression analysis will reduce the interference of hypothesis testing from multiple linked QTLs, but with a possible increase of sampling variance. The last property may be interpreted by the fact that the conditional variance of $x_j^*$ may be much smaller than the unconditional variance. Since the standard error of the estimated $b^*$ is inversely proportional to the standard deviation of $x_j^*$, a smaller variance of $x_j^*$ will lead to a larger standard error of the estimated $b^*$. Anyway, if co-factors are selected with care, the CIM method will increase the precision of QTL mapping and possibly increase the statistical power.

Zeng (Zeng 1994) used a mixture model maximum likelihood method to estimate parameters and a likelihood ratio test to test the null hypothesis. He treated $x_j^*$ as a missing value in order to find the maximum likelihood solution of parameters via an expectation-conditional-maximization (ECM) algorithm. Let $B = \{b_k\}$ be the array of regression coefficients for all co-factors, $X = \{x_{jk}\}$ be the genotype indicator variables for all markers in the list of co-factors and $Y = \{y_j\}$ be the array of phenotypic values. The array of missing genotypes of the putative QTL is denoted by $X^* = \left\{x_j^*\right\}$. Starting from some initial values of the parameters, the ECM updates the parameters by

$$b^* = \frac{(Y - XB)^T \mathrm{E}(X^*)}{\mathrm{E}(X^{*T}X^*)} \tag{18.18}$$

$$B = (X^TX)^{-1} X^T [Y - \mathrm{E}(X^*)b^*] \tag{18.19}$$

$$\sigma^2 = \frac{1}{n} \left[ (Y - XB)^T (Y - XB) - \mathrm{E}(X^{*T}X^*)b^{*2} \right] \tag{18.20}$$

where $\mathrm{E}(X^*) = \left\{ \mathrm{E}\left(x_j^*\right) \right\}$ are the posterior expectations for the missing genotypes and $\mathrm{E}(X^{*T}X^*) = \left\{ \mathrm{E}\left(x_j^{*2}\right) \right\}$ are the posterior expectations of the squared missing genotypes. For an $F_2$ population under the additive model,

$$\mathrm{E}\left(x_j^*\right) = p_j^*(+1) - p_j^*(-1) \tag{18.21}$$

and

$$\mathrm{E}\left(x_j^{*2}\right) = p_j^*(+1) + p_j^*(-1) \tag{18.22}$$

where $p_j^*(+1)$, $p_j^*(0)$, and $p_j^*(-1)$ are the posterior probabilities of the three genotypes of the candidate QTL. The posterior probabilities are presented in Eq. (18.16). The posterior expectations in Eqs. (18.21) and (18.22) represent the expectation step (E-step). Updating the parameters in Eqs. (18.18), (18.19), and (18.20) represents the conditional maximization step (CM-step). The E-step and CM-step are repeated until the parameters converge to a satisfactory criterion. It is easy to understand Eqs. (18.18) and (18.19) for updating the regression coefficients, but may be difficult to understand Eq. (18.20) for updating the residual variance. Below is the derivation for it. Let $R = Y - XB$ and $b^* = R^T X^* (X^{*T}X^*)^{-1}$ when the genotypes of the candidate QTL are observed. The residual variance is

$$\sigma^2 = \frac{1}{n}(Y - XB - X^*b^*)^T(Y - XB - X^*b^*)$$

$$= \frac{1}{n}(R - X^*b^*)^T(R - X^*b^*)$$

$$= \frac{1}{n}\left(R^T R - 2R^T X^* b^* + b^* X^{*T} X^* b^*\right)$$

$$= \frac{1}{n}\left(R^T R - 2R^T X^* \left(X^{*T} X^*\right)^{-1} X^{*T} X^* b^* + b^* X^{*T} X^* b^*\right) \quad (18.23)$$

$$= \frac{1}{n}\left(R^T R - 2b^* X^{*T} X^* b^* + b^* X^{*T} X^* b^*\right)$$

$$= \frac{1}{n}\left(R^T R - b^* X^{*T} X^* b^*\right)$$

Taking the expectation of Eq. (18.23) with respect to the missing values, we have

$$\sigma^2 = \frac{1}{n}\left[R^T R - b^* \mathrm{E}\left(X^{*T} X^*\right) b^*\right]$$

$$= \frac{1}{n}\left[(Y - XB)^T(Y - XB) - \mathrm{E}\left(X^{*T} X^*\right) b^{*2}\right]$$

which is identical to Eq. (18.20). Many software packages are available for composite interval mapping, e.g., QTL Cartographer (Wang et al. 2012). Some CIM programs replace the ML method by the LS method but still keep the cofactor selection process to alleviate interference of hypothesis testing and control the polygenic background (Broman et al. 2003).

## 18.4 Control of Polygenic Background

Composite interval mapping also has its own concerns. The major concern is its sensitiveness to the co-factor selection. For some data, selecting different markers as co-factors may result in different results. Some modifications have been done to improve the stability of CIM, e.g., by leaving one chromosome (the one with the interval under investigation) out when markers are selected as co-factors. Motivated by the mixed model genome-wide association studies (Yu et al. 2006; Kang et al. 2008; Zhang et al. 2010; Lippert et al. 2011, 2012; Zhou and Stephens 2012), Xu (2013) proposed to use a marker inferred kinship matrix to capture the polygenic background in QTL mapping. The polygenic effect plays a role similar to the selected co-factors in controlling the genetic background. With such a polygenic background control, the result is often consistent regardless of the way that the kinship matrix is calculated. Users can even incorporate multiple kinship matrices (e.g., dominance and epistatic effect kinship matrices) into the mixed model (Xu 2013). We now describe the mixed model of QTL mapping. Let $y$ be an $n \times 1$ vector of the phenotypic values from an $F_2$ population. The mixed model is

$$y = X\beta + Z_k\gamma_k + \xi + \varepsilon$$

where $X$ is an $n \times p$ design matrix for some systematic environmental effects, e.g., year and location effects, $\beta$ is a $p \times 1$ vector of the fixed effects, $Z_k$ is an $n \times 1$ vector holding the genotype indicator variable for the $k$th candidate QTL, $\gamma_k$ is the QTL (additive) effect, $\xi$ is an $n \times 1$ vector of polygenic effects (sum of all QTL of the entire genome) assumed to be $N(0, K\phi^2)$ distributed, and $\varepsilon$ is an $n \times 1$ vector of residual errors with an $N(0, I\sigma^2)$ distribution. The polygenic effect is the sum of all marker effects,

$$\xi = \sum_{k'=1}^{m} Z_{k'}\alpha_{k'}$$

where $Z_{k'}$ are the genotype code for marker $k'$ and $\alpha_{k'}$ is the effect of the marker. This term mimics the collection of all co-factors in the composite interval mapping procedure (Zeng 1994). To deal with the large number of markers, we assume $\alpha_{k'} \sim N(0, \phi^2/m)$ so that

$$\text{var}(\xi) = \sum_{k'=1}^{m} Z_{k'} Z_{k'}^T \text{var}(\alpha_{k'}) = \frac{1}{m} \sum_{k'=1}^{m} Z_{k'} Z_{k'}^T \phi^2 = K\phi^2$$

You can see that the covariance structure (kinship matrix) is calculated from all markers of the genome,

$$K = \frac{1}{m} \sum_{k'=1}^{m} Z_{k'} Z_{k'}^T \tag{18.24}$$

If the number of markers is not very large, we can redefine the kinship matrix as

$$K = \frac{1}{m-2} \sum_{k' \neq i, i+1}^{m} Z_{k'} Z_{k'}^T \tag{18.25}$$

to exclude the flanking markers, $i$ and $i + 1$, that define the current interval under study. However, when the number of markers is large, whether the flanking markers are included or excluded does not affect the test of the candidate QTL. The model includes both the fixed effects and the random effects, and thus is called a mixed model. The expectation of the model is

$$\text{E}(y) = X\beta + Z_k \gamma_k = \begin{bmatrix} X & Z_k \end{bmatrix} \begin{bmatrix} \beta \\ \gamma_k \end{bmatrix} = W\theta$$

where $W = \begin{bmatrix} X & Z_k \end{bmatrix}$ and $\theta^T = \begin{bmatrix} \beta^T & \gamma_k^T \end{bmatrix}$. The variance of the model is

$$\text{var}(y) = V = K\phi^2 + I\sigma^2$$

The variance parameters, $\psi = \{\phi^2, \sigma^2\}$, are estimated via the restricted maximum likelihood method (Patterson and Thompson 1971). Note that the fixed effects have been absorbed in the restricted maximum likelihood method and, as a result, they are no longer included in the vector of parameters. The restricted log likelihood function is

$$L(\psi) = -\frac{1}{2} \ln |V| - \frac{1}{2} \ln |W^T V^{-1} W| - \frac{1}{2} \left(y - W\widehat{\theta}\right)^T V^{-1} \left(y - W\widehat{\theta}\right)$$

where

$$\widehat{\theta} = \left(W^T V^{-1} W\right)^{-1} W^T V^{-1} y \tag{18.26}$$

is a function of $\psi = \{\phi^2, \sigma^2\}$. A numerical algorithm is required to find the REML estimates of the variances. The best linear unbiased estimates (BLUE) of the fixed effects are given in Eq. (18.26), which is expanded into

$$\begin{bmatrix} \widehat{\beta} \\ \widehat{\gamma}_k \end{bmatrix} = \begin{bmatrix} X^T V^{-1} X & X^T V^{-1} Z_k \\ Z_k^T V^{-1} X & Z_k^T V^{-1} Z_k \end{bmatrix}^{-1} \begin{bmatrix} X^T V^{-1} y \\ Z_k^T V^{-1} y \end{bmatrix}$$

Let us define a $C$ matrix by

$$\begin{bmatrix} C_{XX} & C_{XZ} \\ C_{ZX} & C_{ZZ} \end{bmatrix} = \begin{bmatrix} X^T V^{-1} X & X^T V^{-1} Z_k \\ Z_k^T V^{-1} X & Z_k^T V^{-1} Z_k \end{bmatrix}^{-1}$$

The variance of the estimated QTL effect is $\text{var}(\widehat{\gamma}_k) = C_{ZZ}$. The Wald test is

$$W_k = \frac{\widehat{\gamma}_k^2}{C_{ZZ}}$$

The p-value is calculated using

$$p_k = 1 - \Pr(\chi_1^2 < W_k)$$

So far we assume that $Z_k = \{Z_{jk}\}$ are observed. In interval mapping, $Z_{jk}$ is missing and the probability distribution is inferred from flanking markers. Let $p_j(+1)$, $p_j(0)$, and $p_j(-1)$ be the conditional probabilities of the putative QTL taking the three genotypes, the expectation of $Z_{jk}$ is

$$E(Z_{jk}) = U_{jk} = p_j(+1) - p_j(-1)$$

Let $U_k = \{U_{jk}\}$ be the array of the expectations. Substituting $Z_k$ in all equations by $U_k$ gives the method of interval mapping with polygenic control (Xu 2013). This is the Haley-Knott (Haley and Knott 1992) version of the interval mapping with the additional feature of polygenic control.

If the kinship matrix is calculated from Eq. (18.24), the method is identical to the efficient mixed model association (EMMA) studies (Kang et al. 2008) provided that the eigenvalue decomposition algorithm has been used to improve the computational speed. If Eq. (18.25) is used to compute the kinship matrix, the method is called decontaminated efficient mixed model association (DEMMA) because the polygenic counterpart of a marker under investigation has been removed (Lippert et al. 2011, 2012). When the number of markers is relatively large, DEMMA and EMMA are indistinguishable. Details of the EMMA and DEMMA will be described in Chap. 19.

## 18.5  Ridge Regression

Is it possible to use ridge regression to map QTL? Ridge regression is a multiple regression method that allows a model to hold more variables than the sample size (Hoerl 1970b; Hoerl and Kennard 1970a). As marker density grows, there is no reason to insert pseudo markers in intervals between consecutive true markers. The situation may be even reversed—rather than inserting pseudo markers, deleting true markers may be needed for QTL mapping. We now try to apply the ridge regression to QTL mapping. The model is a multiple marker model,

$$y = X\beta + \sum_{k=1}^{m} Z_k\gamma_k + \varepsilon$$

When $m$ approaches $n$, the marker effects are not estimable under the usual multiple regression analysis. Let $Z = \{Z_k\}$ be an $n \times m$ matrix holding all the Z vectors. Define $\gamma = \{\gamma_k\}$ as an $m \times 1$ vector for all marker effects. The ridge regression handles this problem by adding a small positive number to the diagonal elements of the normal equation system,

$$\begin{bmatrix} \widehat{\beta} \\ \widehat{\gamma} \end{bmatrix} = \begin{bmatrix} X^TX & X^TZ \\ Z^TX & Z^TZ + \lambda^{-1}I \end{bmatrix}^{-1} \begin{bmatrix} X^Ty \\ Z^Ty \end{bmatrix} \tag{18.27}$$

where $\lambda$ is a ridge factor to shrink the regression coefficients ($\gamma$ not $\beta$) to a level so that the solution of Eq. (18.27) is unique. In the original ridge regression analysis, a fixed value is assigned to $\lambda$ or a value determined via cross validation (Wahba and Wold 1975a; Wahba and Wold 1975b; Golab et al. 1979). However, if $\gamma_k$ is assumed to be normally distributed with an equal variance $\phi^2$ across all markers, $\lambda$ can take the ratio $\lambda = \phi^2/\sigma^2$, where $\sigma^2$ is the residual variance. Of course $\phi^2$ and $\sigma^2$ must be estimated from the data using, e.g., the restricted maximum likelihood method. Once $\lambda$ is estimated from the data, it is used in Eq. (18.27) to solve for the BLUE of $\beta$ and the BLUP of $\gamma$. Strictly speaking, $\widehat{\gamma}$ is called the empirical Bayes estimate if $\lambda$ is replaced by $\widehat{\lambda}$. If the number of markers is very large, the BLUP of $\gamma$ is hard to obtain with Eq. (18.27). Therefore, an alternative expression for the BLUP is

$$\widehat{\gamma} = \widehat{\phi}^2 Z^T \left(ZZ^T\widehat{\phi}^2 + I\widehat{\sigma}^2\right)^{-1}\left(y - X\widehat{\beta}\right)$$
$$= \widehat{\lambda} Z^T \left(ZZ^T\widehat{\lambda} + I\right)^{-1}\left(y - X\widehat{\beta}\right)$$

where

$$\widehat{\beta} = \left[X^T\left(ZZ^T\widehat{\lambda} + I\right)^{-1}X\right]^{-1}X^T\left(ZZ^T\widehat{\lambda} + I\right)^{-1}y$$

The BLUP effect of the $k$th marker is

$$\widehat{\gamma}_k = \widehat{\lambda} Z_k^T\left(ZZ^T\widehat{\lambda} + I\right)^{-1}\left(y - X\widehat{\beta}\right)$$

with a variance of

$$\text{var}(\widehat{\gamma}_k) = \left[\widehat{\lambda} - \widehat{\lambda}Z_k^T\left(ZZ^T\widehat{\lambda} + I\right)^{-1}Z_k\widehat{\lambda}\right]\widehat{\sigma}^2$$

A Wald test for the null hypothesis $H_0 : \gamma_k = 0$ may be defined as

$$W_k = \frac{\widehat{\gamma}_k}{\text{var}(\widehat{\gamma}_k)} \tag{18.28}$$

Unfortunately, this Wald test does not following the usual Chi-square distribution under the null model because $\gamma_k$ is not a fixed effect. Mackay (MacKay 1992) and Tipping (Tipping 2000) showed that the degree of freedom for an estimated random effect is not one but determined by

$$d_k = 1 - \frac{\text{var}(\widehat{\gamma}_k)}{\widehat{\phi}^2}$$

From a Bayesian standpoint of view, $\text{var}(\widehat{\gamma}_k)$ is the posterior variance and $\widehat{\phi}^2$ is the prior variance. This degree of freedom is unity only if $\phi^2 \to \infty$ or $\text{var}(\widehat{\gamma}_k) \to 0$ or both. Using this degree of freedom, we can find the actual distribution of the Wald test given in Eq. (18.28), which is a Gamma distribution with a shape parameter 1/2 and a scale parameter $2d_k$, i.e., $W_k \sim \Gamma(k = 1/2, \theta = 2d_k)$. The p-value is calculated with

$$p_k = 1 - \text{Pr}\left(\Gamma_{1/2,2d_k} < W_k\right) \tag{18.29}$$

For example, if $d_k = 0.25$ and $W_k = 1.5$, the p-value is 0.01430588. The R code to calculate the p-value is

$$p < -1 - \text{pgamma}(1.5, \text{shape} = 0.5, \text{scale} = 2*0.25)$$

Alternatively, the p-value can be calculated from a Chi-square distribution,

$$p_k = 1 - \text{Pr}\left(\chi_1^2 < W_k/d_k\right) \tag{18.30}$$

which is also 0.01430588. The R code for the Chi-square p-value is

$$p < -1 - \text{pchisq}(1.5/0.25, 1)$$

Failure to understand the distribution of the Wald test for an estimated random effect would give a p-value of

$$1 - \text{pchisq}(1.5, 1) = 0.2206714$$

The ridge regression method in its original form is called the ordinary ridge regression (ORR). The effects and Wald test are not altered, except that the p-value is calculated from a Gamma distribution other than a Chi-squared distribution. The effect and the test of the ordinary ridge regression are denoted by $\hat{\gamma}_k^{ORR}$ and $W_k^{ORR}$, respectively.

Alternatively, we can deshrink both the estimated effect and the Wald test back to the level comparable to the effect and test in EMMA so that the p-value can be calculated from a Chi-square distribution. The deshrunk effect is

$$\hat{\gamma}_k^{DRR} = \hat{\gamma}_k^{ORR}/d_k$$

and the deshrunk test is

$$W_k^{DRR} = W_k^{ORR}/d_k$$

Under the null hypothesis, the p-value is calculated from

$$p_k = 1 - \Pr(\chi_1^2 < W_k^{DRR})$$

which is identical to equations (18.29) and (18.30).

## 18.6 An Example (a Mouse Data)

A dataset from an $F_2$ mouse population consisting of 110 individuals was used as an example for illustration. The data were published by Lan et al. (2006) and are freely available from the journal website. Parents of the $F_2$ population were B6 (29 males) and BTBR (31 females). The $F_2$ mice used in this study were measured for various clinical traits related to obesity and diabetes. The framework map consists of 194 microsatellite markers, with an average marker spacing of about 10 cM. The mouse genome has 19 chromosomes (excluding the sex chromosome). The 194 markers cover about 1800 cM of the mouse genome. The trait in this analysis was the tenth week body weight. The marker map, the genotypes of the 110 mice for the 194 markers and the tenth week body weights of the $F_2$ mice are also provided in the author's personal website (www.statgen. ucr.edu). The files stored in our website are not the original data but preprocessed by our laboratory members and are ready for analysis using QTL mapping software packages such as the Win QTL Cartographer software package (Wang et al. 2012). For illustration purpose, we ignored the dominance effects and only presented the results from the additive model.

We inserted one pseudo marker in every 1 cM with a total of 1705 inserted pseudo markers. Including the original 194 markers, the total number of points scanned on the genome is 1899. Three interval mapping procedures were demonstrated, including the least squared (LS) method, the iteratively reweighted least square (IRWLS) method and the maximum likelihood (ML) method. We also demonstrate the results of more advanced methods, including the composite interval mapping (CIM) procedure, the efficient mixed model association (EMMA) procedure with polygenic background control, the decontaminated efficient mixed model association (DEMMA) method, the ordinary ridge regression (ORR), and the deshrunken ridge regression (DRR) method,

### 18.6.1 Technical Detail

Before showing the overall results, let us show some technical details of the analysis for a certain marker from a certain individual mouse. Consider the interval between markers 4 and 5 on chromosome 2 of the mouse genome. The two markers are named D2Mit241 and D2Mit9, respectively. The positions of the two markers are 26.4 cM and 42.9 cM, respectively, counted from the first marker of the second chromosome. To scan QTL in this interval, we inserted 16 pseudo markers. The seventh pseudo marker has a position 33.1941 cM. Let this pseudo marker be the candidate QTL and the two flanking markers be A and B, respectively. The recombination fractions between the three loci (A Q B) are

$$r_{AB} = \frac{1}{2}\{1 - \exp[-2 \times (42.9 - 26.4)/100]\} = 0.140538133$$

$$r_{AQ} = \frac{1}{2}\{1 - \exp[-2 \times (33.1941 - 26.4)/100]\} = 0.063527337$$

$$r_{QB} = \frac{1}{2}\{1 - \exp[-2 \times (42.9 - 33.1941)/100]\} = 0.088219495$$

where the Haldane (Haldane 1919) map function is used. The 93th mouse (male with phenotype $y_{93} = 73.7$) in the population has genotypes $A_1A_2$ (26.4 cM) and $B_2B_2$ (42.9 cM) at the two flanking markers. We want to calculate the probabilities of the three possible genotypes for the putative position Q (33.1941 cM). The three probabilities are

$$\Pr(Q = Q_1Q_1 | A = A_1A_2, B = B_2B_2) = \frac{r_{AQ}(1 - r_{AQ})r_{QB}^2}{r_{AB}(1 - r_{AB})} = 0.003833223$$

$$\Pr(Q = Q_1Q_2 | A = A_1A_2, B = B_2B_2) = \frac{\left[(1 - r_{AQ})^2 + r_{AQ}^2\right](1 - r_{QB})r_{QB}}{r_{AB}(1 - r_{AB})} = 0.586702925$$

$$\Pr(Q = Q_2Q_2 | A = A_1A_2, B = B_2B_2) = \frac{(1 - r_{AQ})r_{AQ}(1 - r_{QB})^2}{r_{AB}(1 - r_{AB})} = 0.409463851$$

In the EM algorithm, these three probabilities are called the prior probabilities, denoted by $p_j(+1) = 0.003833223$, $p_j(0) = 0.586702925 =$, and $p_j(-1) = 0.409463851$. For the LS and IRWLS methods, we need to calculate the conditional expectation of $Z_j$, which is

$$U_j = p_j(+1) - p_j(-1) = 0.003833223 - 0.409463851 = -0.405630627$$

The conditional variance of $Z_j$ is.

$$\Sigma_j = p_j(+1) + p_j(-1) - [p_j(+1) - p_j(-1)]^2 = 0.2487609$$

In the EM algorithm, we need to calculate the posterior probabilities of the three genotypes for each individual of the population. The estimated effects for the intercept and the sex effect are

$$\begin{bmatrix} \beta_1 \\ \beta_2 \end{bmatrix} = \begin{bmatrix} 61.26849 \\ -3.08815 \end{bmatrix}$$

This means that, on average, a female mouse is larger than a male mouse by 3.08815 grams. The additive effect of the putative position is $\gamma_k = 3.412144$. The residual error variance is $\sigma^2 = 27.31962$. We now have enough information to see the posterior probabilities of the three genotypes after we observed the body weight (73.7 g) of this individual. Let $j = 93$ and $y_j = 73.7$. The densities of the mouse body weight under the three genotypes are

$$\begin{aligned} f_1(y_j) &= N(y_j | \beta_1 + \beta_2 + \gamma_k, \sigma^2) \\ &= N(73.7 | 61.26849 - 3.08815 + 3.412144, 27.31962) \\ &= 0.006255615 \end{aligned}$$

$$\begin{aligned} f_2(y_j) &= N(y_j | \beta_1 + \beta_2, \sigma^2) \\ &= N(73.7 | 61.26849 - 3.08815, 27.31962) \\ &= 0.001173757 \end{aligned}$$

$$\begin{aligned} f_3(y_j) &= N(y_j | \beta_1 + \beta_1 - \gamma_k, \sigma^2) \\ &= N(73.7 | 61.26849 - 3.08815 - 3.412144, 27.31962) \\ &= 0.000143815 \end{aligned}$$

**Table 18.3**   The probabilities of three genotypes for the 93th mouse in the $F_2$ population

| Genotype | Before experiment | After collecting markers | After observing phenotype |
|---|---|---|---|
| $Q_1Q_1$ | 0.25 | 0.0038 | 0.03108 |
| $Q_1Q_2$ | 0.50 | 0.5867 | 0.89259 |
| $Q_2Q_2$ | 0.25 | 0.4095 | 0.07633 |

The mixture density (weighted sum of the above three genotype-specific densities) is

$$
\begin{aligned}
f(y_j) &= \sum_{k'=1}^{3} p_j(2-k') f_{k'}(y_j) \\
&= p_j(+1) f_1(y_j) + p_j(0) f_2(y_j) + p_j(-1) f_3(y_j) \\
&= 0.003833223 \times 0.006255615 + 0.5867 \times 0.001173757 + 0.4095 \times 0.000143815 \\
&= 0.0007715126
\end{aligned}
$$

The three posterior probabilities are

$$
p_j^*(+1) = \frac{p_j(+1) f_1(y_j)}{f(y_j)} = \frac{0.003833223 \times 0.006255615}{0.0007715126} = 0.03108072
$$

$$
p_j^*(0) = \frac{p_j(0) f_2(y_j)}{f(y_j)} = \frac{0.5867 \times 0.001173757}{0.0007715126} = 0.89259257
$$

$$
p_j^*(-1) = \frac{p_j(-1) f_3(y_j)}{f(y_j)} = \frac{0.4095 \times 0.000143815}{0.0007715126} = 0.07632672
$$

Table 18.3 compares the genotype probabilities before the experiment (Mendelian expectation), the genotype probabilities after observing marker information and the genotype probabilities after observing the phenotypic value of this individual. After observing markers, the first genotype is almost excluded and the probability is almost split between the heterozygote and the second homozygote. After observing the phenotype, the third genotype is almost excluded, making the posterior probability heavily concentrated on the heterozygote.

Let us look at the IRWLS method. For the same individual at the same location of the genome, the conditional variance of $Z_j$ is $\Sigma_j = 0.2487609$. The estimated intercept and sex effect are $\beta_1 = 61.26559$ and $\beta_2 = -3.09708$, respectively. The estimated QTL effect is $\gamma_k = 3.45015$ and the residual variance is $\sigma^2 = 28.05171$, the weight for this mouse with the IRWLS method is

$$
w_j = \left(\frac{1}{\sigma^2}\Sigma_j \gamma_k^2 + 1\right)^{-1} = \left(\frac{1}{28.05171} \times 0.2487609 \times 3.450153^2 + 1\right)^{-1} = 0.9045193
$$

The weight would be just one if the genotype of the candidate locus is observed ($\Sigma_j = 0$) or the locus has no effect ($\gamma_k = 0$).

## 18.6.2   Results of Different Methods

Let us first compare results of the three interval mapping procedures (LS, IRWLS, and EM). The Wald tests of the LS method and the IRWLS method, and the likelihood ratio test of the EM algorithm are divided by $2 \ln(10) \approx 4.61$ to convert into LOD scores. The LOD scores of all putative positions are plotted against the genome (see Fig. 18.3). The three methods are almost indistinguishable, except in peak regions where the IRWLS method shows higher test statistics than the other two methods. The marker density is relatively high and the ML method implemented via the EM algorithm does not show any advantages over the LS and IRWLS methods. The number of iterations for the IRWLS method is, on average, about 3, while the EM algorithm is about 6. The estimated effects of the three methods are presented in Fig. 18.4. Again, the three methods are almost indistinguishable, just like the LOD score profiles.

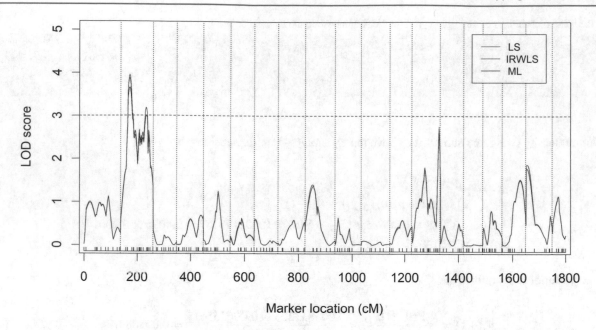

**Fig. 18.3** LOD test statistic profiles for three interval mapping procedures of the mouse data. Chromosomes are separated by the vertical dotted lines. The inwards ticks on the x-axis indicate the positions of the actual markers. The horizontal line at LOD = 3 of the y-axis is the threshold for significance test, above which QTL are declared

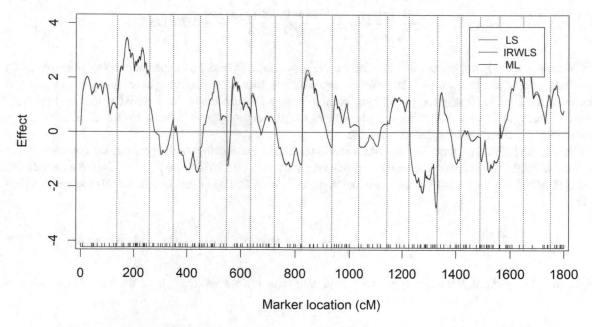

**Fig. 18.4** Estimated effects plotted against genome positions for the three interval mapping procedures (LS, IRWLS and ML)

If LOD = 3 is used as the threshold for declaration of significance, the IM methods detect two QTLs on chromosome 2. The first QTL is located between markers 4 (26.4 cM) and 5 (42.9 cM) with a position at 34.165 (cM). The LOD scores are 3.6581, 3. 8766, and 3.9514, respectively, for the EM, LS, and IRWLS methods. The corresponding QTL effects of the three methods are 3.4121, 3.4366, and 3.4502. The second QTL is located between markers 13 (85.4 cM) and 14 (98.7 cM) with a position (96.8 cM). The QTL effects are about 3.00 for all the three methods.

The three QTL mapping methods with polygenic background control adopted from GWAS are EMMA, DEMMA, and DRR. The LOD scores are shown in Fig. 18.5. The LOD score profiles have the same patterns as the profiles observed from the interval mapping procedures, but they are significantly shrunken towards zero. Therefore, LOD = 3 is too high as the threshold for significance declaration. Empirical threshold may be needed to declare for significance. The DRR methods show

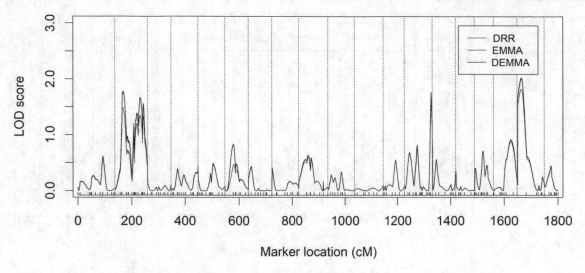

**Fig. 18.5** LOD test statistic profiles of the three polygenic-based QTL mapping procedures (DRR, EMMA, and DEMMA) for the mouse data

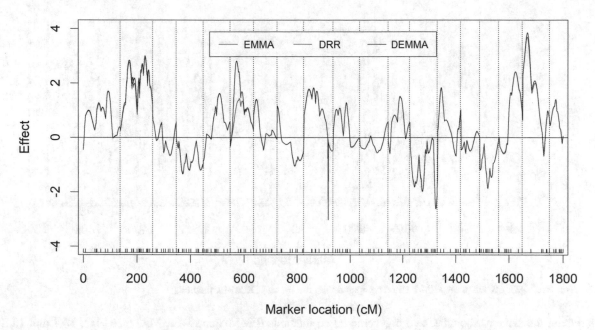

**Fig. 18.6** Estimated effects plotted against genome positions for the three polygenic procedures of QTL mapping (EMMA, DEMMA, and DRR)

peaks slightly lower than the EMMA and DEMMA methods who are not distinguishable themselves. Figure 18.6 shows the estimated effect profiles of the three methods. Again, at the peak areas, DRR is slightly lower than the other two methods.

We took the EM algorithm from the three interval mapping procedures and the DEMMA method from the three polygenic-based methods to compare their tests with the composite interval mapping (CIM) procedure. Figure 18.7 shows the LOD score profiles of the three methods. The interval mapping procedure is consistently higher than the polygenic-based method, but the shapes of the two methods are much the same. At the peak areas, the CIM procedure is indeed higher than both methods. Surprisingly, the peaks on chromosome 2 from the CIM method are different from the other two methods compared. The peak identified by CIM is right at the 12th marker of chromosome 2. Figure 18.8 shows the estimated QTL effects of the three methods (IM, DEMMA, and CIM). In several regions of the genome, the CIM method shows different patterns from the EM and DEMMA methods.

Before leaving this section, let us look at the result of the ordinary ridge regression (ORR). Figure 18.9 shows the ridge estimates and the Wald test of the ORR method in its original form along with the degrees of freedom of the random effects. The ridge effects and the Wald tests are significantly shrunken towards zero. However, the patterns of the effect and test

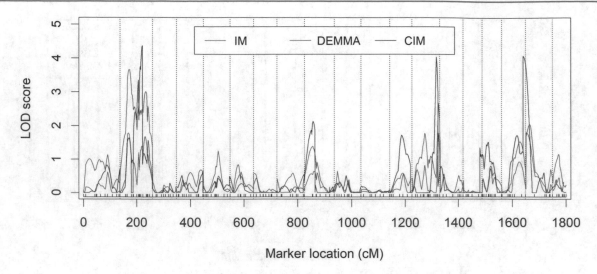

**Fig. 18.7**  LOD score profile of the CIM method compared with IM and DEMMA for the mouse data

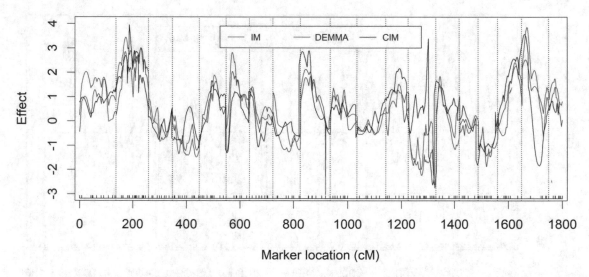

**Fig. 18.8**  Estimated QTL effects of the CIM method compared with the IM and DEMMA methods

profiles remain the same as the other two polygenic-based methods (EMMA and DEMMA) (see Figs. 18.7 and 18.8). The Wald tests of ORR do not follow the usual Chi-square distribution; instead, they follow some Gamma distributions. Therefore, the p-values should be calculated from the Gamma distributions, as shown in Eq. (18.29). Alternatively, the Wald tests should be divided by the degrees of freedom to bring the tests to the levels comparable to the Wald tests of the other polygenic-based methods. Figure 18.10 shows the estimated effects and the Wald tests against the genome location for the original ORR and the DRR methods.

### 18.6.3  Remarks on the Interval Mapping Procedures

The LS method is an approximation of the ML method, aiming to improve the computational speed. The method has been extended substantially to many other situations, e.g., multiple trait QTL mapping and QTL mapping for binary traits. When used for binary and other non-normal traits, the method is no longer called LS. Because of the fast speed, LS remains a popular method, even though the computer power has increased by many orders of magnitude since the LS was developed. In some literature, the LS method is also called the H-K method in honor of the authors. Xu (1995) noticed that the LS method, although a good approximation to ML in terms of estimates of QTL effects and test statistic, may lead to a biased (inflated)

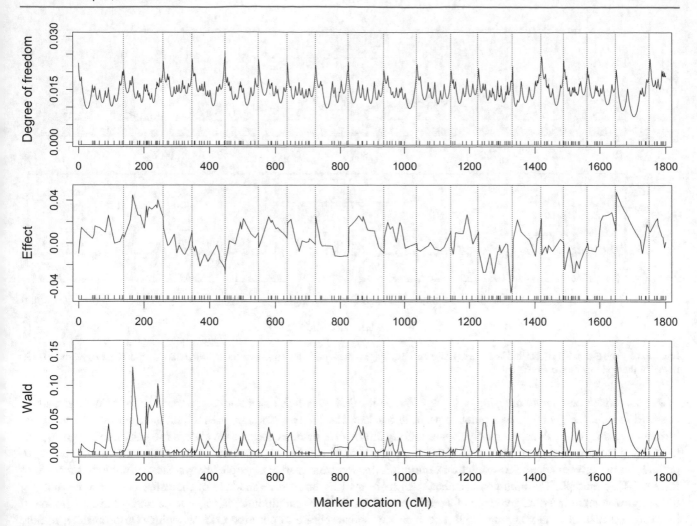

**Fig. 18.9**  The degrees of freedom of the tests (top panel), the estimated effects (the panel in the middle), and the Wald tests (panel at the bottom) of the ordinary ridge regression (ORR) analysis for the mouse data

estimate for the residual error variance. Based on this work, Xu (1998a, b) eventually developed the iteratively reweighted least squares (IRWLS) method. Xu compared LS, IRWLS, and ML in a variety of situations and conclude that IRWLS is always better than LS and as efficient as ML. When the residual error does not have a normal distribution, which is required by the ML method, LS and IRWLS can be better than ML. In other words, LS and IRWLS are more robust than ML to the departure from normality. Kao (2000) and Feenstra et al. (2006) conducted more comprehensive investigation on LS, IRWLS, and ML and found that when epistatic effects exist, LS can generate unsatisfactory results, but IRWLS and ML usually map QTL better than LS. In addition, Feenstra et al. (2006) modified the weighted least squares method by using the estimating equations (EE) algorithm. This algorithm further improved the efficiency of the weighted least squares by maximizing an approximate likelihood function. Most recently, Han and Xu (2008) developed a Fisher scoring algorithm to maximize the approximate likelihood function. Both the EE and Fisher algorithm maximize the same likelihood function, and thus they produce identical results.

The LS method ignores the uncertainty of the QTL genotype. The IRWLS, FISHER (or EE), and ML methods use different ways to extract information from the uncertainty of QTL genotype. If the putative location of QTL overlaps with a fully informative marker, all four methods produce identical result. Therefore, if the marker density is sufficiently high, there is virtually no difference between the four methods. For low marker density, when the putative position is far away from either flanking marker, the four methods will show some difference. This difference will be magnified by large QTL. Han and Xu (2008) compared the four methods in a simulation experiment and showed that when the putative QTL position is fixed in the middle of a 10 cM interval, the four methods generated almost identical results. However, when the interval expands to 20 cM, the differences among the four methods become noticeable.

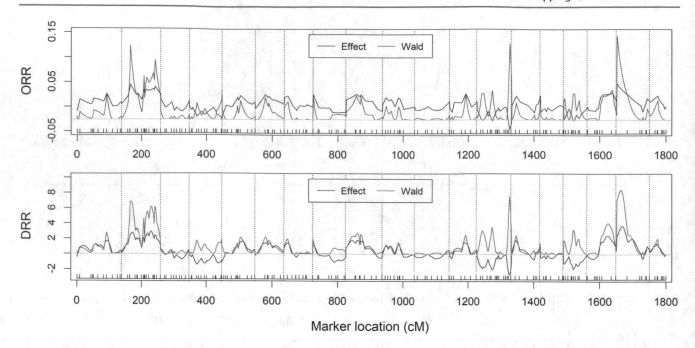

**Fig. 18.10** The estimated effects and Wald tests of the ordinary ridge regression (ORR) (panel at the top) and the deshrink ridge regression (DRR) (panel at the bottom) for the mouse data

A final remark on interval mapping is the way to infer the QTL genotype using flanking markers. If only flanking markers are used to infer the genotype of a putative position bracketed by the two markers, the method is called interval mapping. Strictly speaking, interval mapping only applies to fully informative markers because we always use flanking markers to infer the QTL genotype. However, almost all datasets obtained from real-life experiments contain missing, uninformative, or partially informative markers. To extract maximum information from markers, people always use the multipoint method to infer a QTL genotype. The multipoint method uses more markers or even all markers of the entire chromosome (not just flanking markers) to infer the genotype of a putative position. With the multipoint analysis, we no longer have the notion of interval, and thus interval mapping is no longer an appropriate phrase to describe QTL mapping. Unfortunately, a more appropriate phrase has not been proposed and people are used to the phrase of interval mapping. Therefore, the so-called interval mapping in the current literature means QTL mapping under a single QTL model, regardless whether the genotype of a putative QTL position is inferred from flanking markers or from all markers.

## 18.7  Bonferroni Correction of Threshold for Multiple Tests

Genome scanning involves multiple tests. Sometimes the number of tests may reach hundreds or even thousands. For a single test, the critical value for any test statistic simply takes the 95% or 99% quantile of the distribution that the test statistic follows under the null hypothesis. For example, the F-test statistic follows an F distribution, the likelihood ratio test statistic follows a Chi-square distribution and the Wald-test statistic also follows a Chi-square distribution. When multiple tests are involved, the critical value used for a single test must be adjusted to make the experiment-wise Type I error at a desired level, say 0.05. The Bonferroni correction is a multiple test correction used when multiple statistical tests are being performed in a single experiment. While a given alpha value $\alpha$ may be appropriate for each individual test, it is not for the set of all tests involved in a single experiment. In order to avoid spurious positives, the alpha value needs to be lowered to account for the number of tests being performed. The Bonferroni correction sets the Type I error for the entire set of $m$ tests equal to $\omega$ by taking the alpha value for each test equal to $\alpha$. The $\omega$ is now called the experiment-wise Type I error rate and $\alpha$ is called the test-wise Type I error rate or nominal Type I error rate. The Bonferroni correction states that, in an experiment involving $m$ tests, if you want to control the experiment-wise Type I error rate at $\omega$, the nominal Type I error rate for a single test should be

$$\alpha = \omega/m$$

For example, if an experiment involves 100 tests and the investigator wants to control the experiment-wise Type I error at $\omega = 0.05$, for each of the individual tests, the nominal Type I error rate should be $\alpha = \omega/m = 0.05/100 = 0.0005$. In other words, for any individual test the $p$-value should be less than 0.0005 in order to declare significance for that test. The Bonferroni correction does not require independence of the multiple tests.

When the multiple tests are independent, there is an alternative correction for the Type I error, which is called the Šidák correction. This correction is often confused with the Bonferroni correction. If a test-wise Type I error is $\alpha$, the probability of non-significance is $1 - \alpha$ for this particular test. For $m$ independent tests and none of them is significant, the probability is $(1 - \alpha)^m$. The experiment-wise Type I error is defined as the probability that at least one of the $m$ tests is significant. This probability is

$$\omega = 1 - (1 - \alpha)^m \tag{18.31}$$

To find the nominal $\alpha$ value given the experiment wise value $\omega$, we use the reverse function of (18.31),

$$\alpha = 1 - (1 - \omega)^{1/m}$$

This correction is the Šidák correction. The two corrections are approximately the same when $\omega$ is small because $(1 - \omega)^{1/m} \approx 1 - \omega/m$ and thus $\alpha \approx \omega/m$. Therefore, the Bonferroni correction is an approximation of the Šidák correction for multiple independent tests for small $\omega$.

## 18.8 Permutation Test

When the number of tests is large, the Bonferroni correction tends to be over conservative. In addition, if a test statistic does not follow any standard distribution under the null model, calculation of the $p$-value may be difficult for each individual test. In this case, we can adopt the permutation test to draw an empirical critical value. This method was developed by Churchill and Doerge (Churchill and Doerge 1994) for QTL mapping. The idea is simple, but implementation can be time consuming. When the sample size $n$ is small, we can evaluate all $n!$ different permuted samples of the original phenotypic values while keeping the marker genotype data intact. In other words, we only reshuffle the phenotypes, not the marker genotypes. For each permuted sample, we apply any method of genome scanning to calculate the test statistical values for all markers. In each of the permuted samples, the associations of the phenotype and genotypes of markers have been (purposely) destroyed so that the distribution of the test statistics will mimic the actual distribution under the null model, from which a desirable critical value can be drawn from the empirical null distribution.

The number of permuted samples can be extremely large if the sample size is large. In this case, we can randomly shuffle the data to purposely destroy the association between the phenotype and the marker genotype. By random shuffling the phenotypes, individual $j$ may take the phenotypic value of individual $i$ for $i \neq j$ while the marker genotype of individual $j$ remains unchanged. After shuffling the phenotypes, we analyze the data and scan the entire genome. By chance, we may find some peaks in the test statistic profile. We know that these peaks are false because we have already destroyed the association between markers and phenotypes. We record the value of the test statistic at the highest peak of the profile and denote it by $\lambda_1$. We then reshuffle the data and scan the genome again. We may find some false peaks again. We then record the highest peak and write down the value, denoted by $\lambda_2$, and put it in the dataset. We repeat the reshuffling process many times to form a large sample of $\lambda$ 's, denoted by $\{\lambda_1, \ldots, \lambda_M\}$, where $M$ is a large number, say 1000 or 10,000. These $\lambda$ values will form a distribution, called the null distribution. The 95% or 99% quantile of the null distribution is the empirical critical value for our test statistic. We then compare our test statistic for each marker (from the original data analysis) against this empirical critical value. If the test statistic of a marker is larger than this critical value, we can declare this marker as being significant. Note that the permutation test is time consuming, but it is realistic with the advanced computing system currently available in most laboratories.

We now provide an example to show how to use the permutation test to draw the critical value of a test statistic. Table 18.4 gives a small sample of ten plants (the original dataset). Fifteen randomly reshuffled samples are demonstrated in Table 18.5. We can see that the first observation of sample 1 ($S_1$) takes the phenotype of plant number 6 while the genotypes of the five markers remain unchanged. Another example is that the second observation of sample 2 ($S_2$) takes the phenotype of plant number 8 while the genotypes of the five markers are still the genotypes for plant number 2. The phenotypic values

**Table 18.4**  Phenotypic values of trait $y$ and the genotypes of five markers from ten plants (the original dataset)

| Line | y | $M_1$ | $M_2$ | $M_3$ | $M_4$ | $M_5$ |
|------|------|-------|-------|-------|-------|-------|
| 1 | 55.0 | H | H | H | H | A |
| 2 | 54.2 | H | H | H | H | U |
| 3 | 61.6 | H | H | U | A | A |
| 4 | 66.6 | H | H | H | H | U |
| 5 | 67.4 | A | H | H | B | U |
| 6 | 64.3 | H | H | H | H | H |
| 7 | 54.0 | A | A | B | B | B |
| 8 | 57.2 | H | B | H | H | H |
| 9 | 63.7 | H | H | H | H | H |
| 10 | 55.0 | H | H | A | H | U |

**Table 18.5**  Plant IDs of 15 randomly reshuffled samples, denoted by $S_1$, $S_2$, ..., $S_{15}$, respectively

| S1 | S2 | S3 | S4 | S5 | S6 | S7 | S8 | S9 | S10 | S11 | S12 | S13 | S14 | S15 |
|----|----|----|----|----|----|----|----|----|-----|-----|-----|-----|-----|-----|
| 6 | 5 | 9 | 10 | 5 | 6 | 2 | 10 | 1 | 10 | 7 | 5 | 3 | 10 | 8 |
| 2 | 8 | 6 | 8 | 1 | 2 | 7 | 6 | 4 | 5 | 6 | 8 | 1 | 7 | 2 |
| 4 | 1 | 5 | 7 | 8 | 4 | 1 | 9 | 3 | 9 | 9 | 1 | 8 | 1 | 7 |
| 3 | 10 | 1 | 4 | 10 | 3 | 10 | 8 | 5 | 3 | 8 | 10 | 7 | 2 | 3 |
| 8 | 3 | 2 | 1 | 7 | 8 | 4 | 5 | 8 | 1 | 5 | 3 | 10 | 4 | 6 |
| 9 | 4 | 3 | 5 | 2 | 9 | 3 | 3 | 7 | 8 | 3 | 4 | 2 | 9 | 9 |
| 10 | 7 | 10 | 2 | 4 | 10 | 8 | 2 | 9 | 2 | 2 | 7 | 4 | 8 | 10 |
| 1 | 9 | 7 | 6 | 3 | 1 | 9 | 1 | 10 | 6 | 1 | 9 | 5 | 3 | 1 |
| 7 | 6 | 8 | 3 | 6 | 7 | 6 | 7 | 6 | 7 | 10 | 6 | 6 | 6 | 4 |
| 5 | 2 | 4 | 9 | 9 | 5 | 5 | 4 | 2 | 4 | 4 | 2 | 9 | 5 | 5 |

corresponding to the 15 reshuffled samples are given in Table 18.6. The genotypes of the ten plants remain the same as those given in Table 18.4, but we now have 15 reshuffled datasets (Table 18.6). Therefore, we can perform QTL mapping 15 times, one for each reshuffled dataset. Each sample is subject to genome scanning, i.e., five F-test statistics are calculated, one for each marker. The maximum F-test statistic value for each reshuffled sample is given in the second column of Table 18.7. The 15 F-test statistics from the 15 reshuffled datasets are assumed to be sampled from the null distribution. The last column of Table 18.7 gives the percentile. For example, 93% of the 15 samples have F-test statistics less than or equal to 1.7799. Therefore, 1.7799 should be the empirical threshold to control a Type I error at $1 - 0.93 = 0.07$. The number of reshuffled datasets in the example is not sufficiently large to give an exact 95% quantile for the experimental wise Type I error of 0.05. In practice, the number of randomly reshuffled samples depends on the predetermined Type I error ($\alpha$) due to Monte Carlo error. Nettleton and Doerge (Nettleton and Doerge 2000) recommended that the permutation sample size should be at least $5\alpha^{-1}$, where $\alpha$ is the experiment-wise Type I error rate. In permutation analysis, there is no such a thing as nominal Type I error. In practice, we often choose 1000 as the permutation sample size.

## 18.9  Quantification of QTL Size

After a QTL is detected, the immediate next step is to calculate the size of the QTL. The estimated effect of the QTL is a measurement of the QTL size. However, the size is often reported relative to the trait variation. The most informative measurement of QTL size is the proportion of the phenotypic variance contributed by the QTL. People often call it the QTL heritability although it is not appropriate to call it heritability. When everybody calls it this way, it eventually becomes the norm. Let $\sigma^2_{QTL}$ be the variance of the QTL, $\sigma^2_{POLY}$ be the polygenic variance (variance contributed by the collection of all small polygenic effects) and $\sigma^2$ be the variance of the residuals. The QTL heritability is defined as

**Table 18.6**  Phenotypes of 15 randomly reshuffled samples

| S1 | S2 | S3 | S4 | S5 | S6 | S7 | S8 | S9 | S10 | S11 | S12 | S13 | S14 | S15 |
|----|----|----|----|----|----|----|----|----|-----|-----|-----|-----|-----|-----|
| 64.3 | 67.4 | 63.7 | 55.0 | 67.4 | 64.3 | 54.2 | 55.0 | 55 | 55.0 | 54.0 | 67.4 | 61.6 | 55.0 | 57.2 |
| 54.2 | 57.2 | 64.3 | 57.2 | 55.0 | 54.2 | 54.0 | 64.3 | 66.6 | 67.4 | 64.3 | 57.2 | 55 | 54.0 | 54.2 |
| 66.6 | 55.0 | 67.4 | 54.0 | 57.2 | 66.6 | 55.0 | 63.7 | 61.6 | 63.7 | 63.7 | 55.0 | 57.2 | 55.0 | 54.0 |
| 61.6 | 55.0 | 55.0 | 66.6 | 55.0 | 61.6 | 55.0 | 57.2 | 67.4 | 61.6 | 57.2 | 55.0 | 54 | 54.2 | 61.6 |
| 57.2 | 61.6 | 54.2 | 55.0 | 54.0 | 57.2 | 66.6 | 67.4 | 57.2 | 55.0 | 67.4 | 61.6 | 55 | 66.6 | 64.3 |
| 63.7 | 66.6 | 61.6 | 67.4 | 54.2 | 63.7 | 61.6 | 61.6 | 54.0 | 57.2 | 61.6 | 66.6 | 54.2 | 63.7 | 63.7 |
| 55.0 | 54.0 | 55.0 | 54.2 | 66.6 | 55.0 | 57.2 | 54.2 | 63.7 | 54.2 | 54.2 | 54.0 | 66.6 | 57.2 | 55.0 |
| 55.0 | 63.7 | 54.0 | 64.3 | 61.6 | 55.0 | 63.7 | 55.0 | 55.0 | 64.3 | 55.0 | 63.7 | 67.4 | 61.6 | 55.0 |
| 54.0 | 64.3 | 57.2 | 61.6 | 64.3 | 54.0 | 64.3 | 54.0 | 64.3 | 54.0 | 55.0 | 64.3 | 64.3 | 64.3 | 66.6 |
| 67.4 | 54.2 | 66.6 | 63.7 | 63.7 | 67.4 | 67.4 | 66.6 | 54.2 | 66.6 | 66.6 | 54.2 | 63.7 | 67.4 | 67.4 |

**Table 18.7**  Original F-tests of 15 reshuffled data and the rank of sorted F-tests

| Original sample | | | Sorted sample | | |
|-----------------|--------|------|---------------|--------------|------------|
| Sample | F-test | Rank | Sorted sample | Ranked F-test | Percentile |
| S1 | 1.3906 | 1 | S8 | 2.5477 | 1.0000 |
| S2 | 1.6488 | 2 | S14 | 1.7799 | 0.9333 |
| S3 | 0.6485 | 3 | S11 | 1.7581 | 0.8667 |
| S4 | 0.8870 | 4 | S2 | 1.6488 | 0.8000 |
| S5 | 0.3247 | 5 | S15 | 1.5273 | 0.7333 |
| S6 | 1.2162 | 6 | S1 | 1.3906 | 0.6667 |
| S7 | 0.7994 | 7 | S6 | 1.2162 | 0.6000 |
| S8 | 2.5477 | 8 | S13 | 1.1726 | 0.5333 |
| S9 | 0.5377 | 9 | S4 | 0.8870 | 0.4667 |
| S10 | 0.7429 | 10 | S12 | 0.8216 | 0.4000 |
| S11 | 1.7581 | 11 | S7 | 0.7994 | 0.3333 |
| S12 | 0.8216 | 12 | S10 | 0.7429 | 0.2667 |
| S13 | 1.1726 | 13 | S3 | 0.6485 | 0.2000 |
| S14 | 1.7799 | 14 | S9 | 0.5377 | 0.1333 |
| S15 | 1.5273 | 15 | S5 | 0.3247 | 0.0667 |

$$h^2_{QTL} = \frac{\sigma^2_{QTL}}{\sigma^2_{QTL} + \sigma^2_{POLY} + \sigma^2}$$

where the denominator is the total phenotypic variance assuming absence of dominance and epistatic components in the polygene. The QTL variance is defined as

$$\sigma^2_{QTL} = \mathrm{var}(Z_k\gamma_k) = \mathrm{var}(Z_k)\gamma^2_k = \sigma^2_Z\gamma^2_k \tag{18.32}$$

where $Z_k$ is the genotype indicator variable for the QTL. Recall that the linear mixed model for QTL mapping is

$$y = X\beta + Z_k\gamma_k + \xi + \varepsilon \tag{18.33}$$

where $\gamma_k$ is treated as a fixed effect and $\xi \sim N(0, K\phi^2)$. We can explain Eq. (18.32) only if $\gamma_k$ is treated as a fixed effect. However, when we call QTL variance, we actually treat the QTL effect as a random effect. There is a contradiction between the fixed effect in data analysis and the random effect in concept. When $\gamma_k$ is treated as a fixed effect, the expectation of $y$ is $\mathrm{E}(y) = X\beta + Z_k\gamma_k$ and the variance of $y$ is $\mathrm{var}(y) = K\phi^2 + I\sigma^2$. The phenotypic variance of the trait is

$$\sigma_P^2 = \frac{1}{n} \text{tr}[\text{var}(y)]$$

So, the QTL effect is not included in the variance part. A logically correct model should treat the QTL effect as random, even though it may be just a constant (a parameter). When the QTL effect is treated as a random effect, we must assume $\gamma_k \sim N\left(0, \sigma_\gamma^2\right)$. The expectation of $y$ is $\text{E}(y) = X\beta$ and the variance of $y$ is

$$\text{var}(y) = Z_k Z_k^T \sigma_\gamma^2 + K\phi^2 + I\sigma^2$$

The phenotypic variance is defined as

$$\sigma_P^2 = \frac{1}{n}\text{tr}[\text{var}(y)] = \frac{1}{n}\text{tr}\left(Z_k Z_k^T\right)\sigma_\gamma^2 + \frac{1}{n}\text{tr}(K)\phi^2 + \frac{1}{n}\text{tr}(I)\sigma^2 = \sigma_{QTL}^2 + \sigma_{POLY}^2 + \sigma^2$$

where the QTL variance is

$$\sigma_{QTL}^2 = \frac{1}{n}\text{tr}\left(Z_k Z_k^T\right)\sigma_\gamma^2 = \text{E}(z^2)\sigma_\gamma^2 = \left(\sigma_Z^2 + \mu_Z\right)\sigma_\gamma^2$$

and the polygenic variance is

$$\sigma_{POLY}^2 = \frac{1}{n}\text{tr}(K)\phi^2$$

If $Z_k$ is already standardized, $\sigma_Z^2 = 1$ and $\mu_Z = 0$, and thus $\sigma_{QTL}^2 = \sigma_\gamma^2$. The polygenic variance is $\sigma_{POLY}^2 = \phi^2$ only if $\text{tr}(K) = n$. What is $\sigma_\gamma^2$? Recall that $\gamma_k \sim N\left(0, \sigma_\gamma^2\right)$ and the definition of the variance is

$$\sigma_\gamma^2 = (\gamma_k - 0)^2 = \gamma_k^2$$

for one sampled value. So, the QTL variance is the square of the QTL effect. Under the fixed effect assumption for the QTL effect, we are not estimating the QTL variance or the square of the effect; rather, we estimate the effect itself $(\widehat{\gamma}_k)$ and then take the square $(\widehat{\gamma}_k)^2$ to obtain an estimated QTL variance. It is well known that $\text{E}\left(\gamma_k^2\right) \neq [\text{E}(\gamma_k)]^2$, therefore, $\widehat{\sigma}_\gamma^2 \neq (\widehat{\gamma}_k)^2$. An asymptotically unbiased estimate of the QTL variance should be obtained from the random model,

$$\text{E}(y) = X\beta$$
$$\text{var}(y) = Z_k Z_k^T \sigma_\gamma^2 + K\phi^2 + I\sigma^2$$

with the restricted maximum likelihood method. Let $V = Z_k Z_k^T \sigma_\gamma^2 + K\phi^2 + I\sigma^2$ and define the parameters by $\theta = \left\{\sigma_\gamma^2, \phi^2, \sigma^2\right\}$. The restricted log likelihood function is

$$L(\theta) = -\frac{1}{2}\ln|V| - \frac{1}{2}\ln|X^T V^{-1} X| - \frac{1}{2}\left(y - X\widehat{\beta}\right)^T V^{-1}\left(y - X\widehat{\beta}\right) \tag{18.34}$$

We now have three variances to estimate and QTL mapping under this random model can be very challenging for large number of markers. A convenient method to convert the estimated fixed effect into an estimated variance is (Luo et al. 2003)

$$\widehat{\sigma}_\gamma^2 = \widehat{\gamma}_k^2 - \text{var}(\widehat{\gamma}_k)$$

where $\text{var}(\widehat{\gamma}_k)$ is the square of the standard error of the estimated QTL effect under the usual fixed model. This estimate is actually an unbiased estimate of the QTL variance. If we only take the positive value, i.e.,

$$\widehat{\sigma}_\gamma^2 = \begin{cases} \widehat{\gamma}_k^2 - \text{var}(\widehat{\gamma}_k) & \text{if} \quad \widehat{\gamma}_k^2 > \text{var}(\widehat{\gamma}_k) \\ 0 & \text{if} \quad \widehat{\gamma}_k^2 \le \text{var}(\widehat{\gamma}_k) \end{cases} \tag{18.35}$$

the estimate is identical to the REML estimate obtained by maximizing Eq. (18.34). A statistically more elegant expression of Eq. (18.35) is

$$\widehat{\sigma}_\gamma^2 = \left[ \widehat{\gamma}_k^2 - \text{var}(\widehat{\gamma}_k) \right]^+ \tag{18.36}$$

In summary, the QTL size should be measured by the QTL heritability.

$$\begin{aligned} \widehat{h}_{QTL}^2 &= \frac{n^{-1}\text{tr}(Z_k Z_k^T) \left[ \widehat{\gamma}_k^2 - \text{var}(\widehat{\gamma}_k) \right]^+}{n^{-1}\text{tr}(Z_k Z_k^T) \left[ \widehat{\gamma}_k^2 - \text{var}(\widehat{\gamma}_k) \right]^+ + n^{-1}\text{tr}(K)\widehat{\phi}^2 + \widehat{\sigma}^2} \\ &= \frac{\text{tr}(Z_k Z_k^T) \left[ \widehat{\gamma}_k^2 - \text{var}(\widehat{\gamma}_k) \right]^+}{\text{tr}(Z_k Z_k^T) \left[ \widehat{\gamma}_k^2 - \text{var}(\widehat{\gamma}_k) \right]^+ + \text{tr}(K)\widehat{\phi}^2 + n\widehat{\sigma}^2} \end{aligned}$$

When $\text{tr}(K) = \text{tr}(Z_k Z_k^T) = n$, the estimated QTL heritability is expressed as

$$\widehat{h}_{QTL}^2 = \frac{\left[ \widehat{\gamma}_k^2 - \text{var}(\widehat{\gamma}_k) \right]^+}{\left[ \widehat{\gamma}_k^2 - \text{var}(\widehat{\gamma}_k) \right]^+ + \widehat{\phi}^2 + \widehat{\sigma}^2}$$

For the first QTL on chromosome 2 of the mouse tenth week body weight trait, we have $\text{tr}(Z_k Z_k^T) = 47.1557$ and $\sigma_Z^2 = 0.410934148$. The sample size is $n = 110$. The kinship matrix has a trace of $\text{tr}(K) = 48.5225$. The estimated polygenic variance is $\widehat{\phi}^2 = 22.41959$. The estimated QTL effect is $\widehat{\gamma}_k = 2.835671$ and the variance of the estimate is $\text{var}(\widehat{\gamma}_k) = 1.0092$. The estimated residual variance is $\widehat{\sigma}^2 = 19.12445$. The estimated heritability should be

$$\begin{aligned} \widehat{h}_{QTL}^2 &= \frac{\text{tr}(Z_k Z_k^T) \left[ \widehat{\gamma}_k^2 - \text{var}(\widehat{\gamma}k) \right]^+}{\text{tr}(Z_k Z_k^T) \left[ \widehat{\gamma}_k^2 - \text{var}(\widehat{\gamma}k) \right]^+ + \text{tr}(K)\widehat{\phi}^2 + n\widehat{\sigma}^2} \\ &= \frac{47.1557 \times (2.8356^2 - 1.0092)}{47.1557 \times (2.8356^2 - 1.0092) + 48.5225 \times 22.4196 + 110 \times 19.1245} \\ &= 0.0941 \end{aligned}$$

If we do not correct for the bias in the estimated QTL variance, the heritability would be

$$\begin{aligned} \widehat{h}_{QTL}^2 &= \frac{\sigma_Z^2 \widehat{\gamma}_k^2}{\sigma_Z^2 \widehat{\gamma}_k^2 + n^{-1}\text{tr}(K)\widehat{\phi}^2 + \widehat{\sigma}^2} \\ &= \frac{0.4109 \times 2.8356^2}{0.4109 \times 2.8356^2 + (48.5225/110) \times 22.4196 + 19.1245} \\ &= 0.1022432 \end{aligned}$$

The moment estimate of QTL variance shown in Eq. (18.36) is also the restricted maximum likelihood estimate by maximizing the restricted log likelihood function in (34). We now present an approximate but simplified proof. Let $\widehat{\gamma}_k$ and $\text{var}(\widehat{\gamma}_k | \gamma_k) = s_\gamma^2$ be the estimated QTL effect and the squared estimation error, respectively, from the fixed model shown in

Eq. (18.33). They are sufficient stastistics of the QTL effect. As sufficient statistics, $\widehat{\gamma}_k$ and $s_{\widehat{\gamma}}^2$ provide the same amount of information as the raw data for $\gamma_k$. When the residual error of the fixed model is normally distributed, the estimated QTL effect is also normally distributed, i.e., $\widehat{\gamma}_k \sim N\left(\gamma_k, s_{\widehat{\gamma}}^2\right)$. To estimate the variance of the QTL ($\sigma_\gamma^2$), we can simply obtain it from the sufficient statistics, not necessarily from the original data. Let us propose a random model for $\widehat{\gamma}_k$ (treated as an observed data point),

$$\widehat{\gamma}_k = \gamma_k + e_{\widehat{\gamma}} \tag{18.37}$$

where $\gamma_k$ is the true value with a normal distribution $\gamma_k \sim N\left(0, \sigma_\gamma^2\right)$ and $e_{\widehat{\gamma}} \sim N\left(0, s_{\widehat{\gamma}}^2\right)$ is the residual error with a known error variance. The expectation of model (37) is $E(\widehat{\gamma}_k) = 0$ and the variance is

$$\mathrm{var}(\widehat{\gamma}_k) = \mathrm{var}(\gamma_k) + \mathrm{var}\left(e_{\widehat{\gamma}}\right) = \sigma_\gamma^2 + s_{\widehat{\gamma}}^2$$

The likelihood function is

$$L\left(\sigma_\gamma^2\right) = -\frac{1}{2}\left[\ln\left(\sigma_\gamma^2 + s_{\widehat{\gamma}}^2\right) + \frac{\widehat{\gamma}_k^2}{\sigma_\gamma^2 + s_{\widehat{\gamma}}^2}\right]$$

The derivative of the likelihood function with respect to $\sigma_\gamma^2$ is.

$$\frac{\partial}{\partial \sigma_\gamma^2} L\left(\sigma_\gamma^2\right) = -\frac{1}{2}\left[\frac{1}{\sigma_\gamma^2 + s_{\widehat{\gamma}}^2} - \frac{\widehat{\gamma}_k^2}{\left(\sigma_\gamma^2 + s_{\widehat{\gamma}}^2\right)^2}\right]$$

Setting this derivative to zero, we have

$$-\frac{1}{2}\left[\frac{1}{\sigma_\gamma^2 + s_{\widehat{\gamma}}^2} - \frac{\widehat{\gamma}_k^2}{\left(\sigma_\gamma^2 + s_{\widehat{\gamma}}^2\right)^2}\right] = 0 \tag{18.38}$$

Further simplification of Eq. (18.38) leads to

$$\frac{\widehat{\gamma}_k^2}{\sigma_\gamma^2 + s_{\widehat{\gamma}}^2} = 1$$

The solution is

$$\widehat{\sigma}_\gamma^2 = \begin{cases} \widehat{\gamma}_k^2 - s_{\widehat{\gamma}}^2 & \text{for } \widehat{\gamma}_k^2 > s_{\widehat{\gamma}}^2 \\ 0 & \text{for } \widehat{\gamma}_k^2 < s_{\widehat{\gamma}}^2 \end{cases}$$

which is exactly the MM estimate of $\sigma_\gamma^2$ if negative solution is truncated at zero.

## 18.10  The Beavis Effect

There is an important concept in QTL mapping, the so-called Beavis effect (Beavis 1994; Otto and Jones 2000). In QTL mapping, a QTL is reported only if the test statistics is significant. When a reported effect is associated with a significance test, the effect is doomed to be biased. This is called the winner's curse. The effect of QTL in Beavis's original report (Beavis 1994) means the variance of a detected QTL. So, the Beavis effect describes a phenomenon that the variance of a small QTL detected in a small sample is biased upward due to censored (truncated) reports. A large QTL detected in a large sample will not show the Beavis effect. Many papers of QTL mapping from small samples may have been brutally rejected by reviewers and editors using the Beavis effect as a justification for the rejection. The Beavis effect only shows up in comparative genomics. If you do not compare your result with others, there is no Beavis effect at all. So, those reviewers and editors may have abused the Beavis effect. In this section, we will describe the theoretical basis of the Beavis effect (Xu 2003) and propose a mechanism to correct for the Beavis effect.

Figure 18.11 shows four truncated distributions (shaded areas). Panel A is a truncated normal distribution $N(2, 4)$ at a truncation point of 2. The mean of the truncated distribution is 3.5958, which is greater than 2, the mean of the original normal distribution. Panel B is a truncated non-central Chi-square distribution $\chi^2_{k=2}(\delta = 2)$ at a truncation point of 5. The mean of this truncated non-central Chi-square distribution is 8.2671, larger than 4, the mean of the original distribution. Panel C is a truncated normal distribution $N(2, 4)$ at a truncation point of 4. The mean of the truncated distribution is 5.0502, which is

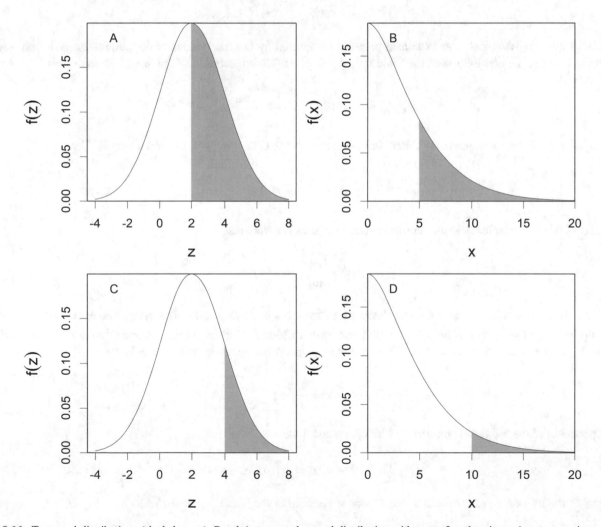

**Fig. 18.11**  Truncated distributions (shaded areas). Panel A: truncated normal distribution with mean 2 and variance 4 at a truncation point of 2. Panel B: truncated non-central Chi-square distribution with degree of freedom 2 and non-centrality parameter 2 at a truncation point of 5. Panel C: truncated normal distribution with mean 2 and variance 4 at a truncation point of 4. Panel D: truncated non-central Chi-square distribution with degree of freedom 2 and non-centrality parameter 2 at a truncation point of 10

greater than 2, the mean of the original normal distribution. Panel D is a truncated non-central Chi-square distribution $\chi^2_{k=2}(\delta = 2)$ at a truncation point of 10. The mean of this truncated non-central Chi-square distribution is 12.9762, larger than 4, the mean of the original distribution.

As illustrated in Fig. 18.11, parameters estimated from a censored data are biased. In this chapter, we only investigate the Beavis effect with a single marker analysis. The Wald test is defined as

$$W_k = \frac{\widehat{\gamma}_k^2}{\text{var}(\widehat{\gamma}_k)}$$

The variance in the denominator is

$$\text{var}(\widehat{\gamma}_k) = \frac{\sigma^2}{n\sigma_Z^2}$$

where $n$ is the sample size, $\sigma^2$ is the residual variance, and $\sigma_Z^2$ is the variance of the genotype indicator variable. The Wald test follows a non-central Chi-square distribution with a non-centrality parameter of

$$\delta = \frac{\gamma_k^2}{\text{var}(\gamma_k)} = \frac{n\sigma_Z^2}{\sigma^2}\gamma_k^2$$

which is simply the Wald test with estimated parameters replaced by the true values of the parameters. We may use this notation to describe the distribution of the Wald test, $W_k \sim \chi_1^2(\delta)$. The expectation of the Wald test statistic is

$$\text{E}(W_k) = \text{E}\left[\chi_1^2(\delta)\right] = \delta + 1 = \frac{n\sigma_Z^2}{\sigma^2}\sigma_\gamma^2 + 1$$

where $\sigma_\gamma^2 = \gamma_k^2$. This equation provides a mechanism to estimate the QTL variance, as shown below,

$$W_k = \frac{n\sigma_Z^2}{\sigma^2}\sigma_\gamma^2 + 1$$

Rearranging this equation leads to the moment estimate of the QTL variance,

$$\widehat{\sigma}_\gamma^2 = \frac{\sigma^2}{n\sigma_Z^2}(W_k - 1)$$

Let $t$ be the critical value of the Wald test above which the locus is claimed to be significant. The estimated $\sigma_\gamma^2$ should be biased upward because $W_k > t$ follows a truncated non-central Chi-square distribution. According to the definition of non-central Chi-square distribution (Li and Yu 2009), the non-centrality parameter is redefined as

$$\Delta = \frac{n\sigma_Z^2}{\sigma^2}\sigma_\gamma^2$$

The expectation of the truncated non-central Chi-square distribution is

$$\text{E}(W_k|W_k > t) = \text{E}\left[\chi_1^2(\delta)|\chi_1^2(\delta) > t\right] = \Delta + 1$$

This non-centrality parameter is related to the genetic variance after the truncation. Now let

$$\frac{n\sigma_Z^2}{\sigma^2}\sigma_\gamma^2 = E\left[\chi_1^2(\delta)|\chi_1^2(\delta) > t\right] - 1$$

Rearranging the above equation leads to

$$\sigma_\gamma^2(\text{biased}) = \frac{\sigma^2}{n\sigma_Z^2}\Delta = \frac{\sigma^2}{n\sigma_Z^2}\left\{E\left[\chi_1^2(\delta)|\chi_1^2(\delta) > t\right] - 1\right\} \tag{18.39}$$

This equation provides a way to evaluate the bias in $\sigma_\gamma^2$ due to significance test, i.e., the Beavis effect.

Now, let us use an example to show how to evaluate the bias. Assume that the mapping population is an $F_2$ population with the genotype ($Z$) coded as 1 for $A_1A_1$, 0 for $A_1A_2$, and $-1$ for $A_2A_2$. Assuming that the locus follows the typical Mendelian segregation ratio, i.e., 1:2:1, the variance of $Z$ is $\sigma_Z^2 = 0.5$. Assume that the genetic variance is $\sigma_\gamma^2 = \gamma_k^2 = 2^2 = 4$ and the heritability of the QTL is $h_{QTL}^2 = 0.05$. The residual variance is obtained by solving the following equation:

$$h_{QTL}^2 = \frac{\sigma_Z^2\sigma_\gamma^2}{\sigma_Z^2\sigma_\gamma^2 + \sigma^2} = \frac{0.5 \times 4}{0.5 \times 4 + \sigma^2} = 0.05$$

The solution is

$$\sigma^2 = \frac{1 - h_{QTL}^2}{h_{QTL}^2}\sigma_Z^2\sigma_\gamma^2 = \frac{1 - 0.05}{0.05} \times 0.5 \times 4 = 38 \tag{18.40}$$

When the sample size is $n = 100$, the non-centrality parameter is

$$\delta = \frac{n\sigma_Z^2}{\sigma^2}\sigma_\gamma^2 = \frac{100 \times 0.5}{38} \times 4 = 5.263158 \tag{18.41}$$

Assume that $t = 3.84$ so that the Wald test statistic follows a truncated non-central Chi-square distribution. The expectation of the truncated Wald test is

$$E\left[\chi_1^2(\delta)|\chi_1^2(\delta) > t\right] = 8.805996$$

The biased estimate of the QTL variance is

$$\begin{aligned}
\sigma_\gamma^2(\text{biased}) &= \frac{\sigma^2}{n\sigma_Z^2}\left\{E\left[\chi_1^2(\delta)|\chi_1^2(\delta) > t\right] - 1\right\} \\
&= \frac{38}{0.5 \times 100} \times (8.805996 - 1) \\
&= 5.932557
\end{aligned} \tag{18.42}$$

The biased QTL heritability is

$$h_{QTL}^2(\text{biased}) = \frac{\sigma_Z^2\sigma_\gamma^2}{\sigma_Z^2\sigma_\gamma^2 + \sigma^2} = \frac{0.5 \times 5.9325}{0.5 \times 5.9325 + 38} = 0.0724$$

The true QTL heritability is 0.05, but the biased heritability is 0.0724. If the truncation point is $t = 12.00$, the biased QTL variance will be 11.27718 and the biased heritability will be 0.1292. For the same truncation point of $t = 12$, if the sample size is $n = 500$, the biased QTL variance will be 4.13668, and the biased QTL heritability will be just 0.0516203, which is very close to the true value of 0.05.

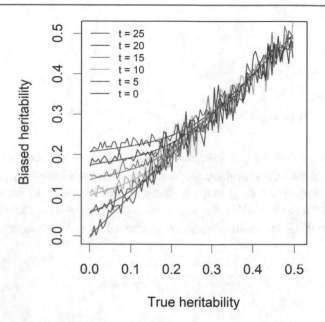

**Fig. 18.12** Bias in heritability due to significance test (the Beavis effect). The smooth curves are the theoretical biased heritability values and the noisy curves are averages of estimated values from 10 simulated samples

Let us perform some simulation studies to validate the theoretical value of the biased QTL variance. The true QTL variance is set to $\sigma_\gamma^2 = \gamma_k^2 = 4$. The genotype indicator variable of the locus is $Z$ with a theoretical variance of $\sigma_Z^2 = 0.5$. The sample size is fixed at $n = 100$. The residual variance is determined by the QTL heritability, as shown in Eq. (18.40). The QTL heritability ($h_{QTL}^2$) varies from 0 to 0.5 incremented by 0.005. The threshold of the test statistic varies from $t = 0$ to $t = 25$ incremented by 5. The biased heritability is plotted against the true heritability (see Fig. 18.12). Smaller heritability with higher threshold shows greater bias. When the actual heritability is greater than 0.3, the bias is barely noticeable.

Given the biased estimate of the QTL variance, we can correct the bias. The following equation serves as theoretical evaluation of the bias and the bias correction.

$$\sigma_\gamma^2(\text{biased}) = \frac{\sigma^2}{n\sigma_Z^2}\left\{ \mathrm{E}\left[\chi_1^2(\delta)\middle|\chi_1^2(\delta) > t\right] - 1\right\} \tag{18.43}$$

If $\sigma_\gamma^2$ is known then $\delta = \left(n\sigma_Z^2/\sigma^2\right)\sigma_\gamma^2$ is known and the theoretical bias is evaluated using Eq. (18.43). If the biased QTL variance is obtained, we can use Eq. (18.43) to solve for $\delta$ backward and thus obtained a bias corrected $\sigma_\gamma^2$. Now, let us show an example. If an estimated QTL variance is 5.9325 with a truncation point of $t = 3.84$, from Eq. (18.43), we can solve for $\mathrm{E}\left[\chi_1^2(\delta)\middle|\chi_1^2(\delta) > t\right]$ using

$$\begin{aligned} \mathrm{E}\left[\chi_1^2(\delta)\middle|\chi_1^2(\delta) > t\right] &= \frac{n\sigma_Z^2}{\sigma^2}\sigma_\gamma^2(\text{biased}) + 1 \\ &= \frac{0.5 \times 100}{38} \times 5.932557 + 1 \\ &= 8.805996 \end{aligned}$$

where we assume $\sigma_Z^2 = 0.5$, $n = 100$ and $\sigma^2 = 38$. In reality, these parameters should be replaced by their estimates. What is the non-centrality parameter to give the above expectation of the truncated non-central Chi-square distribution? Let

$$f(\delta) = \mathrm{E}\left[\chi_1^2(\delta)\middle|\chi_1^2(\delta) > 3.84\right] = 8.805996$$

We can solve for $\delta$ numerically, which is 5.263158. Using Eq. (18.41), we can solve for the unbiased QTL heritability

$$\sigma_\gamma^2 = \frac{\sigma^2}{n\sigma_Z^2}\delta = \frac{38}{100 \times 0.5} \times 5.263158 = 4$$

In real data analysis, correction for the bias is not as easy as shown in the above example. To calculate the non-centrality parameter after truncation on the Wald test statistic, Li and Yu (Li and Yu 2009) proposed a modified moment method of estimation using a single value of the Wald test $W_k$. Let $\beta$ be a number $0 < \beta < 1$ and

$$\beta = (t + 2)/(t + 2 + df)$$

is recommended (Li and Yu 2009), where $df = 1$ is the degree of freedom of the Wald test. Define

$$J(0) = \mathrm{E}\left[\chi_1^2(0)|\chi_1^2(0) > t\right]$$

as the expectation of the truncated central Chi-square distribution with 1 degree of freedom. Let

$$\kappa_\beta = \min\left\{\delta_{TM}(\kappa) = \beta(\kappa - t), \kappa \geq J(0)\right\}$$

i.e., $\kappa_\beta$ is the minimum solution of $\delta_{TM}(\kappa) = \beta(\kappa - t)$ when $\kappa \geq J(0)$. The modified estimate of the non-centrality parameter is

$$\delta_\beta(W_k) = \begin{cases} \beta(W_k - t) & t < W_k \leq \kappa_\beta \\ \delta_{TM}(W_k) & W_k > \kappa_\beta \end{cases}$$

where $\delta_{TM}(W_k)$ is the moment estimate of the non-centrality parameter satisfying

$$\mathrm{E}\left[\chi_1^2(\delta)|\chi_1^2(\delta) > t\right] = W_k$$

The bias corrected estimate of the QTL variance is

$$\sigma_\gamma^2(\text{unbiased}) = \frac{\sigma^2}{n\sigma_Z^2}\delta_\beta(W_k)$$

We performed another simulation experiment to show the bias corrected heritability estimates under $t = \{5, 10, 15, 20\}$. The sample size was $n = 100$ for both scenarios. The simulation was replicated 20 times. Figure 18.13 shows the result of this experiment. The bias-corrected heritability estimate varies around the diagonal line. The theoretical biased heritability (smooth curve) and empirical biased heritability are shown in the figure also for comparison with the bias-corrected heritability.

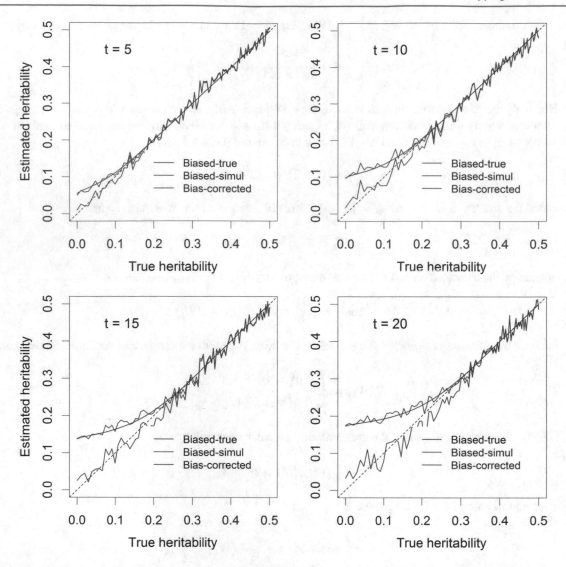

**Fig. 18.13** Result of bias-corrected heritability estimated from 20 replicated simulations. The upper panels show the results of simulation when the truncation points of the Wald test are $t = 5$ and $t = 10$. The lower panels show the results of simulation when the truncation point of the Wald test are $t = 15$ and $t = 20$

# References

Beavis WD. The power and deceit of QTL experiments: lessons from comparative QTL studies. In: the forty-ninth annual corn & sorghum industry research conference (ed. AST Association), 1994, pp. 250–266. American Seed Trade Association, Washington

Broman KW, Wu H, Sen Ś, Churchill GA. R/qtl: QTL mapping in experimental crosses. Bioinformatics. 2003;19:889–90.

Churchill GA, Doerge RW. Empirical threshold values for quantitative trait mapping. Genetics. 1994;138:963–71.

Dempster AP, Laird MN, Rubin DB. Maximum likelihood from incomplete data via the EM algorithm. J R Stat Soc B Methodol. 1977;39:1–38.

Feenstra B, Skovgaard IM, Broman KW. Mapping quantitative trait loci by an extension of the Haley-Knott regression method using estimating equations. Genetics. 2006;173:2269–82.

Golab GH, Heath M, Wahba G. Generalized cross-validation as a method for choosig a good ridge parameter. Dent Tech. 1979;21:215–23.

Haldane JBS. The combination of linkage values, and the calculation of distances between the loci of linked factors. J Genet. 1919;8:299–309.

Haley CS, Knott SA. A simple regression method for mapping quantitative trait loci in line crosses using flanking markers. Heredity. 1992;69:315–24.

Han L, Xu S. A Fisher scoring algorithm for the weighted regression method of QTL mapping. Heredity. 2008;101:453–64.

Hoerl AE, Kennard RW. Ridge regression: applications to nonorthogonal problems. Dent Tech. 1970a;12:69–82.

Hoerl AE, Kennard RW. Ridge regression: biased estimation for nonorthogonal problems. Dent Tech. 1970b;12:55–67.

Jansen RC, Stam P. High resolution of quantitative traits into multiple loci via interval mapping. Genetics. 1994;136:1447–55.

Jiang C, Zeng ZB. Mapping quantitative trait loci with dominant and missing markers in various crosses from two inbred lines. Genetica. 1997;101: 47–58.

Kang HM, Zaitlen NA, Wade CM, Kirby A, Heckerman D, Daly MJ, Eskin E. Efficient control of population structure in model organism association mapping. Genetics. 2008;178:1709–23.

Kao CH. On the differences between maximum likelihood and regression interval mapping in the analysis of quantitative trait loci. Genetics. 2000;156:855–65.

Lan H, Chen M, Flowers JB, Yandell BS, Stapleton DS, Mata CM, Mui ET, Flowers MT, Schueler KL, Manly KF, et al. Combined expression trait correlations and expression quantitative trait locus mapping. Public Lib Sci Genet. 2006;2:e6.

Lander ES, Botstein D. Mapping Mendelian factors underlying quantitative traits using RFLP linkage maps. Genetics. 1989;121:185–99.

Li Q, Yu K. Inference of non-centrality parameter of a truncated non-central chi-squared distribution. J Statist Plann Inf. 2009;139:2431–44.

Lippert C, Listgarten J, Liu Y, Kadie CM, Davidson RI, Heckerman D. FaST linear mixed models for genome-wide association studies. Nat Methods. 2011;8:833–5.

Lippert C, Listgarten J, Liu Y, Kadie CM, Davidson RI, Heckerman D. Improved linear mixed models for genome-wide association studies. Nat Methods. 2012;9:525–6.

Louis T. Finding the observed information matrix when using the EM algorithm. J R Stat Soc B Methodol. 1982;44:226–33.

Luo L, Mao Y, Xu S. Correcting the bias in estimation of genetic variances contributed by individual QTL. Genetica. 2003;119:107–14.

MacKay DJC. Bayesian interpolation. Neural Comput. 1992;4:415–47.

Nettleton D, Doerge RW. Accounting for variability in the use of permutation testing to detect quantitative trait loci. Biometrics. 2000;56:52–8.

Otto SP, Jones CD. Detecting the undetected: estimating the total number of loci underlying a quantitative trait. Genetics. 2000;156:2093–107.

Patterson HD, Thompson R. Recovery of inter-block information when block sizes are unequal. Biometrika. 1971;58:545–54.

Tipping ME. The relevance vector machine. In: Solla SA, et al., editors. Advances in Neural Information Processing Systems. Cambidge: MIT Press; 2000. p. 652–8.

Wahba G, Wold S. A completely automatic french curve: fitting spline functions by cross validation. Commun Statist. 1975a;4:1–17.

Wahba G, Wold S. Periodic splines for spectral density estimation: the use of cross-validation for determining the degree of smoothing. Commun Statist. 1975b;4:125–41.

Wang S, Basten CJ, Zeng Z-B. Windows QTL cartographer 2.5. Raleigh: Department of Statistics, North Carolina State University; 2012.

Xu S. A comment on the simple regression method for interval mapping. Genetics. 1995;141:1657–9.

Xu S. Further investigation on the regression method of mapping quantitative trait loci. Heredity. 1998a;80:364–73.

Xu S. Iteratively reweighted least squares mapping of quantitative trait loci. Behav Genet. 1998b;28:341–55.

Xu S. Theoretical basis of the Beavis effect. Genetics. 2003;165:259–2268.

Xu S. Mapping quantitative trait loci by controlling polygenic background effects. Genetics. 2013;195:1209–22.

Yu J, Pressoir G, Briggs WH, Vroh Bi I, Yamasaki M, Doebley JF, McMullen MD, Gaut BS, Nielsen DM, Holland JB, et al. A unified mixed-model method for association mapping that accounts for multiple levels of relatedness. Nat Genet. 2006;38:203–8.

Zeng ZB. Theoretical basis for separation of multiple linked gene effects in mapping quantitative trait loci. Proc Natl Acad Sci U S A. 1993;90: 10972–6.

Zeng ZB. Precision mapping of quantitative trait loci. Genetics. 1994;136:1457–68.

Zhang Z, Ersoz E, Lai CQ, Todhunter RJ, Tiwari HK, Gore MA, Bradbury PJ, Yu J, Arnett DK, Ordovas JM, et al. Mixed linear model approach adapted for genome-wide association studies. Nat Genet. 2010;42:355–60.

Zhou X, Stephens M. Genome-wide efficient mixed-model analysis for association studies. Nat Genet. 2012;44:821–4.

# Genome-Wide Association Studies

## 19.1 Introduction

There are many different technologies, study designs, and analytical tools for identifying genetic risk factors in human and quantitative trait loci in agriculture species. Genome-wide association studies (GWAS) are statistical approaches that involve rapidly scanning markers across the complete sets of DNA, or genomes, of a randomly selected population to find genetic variants associated with a particular trait. Once new genetic associations are identified, researchers can use the information to develop better strategies to detect, treat, and prevent diseases. Such studies are particularly useful in finding genetic variants that contribute to common, complex diseases, such as asthma, cancer, diabetes, heart disease, and mental illnesses. For animal and plant geneticists and breeders, genetic associations can help identify quantitative trait loci, clone genes, and facilitate marker-assisted selection. GWAS typically focus on associations between single-nucleotide polymorphisms (SNPs) and traits like some serious human diseases and grain yield in plants, but can equally be applied to any other organisms. Genome-wide association studies have evolved over the last 10 years into a powerful tool for investigating the genetic architectures of human diseases and complex traits in agricultural species. In this chapter, we will review the key concepts underlying GWAS and introduce various statistical methods used to perform GWAS. We will start with the simple method and then introduce the most advanced ones. The target traits for GWAS can be case-control (binary) outcomes or continuous traits. We will only focus on continuous traits.

Note that GWAS in random populations and QTL mapping in line crosses can use exactly the same set of statistical methods (Xu 2013). One difference between the two is that there is no need to fit population structures (covariates) in QTL mapping because a family of line crosses is often homogenous. However, if the QTL mapping population consists of recombinant inbred lines (RIL), the experiment can be replicated in multiple years and multiple locations. In this case, the year and location effects can be included in the model as fixed effects. These fixed effects are treated exactly the same as the population structure effects in GWAS. Therefore, the statistical methods introduced in this chapter also apply to QTL mapping. One fundamental difference between GWAS and the traditional QTL mapping is that GWAS analyzes markers only while the traditional QTL mapping also analyzes putative positions between markers.

## 19.2 Simple Regression Analysis

Simple regression analysis is an approach of genome scanning for testing one marker at a time. The model may contain other fixed effects such as covariates drawn from population structures, but there is no polygenic control. As a result, the simple regression method cannot properly control false positives. The popular GWAS software Plink (Purcell et al. 2007) actually performs such a simple linear regression analysis. Let $y$ be the phenotypic values of $n$ individuals in the population of interest. The simple linear model is

$$y = X\beta + Z_k\gamma_k + e$$

© Springer Nature Switzerland AG 2022
S. Xu, *Quantitative Genetics*, https://doi.org/10.1007/978-3-030-83940-6_19

where $X$ is a design matrix of covariates, $\beta$ is a vector of effects for the covariates to reduce interference from other non-genetic factors, $\gamma_k$ is the effect of marker $k$ for $k = 1, \ldots, m$ where $m$ is the total number of markers and $e$ is a vector of residual errors with an assumed $N(0, \sigma^2)$ distribution. For SNP data, $Z_k$ is an $n \times 1$ vector of genotype indicator variables and the $j$th element (individual) is defined as

$$
Z_{jk} = \begin{cases} +1 & \text{for } A_1A_1 \\ 0 & \text{for } A_1A_2 \\ -1 & \text{for } A_2A_2 \end{cases}
$$

The size of this locus is measured by the total phenotypic variance explained by this locus, which is defined as

$$
h_k^2 = \frac{\sigma_{G_k}^2}{\sigma_{G_k}^2 + \sigma^2} = \frac{\mathrm{var}(Z_k)\gamma_k^2}{\mathrm{var}(Z_k)\gamma_k^2 + \sigma^2} = \frac{\sigma_{Z_k}^2 \gamma_k^2}{\sigma_{Z_k}^2 \gamma_k^2 + \sigma^2}
$$

where $\mathrm{var}(Z_k) = \sigma_{Z_k}^2$ represents the variance of the $k$th marker and $\sigma_{G_k}^2 = \sigma_{Z_k}^2 \gamma_k^2$ is the genetic variance of the $k$th marker. Estimation and hypothesis test for the simple regression analysis is straightforward. In matrix notation, the estimates are

$$
\begin{bmatrix} \widehat{\beta} \\ \widehat{\gamma}_k \end{bmatrix} = \begin{bmatrix} X^TX & X^TZ_k \\ Z_k^TX & Z_k^TZ_k \end{bmatrix}^{-1} \begin{bmatrix} X^Ty \\ Z_k^Ty \end{bmatrix}
$$

The residual error variance is estimated using

$$
\widehat{\sigma}^2 = \frac{1}{n-q-1} \left( y - X\widehat{\beta} - Z_k\widehat{\gamma}_k \right)^T \left( y - X\widehat{\beta} - Z_k\widehat{\gamma}_k \right) \tag{19.1}
$$

where $q$ is the number of columns (covariates including the intercept) of matrix $X$. When the sample size is sufficiently large, the numerator of Eq. (19.1) can be replaced by $n$. The variance-covariance matrix of the estimates is

$$
\mathrm{var}\begin{bmatrix} \widehat{\beta} \\ \widehat{\gamma}_k \end{bmatrix} = \begin{bmatrix} \mathrm{var}\left(\widehat{\beta}\right) & \mathrm{cov}\left(\widehat{\beta}, \widehat{\gamma}_k\right) \\ \mathrm{cov}\left(\widehat{\gamma}_k, \widehat{\beta}\right) & \mathrm{var}(\widehat{\gamma}_k) \end{bmatrix} = \begin{bmatrix} X^TX & X^TZ_k \\ Z_k^TX & Z_k^TZ_k \end{bmatrix}^{-1} \widehat{\sigma}^2 = \begin{bmatrix} C_{XX} & C_{XZ} \\ C_{ZX} & C_{ZZ} \end{bmatrix} \widehat{\sigma}^2
$$

Note that $\mathrm{var}(\widehat{\gamma}_k) = C_{ZZ}\widehat{\sigma}^2$ and $C_{ZZ}$ is the lower left diagonal element of matrix $C$ and $C_{ZZ} \neq \left(Z_k^TZ_k\right)^{-1}$. The Wald test statistic for $H_0 : \gamma_k = 0$ is

$$
W_k = \frac{\widehat{\gamma}_k^2}{\mathrm{var}(\widehat{\gamma}_k)}
$$

Under the null hypothesis, $W_k$ follows a Chi-square distribution with 1 degree of freedom (exact distribution is $F$ with 1 and $n - q - 1$ degrees of freedom). The $p$-value is calculated using

$$
p_k = 1 - \mathrm{Pr}\left(\chi_1^2 < W_k\right)
$$

where $\chi_1^2$ is a Chi-square variable with 1 degree of freedom. In GWAS, people often convert the $p$-value into

$$
\eta_k = -\log_{10}(p_k)
$$

and report $\eta_k$ as the test statistic. The entire genome is then scanned. Any markers with $\eta_k$ larger than a predetermined critical value are declared as being associated with the trait of interest. The major flaw of the simple regression analysis is the inability to control false-positive rate. Therefore, this method is often used as an example to demonstrate the advantages of more advanced methods of GWAS.

## 19.3 Mixed Model Methodology Incorporating Pedigree Information

Zhang et al. (2005) developed a mixed model methodology for mapping QTL using multiple inbred lines of maize. They collected over 400 varieties (inbred lines) of maize derived from the past 80 years in the USA. The pedigree relationship of the lines was known exactly. Their mixed model included covariates (fixed effects), a marker (treated as a random effect), and a polygenic effect (random effect) whose covariance structure is determined by the additive relationship matrix inferred from the pedigree. The model is described as

$$y = X\beta + Z_k\gamma_k + W\alpha + e$$

The following defines entries of the above model. $X\beta$ remains the same as defined earlier in the simple regression analysis. $Z_k$ is a design matrix with $n$ rows and $s$ columns, where $s$ is the number of founders in the population. This matrix is inferred from marker information. It contains the probabilities of lines inheriting from the founder alleles given the marker genotypes. Therefore, $\gamma_k$ is a $s \times 1$ vector of founder effects. $W$ has the same dimension as $Z_k$ but it is the expectation of $Z_k$ across all loci. $\alpha$ is an $s \times 1$ vector of polygenic effects of all founders. $e$ is a vector of residual errors with an assumed $N(0, I\sigma^2)$ distribution (multivariate normal distribution). Both the marker specific effects and polygenic effects are assumed to be random effects with $\gamma_k \sim N(0, I\sigma_k^2)$ and $\alpha \sim N(0, I\sigma_\alpha^2)$ distributions, respectively. Under these assumptions, the expectation of $y$ is $E(y) = X\beta$ and the variance is

$$\text{var}(y) = V = Z_k Z_k^T \sigma_k^2 + WW^T \sigma_\alpha^2 + I\sigma^2$$

Zhang et al. (2005) defined $\Pi_k = Z_k Z_k^T$ and called it the marker-specific IBD matrix, $\Pi = WW^T$ and called it the additive relationship matrix, and rewrote the variance as

$$\text{var}(y) = V = \Pi_k \sigma_k^2 + \Pi\sigma_\alpha^2 + I\sigma^2$$

They used the restricted maximum likelihood (REML) method to estimate the variance components and performed a likelihood ratio test for $H_0 : \sigma_k^2 = 0$,

$$\lambda_k = -2\left[L(0) - L(\widehat{\sigma_k^2})\right]$$

where $L(0)$ is the log likelihood function evaluated under the null model (the pure polygenic model) and $L(\widehat{\sigma_k^2})$ is the log likelihood function evaluated under the alternative model. The method was implemented using the MIXED procedure in SAS (PROC MIXED). Since the title of the paper did not include the key word "GWAS," the paper was published in Genetics (not Nature Genetics). Nobody knows that the method is a prototype of the mixed linear model (MLM) published a year later by Yu et al. (2006). Zhang et al. (2005) did not provide a user friendly program to implement the mixed model procedure, which partly explains why the method has not been cited 16 years after its publication.

## 19.4 Mixed Linear Model (MLM) Using Marker-Inferred Kinship

A year later after Zhang et al. (2005), Yu et al. (2006) published a similar mixed model procedure for GWAS, but they used a marker-inferred kinship matrix $K$ in place of Zhang's $\Pi$ matrix. They replaced the $X$ matrix by a $Q$ matrix (population structure) and treated $\gamma_k$ as a fixed effect and directly estimated and tested this marker effect. The model was written in the following form

$$y = Q\alpha + X\beta + Zu + e$$

where $Q\alpha$ is the fixed effect part to control the population structure, $X\beta$ is the marker genotype indicator multiplied by the marker effect (replacing our $Z_k\gamma_k$), $Zu$ captures the polygene (replacing our $W\alpha$). The expectation of $y$ is $\mathrm{E}(y) = Q\alpha + X\beta$ and the variance is

$$\mathrm{var}(y) = V = ZKZ^T \sigma_u^2 + I\sigma^2$$

where $K$ is the kinship matrix, which is the covariance structure in the form of $\mathrm{var}(u) = K\sigma_u^2$, where $\sigma_u^2$ is the polygenic variance. The $Z$ matrix depends on the type of populations and often takes the form of identity if $u$ is an $n \times 1$ polygenic effects. The "kinship" matrix is calculated from markers of the whole genome. The genotype indicator for marker $k$ is $X$ and the marker effect is $\beta$ (marker index is ignored). The test statistics is again the Wald test defined as

$$W = \frac{\widehat{\beta}^2}{\mathrm{var}\left(\widehat{\beta}\right)}$$

Such a test is done for every marker and the entire genome is scanned to complete the GWAS. This mixed model is called the Q + K model and is considered the first method of this kind. Yu was eventually honored the "father" of the modern GWAS procedure. The linear mixed model procedure (they call it mixed linear model) is implemented in a software package called TASSEL (Bradbury et al. 2007).

## 19.5  Efficient Mixed Model Association (EMMA)

The mixed model in the original form is computationally very expensive because the inclusion of the kinship matrix. Kang et al. (2008) proposed an eigenvalue decomposition method to improve the computational speed. They first decomposed the $K$ matrix to simplify the computation and then estimated the polygenic and residual variance components for each marker screened by treating the variance ratio $\sigma^2/\sigma_u^2$ as the parameter. The method is called efficient mixed model association (EMMA). Zhou and Stevens (2012) further investigated the method and gave a new name called genome-wide efficient mixed model association (GEMMA). Other than the recursive algorithm to evaluate the likelihood function and the beautiful presentation, I do not see much value of GEMMA compared with EMMA. I originally thought that EMMA treated the estimated $\sigma^2/\sigma_u^2$ from the pure polygenic model (no marker effect is included) as a constant and used it to scan the genome without further estimation of the ratio. I thought that the re-estimation of the ratio for each marker was the improvement of GEMMA over EMMA. In fact, EMMA is also re-estimating the ratio for each marker. This puzzled me how GEMMA was published in Nature Genetics while the original idea of eigenvalue decomposition is already presented in Kang et al. (2008).

I now introduce these methods using my notation with some modification to help readers understand EMMA. The linear mixed model for the $k$th marker ($k = 1, \ldots, m$) is

$$y = X\beta + Z_k\gamma_k + \xi + \varepsilon \tag{19.2}$$

where $X$ is a design matrix for $q$ fixed effects and $\beta$ are the fixed effects themselves. The fixed effects are included in the model to capture effects of any covariates that are not relevant to the marker effects, e.g., population structure and age effect. The remaining variables in the model are defined as follows. The $Z_k$ vector stores the genotype indicator variable for all individuals and the $j$th element of it is defined as $Z_{jk} = 1$ for the homozygote of the major allele, $Z_{jk} = 0$ for the heterozygote and $Z_{jk} = -1$ for the homozygote of the minor allele. The effect of the $k$th marker is denoted by $\gamma_k$ and treated as fixed effect. The polygenic term $\xi$ is assumed to be random with mean 0 and variance

$$\mathrm{var}(\xi) = K\phi^2$$

where $K$ is the covariance structure (kinship matrix) and $\sigma_\xi^2$ is the polygenic variance. You may wonder how the $K$ matrix is derived from the markers. We can write $\xi$ explicitly using

$$\xi = \sum_{k=1}^{m} Z_k \delta_k \tag{19.3}$$

The $Z_k$'s in Eqs. (19.3) and (19.2) are the same, but $\gamma_k$ and $\delta_k$ in the two equations are different because they are from different distributions; the former is a fixed effect (no distribution) and the latter is a random variable from a normal distribution, i.e., $\delta_k \sim N(0, \phi^2)$ for all $k = 1, \ldots, m$. So, each marker is assumed to have two effects, one being a major effect and the other being a component of the polygene. The major effect determines the association but the polygenic component is small and assume to be sampled from the same normal distribution for all markers. The variance of $\xi$ is

$$\text{var}(\xi) = \sum_{k=1}^{m} Z_k \text{var}(\delta_k) Z_k^T = \sum_{k=1}^{m} Z_k Z_k^T \phi^2 = K \phi^2$$

Therefore,

$$K = \sum_{k=1}^{m} Z_k Z_k^T$$

If $m$ is small, we can directly use matrix multiplication to calculate $K$ without using the summation. In other words, $K = ZZ^T$, where $Z$ is an $n \times m$ matrix. There are more than a dozen different forms of $K$ in the literature, but this is the only one directly derived from the properties of variance. The residual errors are assumed to be $\varepsilon \sim N(0, I\sigma^2)$. Going back to model (19.2), the expectation is

$$\text{E}(y) = X\beta + Z_k \gamma_k$$

and the variance is

$$\text{var}(y) = K\phi^2 + I\sigma^2 = (K\lambda + I)\sigma^2$$

where $\lambda = \phi^2/\sigma^2$ is the variance ratio. This ratio is marker specific because it depends on the fixed effects, which include the $k$th marker effect. We use the restricted maximum likelihood (REML) to estimate parameters. Evaluation of the restricted log likelihood function can be very costly when the sample size is very large. Let $T_k = (X\|Z_k)$ be the horizontal concatenation of the two matrices, equivalent to the `cbind(X, Zk)` in R. The log likelihood function for REML is

$$L(\lambda) = -\frac{1}{2} \ln |H| - \frac{1}{2} \ln |T_k^T H^{-1} T_k| - \frac{n-q-1}{2} \ln (y^T P_k y) \tag{19.4}$$

where

$$H = K\lambda + I$$

and

$$P_k = H^{-1} - H^{-1} T_k (T_k^T H^{-1} T_k)^{-1} T_k^T H^{-1}$$

The fixed effect and the residual variance have been absorbed (profiled likelihood function). Therefore, this likelihood function only involves one parameter $\lambda$. Any numerical algorithm can be used to search for the REML estimate, denoted by $\widehat{\lambda}$. Once $\lambda$ is replaced by $\widehat{\lambda}$, we can find $\beta$ and $\gamma_k$ using

$$\begin{bmatrix} \widehat{\beta} \\ \widehat{\gamma}_k \end{bmatrix} = \begin{bmatrix} X^T H^{-1} X & X^T H^{-1} Z_k \\ Z_k^T H^{-1} X & Z_k^T H^{-1} Z_k \end{bmatrix}^{-1} \begin{bmatrix} X^T H^{-1} y \\ Z_k^T H^{-1} y \end{bmatrix}$$

which is called the mixed model equation (Henderson 1953). The residual error variance is estimated using

$$\widehat{\sigma}^2 = \frac{1}{n-q-1} \left( y - X\widehat{\beta} - Z_k\widehat{\gamma}_k \right)^T H^{-1} \left( y - X\widehat{\beta} - Z_k\widehat{\gamma}_k \right)$$

The variance matrix of the estimated fixed effects is

$$\mathrm{var}\begin{bmatrix} \widehat{\beta} \\ \widehat{\gamma}_k \end{bmatrix} = \begin{bmatrix} X^T H^{-1} X & X^T H^{-1} Z_k \\ Z_k^T H^{-1} X & Z_k^T H^{-1} Z_k \end{bmatrix}^{-1} \widehat{\sigma}^2 = \begin{bmatrix} \mathrm{var}\left(\widehat{\beta}\right) & \mathrm{cov}\left(\widehat{\beta},\widehat{\gamma}_k\right) \\ \mathrm{cov}\left(\widehat{\gamma}_k,\widehat{\beta}\right) & \mathrm{var}(\widehat{\gamma}_k) \end{bmatrix}$$

The Wald test statistic for $H_0 : \gamma_k = 0$ is

$$W_k = \frac{\widehat{\gamma}_k^2}{\mathrm{var}(\widehat{\gamma}_k)}$$

Under the null hypothesis, $W_k$ follows a Chi-square distribution with 1 degree of freedom (exact distribution is $F$ with 1 and $n - q - 1$ degrees of freedom). The analysis is identical to the simple regression analysis except that there is a weight matrix $H^{-1}$ involved in the linear mixed model. The mixed model requires multiple iterations for estimating $\lambda$, which is marker specific, and the entire genome has to be scanned. This method is called the exact method by Zhou and Stevens (2012).

Corresponding to the exact method, there is an approximate method, in which $\lambda = \widehat{\lambda}$ where $\widehat{\lambda}$ is the estimated variance ratio from the pure polygenic model. In other words, the variance ratio is fixed for all loci. Such an approximated method is a typical weighted least square method. The weight matrix $H^{-1}$ is only calculated once from the pure polygenic model and will be kept as constant when the genome is scanned. The efficient mixed model association expedited (EMMAX) method (Kang et al. 2010) is such an approximate method. Zhang et al. (2010) called the approximate method population parameters predefined (P3D).

The exact method does not cost too much additional computing time compared with the approximate method because of the implementation of the eigenvalue decomposition algorithm by both the EMMA and GEMMA methods. The eigenvalue decomposition is performed on the kinship matrix $K$ as

$$K = UDU^T$$

where $U$ is an $n \times n$ matrix of eigenvectors and $D$ is a diagonal matrix holding the eigenvalues, $D = \mathrm{diag}\{d_1, \ldots, d_n\}$. The decomposition is only conducted once prior to the GWAS. The $H$ matrix is

$$H = K\lambda + I = UDU^T + I = U(D\lambda + I)U^T$$

Therefore, the log determinant of matrix $H$ involved in Eq. (19.4) is

$$\ln |H| = \ln |D\lambda + I| = \sum_{j=1}^n \ln(d_j\lambda + 1)$$

because $D$ is a diagonal matrix and $UU^T = I$. Various quadratic forms are involved in the likelihood function, Eq. (19.4) in the form of $a^T H^{-1} b$, for example, $X^T H^{-1} X$, $X^T H^{-1} y$, and $y^T H^{-1} y$. Using eigenvalue decomposition, we can rewrite the quadratic form by

$$a^T H^{-1} b = a^T U (D\lambda + I)^{-1} U^T b = a^{*T}(D\lambda + I)^{-1} b^* = \sum_{j=1}^{n} a_j^{*T} b_j^* \left( d_j \lambda + 1 \right)^{-1} \tag{19.5}$$

With such a decomposition, matrix inversion has been replaced by the above summations and the computational speed has been improved significantly. As a result, the exact method by updating $\lambda$ for each locus is possible (Zhou and Stephens 2012) although the improvement of the exact method over the approximate method has never been shown to be sufficiently significant to change the conclusions of a study.

A technical remark on computing the "quadratic form" (19.5), looping in R is slow and, therefore, you have to avoid directly using the summation in (19.5) to compute the "quadratic form." Instead, take advantage of built-in R functions. Let $d = \{d_j\}$ for $j = 1, \cdots, n$ be an $n \times 1$ array and define $w = 1/(d\lambda + 1)$ as an $n \times 1$ array. Let $A = U^T a$ and $B = U^T b$. The R code to calculate the "quadratic form" is.

```
aWb = crossprod(A*sqrt(w),B*sqrt(w))
```

If $b = a$, Eq. (5) is a true quadratic form, the R code is.

```
aWa = crossprod(A*sqrt(w))
```

The `crossprod()` function is a built-in R function written in C++ or FORTRAN, which is very fast.

## 19.6 Decontaminated Efficient Mixed Model Association (DEMMA)

The polygenic background control is similar to the composite interval mapping using co-factors to control the background effects in QTL mapping. However, if the number of markers used to calculate the kinship matrix is small, the effect from the polygene ($\delta_k$) will compete with the effect of the major effect ($\gamma_k$). This will lead to a decreased power. The decreased false-positive rate of EMMA will be compromised by a decreased power, which is not a desired property of the mixed model procedure. Can we avoid or reduce the competition? Wang et al. (2016) and Wei and Xu (2016) proposed a method that can boost the power but still keep the low false-positive rate. The method first calculates the BLUP value for each $\delta_k$ using.

$$\widehat{\delta}_k = \mathrm{E}(\delta_k | y) = Z_k \widehat{\lambda} \left( K\widehat{\lambda} + I \right)^{-1} \left( y - X\widehat{\beta} \right) \tag{19.6}$$

From this BLUP value, they defined $\widehat{\xi}_k = Z_k \widehat{\delta}_k$ as the polygenic component contributed by marker $k$. Using this term, they adjusted the original model (19.2) by

$$y = X\beta + Z_k \gamma_k + \xi - \widehat{\xi}_k + \varepsilon$$

A rearrangement of this equation leads to

$$y + \widehat{\xi}_k = X\beta + Z_k \gamma_k + \xi + \varepsilon$$

Let $y_k = y + \widehat{\xi}_k$, we have

$$y_k = X\beta + Z_k \gamma_k + \xi + \varepsilon$$

The right hand side of the equation is the same as that of the original model (2), but the left hand side is an adjusted phenotype. Such an adjustment is not necessary when the number of markers used to calculate the kinship matrix is large, say >5000. Since the kinship matrix defined this way, $K = ZZ^T$, depends on the number of loci, a very large number of loci will cause a very large value of the $K$ matrix, which eventually leads to an extremely small $\widehat{\delta}_k$, as shown in Eq. (19.6). A small $\widehat{\delta}_k$ will not compete with $\gamma_k$. However, if the number of loci used to calculate the kinship matrix is small, the competition can be serious and the adjustment is necessary.

The method of Wang et al. (2016) and Wei and Xu (2016) is an ad hoc approach. Although it works satisfactorily based on their empirical data analysis, the ad hoc method lacks the beauty of rigorous statistical derivation. A method with a solid statistical foundation was proposed by Lippert et al. (2011, 2012), which is presented below.

The polygenic effect in the original EMMA model is the sum of all marker effects that are modeled as random effects,

$$\xi = \sum_{k'=1}^{m} Z_{k'}\delta_{k'} = Z_k\delta_k + \sum_{k'\neq k}^{m} Z_{k'}\delta_{k'}$$

The linear mixed model already includes effect of marker $k$ as a fixed effect, denoted by $\gamma_k$. So there are two effects for each marker scanned, $\gamma_k$ and $\delta_k$, and these two effects compete each other, leading to a reduced power for testing $H_0 : \gamma_k = 0$. This phenomenon is called proximal contamination (Listgarten et al. 2012). To decontaminate EMMA, they excluded $Z_k\delta_k$ from the polygene and redefine the polygene as

$$\xi_{-k} = \sum_{k'\neq k}^{m} Z_{k'}\delta_{k'}$$

The revised mixed model is

$$y = X\beta + Z_k\gamma_k + \xi_{-k} + \varepsilon$$

We assume that $\delta_{k'}\sim N(0, \phi^2)$ for all $k' = 1, \cdots, m$, where $m$ is the total number of markers. The expectation of $y$ is $E(y) = X_k\beta_k = X\beta + Z_k\gamma_k$ and the variance matrix is $\text{var}(y) = V_k$, where.

$$V_k = \sum_{k'\neq k}^{m} Z_{k'}Z_{k'}^T\phi^2 + I\sigma^2 = K_{-k}\phi^2 + I\sigma^2$$

and

$$K_{-k} = \sum_{k'\neq k}^{m} Z_{k'}Z_{k'}^T = K - Z_kZ_k^T$$

which is a kinship matrix inferred from $m - 1$ markers that excludes marker $k$. This marker-specific kinship matrix will change for every marker scanned, presenting a tremendous increase in computational burden. Therefore, the marker-specific kinship matrix is replaced by an adjusted kinship matrix inferred from all $m$ markers of the entire genome. This allows us to deal with $K$ and modify it to get $K_{-k}$. The $V_k$ matrix can be written as

$$V_k = K_{-k}\phi^2 + I\sigma^2 = K\phi^2 - Z_kZ_k^T\phi^2 + I\sigma^2 = K\phi^2 + I\sigma^2 - Z_kZ_k^T\phi^2$$

which is further simplified into

$$V_k = [(K\lambda + I) - Z_kZ_k^T\lambda]\sigma^2 = (H - Z_kZ_k^T\lambda)\sigma^2 = H_k\sigma^2$$

where $\lambda = \phi^2/\sigma^2$ and $H_k = H - Z_kZ_k^T\lambda$. The restricted likelihood function involves $H_k^{-1}$ and $|H_k|$. Since matrix $H_k$ is a highly structured matrix, we can use the Sherman-Morris-Woodbury matrix identity (Woodbury 1950; Golub and Van Loan 1996) to update $H$,

$$H_k^{-1} = H^{-1} - H^{-1}Z_k\lambda(Z_k^TH^{-1}Z_k\lambda - I)^{-1}Z_k^TH^{-1}$$

and

$$| H_k | = | H || Z_k^T H^{-1} Z_k - \lambda^{-1} || - \lambda | = | H || I - Z_k^T H^{-1} Z_k \lambda |$$

where the inverse and determinant of $H$ are already simplified via eigenvalue decomposition. Estimated effects and their variance matrix are obtained using the same formulas as those of the EMMA method. The Wald test for marker $k$ is defined as

$$W_k^{\text{DEMMA}} = \frac{\left(\hat{\gamma}_k^{\text{DEMMA}}\right)^2}{\text{var}\left(\hat{\gamma}_k^{\text{DEMMA}}\right)}$$

Under the null model $H_0 : \gamma_k = 0$, $W_k^{\text{DEMMA}}$ follows a Chi-square distribution with one degree of freedom so that the $p$-value can be easily obtained from this Chi-square distribution.

## 19.7 Manhattan Plot and Q-Q Plot

The result of a GWAS is often presented graphically with the test statistic being the $y$-variable and the genome location being the $x$-variable. The test statistic we have learned so far is the Wald test statistic, from which a p-value is calculated. Therefore, the p-value is often presented, rather than the Wald test statistic. If the p-value of a marker is smaller than a predetermined cut-off value, say $10^{-7}$, the marker is claimed to be associated with the trait. The p-values themselves, however, cannot be visualized graphically. Therefore, we often present the log transformation of the p-value, $\eta_k = -\log_{10}(p_k)$. The reverse relationship is $p_k = 10^{-\eta_k}$, so $\eta_k$ has a very intuitive interpretation. The plot of $\eta_k$ against the marker position is called the Manhattan plot because the graph can look like a picture of the Manhattan in New York City. Figure 19.1 is an example of the Manhattan plot of the mouse genome for testing of selection signature (Xu and Garland 2017). The majority of markers have $\eta_k \leq 1$. Some have values slightly above 1. Occasionally, one or two markers have very large $\eta_k$, say $\eta_k = 7$. The markers with very large $\eta_k$ values may not be explained by chance and thus can be claimed to be associated with the trait of interest. Figure 19.2 is a picture of Manhattan in New York City of the USA. Comparing the Manhattan plot (Fig. 19.1) with the Manhattan district (Fig. 19.2), we do find some similarities. A necessary condition to make the Manhattan plot look like Manhattan is the linkage disequilibrium (LD) between neighboring markers. If all loci are independent (no LD), the Manhattan plot will look like the one shown in Fig. 19.3, where a high peak is just an isolated point.

In addition to the Manhattan plot, a quantile-quantile plot, i.e., a Q-Q plot, is also commonly seen in GWAS reports. In statistics, a Q-Q plot is a probability plot, which is a graphical method for comparing two probability distributions by plotting their quantiles against each other. First, the set of intervals for the quantiles is chosen. A point $(x, y)$ on the plot corresponds to one of the quantiles of the second distribution ($y$-coordinate) plotted against the same quantile of the first distribution

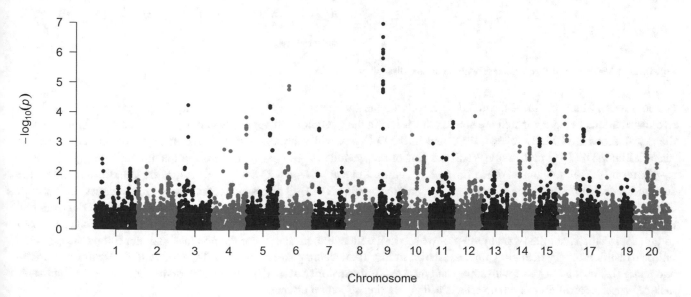

**Fig. 19.1** Manhattan plot of the mouse genome for GWAS of selection signature (Xu and Garland 2017)

**Fig. 19.2** A picture of Manhattan of the New York City (Tony Jin, https://en.wikipedia.org/wiki/Manhattan#/media/File:Lower_Manhattan_from_Jersey_City_November_2014_panorama_3.jpg)

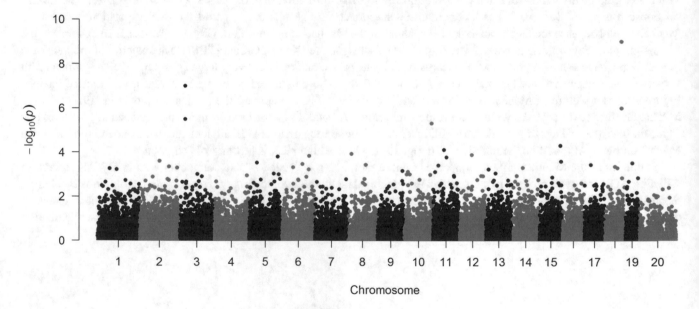

**Fig. 19.3** A Manhattan plot for a genome with no linkage disequilibrium (LD)

(*x*-coordinate). Thus, the line is a parametric curve with the parameter being the number of the interval for the quantile. If the two distributions being compared are similar, the points in the Q-Q plot will approximately lie on the diagonal line, $y = x$. If the distributions are linearly related, the points in the Q-Q plot will approximately lie on a line, but not necessarily on the diagonal line. The Q-Q plot in a GWAS is a plot of the quantile of $-\log_{10}(p_k)$ against the quantile of $-\log_{10}(u_k)$, where $u_k \sim \text{Uniform}(0, 1)$ is random variable for $k = 1, \ldots, m$. Let $u_{(k)} = k/(n + 1)$ for $k = 1, \ldots, m$ be the $k$th quantile of the standardized uniform distribution. Let $p_{(k)}$ be the ordered p-value. The Q-Q plot is the plot of $-\log_{10}(p_{(k)})$ against $-\log_{10}(u_{(k)})$. Table 19.1 shows 10 ordered p-values and the corresponding theoretical quantiles of a uniform (*u*) distribution. For example, $u_1 = 1/(10 + 1) = 0.0909$ and $u_5 = 5/(10 + 1) = 0.4545$. Figure 19.4 shows the Q-Q plot of the mouse GWAS data presented in Fig. 19.1. This is not the typical Q-Q plot we see in the literature, meaning that the p-values may not follow the expected uniform distribution. Typical Q-Q plots are shown in Fig. 19.5, where panel A shows a Q-Q plot with no significant markers and panel B shows a Q-Q plot with significant markers. If the majority of markers deviate from the diagonal line, it indicates lack of proper control for polygenic effects and population structural effects.

**Table 19.1** A sample of ten ordered p-values against ten quantiles of a standardized uniform random variable

| Order ($k$) | 1 | 2 | 3 | 4 | 5 | 6 | 7 | 8 | 9 | 10 |
|---|---|---|---|---|---|---|---|---|---|---|
| $p_{(k)}$ | 0.01 | 0.15 | 0.20 | 0.25 | 0.41 | 0.52 | 0.69 | 0.82 | 0.95 | 0.99 |
| $u_{(k)}$ | 0.0909 | 0.1818 | 0.2727 | 0.3636 | 0.4545 | 0.5455 | 0.6364 | 0.7273 | 0.8182 | 0.9091 |

**Fig. 19.4** Q-Q plot of the mouse GWAS data for selection signature (Xu and Garland 2017)

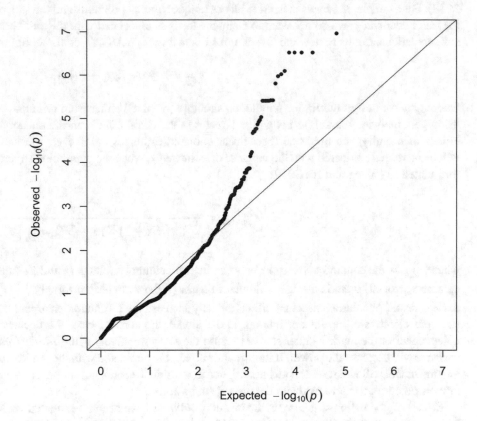

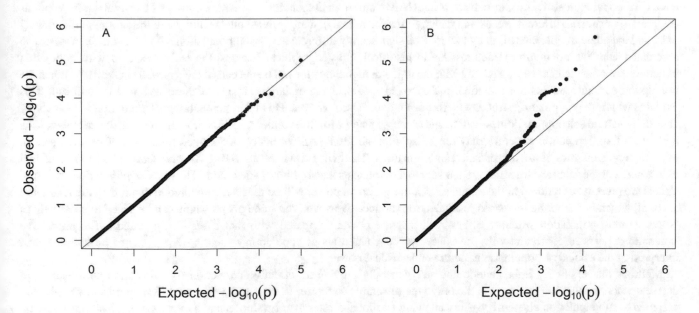

**Fig. 19.5** Typical Q-Q plots of GWAS results, where panel A shows no loci are detected and panel B shows some significant loci

## 19.8    Population Structure

Population structure is often caused by population heterogeneity (or admixture) represented by multiple ethnic groups or subpopulations within the association population (Pritchard et al. 2000a, b). The purpose of fitting population structure effects into the linear mixed model is to reduce false positives for loci that are confounded with population structure (Toosi et al. 2018). For example, if a locus is fixed to alleles unique to subpopulations and the subpopulations are strongly associated with the trait under study, we do not want to claim the locus as associated with the trait because the association may be caused by subpopulations. Let us review the Q + K mixed model for GWAS (Yu et al. 2006),

$$y = \mu + Q\alpha + Z_k\gamma_k + \xi + e$$

where $Q$ is the design matrix for population structure (obtained either from principal component analysis or cluster analysis using genome-wide markers), $\alpha$ is a $q \times 1$ vector of structural effects on the phenotype. If the model is true, the estimated effects of $\alpha$ and $\gamma_k$ are unbiased (best linear unbiased estimates). However, the variance of $\widehat{\gamma}_k$ with population structure will be increased compared with the variance of estimated $\gamma_k$ when the population structure is absent. The increased variance is formulated as (Wang and Xu 2019)

$$\text{var}(\widehat{\gamma}_k) = \frac{\sigma^2}{\sum_{j=1}^{n}(d_j\lambda + 1)^{-1}\left(1 - \sum_{i=1}^{q} r_{ZQ_i}^2\right)\sigma_Z^2}$$

where $r_{ZQ_i}^2$ is the squared correlation between the $i$th column of matrix $Q$ and $Z_k$ (under the additive model, $Z_k$ is a single column vector). If there is a single column of matrix $Q$, the extra term is simply $1 - r_{ZQ}^2$, which is a fraction between 0 and 1. As a result, population structure effects actually increase the estimation error and thus lower the power. If the population structure effects are present but ignored in the model, the consequence is a decreased power (if the structure effects are independent of the marker under study) because the structure effects will go to the residual error. An inflated residual error variance will decrease the power. If the structure effects are correlated with the marker under study, failure to incorporate them into the model will violate the model assumption that residual error is not supposed to correlate with the model effects and thus there is no correct way to evaluate the theoretical power.

Effects of population structure on the powers of GWAS have been investigated via Monte Carlo simulations (Atwell et al. 2010; Platt et al. 2010; Korte and Farlow 2013; Shin and Lee 2015; Toosi et al. 2018). A consensus conclusion is that proper control of population structure can reduce false-positive rate. If an association population consists of several subpopulations (in human) or several different breeds and their hybrids (in animals), many private alleles (unique to the subpopulations) may exist and the allele frequencies of many loci may be significantly different across subpopulations. If the trait of interest is also associated with the population structures due to historical and geographical reasons, the loci associated with population structures are often detected as associated with the trait although they may not be the causal loci (Atwell et al. 2010). When the population structure effects are included in the mixed model, the association signals of these loci will be reduced. This explains why fitting population structure effects can reduce false positives. However, population differentiation is most likely caused by natural selection or domestication and loci associated with traits under selection pressure may be the causal loci. As a result, fitting population structure may not be appropriate in GWAS for adaptation-related traits. A well-studied area in evolutionary genomics is to detect selection signatures (Baldwin-Brown et al. 2014; Xu and Garland 2017). The loci associated with population structures are the very loci of interests in evolutionary genomics. However, if population structural effects are present but ignored in the mixed model, the statistical power will be reduced compared to that if they are taken into account, which is due to the increased residual error variance. However, the same phenomenon can be stated alternatively as "Incorporating population structure effects will increase power compared with that if they are ignored." The alternative statement appears to contradict with the consensus conclusion about population structure. One needs to be careful when interpreting the effects of population structure on statistical power.

Methods that explicitly infer genetic ancestry generally provide an effective correction for population stratification in datasets where population structure is the only type of sample structure. In the structured association approach, samples are assigned to subpopulation clusters (possibly allowing fractional cluster membership) using a model-based clustering program such as STRUCTURE (Pritchard et al. 2000a, b). After a structured (or cluster) analysis of genomic data, individuals are divided into subgroups, say three groups. To incorporate the three groups into the GWAS model, one simply add this group ID

**Table 19.2** A toy example of six individuals divided into three groups with ten markers

| Individual | Y | Group | $Z_1$ | $Z_2$ | $Z_3$ | $Z_4$ | $Z_5$ | $Z_6$ | $Z_7$ | $Z_8$ | $Z_9$ | $Z_{10}$ |
|---|---|---|---|---|---|---|---|---|---|---|---|---|
| 1 | 9.794 | 1 | −1 | −1 | 0 | 0 | 0 | 0 | 0 | 0 | 0 | 0 |
| 2 | 8.749 | 1 | 0 | 0 | 0 | 0 | 0 | 0 | 0 | 0 | 0 | 0 |
| 3 | 9.772 | 2 | 0 | 0 | 0 | 0 | −1 | −1 | 0 | 0 | 0 | 0 |
| 4 | 8.353 | 2 | −1 | 0 | 0 | 0 | 0 | 0 | 0 | 0 | −1 | 0 |
| 5 | 10.695 | 3 | 0 | 0 | 0 | 0 | 0 | 0 | 0 | 0 | 0 | 0 |
| 6 | 7.526 | 3 | 0 | −1 | 0 | −1 | 0 | 0 | 0 | 0 | 0 | 0 |

**Table 19.3** A toy example of six individuals divided into three groups (dummy variables) with ten markers

| Individual | Y | $Q_1$ | $Q_2$ | $Q_3$ | $Z_1$ | $Z_2$ | $Z_3$ | $Z_4$ | $Z_5$ | $Z_6$ | $Z_7$ | $Z_8$ | $Z_9$ | $Z_{10}$ |
|---|---|---|---|---|---|---|---|---|---|---|---|---|---|---|
| 1 | 9.794 | 1 | 0 | 0 | −1 | −1 | 0 | 0 | 0 | 0 | 0 | 0 | 0 | 0 |
| 2 | 8.749 | 1 | 0 | 0 | 0 | 0 | 0 | 0 | 0 | 0 | 0 | 0 | 0 | 0 |
| 3 | 9.772 | 0 | 1 | 0 | 0 | 0 | 0 | 0 | −1 | −1 | 0 | 0 | 0 | 0 |
| 4 | 8.353 | 0 | 1 | 0 | −1 | 0 | 0 | 0 | 0 | 0 | 0 | 0 | −1 | 0 |
| 5 | 10.695 | 0 | 0 | 1 | 0 | 0 | 0 | 0 | 0 | 0 | 0 | 0 | 0 | 0 |
| 6 | 7.526 | 0 | 0 | 1 | 0 | −1 | 0 | −1 | 0 | 0 | 0 | 0 | 0 | 0 |

**Table 19.4** A toy example of six individuals divided into three groups (mixture proportions) with ten markers

| Individual | Y | $Q_1$ | $Q_2$ | $Q_3$ | $Z_1$ | $Z_2$ | $Z_3$ | $Z_4$ | $Z_5$ | $Z_6$ | $Z_7$ | $Z_8$ | $Z_9$ | $Z_{10}$ |
|---|---|---|---|---|---|---|---|---|---|---|---|---|---|---|
| 1 | 9.794 | 0.85 | 0.05 | 0.10 | −1 | −1 | 0 | 0 | 0 | 0 | 0 | 0 | 0 | 0 |
| 2 | 8.749 | 0.78 | 0.12 | 0.10 | 0 | 0 | 0 | 0 | 0 | 0 | 0 | 0 | 0 | 0 |
| 3 | 9.772 | 0.08 | 0.75 | 0.17 | 0 | 0 | 0 | 0 | −1 | −1 | 0 | 0 | 0 | 0 |
| 4 | 8.353 | 0.20 | 0.69 | 0.11 | −1 | 0 | 0 | 0 | 0 | 0 | 0 | 0 | −1 | 0 |
| 5 | 10.695 | 0.05 | 0.00 | 0.95 | 0 | 0 | 0 | 0 | 0 | 0 | 0 | 0 | 0 | 0 |
| 6 | 7.526 | 0.17 | 0.10 | 0.73 | 0 | −1 | 0 | −1 | 0 | 0 | 0 | 0 | 0 | 0 |

into the model as a class variable. Table 19.2 shows a toy example of six observations with ten markers and the six individuals are divided by three groups (structure). To capture the population structure, we first convert the class variable into three dummy variables as shown in Table 19.3. The sum of the three columns equal a unity vector (all one's), which is the design vector of intercept. Therefore, you cannot include the intercept in the model.

If you use the STRUCTURE program to infer the groups, you will get the group proportions. For example, if the first individual has a 0.85 probability of being from group 1, 0.05 probability from group 2 and 0.10 probability from group 3, then you should enter the group probabilities into the data, as shown in Table 19.4. The three columns of group information are collectively called the Q matrix in the original GWAS (Yu et al. 2006). This Q matrix is the design matrix for the fixed effects in the linear mixed model.

Principal component analysis (PCA) is a tool that has been used to infer population structure in genetic data for several decades, long before the GWAS era. It should be noted that top PCs do not always reflect population structure: they may reflect family relatedness, long-range LD (e.g., due to inversion polymorphisms), or assay artifacts; these effects can often be eliminated by removing related samples, regions of long-range LD, or low-quality data, respectively, from the data used to compute PCs. In addition, PCA can highlight effects of differential bias that require additional quality control. For PCA control of population structure, we first perform principal components analysis on the genomic data (an $n \times m$ matrix $Z$) where $n$ is the sample size and $m$ is the number of markers. We first find the eigenvector (a matrix) for matrix $Z^T Z$ (this is an $m \times m$ matrix and can be very large if $m$ is very large). The eigenvalue decomposition of this matrix is expressed by

$$Z^T Z = V D V^T$$

where $D$ is a diagonal matrix holding $m$ eigenvalues, $V$ is the left eigenvectors with dimension $m \times m$, and $V^T$ is the right eigenvector (transposition of the same matrix). Matrix $V$ are often called the $Z$ loadings (or rotations). The principal components (matrix $S$) are calculated via

**Table 19.5** The $Z^TZ$ matrix from the toy data

| $Z^TZ$ | $Z_1$ | $Z_2$ | $Z_3$ | $Z_4$ | $Z_5$ | $Z_6$ | $Z_7$ | $Z_8$ | $Z_9$ | $Z_{10}$ |
|---|---|---|---|---|---|---|---|---|---|---|
| $Z_1$ | 2 | 1 | 0 | 0 | 0 | 0 | 0 | 0 | 1 | 0 |
| $Z_2$ | 1 | 2 | 0 | 1 | 0 | 0 | 0 | 0 | 0 | 0 |
| $Z_3$ | 0 | 0 | 0 | 0 | 0 | 0 | 0 | 0 | 0 | 0 |
| $Z_4$ | 0 | 1 | 0 | 1 | 0 | 0 | 0 | 0 | 0 | 0 |
| $Z_5$ | 0 | 0 | 0 | 0 | 1 | 1 | 0 | 0 | 0 | 0 |
| $Z_6$ | 0 | 0 | 0 | 0 | 1 | 1 | 0 | 0 | 0 | 0 |
| $Z_7$ | 0 | 0 | 0 | 0 | 0 | 0 | 0 | 0 | 0 | 0 |
| $Z_8$ | 0 | 0 | 0 | 0 | 0 | 0 | 0 | 0 | 0 | 0 |
| $Z_9$ | 1 | 0 | 0 | 0 | 0 | 0 | 0 | 0 | 1 | 0 |
| $Z_{10}$ | 0 | 0 | 0 | 0 | 0 | 0 | 0 | 0 | 0 | 0 |

**Table 19.6** The eigenvector (an $m \times m$ matrix) of matrix $Z^TZ$ from the toy example

| V | $Z_1$ | $Z_2$ | $Z_3$ | $Z_4$ | $Z_5$ | $Z_6$ | $Z_7$ | $Z_8$ | $Z_9$ | $Z_{10}$ |
|---|---|---|---|---|---|---|---|---|---|---|
| $Z_1$ | −0.65328 | 0 | 0.5 | 0.270598 | 0 | 0 | 0 | 0 | 0 | 0.5 |
| $Z_2$ | −0.65328 | 0 | −0.5 | 0.270598 | 0 | 0 | 0 | 0 | 0 | −0.5 |
| $Z_3$ | 0 | 0 | 0 | 0 | 0 | 1 | 0 | 0 | 0 | 0 |
| $Z_4$ | −0.2706 | 0 | −0.5 | −0.65328 | 0 | 0 | 0 | 0 | 0 | 0.5 |
| $Z_5$ | 0 | −0.70711 | 0 | 0 | 0.707107 | 0 | 0 | 0 | 0 | 0 |
| $Z_6$ | 0 | −0.70711 | 0 | 0 | −0.70711 | 0 | 0 | 0 | 0 | 0 |
| $Z_7$ | 0 | 0 | 0 | 0 | 0 | 0 | 0 | 1 | 0 | 0 |
| $Z_8$ | 0 | 0 | 0 | 0 | 0 | 0 | 0 | 0 | 1 | 0 |
| $Z_9$ | −0.2706 | 0 | 0.5 | −0.65328 | 0 | 0 | 0 | 0 | 0 | −0.5 |
| $Z_{10}$ | 0 | 0 | 0 | 0 | 0 | 0 | 1 | 0 | 0 | 0 |

$$S = ZV$$

which is an $n \times m$ matrix. If $n < m$ (often the case), only the first $n$ eigenvalues are meaningful and the remaining $m - n$ eigenvalues must be zero. The columns of matrix $V^T$ corresponding to the zero eigenvalues are meaningless and should be deleted from the matrix. We often choose the first few principal components for inclusion to control populations structure. For example, the first three principal components (the first three columns of $S$) may be included in the model. Let us consider the $Z$ matrix in Table 19.2 of the toy example. The $Z$ matrix has $n = 6$ rows and $m = 10$ columns. The $Z^TZ$ matrix is shown in Table 19.5. The right eigenvector is shown in Table 19.6. The diagonal matrix of eigenvalues ($D$) is shown in Table 19.7, where only the first four eigenvalues are non-zero (the rank is 4). The score matrix or the principal components are shown in Table 19.8. You can see that the last six columns are all zero. If we use the first three principal components to control the population structure, the $Q$ matrix should take the first three columns of Table 19.8. The original toy data including the first three components are shown in Table 19.9. So, if you want to use PCA to control population structure, the input data will have a format like Table 19.9.

## 19.9　Genome-Wide Association Study in Rice—A Case Study

The rice data of Chen et al. (2014) were used as an example to demonstrate the linear mixed model GWAS procedure. The rice population consists of 524 varieties collected from a wide range of geographical locations with a diverse genetic background, including *indica*, *japonica*, and some intermediate type between the two subspecies. The trait analyzed is the grain width (GW). The number of SNP markers is 314,393, covering the 12 chromosomes of the rice genome. We first performed principal component analysis (PCA) and took the first five principal components (scores) as co-factors to control potential population structures. Figure 19.6 shows the plot of PC2 against PC1 for the 524 varieties of rice, where the PCs are calculated from all SNP data. Three major groups were found, *indica*, *japonica*, and *aus*. We also used six cluster membership identified from STRUCTURE as the co-factors to control the population structural effects. The Bonferroni corrected threshold was used as the cut-off point for declaration of statistical significance, which is $-\log_{10}(0.05/314393) = 6.7985$. The LMM procedure

**Table 19.7** Eigenvalues (an $m \times m$ matrix D) from $Z^T Z$ of the toy data

| D | $Z_1$ | $Z_2$ | $Z_3$ | $Z_4$ | $Z_5$ | $Z_6$ | $Z_7$ | $Z_8$ | $Z_9$ | $Z_{10}$ |
|---|---|---|---|---|---|---|---|---|---|---|
| $Z_1$ | 3.4142 | 0 | 0 | 0 | 0 | 0 | 0 | 0 | 0 | 0 |
| $Z_2$ | 0 | 2 | 0 | 0 | 0 | 0 | 0 | 0 | 0 | 0 |
| $Z_3$ | 0 | 0 | 2 | 0 | 0 | 0 | 0 | 0 | 0 | 0 |
| $Z_4$ | 0 | 0 | 0 | 0.5858 | 0 | 0 | 0 | 0 | 0 | 0 |
| $Z_5$ | 0 | 0 | 0 | 0 | 0 | 0 | 0 | 0 | 0 | 0 |
| $Z_6$ | 0 | 0 | 0 | 0 | 0 | 0 | 0 | 0 | 0 | 0 |
| $Z_7$ | 0 | 0 | 0 | 0 | 0 | 0 | 0 | 0 | 0 | 0 |
| $Z_8$ | 0 | 0 | 0 | 0 | 0 | 0 | 0 | 0 | 0 | 0 |
| $Z_9$ | 0 | 0 | 0 | 0 | 0 | 0 | 0 | 0 | 0 | 0 |
| $Z_{10}$ | 0 | 0 | 0 | 0 | 0 | 0 | 0 | 0 | 0 | 0 |

**Table 19.8** The score matrix (also called principal components) of the toy data

| Obs | PC1 | PC2 | PC3 | PC4 | PC5 | PC6 | PC7 | PC8 | PC9 | PC10 |
|---|---|---|---|---|---|---|---|---|---|---|
| 1 | 1.3066 | 0 | 0 | −0.5412 | 0 | 0 | 0 | 0 | 0 | 0 |
| 2 | 0 | 0 | 0 | 0 | 0 | 0 | 0 | 0 | 0 | 0 |
| 3 | 0 | 1.4142 | 0 | 0 | 0 | 0 | 0 | 0 | 0 | 0 |
| 4 | 0.9239 | 0 | −1 | 0.3827 | 0 | 0 | 0 | 0 | 0 | 0 |
| 5 | 0 | 0 | 0 | 0 | 0 | 0 | 0 | 0 | 0 | 0 |
| 6 | 0.9239 | 0 | 1 | 0.3827 | 0 | 0 | 0 | 0 | 0 | 0 |

**Table 19.9** The toy example of 6 individuals with 10 markers including three principal components

| Individual | Y | PC1 | PC2 | PC3 | $Z_1$ | $Z_2$ | $Z_3$ | $Z_4$ | $Z_5$ | $Z_6$ | $Z_7$ | $Z_8$ | $Z_9$ | $Z_{10}$ |
|---|---|---|---|---|---|---|---|---|---|---|---|---|---|---|
| 1 | 9.794 | 1.3066 | 0 | 0 | −1 | −1 | 0 | 0 | 0 | 0 | 0 | 0 | 0 | 0 |
| 2 | 8.749 | 0 | 0 | 0 | 0 | 0 | 0 | 0 | 0 | 0 | 0 | 0 | 0 | 0 |
| 3 | 9.772 | 0 | 1.4142 | 0 | 0 | 0 | 0 | 0 | −1 | −1 | 0 | 0 | 0 | 0 |
| 4 | 8.353 | 0.9239 | 0 | −1 | −1 | 0 | 0 | 0 | 0 | 0 | 0 | 0 | −1 | 0 |
| 5 | 10.695 | 0 | 0 | 0 | 0 | 0 | 0 | 0 | 0 | 0 | 0 | 0 | 0 | 0 |
| 6 | 7.526 | 0.9239 | 0 | 1 | 0 | −1 | 0 | −1 | 0 | 0 | 0 | 0 | 0 | 0 |

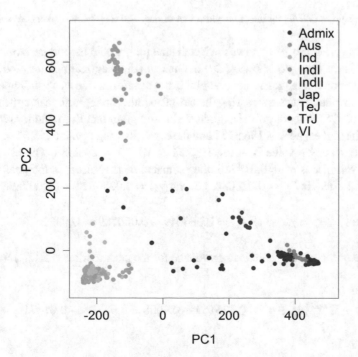

**Fig. 19.6** Principal component plot of the SNP data for the 524 varieties of rice

**Fig. 19.7** Manhattan plot of the GWAS result of the rice data for grain width (GW). The red horizontal line is the Bonferroni corrected threshold $-\log_{10}(0.05/314393) = 6.7985$

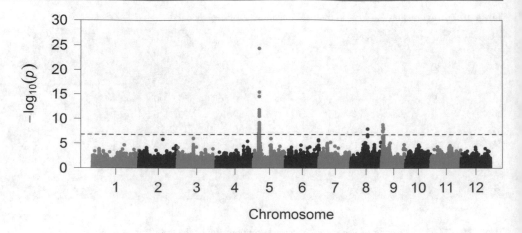

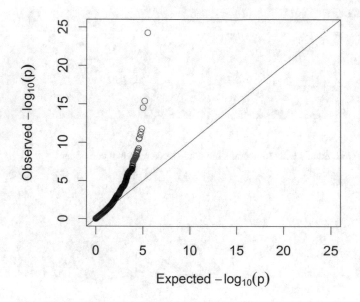

**Fig. 19.8** Q-Q plot of the test statistic of GWAS for the grain width trait of rice. The significant markers are annotated in blue

was coded in R by our laboratory. Figure 19.7 shows the Manhattan plot, where the red horizontal line is the cut-off point for significance test. Figure 19.8 is the Q-Q plot showing deviations from the diagonal line for a number of markers. In fact, the majority of markers (98.4%) are on the diagonal line (the plot is not as informative as we anticipated). Back to the Manhattan plot, three regions of the genome have markers passing the cut-off point. The highest peak appears on chromosome 5, which overlaps with a cloned gene (GW5) that controls rice grain width and grain length to width ratio (Wan et al. 2008).

The Wald test statistic of the highest peak is 106.5774 and the corresponding p-value is $5.51 \times 10^{-25}$. The peak value in the Manhattan plot corresponding to this p-value is $-\log 10(5.51 \times 10^{-25}) = 24.2588$. The estimated polygenic variance is $\phi^2 = 0.126159$, the residual variance is $\sigma^2 = 0.014716$, the estimated marker effect of the peak is $\gamma_k = -0.28991$, and the standard error of the estimated effect is $S_{\gamma_k} = 0.028082$, i.e., $\text{var}(\gamma_k) = 0.028082^2 = 0.000789$. The QTL variance is

$$\sigma_\gamma^2 = \gamma_k^2 - \text{var}(\gamma_k) = 0.084049 - 0.000789 = 0.08326$$

The expectation of the square of the genotype indicator variance for the peak marker is $\text{E}(Z_k^2) = 0.4008$. Therefore, the total phenotypic variance is

$$\sigma_P^2 = \phi^2 + \text{E}(Z_k^2)\sigma_\gamma^2 + \sigma^2 = 0.126159 + 0.4008 \times 0.08326 + 0.014716 = 0.1742$$

The estimated QTL heritability is

$$h_{QTL}^2 = \frac{\mathrm{E}\left(Z_k^2\right)\sigma_\gamma^2}{\phi^2 + \mathrm{E}\left(Z_k^2\right)\sigma_\gamma^2 + \sigma^2} = \frac{0.4008 \times 0.08326}{0.126159 + 0.4008 \times 0.08326 + 0.014716} = 0.1915$$

Grain width in rice is a complex quantitative trait. A single locus explaining 19% of the phenotypic variance is really a major gene.

## 19.10  Efficient Mixed Model Association Studies Expedited (EMMAX)

Before we close this chapter, let us introduce a fast version of the efficient mixed model association studies called efficient mixed model association studies expedited (EMMAX). This method was developed by Kang et al. (2010). Assume that the QTL effects are relatively small for all loci. Under this assumption, the polygenic variance $\phi^2$ and the ratio of the polygenic variance to the residual variance $\lambda = \phi^2/\sigma^2$ will not change too much across loci. Instead of estimating $\lambda$ for every locus scanned, we can take the estimated value from the null model and then use it as a constant when loci are scanned. This is an approximate but fast version of EMMA. Zhang et al. (2010) called such a method population parameters previously determined (P3D). The approximation is extremely fast and it can short the computational time from years to hours for extremely large population with millions of markers (Kang et al. 2010).

Here, we introduce a weighted least squares method to implement EMMAX. With this method, we run the mixed model just once under the null model (no QTL effect) to estimate $\lambda$. Computation of the null model does not depend on the number of markers but is merely determined by the sample size. The estimated $\lambda$ from the null model is then treated as a constant later when the genome is scanned. The polygenic model under the null hypothesis (no QTL effect) is

$$y = X\beta + \xi + \varepsilon$$

where the expectation of $y$ is $\mathrm{E}(y) = X\beta$ and the variance matrix of $y$ is

$$\mathrm{var}(y) = V = K\phi^2 + I\sigma^2 = \left(K\phi^2/\sigma^2 + I\right)\sigma^2 = (K\lambda + I)\sigma^2$$

Let $K = UDU^T$ be the eigenvalue decomposition of the kinship matrix so that

$$(K\lambda + I)^{-1} = U(D\lambda + I)^{-1}U^T$$

where

$$(D\lambda + I)^{-1} = W = \mathrm{diag}\left\{\left(d_j\lambda + 1\right)^{-1}\right\}$$

Let $\widehat{\lambda}$ be the REML estimate of $\lambda$ and it is treated as a known value so that the weight matrix is a given diagonal matrix of constants. Let us define a general error term by $e = \xi + \varepsilon$. The mixed model of GWAS is

$$y = X\beta + Z_k\gamma_k + e$$

Let us take linear transformations of $X$, $Z_k$ and $y$, $X^* = U^T X$, $Z_k^* = U^T Z_k$ and $y^* = U^T y$, so that

$$y^* = X^*\beta + Z_k^*\gamma_k + e^*$$

where $e^* = U^T e$. The expectation of $y^*$ is.

$$\mathrm{E}(y^*) - X^*\beta + Z_k^*\gamma_k$$

and the variance is

$$\text{var}(y^*) = U^T \text{var}(e) U = U^T (K\lambda + I) U \sigma^2 = U^T \left( U D U^T \lambda + I \right) U \sigma^2 = (D\lambda + I)\sigma^2 = W^{-1}\sigma^2$$

The weighted least squares estimates of the model effects are

$$\begin{bmatrix} \widehat{\beta} \\ \widehat{\gamma}_k \end{bmatrix} = \begin{bmatrix} X^{*T} W X^* & X^{*T} W Z_k^* \\ Z_k^{*T} W X^* & Z_k^{*T} W X^* \end{bmatrix}^{-1} \begin{bmatrix} X^{*T} W y^* \\ Z_k^{*T} W y^* \end{bmatrix}$$

The estimated residual variance is

$$\widehat{\sigma}^2 = \frac{1}{n-q-1} \left( y^* - X^*\widehat{\beta} - Z_k^*\gamma_k \right)^T W \left( y^* - X^*\widehat{\beta} - Z_k^*\gamma_k \right)$$

The variance matrix of the estimates is

$$\text{var} \begin{bmatrix} \widehat{\beta} \\ \widehat{\gamma}_k \end{bmatrix} = \begin{bmatrix} X^{*T} W X^* & X^{*T} W Z_k^* \\ Z_k^{*T} W X^* & Z_k^{*T} W X^* \end{bmatrix}^{1} \widehat{\sigma}^2 = \begin{bmatrix} C_{XX} & C_{XZ} \\ C_{ZX} & C_{ZZ} \end{bmatrix} \widehat{\sigma}^2$$

where $C_{ZZ}$ is the lower right block of the inverse of the coefficient matrix, i.e., $\text{var}(\widehat{\gamma}_k) = C_{ZZ}\sigma^2$. The Wald test for the $k$th marker is

$$W_k = \frac{\widehat{\gamma}_k^2}{\text{var}(\widehat{\gamma}_k)} = \frac{\widehat{\gamma}_k^2}{C_{ZZ}\widehat{\sigma}^2} = C_{ZZ}^{-1} \frac{\widehat{\gamma}_k^2}{\widehat{\sigma}^2}$$

Note that $W_k$ is the Wald test statistic, not an element of the weight matrix. The special case of no polygenic background control becomes a special case when the weight matrix is replaced by an identity matrix.

For multiple variance components, the model under the null hypothesis (no QTL) is

$$y = X\beta + \sum_{l=1}^{L} \xi_l + \varepsilon$$

where $\xi_l$ is the $l$th components of the polygene for $l = 1, \ldots, L$ and $L$ is the total number of polygenic components. For example, in the additive and dominance model, there are two components, $\xi_1 = \xi_a$ and $\xi_2 = \xi_d$. In the completely full genetic model, there are six components, $\xi_1 = \xi_a$, $\xi_2 = \xi_d$, $\xi_3 = \xi_{aa}$, $\xi_4 = \xi_{ad}$, $\xi_5 = \xi_{da}$, and $\xi_6 = \xi_{dd}$. The variance of $y$ is

$$\text{var}(y) = V = \sum_{l=1}^{L} K_l \phi_l^2 + I\sigma^2 = \left( \sum_{l=1}^{L} K_l \lambda_l + I \right) \sigma^2 = (K + I)\sigma^2$$

where

$$K = \sum_{l=1}^{L} K_l \lambda_l$$

and $\lambda_l = \phi_l^2/\sigma^2$ are estimated from the REML method. Let $K = U D U^T$ so that

$$V = (K + I)\sigma^2 = U(D + I)U^T\sigma^2$$

When the $k$th marker is included in the model as a fixed effect, the model becomes

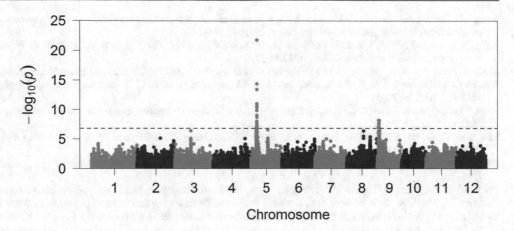

**Fig. 19.9** Manhattan plot of the rice data using the weighted least squares method (EMMAX)

$$y = X\beta + Z_k\gamma_k + \sum_{l=1}^{L} \xi_l + \varepsilon$$

Linear transformation of the original variables by the eigenvector is

$$y^* = X^*\beta + Z_k^*\gamma_k + \sum_{l=1}^{L} \xi_l^* + \varepsilon^*$$

The variance of $y^*$ is.

$$\mathrm{var}(y^*) = U^T(K+I)U\sigma^2 = U^T(UDU^T + I)U\sigma^2 = (D+I)\sigma^2$$

Let the weight matrix be

$$W = (D+I)^{-1} = \mathrm{diag}\left\{(d_j + 1)^{-1}\right\}$$

where $d_j$ is the $j$th ordered eigenvalue. Parameter estimation and hypothesis test follows exactly what is done for a single polygenic effect. Once we calculate the weights matrix, we can call the lm() function of R to perform a weighted least square analysis for significance test. For the rice data with 524 varieties and over 300 k markers, the lm() scanning of all markers only took about 6 min. Figure 19.9 shows the Manhattan plot of the result (EMMAX).

Compared with the Manhattan plot of EMMA shown in Fig. 19.7, the Manhattan plot of EMMAX shares exactly the same pattern. The peak values for chromosomes 5 and 9 are indeed lower than the peaks of EMMA but they are still significant (passed the Bonferroni corrected threshold). The peak of chromosome 8 now is below the threshold while it is above the threshold in EMMA (Fig. 19.7).

# References

Atwell S, Huang YS, Vilhjálmsson BJ, Willems G, Horton M, Li Y, Meng D, Platt A, Tarone AM, Hu TT, et al. Genome-wide association study of 107 phenotypes in Arabidopsis thaliana inbred lines. Nature. 2010;465:627.

Baldwin-Brown JG, Long AD, Thornton KR. The power to detect quantitative trait loci using Resequenced, experimentally evolved populations of diploid, sexual organisms. Mol Biol Evol. 2014;31:1040–55.

Bradbury PJ, Zhang Z, Kroon DE, Casstevens TM, Ramdoss Y, Buckler ES. TASSEL: software for association mapping of complex traits in diverse samples. Bioinformatics. 2007;23:2633–5.

Chen W, Gao Y, Xie W, Gong L, Lu K, Wang W, Li Y, Liu X, Zhang H, Dong H, et al. Genome-wide association analyses provide genetic and biochemical insights into natural variation in rice metabolism. Nat Genet. 2014;46:714.

Golub GH, Van Loan CF. Matrix computations. Baltimore: Johns Hopkins University Press; 1996.

Henderson CR. Estimation of variance and covariance components. Biometrics. 1953;9:226–52.

Kang HM, Sul JH, Service SK, Zaitlen NA, Kong S-Y, Freimer NB, Sabatti C, Eskin E. Variance component model to account for sample structure in genome-wide association studies. Nat Genet. 2010;42:348–54.

Kang HM, Zaitlen NA, Wade CM, Kirby A, Heckerman D, Daly MJ, Eskin E. Efficient control of population structure in model organism association mapping. Genetics. 2008;178:1709–23.

Korte A, Farlow A. The advantages and limitations of trait analysis with GWAS: a review. Plant Methods. 2013;9:29.

Lippert C, Listgarten J, Liu Y, Kadie CM, Davidson RI, Heckerman D. FaST linear mixed models for genome-wide association studies. Nat Methods. 2011;8:833–5.

Lippert C, Listgarten J, Liu Y, Kadie CM, Davidson RI, Heckerman D. Improved linear mixed models for genome-wide association studies. Nat Methods. 2012;9:525–6.

Listgarten J, Lippert C, Kadie CM, Davidson RI, Eskin E, Heckerman D. Improved linear mixed models for genome-wide association studies. Nat Methods. 2012;9:525.

Platt A, Vilhjálmsson BJ, Nordborg M. Conditions under which genome-wide association studies will be positively misleading. Genetics. 2010;186: 1045–52.

Pritchard JK, Stephens M, Donnelly P. Inference of population structure using multilocus genotype data. Genetics. 2000a;155:945–59.

Pritchard JK, Stephens M, Rosenberg NA, Donnelly P. Association mapping in structured populations. Am J Hum Genet. 2000b;67:170–81.

Purcell S, Neale B, Todd-Brown K, Thomas L, Ferreira Manuel AR, Bender D, Maller J, Sklar P, de Bakker PIW, Daly Mark J, et al. PLINK: a tool set for whole-genome association and population-based linkage analyses. Am J Hum Genet. 2007;81:559–75.

Shin J, Lee C. Statistical power for identifying nucleotide markers associated with quantitative traits in genome-wide association analysis using a mixed model. Genomics. 2015;105:1–4.

Toosi A, Fernando RL, Dekkers JCM. Genome-wide mapping of quantitative trait loci in admixed populations using mixed linear model and Bayesian multiple regression analysis. Genet Sel Evol. 2018;50:32.

Wan X, Weng J, Zhai H, Wang J, Lei C, Liu X, Guo T, Jiang L, Su N, Wan J. Quantitative trait loci (QTL) analysis for Rice grain width and fine mapping of an identified QTL allele gw-5 in a recombination hotspot region on chromosome 5. Genetics. 2008;179:2239–52.

Wang M, Xu S. Statistical power in genome-wide association studies and quantitative trait locus mapping. Heredity. 2019;123:287–306.

Wang Q, Wei J, Pan Y, Xu S. An efficient empirical Bayes method for genomewide association studies. J Anim Breed Genet. 2016;133:253–63.

Wei J, Xu S. A random model approach to QTL mapping in multi-parent advanced generation inter-cross (MAGIC) populations. Genetics. 2016;202:471–86.

Woodbury MA. Inverting modified matrices. Memorandum Rep. 1950;42:336.

Xu S. Mapping quantitative trait loci by controlling polygenic background effects. Genetics. 2013;195:1209–22.

Xu S, Garland T. A mixed model approach to genome-wide association studies for selection signatures, with application to mice bred for voluntary exercise behavior. Genetics. 2017;207:785–99.

Yu J, Pressoir G, Briggs WH, Vroh Bi I, Yamasaki M, Doebley JF, McMullen MD, Gaut BS, Nielsen DM, Holland JB, et al. A unified mixed-model method for association mapping that accounts for multiple levels of relatedness. Nat Genet. 2006;38:203–8.

Zhang Y-M, Mao Y, Xie C, Smith H, Luo L, Xu S. Mapping quantitative trait loci using naturally occurring genetic variance among commercial inbred lines of maize (Zea mays L.). Genetics. 2005;169:2267–75.

Zhang Z, Ersoz E, Lai CQ, Todhunter RJ, Tiwari HK, Gore MA, Bradbury PJ, Yu J, Arnett DK, Ordovas JM, et al. Mixed linear model approach adapted for genome-wide association studies. Nat Genet. 2010;42:355–60.

Zhou X, Stephens M. Genome-wide efficient mixed-model analysis for association studies. Nat Genet. 2012;44:821–4.

# Genomic Selection

**20**

Genomic selection is an alternative form of marker-assisted selection. It was first introduced by Meuwissen et al. (2001) to the animal breeding world. Genomic selection differs from the traditional marker-assisted selection (Lande and Thompson 1990) in that all markers of the genome are used to help predict the breeding value of a candidate, regardless of the sizes (i.e., effects) of the markers. The traditional marker-assisted selection, however, requires a marker detection step to select all significant markers and only the significant markers are used for prediction of the breeding value of a candidate. Figure 20.1 shows the difference between genomic selection and traditional marker-assisted selection. Genomic best linear unbiased prediction (GBLUP) is the most commonly used method for genomic prediction (VanRaden 2008). Although the Bayesian Alphabet series are considered the most popular methods of genomic selection (Gianola 2013), they are not robust in the sense that they may perform well in one dataset but poorly in other datasets. These Bayesian methods require detailed prior distributions and the priors are often selected based on the natures of the data. The GBLUP method, however, only uses the simple maximum likelihood method for parameter estimation and requires only the well-known Henderson's mixed model equation to calculate the BLUP for random effects. Therefore, it is the most robust method in the sense that it performs well in most types of data. For some specific data, it may not be optimal, but its performance has never been too low. On average, its performance is better than all other methods. A nonparametric method called reproducing kernel Hilbert spaces regression (Gianola and van Kaam 2008; de los Campos et al. 2009, 2010, 2013) is often more efficient than BLUP. The least absolute shrinkage and selection operator (Lasso) (Tibshirani 1996) is a very popular method for variable selection, and it is often used in genomic selection and GWAS (Ithnin et al. 2017). In this chapter, we will focus only on the BLUP and RKHS methods for genomic selection. The BLUP method is identical to the ridge regression if the ridge factor is estimated from the data. Therefore, we will introduce the ridge regression and then compare it with the BLUP method.

## 20.1 Genomic Best Linear Unbiased Prediction

### 20.1.1 Ridge Regression

The linear model for a centered response variable $y$ ($n \times 1$ vector) fit by a scaled matrix of predictors $Z$ ($n \times m$ matrix) is

$$y = Z\gamma + \varepsilon$$

where $\gamma$ ($m \times 1$ vector) are the regression coefficients and $\varepsilon$ ($n \times 1$ vector) are the residual errors with $\mathrm{E}(\varepsilon) = 0$ and var $(\varepsilon) = I_n\sigma^2$. The notation is different from the conventional ridge regression where the independent variables are often denoted by $X$ and the regression coefficients are often denoted by $\beta$. Let $\omega$ be the ridge parameter. The ridge estimates of the regression coefficients are

$$\gamma^{Ridge} = \left(Z^T Z + \omega I_m\right)^{-1} Z^T y \tag{20.1}$$

Here, the inverse of the ridge factor appears in the ridge regression coefficients.

© Springer Nature Switzerland AG 2022
S. Xu, *Quantitative Genetics*, https://doi.org/10.1007/978-3-030-83940-6_20

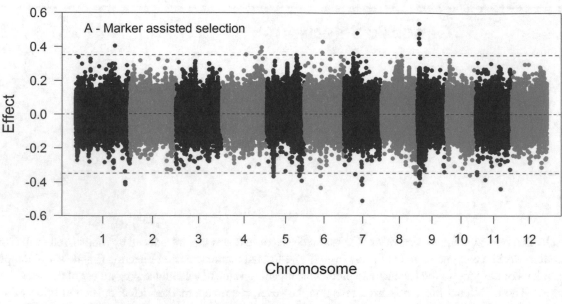

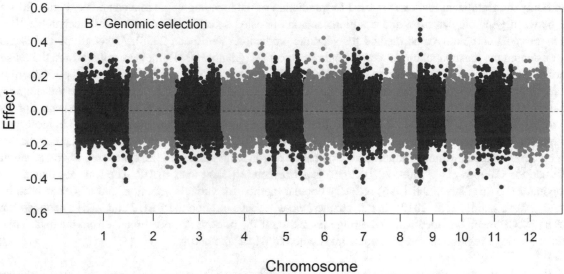

**Fig. 20.1** Plots of marker effect against genome location. The upper panel (**a**) shows markers with significant effects (beyond the two cut-off points) on the trait of interest, and these selected markers are used in the traditional marker-assisted selection. The lower panel (**b**) shows genomic selection where all markers are used to predict genomic value of a candidate

The ridge regression problem can be formulated as a Bayesian regression problem by assigning each regression coefficient a prior normal distribution with mean zero and variance $\mathrm{var}(\gamma_k) = \phi^2 = \sigma^2/\omega$ where $\omega = \sigma^2/\phi^2$. Since $y$ is already centered, the expectation of $y$ is 0 and the variance of $y$ is

$$\mathrm{var}(y) = ZZ^T\phi^2 + I_n\sigma^2 = (ZZ^T + \omega I_n)\phi^2$$

Given $\omega$, the best linear unbiased predictor of $\gamma$ has two forms, one being the same as the ridge estimate given in Eq. (20.1) and the other being derived from the conditional expectation of $\gamma$ given $y$. Let.

$$\mathrm{E}\begin{bmatrix} y \\ \gamma \end{bmatrix} = \begin{bmatrix} 0 \\ 0 \end{bmatrix}$$

and

$$\text{var}\begin{bmatrix} y \\ \gamma \end{bmatrix} = \begin{bmatrix} ZZ^T\phi^2 + I_n\sigma^2 & Z\phi^2 \\ Z^T\phi^2 & I_m\phi^2 \end{bmatrix} = \begin{bmatrix} ZZ^T + \omega I_n & Z \\ Z^T & I_m \end{bmatrix}\phi^2$$

The conditional expectation of $\gamma$ given $y$ is.

$$\begin{aligned} E(\gamma|y) &= E(\gamma) + \text{cov}(\gamma, y)[\text{var}(y)]^{-1}[y - E(y)] \\ &= \text{cov}(\gamma, y)[\text{var}(y)]^{-1}y \\ &= Z^T(ZZ^T + \omega I_n)^{-1}y \end{aligned}$$

This is called the best linear unbiased predictor (BLUP) and denoted by

$$\gamma^{Blup} = Z^T(ZZ^T + \omega I_n)^{-1}y$$

The two different forms of estimate for $\gamma$ appear to be quite different, but they are exactly the same, as proved in the next paragraph. However, the time complexities of computation are different. If $n > m$, $\gamma^{Ridge}$ is more efficient because the estimation involves the inverse of an $m \times m$ matrix with a time complexity of $O(m^3)$. If $n < m$, $\gamma^{Blup}$ is more efficient because the estimation involves the inverse of an $n \times n$ matrix with a time complexity of $O(n^3)$. Let us put the two forms together,

$$\gamma^{Ridge} = (Z^TZ + \omega I_m)^{-1}Z^Ty$$
$$\gamma^{Blup} = Z^T(ZZ^T + \omega I_n)^{-1}y$$

We now prove that the two forms are equivalent. The proof requires the following Woodbury matrix identity,

$$(ZZ^T + \omega I_n)^{-1} = \omega^{-1}I_n - \omega^{-1}Z(Z^TZ + \omega I_m)^{-1}Z^T \qquad (20.2)$$

The original form of the Woodbury matrix identifies are called Sherman-Morrison-Woodbury formulas (Golub and Van Loan 1996), represented by

$$\underbrace{(XHX^T + R)^{-1}}_{n \times n} = R^{-1} - R^{-1}X\underbrace{(X^TR^{-1}X + H^{-1})^{-1}}_{q \times q}X^TR^{-1}$$

and

$$|XHX^T + R| = |R|\,|H|\,|X^TR^{-1}X + H^{-1}|$$

Substituting the inverse matrix in $\gamma^{Blup}$ by the above identity, Eq. (20.2), we have

$$\begin{aligned} Z^T(ZZ^T + \omega I_n)^{-1}y &= \omega^{-1}Z^Ty - \omega^{-1}Z^TZ(Z^TZ + \omega I_m)^{-1}Z^Ty \\ &= \omega^{-1}(Z^TZ + \omega I_m)(Z^TZ + \omega I_m)^{-1}Z^Ty - \omega^{-1}Z^TZ(Z^TZ + \omega I_m)^{-1}Z^Ty \\ &= [\omega^{-1}(Z^TZ + \omega I_m) - \omega^{-1}Z^TZ](Z^TZ + \omega I_m)^{-1}Z^Ty \\ &= [\omega^{-1}Z^TZ + I_m - \omega^{-1}Z^TZ](Z^TZ + \omega I_m)^{-1}Z^Ty \\ &= (Z^TZ + \omega I_m)^{-1}Z^Ty \end{aligned}$$

which is the same as $\gamma^{Ridge}$. Note that we used a very simple trick to complete the proof, that is to insert $(Z^TZ + \omega I_m)(Z^TZ + \omega I_m)^{-1}$ between $\omega^{-1}$ and $Z^Ty$, i.e.,

$$\omega^{-1}Z^T y = \omega^{-1}\left(Z^T Z + \omega I_m\right)\left(Z^T Z + \omega I_m\right)^{-1} Z^T y$$

Let us go back to the two forms of the regression coefficients. The predicted responses corresponding to the two forms are

$$\hat{y}^{\text{Ridge}} = Z\gamma^{\text{Ridge}} = Z\left(Z^T Z + \omega I_m\right)^{-1} Z^T y$$

$$\hat{y}^{\text{Blup}} = Z\gamma^{\text{Blup}} = ZZ^T\left(ZZ^T + \omega I_n\right)^{-1} y$$

In Chap. 18, we learned marker inferred kinship matrix, $K = ZZ^T$. Using the kinship matrix, the BLUP of the trait is

$$\hat{y}^{\text{Blup}} = K(K + \omega I_n)^{-1} y$$

### 20.1.2 Best Linear Unbiased Prediction of Random Effects

Let us review the linear mixed model

$$y = X\beta + Z\gamma + \varepsilon \tag{20.3}$$

where $X$ is a design matrix for the fixed effects, e.g., year and location effects, $\beta$ is a collection of fixed effects, $Z$ is an $n \times m$ matrix of marker genotype indicators, $\gamma$ is an $m \times 1$ vector of marker effects treated as random effects, $\gamma \sim N(0, G\phi^2)$, and $\varepsilon \sim N(0, R\sigma^2)$ are the residual errors. In genomic selection, we assume that marker effects are i.i.d so that $G = I_m$. We also assume that the residuals are i.i.d. so that $R = I_n$. Given the variance parameters, Henderson's mixed model equations are

$$\begin{bmatrix} X^T R^{-1} X & X^T R^{-1} Z \\ Z^T R^{-1} X & Z^T R^{-1} Z + G^{-1}/\lambda \end{bmatrix}\begin{bmatrix} \beta \\ \gamma \end{bmatrix} = \begin{bmatrix} X^T R^{-1} y \\ Z^T R^{-1} y \end{bmatrix}$$

where $\lambda = \omega^{-1} = \phi^2/\sigma^2$ is the variance ratio. In genomic selection, we do not estimate the effects of markers; rather, we are interested in $\xi = Z\gamma$ as a whole, which is interpreted as the polygenic effect. The expectation of $\xi$ is zero and the variance is $\text{var}(\xi) = ZZ^T\phi^2 = K\phi^2$, where $K = ZZ^T$ is a marker-inferred kinship matrix. The mixed model in Eq. (20.3) is rewritten as

$$y = X\beta + \xi + \varepsilon$$

where the expectation is $E(y) = X\beta$ and the variance is

$$\text{var}(y) = V = K\phi^2 + I\sigma^2 = (K\lambda + I)\sigma^2 = H\sigma^2$$

Henderson's BLUP equations under the genomic selection model are

$$\begin{bmatrix} X^T X & X^T \\ X & I + K^{-1}/\lambda \end{bmatrix}\begin{bmatrix} \beta \\ \xi \end{bmatrix} = \begin{bmatrix} X^T y \\ y \end{bmatrix} \tag{20.4}$$

This leads to

$$\begin{bmatrix} \hat{\beta} \\ \hat{\xi} \end{bmatrix} = \begin{bmatrix} X^T X & X^T \\ X & I + K^{-1}/\lambda \end{bmatrix}^{-1}\begin{bmatrix} X^T y \\ y \end{bmatrix} = \begin{bmatrix} C_{11} & C_{12} \\ C_{21} & C_{22} \end{bmatrix}\begin{bmatrix} Q_1 \\ Q_2 \end{bmatrix}$$

Therefore, the BLUP of the polygenic effects are

$$\widehat{\xi} = C_{21}Q_1 + C_{22}Q_2$$

and the variance-covariance matrix of the BLUP is

$$\mathrm{var}\left(\widehat{\xi}\right) = C_{22}\sigma^2$$

In Henderson's notation, this variance is actually denoted by $\mathrm{var}\left(\widehat{\xi} - \xi\right)$. An alternative expression of the BLUP is

$$\widehat{\xi} = \phi^2 K V^{-1}\left(y - X\widehat{\beta}\right) = \lambda K H^{-1}\left(y - X\widehat{\beta}\right)$$

where

$$\widehat{\beta} = \left(X^T H^{-1} X\right)^{-1} X^T H^{-1} y \tag{20.5}$$

Let $K = UDU^T$ be the eigenvalue decomposition of $K$, where $U$ is the eigenvector (an $n \times n$ matrix) and $D$ is an $n \times n$ diagonal matrix holding the eigenvalues. The eigenvector matrix is orthogonal so that $U^T = U^{-1}$ and thus $UU^T = I$. The inverse of the $H$ matrix is

$$H^{-1} = (K\lambda + I)^{-1} = \left(UDU^T\lambda + I\right)^{-1} = U(D\lambda + I)^{-1}U^T$$

where $D\lambda + I$ is a diagonal matrix whose inverse is calculated instantly by taking the inverse of each element of the original matrix. The BLUP of the polygenic effect can be rewritten as

$$\widehat{\xi} = \lambda K U(\lambda D + I)^{-1}U^T\left(y - X\widehat{\beta}\right)$$

This is the array of predicted genomic values. Individuals can be ranked based on $\widehat{\xi}_j$ for $j = 1, \cdots, n$ and selection can be made based on the ranking of the predicted genomic values.

### 20.1.3 Predicting Genomic Values of Future Individuals

Let $y_1$ be the phenotypic values of $n_1$ individuals in the training sample and $y_2$ be the phenotypic values of $n_2$ individuals in the test sample (future individuals). A training sample is a collection of individuals with observed phenotypic values and a test sample contains all individuals whose phenotypic values yet to be measured. Genomic selection is defined as prediction of genomic values for individuals in the test sample using phenotypic values of individuals in the training sample. The linear models for the two samples are

$$\begin{bmatrix} y_1 \\ y_2 \end{bmatrix} = \begin{bmatrix} X_1\beta \\ X_2\beta \end{bmatrix} + \begin{bmatrix} \xi_1 \\ \xi_2 \end{bmatrix} + \begin{bmatrix} \varepsilon_1 \\ \varepsilon_2 \end{bmatrix}$$

The expectation of the model is

$$\mathrm{E}\begin{bmatrix} y_1 \\ y_2 \end{bmatrix} = \begin{bmatrix} X_1\beta \\ X_2\beta \end{bmatrix}$$

and the variance matrix is

$$\mathrm{var}\begin{bmatrix} y_1 \\ y_2 \end{bmatrix} = \begin{bmatrix} K_{11} & K_{12} \\ K_{21} & K_{22} \end{bmatrix} \phi^2 + \begin{bmatrix} I & 0 \\ 0 & I \end{bmatrix} \sigma^2$$

where $K_{11}$ is the kinship matrix of the training sample, $K_{21} = K_{12}^T$ is the genomic relationship between individuals in the test sample and individuals in the training sample and $K_{22}$ is the kinship matrix of the test sample. From the multivariate normal theorem (Giri 1996), we predict the genomic value of individuals in the test sample by

$$\widehat{\xi}_2 = \widehat{\phi}^2 K_{21} \left( K_{11} \widehat{\phi}^2 + I \widehat{\sigma}^2 \right)^{-1} \left( y_1 - X_1 \widehat{\beta} \right) \tag{20.6}$$

where the parameters are assumed to be estimated from the training sample. It is important to remember that individuals in the test sample are ranked based on $\widehat{\xi}_2$, the predicted genomic values, not based on the predicted phenotypic values, which are

$$\widehat{y}_2 = X_2 \widehat{\beta} + \widehat{\phi}^2 K_{21} \left( K_{11} \widehat{\phi}^2 + I \widehat{\sigma}^2 \right)^{-1} \left( y_1 - X_1 \widehat{\beta} \right)$$

because $\widehat{y}_2$ contain the fixed environmental effects. The predicted genomic values are subject to prediction errors, which are defined as

$$\mathrm{var}\left( \widehat{\xi}_2 \right) = K_{22} \widehat{\phi}^2 - \widehat{\phi}^2 K_{21} \left( K_{11} \widehat{\phi}^2 + I \widehat{\sigma}^2 \right)^{-1} K_{12} \widehat{\phi}^2$$

Square roots of the diagonal elements of $\mathrm{var}\left( \widehat{\xi}_2 \right)$ are the standard errors of the prediction, i.e., the prediction errors. The predicted genomic values for future individuals are also called the selection indices. A selection index is a linear combination of phenotypic values of all individuals in the training sample. Substituting the fixed effects in Eq. (20.6) by the estimated values from Eq. (20.5), we have

$$\begin{aligned} \widehat{\xi}_2 &= \widehat{\phi}^2 K_{21} \left( K_{11} \widehat{\phi}^2 + I \widehat{\sigma}^2 \right)^{-1} \left[ y_1 - X_1 \left( X_1^T H^{-1} X_1 \right)^{-1} X_1^T H^{-1} y_1 \right] \\ &= \widehat{\phi}^2 K_{21} \left( K_{11} \widehat{\phi}^2 + I \widehat{\sigma}^2 \right)^{-1} \left[ I - X_1 \left( X_1^T H^{-1} X_1 \right)^{-1} X_1^T H^{-1} \right] y_1 \\ &= B y_1 \end{aligned}$$

which are indeed linear combinations of the phenotypic values in the training sample. The weight matrix of the indices is

$$B = \widehat{\phi}^2 K_{21} \left( K_{11} \widehat{\phi}^2 + I \widehat{\sigma}^2 \right)^{-1} \left[ I - X_1 \left( X_1^T H^{-1} X_1 \right)^{-1} X_1^T H^{-1} \right]$$

which has $n_2$ rows and $n_1$ columns.

## 20.1.4 Estimating Variance Parameters

The parameter vector is $\theta = [\beta \ \lambda \ \sigma^2]^T$. The data are represented by $y$. Given the parameters, the distribution of $y$ is multivariate normal. From the above three pieces of information, we can construct the log likelihood function,

$$\begin{aligned} L(\theta) &= -\frac{1}{2} \ln |V| - \frac{1}{2} (y - X\beta)^T V^{-1} (y - X\beta) \\ &= -\frac{1}{2} \ln |H| - \frac{n}{2} \ln \left( \sigma^2 \right) - \frac{1}{2\sigma^2} (y - X\beta)^T H^{-1} (y - X\beta) \end{aligned}$$

where $H = K\lambda + I$ and $\lambda = \phi^2/\sigma^2$. The restricted maximum likelihood (REML) method is often better than the maximum likelihood (ML) method for estimation of variance components. With the REML method, $\beta$ is eliminated from the data and thus the parameter vector, $\theta_R = \{\lambda, \sigma^2\}$, does not contain $\beta$. The log likelihood function for REML is

$$L_R(\theta_R) = -\frac{1}{2} \ln |V| - \frac{1}{2}\left(y - X\widehat{\beta}\right)^T V^{-1}\left(y - X\widehat{\beta}\right) - \frac{1}{2} \ln |X^T V^{-1} X|$$

$$= -\frac{1}{2} \ln |H| - \frac{n}{2} \ln (\sigma^2) - \frac{1}{2\sigma^2}\left(y - X\widehat{\beta}\right)^T H^{-1}\left(y - X\widehat{\beta}\right) - \frac{1}{2} \ln |X^T H^{-1} X| + \frac{q}{2} \ln (\sigma^2)$$

where $q = \text{rank}(X)$, i.e., the number of independent columns of matrix $X$ and $\widehat{\beta}$ is a function of the data ($y$) and the parameters ($\theta_R$),

$$\widehat{\beta} = \left(X^T H^{-1} X\right)^{-1} X^T H^{-1} y$$

We now present the two log likelihood functions together,

$$L(\theta) = -\frac{1}{2} \ln |H| - \frac{n}{2} \ln (\sigma^2) - \frac{1}{2\sigma^2}(y - X\beta)^T H^{-1}(y - X\beta)$$

$$L_R(\theta_R) = -\frac{1}{2} \ln |H| - \frac{n}{2} \ln (\sigma^2) - \frac{1}{2\sigma^2}\left(y - X\widehat{\beta}\right)^T H^{-1}\left(y - X\widehat{\beta}\right) - \frac{1}{2} \ln |X^T H^{-1} X| + \frac{q}{2} \ln (\sigma^2)$$

The difference between the two likelihood functions is obvious. Hereafter, we only discuss the REML method for estimation of variance components. The REML likelihood function can be simplified into

$$L_R(\theta_R) = -\frac{1}{2} \ln |H| - \frac{1}{2\sigma^2}\left(y - X\widehat{\beta}\right)^T H^{-1}\left(y - X\widehat{\beta}\right) - \frac{1}{2} \ln |X^T H^{-1} X| - \frac{n-q}{2} \ln (\sigma^2)$$

The REML estimate of parameter vector $\theta_R$ is obtained by maximizing the restricted likelihood function. The partial derivatives of $L_R(\theta_R)$ with respect to $\sigma^2$ is

$$\frac{\partial}{\partial \sigma^2} L_R(\theta_R) = \frac{1}{2\sigma^4}\left(y - X\widehat{\beta}\right)^T H^{-1}\left(y - X\widehat{\beta}\right) - \frac{n-q}{2\sigma^2}$$

Let the partial derivative to zero and solve for $\sigma^2$, we have

$$\widehat{\sigma}^2 = \frac{1}{n-q}\left(y - X\widehat{\beta}\right)^T H^{-1}\left(y - X\widehat{\beta}\right)$$

The solution is explicit and it is a function of $\lambda$ because $H = K\lambda + I$. Substituting this solution back to the original restricted likelihood function, we have

$$L_R(\lambda) = -\frac{1}{2} \ln |H| - \frac{1}{2\widehat{\sigma}^2}\left(y - X\widehat{\beta}\right)^T H^{-1}\left(y - X\widehat{\beta}\right) - \frac{1}{2} \ln |X^T H^{-1} X| - \frac{n-q}{2} \ln (\widehat{\sigma}^2)$$

which is only a function of $\lambda$ and thus a simple Newton iteration suffices to find the solution for $\lambda$. This likelihood function can be simplified into

$$L_R(\lambda) = -\frac{1}{2} \ln |H| - \frac{1}{2} \ln |X^T H^{-1} X| - \frac{1}{2}(n-q) \ln (\widehat{\sigma}^2) - \frac{1}{2}(n-q)$$

Given $\widehat{\beta} = \left(X^T H^{-1} X\right)^{-1} X^T H^{-1} y$, $\widehat{\sigma}^2$ can be expressed as

$$\hat{\sigma}^2 = \frac{1}{n-q}(y - X\beta)^T H^{-1}(y - X\beta)$$

$$= \frac{1}{n-q}\left[y^T H^{-1}y - y^T H^{-1}X(X^T H^{-1}X)^{-1}X^T H^{-1}y\right]$$

$$= \frac{1}{n-q}y^T Py$$

where

$$P = H^{-1} - H^{-1}X(X^T H^{-1}X)^{-1}X^T H^{-1}$$

Therefore, the restricted log likelihood function can also be expressed as

$$L_R(\lambda) = -\frac{1}{2}\ln|H| - \frac{1}{2}\ln|X^T H^{-1}X| - \frac{1}{2}(n-q)\ln(y^T Py) + \frac{1}{2}(n-q)[\ln(n-q) - 1]$$

The last term (in blue) is a constant in the sense that it has nothing to do with $\lambda$ and thus can be ignored, leading to

$$L_R(\lambda) = -\frac{1}{2}\ln|H| - \frac{1}{2}\ln|X^T H^{-1}X| - \frac{1}{2}(n-q)\ln(y^T Py)$$

The Newton iteration is

$$\lambda^{(t+1)} = \lambda^{(t)} - \left[\frac{\partial^2 L_R(\lambda^{(t)})}{\partial\lambda^2}\right]^{-1}\left[\frac{\partial L_R(\lambda^{(t)})}{\partial\lambda}\right]$$

Once the iteration process converges, we get the REML estimate of $\lambda$, denoted by $\hat{\lambda}$.

### 20.1.5 Eigenvalue Decomposition for Fast Computing

The restricted log likelihood function involving the inverse and determinant of matrix $H$, which is an $n \times n$ matrix. If the sample size is very large, repeatedly inverting such a large matrix can be very costly. The cost can be substantially reduced using the eigenvalue decomposition algorithm. Recall that the $H$ matrix only appears in the forms of $\ln|H|$, $y^T H^{-1}y$, $y^T H^{-1}X$, and $X^T H^{-1}X$, where $H = K\lambda + I$. We can decompose the kinship matrix into $K = UDU^T$ where $U$ is the eigenvector (a matrix) and $D = \text{diag}(d_1 d_2 \cdots d_n)$ is a diagonal matrix containing all the eigenvalues. This allows us to write matrix $H$ in the following form,

$$H = U(\lambda D + I)U^T$$

where $\lambda D + I$ is a diagonal matrix. The determinant of matrix $H$ is expressed as

$$|H| = |U(\lambda D + I)U^T| = |U||\lambda D + I||U^{-1}| = |U||U^{-1}||\lambda D + I| = |\lambda D + I|$$

The inverse of matrix $H$ is expressed as

$$H^{-1} = U(\lambda D + I)^{-1}U^T$$

Note that matrix $\lambda D + I$ is diagonal and thus

$$\ln |H| = \ln |\lambda D + 1| = \sum_{j=1}^{n} \ln (\lambda d_j + 1)$$

The diagonal structure of $\lambda D + I$ also leads to an easy way to calculate the inverse, which is also diagonal with the $j$th diagonal element being $1/(\lambda d_j + 1)$. We now write the log likelihood function as

$$L_R(\lambda) = -\frac{1}{2} \ln |H| - \frac{1}{2} \ln |X^T H^{-1} X| - \frac{1}{2}(n-q) \ln (y^T P y)$$

where

$$y^T P y = y^T H^{-1} y - y^T H^{-1} X (X^T H^{-1} X)^{-1} X^T H^{-1} y$$

The $H^{-1}$ never occurs alone but in the form of $a^T H^{-1} b$, which is then expressed as

$$a^T H^{-1} b = a^T U (D\lambda + I)^{-1} U^T b$$

Define $a^* = U^T a$ and $b^* = U^T b$, we have

$$a^T H^{-1} b = a^{*T} (D\lambda + I)^{-1} b^*$$

Let $a_j^*$ and $b_j^*$ be the $j$th rows of the corresponding matrices, the above quadratic form can be obtained through simple summation,

$$a^T H^{-1} b = \sum_{j=1}^{n} a_j^* (\lambda d_j + 1)^{-1} b_j^{*T}$$

Since matrix inverse and determinant have been replaced by simple summations of $n$ terms involving $a_j^*$, $b_j^*$, $\delta_j$, and parameter $\lambda$, evaluation of the log likelihood can be very efficient because $a^* = U^T a$ and $b^* = U^T b$ are only calculated once prior to the maximum likelihood analysis. More computing time can be saved if $m < n$ because $d_j = 0$ for all $j > m$ and the eigenvector $U$ is an $m \times n$ matrix. The upper limit of the summation is $\min(n, m)$.

### 20.1.6 Kinship Matrix

The kinship matrix has been defined as

$$K = ZZ^T \tag{20.7}$$

where $Z$ is an $n \times m$ matrix of additive genotype indicators that appears in the following mixed model

$$y = X\beta + Z\gamma + \varepsilon \tag{20.8}$$

Let $\gamma \sim N(0, I\phi^2)$ where $\phi^2$ is called the "polygenic variance." This means that all $\gamma_k$ for $k = 1, \cdots, m$ share the same variance $\phi^2$. We now define the polygene as $\xi = Z\gamma$. The variance-covariance matrix of the polygene is

$$\mathrm{var}(\xi) = \mathrm{var}(Z\gamma) = Z\mathrm{var}(\gamma)Z^T = ZZ^T \phi^2 = K\phi^2$$

This matrix is the primary kinship matrix and all other kinship matrices seen in literature are derived from this primary matrix. An alternative expression of the primary kinship matrix is

$$K = \sum_{k=1}^{m} Z_k Z_k^T$$

This formula does not use as much computer memory as the formula given in Eq. (20.7) and thus is often used when computer memory is limited, e.g., desktop and laptop PCs. The variance-covariance matrix of $y$ is

$$\text{var}(y) = \text{var}(\xi) + \text{var}(\varepsilon) = K\phi^2 + I\sigma^2 \tag{20.9}$$

The primary kinship matrix may cause a problem if $m$ is extremely large, say several millions. The reason is that $\phi^2$ and $\sigma^2$ may not be in the same scale, leading to a convergence problem in any optimization package. For example, if $m$ is extremely large, the kinship matrix that is a summation of $m$ "marker-specific kinship matrix $K_k = Z_k Z_k^T$ can be extremely large. The solution for $\phi^2$ will be extremely small, say $\phi^2 = 10^{-8}$, whereas $\sigma^2$ may be in the order of 10 or 100. The polygenic variance may well hit the lower boundary of the parameter space in an optimization algorithm. Therefore, we often normalize the primary kinship matrix by $m$ as shown below,

$$K = \frac{1}{m} \sum_{k=1}^{m} Z_k Z_k^T$$

Another normalization factor is $d$ so that

$$K = \frac{1}{d} \sum_{k=1}^{m} Z_k Z_k^T \tag{20.10}$$

where

$$d = \frac{1}{m} \text{tr} \left( \sum_{k=1}^{m} Z_k Z_k^T \right)$$

is simply the average of the diagonal elements of the primary kinship matrix. This normalized kinship matrix will bring the diagonal elements near unity and thus the estimated polygenic variance is in the same level as the residual error variance. This can be easily understood if we look at the variance matrix given in Eq. (20.9) again, where matrix $K$ and $I$ are comparable because all diagonal elements in matrix $I$ are unity. Recall that the phenotypic variance is partitioned into the polygenic variance and the residual variance using

$$\sigma_P^2 = \frac{1}{n} \text{tr}[\text{var}(y)] = \frac{1}{n} \text{tr}(K)\phi^2 + \frac{1}{n} \text{tr}(I)\sigma^2 = \phi^2 + \sigma^2$$

This partitioning is only true when the kinship matrix is normalized using Eq. (20.10).

VanRaden (2008) proposed three rescaled kinship matrices, which are introduced here. Let the numerically coded genotype indicator be

$$Z_{jk} = \begin{cases} +1 & \text{for } A_1A_1 \text{ with frequency } p_k^2 \\ 0 & \text{for } A_1A_2 \text{ with frequency } 2p_kq_k \\ -1 & \text{for } A_2A_2 \text{ with frequency } q_k^2 \end{cases}$$

where $p_k$ is the frequency of allele $A_1$ and $q_k = 1 - p_k$ is the frequency of allele $A_2$ for locus $k$. The genotype frequency is defined under the assumption of HW equilibrium. Based on the above definition, the expectation and variance of $Z_{jk}$ are.

$$E(Z_{jk}) = p_k^2 - q_k^2 = (p_k + q_k)(p_k - q_k) = p_k - q_k$$

and

$$\text{var}(Z_{jk}) = (p_k^2 + q_k^2) - (p_k - q_k)^2 = (p_k^2 + q_k^2) - (p_k^2 - 2p_kq_k - q_k^2) = 2p_kq_k$$

respectively. VanRaden (2008) define

$$P_k = E(Z_k) = p_k - q_k = 2p_k - 1 = 2(p_k - 0.5)$$

as an $n \times 1$ vector of expectation of $Z_k$. He then centered $Z_k$ by $Z_k = Z_k - P_k$ and used the centered $Z_k$ to construct the kinship matrix,

$$K = \frac{1}{2\sum_{k=1}^{m} p_kq_k} \sum_{k=1}^{m} (Z_k - P_k)(Z_k - P_k)^T$$

He claimed that this $K$ matrix is analogous to the numerator relationship matrix $A$ (two times the coancestry matrix) in the Henderson's mixed model equations. This is the first VanRaden kinship matrix. The second VanRaden kinship matrix is

$$K = \frac{1}{m} \sum_{k=1}^{m} \frac{1}{2p_kq_k} (Z_k - P_k)(Z_k - P_k)^T$$

This kinship matrix is equivalent to using a "standardized" $Z$ matrix to construct the kinship matrix. In other words, he standardized each $Z_{jk}$ using.

$$Z_{jk}^* = \frac{Z_{jk} - E(Z_{jk})}{\sqrt{\text{var}(Z_{jk})}}$$

and then construct the second kinship matrix using

$$K = \frac{1}{m} \sum_{k=1}^{m} Z_k^* Z_k^{*T}$$

VanRaden's third kinship matrix is to perform a matrix regression analysis to incorporate a pedigree-inferred additive relationship matrix into the calculation of $K$. Let $A$ be the additive relationship matrix inferred from pedigree information. VanRaden fitted the following regression model

$$ZZ^T = \alpha(J_{n\times1}J_{1\times n}^T) + \beta A_{n\times n} + \varepsilon_{n\times n}$$

where $J_{n\times1}$ is an $n \times 1$ vector of unity (all elements are 1), $\alpha$ is the intercept, and $\beta$ is the regression coefficient and $\varepsilon_{n\times n}$ is a matrix of residual errors. The intercept and regression coefficient are solved using

$$\begin{bmatrix} \hat{\alpha} \\ \hat{\beta} \end{bmatrix} = \begin{bmatrix} n^2 & \sum_{ij}^{n} A_{ij} \\ \sum_{ij}^{n} A_{ij} & \sum_{ij}^{n} A_{ij}^2 \end{bmatrix}^{-1} \begin{bmatrix} \sum_{ij}^{n} (ZZ^T)_{ij} \\ \sum_{ij}^{n} (ZZ^T)_{ij} A_{ij} \end{bmatrix}$$

The kinship matrix is then obtained via

$$K = \frac{1}{\widehat{\beta}}\left(ZZ - \widehat{\alpha}JJ^T\right)$$

When calculating the expectation and variance of $Z_{jk}$, VanRaden (VanRaden 2008) assumed all markers to be in Hardy-Weinberg equilibrium. This assumption is absolutely unnecessary! One can simply replace the expectation and variance by the sample mean and sample variance. For example,

$$\mathrm{E}(Z_k) = \overline{Z}_k = \frac{1}{n}\sum_{j=1}^{n} Z_{jk}$$

and

$$\mathrm{var}(Z_k) = S_k^2 = \frac{1}{n-1}\sum_{j=1}^{n}\left(Z_{jk} - \overline{Z}_k\right)^2$$

We now summary all the methods for kinship matrix calculation in Table 20.1.

$n$—number of observations (population size).

$m$—number of markers of the genome.

$K$—an $n \times n$ kinship matrix.

$A$—an $n \times n$ numerator relationship matrix.

$k$—index marker for $k = 1, \ldots, m$.

$p_k$—frequency of allele $A_1$.

$q_k$—frequency of allele $A_2$ and $q_k = 1 - p_k$.

$Z_{jk}$—genotype indicator for individual $j$ at marker $k$

$$Z_{jk} = \begin{cases} +1 & \text{for } A_1A_1 \text{ with frequency } p_k^2 \\ 0 & \text{for } A_1A_2 \text{ with frequency } 2p_kq_k \\ -1 & \text{for } A_2A_2 \text{ with frequency } q_k^2 \end{cases}$$

$Z_k = \{Z_{jk}\}_{n \times 1}$ is an $n \times 1$ vector of all $Z_{jk}$ for marker $k$.

$Z = \{Z_k\}_{n \times m}$ is an $n \times m$ matrix of all $Z_{jk}$.

$J$—an $n \times 1$ vector of unity and $JJ^T$ is an $n \times n$ matrix of unity.

**Table 20.1** Summary of all kinship matrices introduced in the text

| Kinship | Formula | Scale | Center |
|---------|---------|-------|--------|
| Primary (base) | $K = \sum_{k=1}^{m} Z_k Z_k^T$ | | |
| Averaged | $K = \frac{1}{m}\sum_{k=1}^{m} Z_k Z_k^T$ | $m$ | |
| Normalized | $K = \frac{1}{d}\sum_{k=1}^{m} Z_k Z_k^T$ | $d = \frac{1}{m}\mathrm{tr}\left(\sum_{k=1}^{m} Z_k Z_k^T\right)$ | |
| VanRaden 1 | $K = \frac{1}{d}\sum_{k=1}^{m}(Z_k - P_k)(Z_k - P_k)^T$ | $d = 2\sum_{k=1}^{m} p_k q_k$ | $P_k = 2p_k - 1$ |
| VanRaden 2 | $K = \frac{1}{m}\sum_{k=1}^{m}\frac{1}{d_k}(Z_k - P_k)(Z_k - P_k)^T$ | $d_k = 2p_k q_k$ | $P_k = 2p_k - 1$ |
| VanRaden 3 | $K = \frac{1}{\widehat{\beta}}(ZZ - \widehat{\alpha}JJ^T)$ | $\widehat{\beta}$ | $\widehat{\alpha}$ |
| Scaled | $K = \frac{1}{m}\sum_{k=1}^{m}\frac{1}{S_k^2}(Z_k - \overline{Z}_k)(Z_k - \overline{Z}_k)^T$ | $S_k^2 = \frac{1}{n-1}\sum_{j=1}^{n}(Z_{jk} - \overline{Z}_k)^2$ | $\overline{Z}_k = \frac{1}{n}\sum_{j=1}^{n} Z_{jk}$ |

**Table 20.2** Numerical codes of the additive and dominance indicators for three loci of ten plants

| plant | marker1 | marker2 | marker3 | z1 | z2 | z3 | w1 | w2 | w3 |
|---|---|---|---|---|---|---|---|---|---|
| 1 | $A_1A_2$ | $A_1A_1$ | $A_1A_2$ | 0 | 1 | 0 | 1 | −1 | 1 |
| 2 | $A_1A_1$ | $A_2A_2$ | $A_1A_1$ | 1 | −1 | 1 | −1 | −1 | −1 |
| 3 | $A_1A_2$ | $A_1A_2$ | $A_1A_2$ | 0 | 0 | 0 | 1 | 1 | 1 |
| 4 | $A_1A_2$ | $A_2A_2$ | $A_1A_1$ | 0 | −1 | 1 | 1 | −1 | −1 |
| 5 | $A_1A_1$ | $A_1A_2$ | $A_1A_2$ | 1 | 0 | 0 | −1 | 1 | 1 |
| 6 | $A_1A_1$ | $A_1A_2$ | $A_2A_2$ | 1 | 0 | −1 | −1 | 1 | −1 |
| 7 | $A_2A_2$ | $A_2A_2$ | $A_1A_2$ | −1 | −1 | 0 | −1 | −1 | 1 |
| 8 | $A_1A_1$ | $A_1A_2$ | $A_2A_2$ | 1 | 0 | −1 | −1 | 1 | −1 |
| 9 | $A_1A_1$ | $A_1A_2$ | $A_1A_1$ | 1 | 0 | 1 | −1 | 1 | −1 |
| 10 | $A_2A_2$ | $A_1A_2$ | $A_1A_2$ | −1 | 0 | 0 | −1 | 1 | 1 |

$$
\begin{bmatrix} \widehat{\alpha} \\ \widehat{\beta} \end{bmatrix} = \begin{bmatrix} n^2 & \sum_{ij}^n A_{ij} \\ \sum_{ij}^n A_{ij} & \sum_{ij}^n A_{ij}^2 \end{bmatrix}^{-1} \begin{bmatrix} \sum_{ij}^n (ZZ^T)_{ij} \\ \sum_{ij}^n (ZZ^T)_{ij} A_{ij} \end{bmatrix}
$$

### 20.1.6.1 Kinship Matrices for Dominance and Epistasis

The models discussed so are still the additive models. A general model that covers dominance and epistasis is

$$
y = X\beta + \xi_a + \xi_d + \xi_{aa} + \xi_{ad} + \xi_{da} + \xi_{dd} + \varepsilon
$$

where $\xi_a, \ldots, \xi_{dd}$ are polygenic effects for additive, dominance, additive by additive, additive by dominance, dominance by additive, and dominance by dominance (Xu 2013). The covariance structure of the model is defined as

$$
\begin{aligned}
\mathrm{var}(y) &= \mathrm{var}(\xi_a) + \mathrm{var}(\xi_d) + \mathrm{var}(\xi_{aa}) + \mathrm{var}(\xi_{ad}) + \mathrm{var}(\xi_{da}) + \mathrm{var}(\xi_{dd}) + \mathrm{var}(\varepsilon) \\
&= K_a\phi_a^2 + K_d\phi_d^2 + K_{aa}\phi_{aa}^2 + K_{ad}\phi_{ad}^2 + K_{da}\phi_{da}^2 + K_{dd}\phi_{dd}^2 + I\sigma^2
\end{aligned}
$$

where $K_a, \cdots, K_{dd}$ are the kinship matrices for additive, dominance, and so on. Corresponding to each type of polygenic effect, there is a variance component, denoted by $\phi_a^2$ and so on. Given these kinship matrices, the variance parameters are estimated from the REML method. We now focus on calculation of various kinship matrices.

We now use three markers from ten individuals to demonstrate the method of kinship matrix calculation. Table 20.2 shows the genotypes and the numerical codes of the genotypes. The additive genotype coding ($Z$) and dominance genotypic coding ($W$) are defined as.

$$
Z_{jk} = \begin{cases} 1 & \text{for } A_1A_1 \\ 0 & \text{for } A_1A_2 \\ -1 & \text{for } A_2A_2 \end{cases} \quad \text{and} \quad W_{jk} = \begin{cases} -1 & \text{for } A_1A_1 \\ 1 & \text{for } A_1A_2 \\ -1 & \text{for } A_2A_2 \end{cases}
$$

This numerical coding system is somehow arbitrary. The proposed method of kinship matrix calculation applies to any coding system. The first three columns of Table 20.2 (marker1, marker2, and marker3) show the marker genotypes, the three columns in the middle (z1, z2, and z3) show the additive codes and the last three columns (w1, w2, and w3) hold the dominance codes. All kinship matrices will be calculated from the $Z$ and $W$ matrices.

Let $Z$ be an $n \times m$ matrix for the additive codes (shown in Table 20.2), where $n = 10$ (sample size) and $m = 3$ (number of markers) in this example. Define $W$ as an $n \times m$ matrix for the dominance codes (shown in Table 20.2). The additive and dominance kinship matrices are defined as

$$
K_a = ZZ^T \text{ and } K_d = WW^T.
$$

**Table 20.3** Numerical codes of epistatic indicators for three loci of 10 plants

| Plant | z1z2 | z1z3 | z2z3 | z1w2 | z1w3 | z2w3 | w1z2 | w1z3 | w2z3 | w1w2 | w1w3 | w2w3 |
|---|---|---|---|---|---|---|---|---|---|---|---|---|
| 1 | 0 | 0 | 0 | 0 | 0 | 1 | −1 | 0 | 0 | −1 | −1 | 1 |
| 2 | −1 | 1 | −1 | 1 | 0 | 0 | 0 | 0 | 1 | 0 | 0 | 0 |
| 3 | 0 | 0 | 0 | 0 | 0 | 0 | 0 | 0 | 0 | 1 | 1 | 1 |
| 4 | 0 | 0 | −1 | 0 | 0 | −1 | 1 | −1 | 0 | 0 | −1 | 0 |
| 5 | 0 | 0 | 0 | −1 | 0 | 0 | 0 | 0 | 0 | −1 | 0 | 0 |
| 6 | 0 | −1 | 0 | 1 | 0 | 0 | 0 | −1 | −1 | 1 | 0 | 0 |
| 7 | 1 | 0 | 0 | −1 | 0 | 0 | 0 | 0 | 0 | 0 | 0 | 0 |
| 8 | 0 | −1 | 0 | −1 | 0 | 0 | 0 | −1 | 1 | −1 | 0 | 0 |
| 9 | 0 | 1 | 0 | −1 | 0 | 0 | 0 | 1 | −1 | −1 | 0 | 0 |
| 10 | 0 | 0 | 0 | 1 | 0 | 0 | 0 | 0 | 0 | −1 | 0 | 0 |

**Table 20.4** Numerical codes for all effects with three loci in 10 plants, where each column represents a plant and each row represents an effect (a, d, aa, ad, da, and dd)

|  | plant1 | plant2 | plant3 | plant4 | plant5 | plant6 | plant7 | plant8 | plant9 | plant10 |
|---|---|---|---|---|---|---|---|---|---|---|
| z1 | 0 | 1 | 0 | 0 | 1 | 1 | −1 | 1 | 1 | −1 |
| z2 | 1 | −1 | 0 | −1 | 0 | 0 | −1 | 0 | 0 | 0 |
| z3 | 0 | 1 | 0 | 1 | 0 | −1 | 0 | −1 | 1 | 0 |
| w1 | 1 | −1 | 1 | 1 | −1 | −1 | −1 | −1 | −1 | −1 |
| w2 | −1 | −1 | 1 | −1 | 1 | 1 | −1 | 1 | 1 | 1 |
| w3 | 1 | −1 | 1 | −1 | 1 | −1 | 1 | −1 | −1 | 1 |
| z1z2 | 0 | −1 | 0 | 0 | 0 | 0 | 1 | 0 | 0 | 0 |
| z1z3 | 0 | 1 | 0 | 0 | 0 | −1 | 0 | −1 | 1 | 0 |
| z2z3 | 0 | −1 | 0 | −1 | 0 | 0 | 0 | 0 | 0 | 0 |
| z1w2 | 0 | 1 | 0 | 0 | −1 | 1 | −1 | −1 | −1 | 1 |
| z1w3 | 0 | 0 | 0 | 0 | 0 | 0 | 0 | 0 | 0 | 0 |
| z2w3 | 1 | 0 | 0 | −1 | 0 | 0 | 0 | 0 | 0 | 0 |
| w1z2 | −1 | 0 | 0 | 1 | 0 | 0 | 0 | 0 | 0 | 0 |
| w1z3 | 0 | 0 | 0 | −1 | 0 | −1 | 0 | −1 | 1 | 0 |
| w2z3 | 0 | 1 | 0 | 0 | 0 | 0 | −1 | 1 | −1 | 0 |
| w1w2 | −1 | 0 | 1 | 0 | −1 | 1 | 0 | −1 | −1 | −1 |
| w1w3 | −1 | 0 | 1 | −1 | 0 | 0 | 0 | 0 | 0 | 0 |
| w2w3 | 1 | 0 | 1 | 0 | 0 | 0 | 0 | 0 | 0 | 0 |

respectively. Do not normalize these matrices at this stage! To calculate the kinship matrices for epistatic effects, we first need to define the numerical codes for these epistatic effects. For the additive by additive effects, the numerical codes are obtained through multiplications of all pairs of columns of the Z matrix, producing a matrix with $n$ rows and $m(m − 1)/2$ columns. In this case, the additive by additive codes are represented by a $10 \times 3$ matrix because $3(3 − 1)/2 = 3$. Similarly, the additive by dominance codes are obtained through multiplications of all columns of Z paired with all columns of W. The dominance by additive codes are obtained from multiplications of all columns of W paired with columns of Z. Finally, the dominance by dominance codes are obtained from all pairwise multiplications of columns of W. These numerical codes are provided in Table 20.3, where the first three columns (z1z2, z1z3, and z2z3) show the additive by additive codes, the six columns in the middle (z1w2, z1w3, z2w3, w1z2, w1z3, and w2z3) show the additive by dominance and the dominance by additive codes and the last three columns (w1w2, w1w3, and w2w3) show the dominance by dominance codes.

### 20.1.6.2 Direct Method

The numerically coded genotypes in Tables 20.4 and 20.5 (transposition of Table 20.4) are partitioned into six blocks. Each block represents the codes for a type of effects, i.e., a, d, aa, ad, da, and dd. So, the kinship matrix for each type of effects is calculated using the codes corresponding to that block. Let Z be the first block for additive effect and it is used to calculate the additive kinship matrix,

**Table 20.5** Numerical codes for all effects with three loci in ten plants, where each row represents a plant and each column represents an effect (a, d, aa, ad, da, and dd)

| Plant | z1 | z2 | z3 | w1 | w2 | w3 | z1z2 | z1z3 | z2z3 | z1w2 | z1w3 | z2w3 | w1z2 | w1z3 | w2z3 | w1w2 | w1w3 | w2w3 |
|---|---|---|---|---|---|---|---|---|---|---|---|---|---|---|---|---|---|---|
| 1 | 0 | 1 | 0 | −1 | −1 | −1 | −1 | 0 | 0 | 0 | 0 | −1 | −1 | 0 | 0 | −1 | −1 | −1 |
| 2 | 1 | −1 | −1 | −1 | −1 | −1 | 0 | −1 | −1 | −1 | 0 | 0 | 0 | 0 | −1 | 0 | 0 | 0 |
| 3 | 0 | 0 | 0 | −1 | −1 | −1 | 0 | 0 | 0 | 0 | 0 | 0 | 0 | 0 | 0 | −1 | −1 | −1 |
| 4 | 0 | −1 | 0 | −1 | −1 | −1 | 0 | 0 | −1 | 0 | 0 | −1 | −1 | −1 | 0 | 0 | 0 | 0 |
| 5 | 1 | 0 | −1 | −1 | −1 | −1 | 0 | 0 | 0 | −1 | 0 | 0 | 0 | 0 | 0 | −1 | 0 | 0 |
| 6 | −1 | 0 | 0 | −1 | −1 | −1 | 0 | −1 | 0 | −1 | 0 | 0 | 0 | −1 | −1 | −1 | 0 | 0 |
| 7 | −1 | −1 | 0 | −1 | −1 | −1 | −1 | 0 | 0 | −1 | 0 | −1 | 0 | 0 | 0 | 0 | 0 | 0 |
| 8 | −1 | 0 | −1 | −1 | −1 | −1 | 0 | −1 | 0 | −1 | 0 | 0 | −1 | −1 | −1 | −1 | 0 | 0 |
| 9 | 1 | 0 | −1 | −1 | −1 | −1 | 0 | −1 | 0 | −1 | 0 | −1 | 0 | 0 | −1 | −1 | 0 | 0 |
| 10 | −1 | 0 | 0 | −1 | −1 | −1 | 0 | 0 | 0 | −1 | 0 | 0 | 0 | 0 | 0 | −1 | 0 | 0 |

$$K_a = ZZ^T = \sum_{k=1}^{m} Z_k Z_k^T$$

The two different expressions are used according to the size of the genomic data. When $m$ is small, it is straightforward to use the first expression. If the $m$ is very large, however, it is more convenient to use the second expression. Let $W$ be the second block representing the code for the dominance effect, the kinship matrix is

$$K_d = WW^T = \sum_{k=1}^{m} W_k W_k^T$$

Let $Z \# Z$ be the third block of the genotype codes that represent the additive by additive epistatic effects. It is obtained from matrix $Z$ using

$$Z\#Z = \text{horzcat}_{k=1}^{m-1} \; \text{horzcat}_{k'=k+1}^{m} (Z_k * Z_{k'})$$

where $Z_k * Z_{k'}$ represents element-wise multiplication of the two vectors and horzcat() represents horizontal concatenation of matrices. The dimension of $Z \# Z$ is $n$ by $m(m-1)/2$. The kinship matrix for the additive by additive effects is

$$K_{aa} = (Z\#Z)(Z\#Z)^T$$

Let us define

$$Z\#W = \text{horzcat}_{k=1}^{m-1} \; \text{horzcat}_{k'=k+1}^{m} (Z_k * W_{k'}) \text{ and } W\#Z = \text{horzcat}_{k=1}^{m-1} \; \text{horzcat}_{k'=k+1}^{m} (W_k * Z_{k'})$$

The additive by dominance and dominance by additive matrices are

$$K_{ad} = (Z\#W)(Z\#W)^T \text{ and } K_{da} = (W\#Z)(W\#Z)^T$$

Note that $Z \# W$ is different from $W \# Z$ and thus $K_{ad}$ is different from $K_{da}$. However, in variance component analysis, we often define the sum of the two as a composite additive by dominance matrix,

$$K_{(ad)} = K_{ad} + K_{da} = (Z\#W)(Z\#W)^T + (W\#Z)(W\#Z)^T$$

Finally, the dominance by dominance kinship matrix is

$$K_{dd} = (W\#W)(W\#W)^T$$

Table 20.6 summarizes the kinship matrices using the direct methods. The five kinship matrices ($K_a$, $K_d$, $K_{aa}$, $K_{(ad)}$, and $K_{dd}$) for the sample data are given in Tables 20.7, 20.8, 20.9, 20.10, 20.11.

### 20.1.6.3 Fast Algorithm for Epistatic Kinship Matrix Calculation

Calculation of the epistatic kinship matrices using the direct method can be time consuming for a large number of markers. We first have to generate the numerical codes for the epistatic effects, and these matrices have $n$ rows and $m(m-1)/2$ columns for each of the four epistatic terms. If the number of markers is 10,000, for example, the number of columns for each of the four epistatic terms will be $10000(10000-1)/2 = 49995000$ columns. From these matrices, we then calculate the kinship matrices using regular matrix multiplication. The Hadamard matrix multiplication has been used to calculate these epistatic kinship matrices (Ober et al. 2012). With this special element-wise matrix multiplication, the additive by addition kinship matrix is simply calculated as $K_{aa}^{Had} = K_a * K_a$ (element-wise multiplication), $K_{ad}^{Had} = K_a * K_d$, and $K_{dd}^{Had} = K_d * K_d$. Unfortunately,

**Table 20.6** Summary of all five types of kinship matrices using the direct method

| Kinship matrix | Expression |
|---|---|
| $K_a$ | $ZZ^T$ |
| $K_d$ | $WW^T$ |
| $K_{aa}$ | $(Z \# Z)(Z \# Z)^T$ |
| $K_{(ad)}$ | $(Z \# W)(Z \# W)^T + (W \# Z)(W \# Z)^T$ |
| $K_{dd}$ | $(W \# W)(W \# W)^T$ |

**Table 20.7** Additive kinship matrix ($K_a$) for ten plants calculated from three loci

| $K_a$ | p1 | p2 | p3 | p4 | p5 | p6 | p7 | p8 | p9 | p10 |
|---|---|---|---|---|---|---|---|---|---|---|
| p1 | 1 | −1 | 0 | −1 | 0 | 0 | −1 | 0 | 0 | 0 |
| p2 | −1 | 3 | 0 | 2 | 1 | 0 | 0 | 0 | 2 | −1 |
| p3 | 0 | 0 | 0 | 0 | 0 | 0 | 0 | 0 | 0 | 0 |
| p4 | −1 | 2 | 0 | 2 | 0 | −1 | 1 | −1 | 1 | 0 |
| p5 | 0 | 1 | 0 | 0 | 1 | 1 | −1 | 1 | 1 | −1 |
| p6 | 0 | 0 | 0 | −1 | 1 | 2 | −1 | 2 | 0 | −1 |
| p7 | −1 | 0 | 0 | 1 | −1 | −1 | 2 | −1 | −1 | 1 |
| p8 | 0 | 0 | 0 | −1 | 1 | 2 | −1 | 2 | 0 | −1 |
| p9 | 0 | 2 | 0 | 1 | 1 | 0 | −1 | 0 | 2 | −1 |
| p10 | 0 | −1 | 0 | 0 | −1 | −1 | 1 | −1 | −1 | 1 |

**Table 20.8** Dominance kinship matrix ($K_d$) for ten plants calculated from three loci

| $K_d$ | p1 | p2 | p3 | p4 | p5 | p6 | p7 | p8 | p9 | p10 |
|---|---|---|---|---|---|---|---|---|---|---|
| p1 | 3 | 1 | 1 | 2 | −2 | 0 | 1 | −2 | −2 | −2 |
| p2 | 1 | 1 | 1 | 0 | −1 | 1 | 1 | −1 | −1 | −1 |
| p3 | 1 | 1 | 3 | 0 | 0 | 2 | 1 | 0 | 0 | 0 |
| p4 | 2 | 0 | 0 | 2 | −1 | −1 | 0 | −1 | −1 | −1 |
| p5 | −2 | −1 | 0 | −1 | 2 | 0 | −1 | 2 | 2 | 2 |
| p6 | 0 | 1 | 2 | −1 | 0 | 2 | 1 | 0 | 0 | 0 |
| p7 | 1 | 1 | 1 | 0 | −1 | 1 | 1 | −1 | −1 | −1 |
| p8 | −2 | −1 | 0 | −1 | 2 | 0 | −1 | 2 | 2 | 2 |
| p9 | −2 | −1 | 0 | −1 | 2 | 0 | −1 | 2 | 2 | 2 |
| p10 | −2 | −1 | 0 | −1 | 2 | 0 | −1 | 2 | 2 | 2 |

**Table 20.9** Additive by additive kinship matrix ($K_{aa}$) for ten plants calculated from three loci

| $K_{aa}$ | p1 | p2 | p3 | p4 | p5 | p6 | p7 | p8 | p9 | p10 |
|---|---|---|---|---|---|---|---|---|---|---|
| p1 | 0 | 0 | 0 | 0 | 0 | 0 | 0 | 0 | 0 | 0 |
| p2 | 0 | 3 | 0 | 1 | 0 | −1 | −1 | −1 | 1 | 0 |
| p3 | 0 | 0 | 0 | 0 | 0 | 0 | 0 | 0 | 0 | 0 |
| p4 | 0 | 1 | 0 | 1 | 0 | 0 | 0 | 0 | 0 | 0 |
| p5 | 0 | 0 | 0 | 0 | 0 | 0 | 0 | 0 | 0 | 0 |
| p6 | 0 | −1 | 0 | 0 | 0 | 1 | 0 | 1 | −1 | 0 |
| p7 | 0 | −1 | 0 | 0 | 0 | 0 | 1 | 0 | 0 | 0 |
| p8 | 0 | −1 | 0 | 0 | 0 | 1 | 0 | 1 | −1 | 0 |
| p9 | 0 | 1 | 0 | 0 | 0 | −1 | 0 | −1 | 1 | 0 |
| p10 | 0 | 0 | 0 | 0 | 0 | 0 | 0 | 0 | 0 | 0 |

the Hadamard matrix multiplication does not generate the correct kinship matrices for the epistatic effects. After careful examination of the kinship matrices calculated from the Hadamard matrix multiplication, we realized that they can be obtained via a direct method by coding the epistatic effects using column wise matrix multiplication and the epistatic effect codes are then used to calculate the kinship matrix using the usual method of matrix multiplication.

**Table 20.10**  Additive by dominance kinship matrix ($K_{(ad)}$)for ten plants calculatedfrom three loci

| $K_{(ad)}$ | p1 | p2 | p3 | p4 | p5 | p6 | p7 | p8 | p9 | p10 |
|---|---|---|---|---|---|---|---|---|---|---|
| p1 | 2 | 0 | 0 | −2 | 0 | 0 | 0 | 0 | 0 | 0 |
| p2 | 0 | 2 | 0 | 0 | −1 | 0 | −1 | 0 | −2 | 1 |
| p3 | 0 | 0 | 0 | 0 | 0 | 0 | 0 | 0 | 0 | 0 |
| p4 | −2 | 0 | 0 | 3 | 0 | 1 | 0 | 1 | −1 | 0 |
| p5 | 0 | −1 | 0 | 0 | 1 | −1 | 1 | 1 | 1 | −1 |
| p6 | 0 | 0 | 0 | 1 | −1 | 3 | −1 | −1 | −1 | 1 |
| p7 | 0 | −1 | 0 | 0 | 1 | −1 | 1 | 1 | 1 | −1 |
| p8 | 0 | 0 | 0 | 1 | 1 | −1 | 1 | 3 | −1 | −1 |
| p9 | 0 | −2 | 0 | −1 | 1 | −1 | 1 | −1 | 3 | −1 |
| p10 | 0 | 1 | 0 | 0 | −1 | 1 | −1 | −1 | −1 | 1 |

**Table 20.11**  Dominance by dominance kinship matrix ($K_{dd}$) for ten plants calculatedfrom three loci

| $K_{dd}$ | p1 | p2 | p3 | p4 | p5 | p6 | p7 | p8 | p9 | p10 |
|---|---|---|---|---|---|---|---|---|---|---|
| p1 | 3 | 0 | −1 | 1 | 1 | −1 | 0 | 1 | 1 | 1 |
| p2 | 0 | 0 | 0 | 0 | 0 | 0 | 0 | 0 | 0 | 0 |
| p3 | −1 | 0 | 3 | −1 | −1 | 1 | 0 | −1 | −1 | −1 |
| p4 | 1 | 0 | −1 | 1 | 0 | 0 | 0 | 0 | 0 | 0 |
| p5 | 1 | 0 | −1 | 0 | 1 | −1 | 0 | 1 | 1 | 1 |
| p6 | −1 | 0 | 1 | 0 | −1 | 1 | 0 | −1 | −1 | −1 |
| p7 | 0 | 0 | 0 | 0 | 0 | 0 | 0 | 0 | 0 | 0 |
| p8 | 1 | 0 | −1 | 0 | 1 | −1 | 0 | 1 | 1 | 1 |
| p9 | 1 | 0 | −1 | 0 | 1 | −1 | 0 | 1 | 1 | 1 |
| p10 | 1 | 0 | −1 | 0 | 1 | −1 | 0 | 1 | 1 | 1 |

Therefore, we can modify the $K_{aa}^{Had}$ matrix by removing the contents generated from the extra terms and adjusting for the duplications to obtain the correct $K_{aa}$ matrix,

$$K_{aa} = \frac{1}{2}\left[K_{aa}^{Had} - (Z*Z)(Z*Z)^T\right] = \frac{1}{2}\left[K_a * K_a - (Z*Z)(Z*Z)^T\right]$$

where $Z*Z$ is the element-wise squares of matrix $Z$. We can also obtain the correct additive by dominance kinship matrix by modifying the kinship matrix calculated from the Hadamard matrix multiplication using

$$K_{(ad)} = K_{ad} + K_{da} = K_{ad}^{Had} - (Z*W)(Z*W)^T = K_a * K_d - (Z*W)(Z*W)^T$$

where $Z*W$ are element-wise multiplication of $Z$ and $W$. We can modify the $K_{dd}^{Had}$ matrix by removing the contents generated from the extra terms and adjusting for the duplications to obtain the correct $K_{dd}$ matrix,

$$K_{dd} = \frac{1}{2}\left[K_{dd}^{Had} - (W*W)(W*W)^T\right] = \frac{1}{2}\left[K_d * K_d - (W*W)(W*W)^T\right]$$

where $W*W$ is element-wise square of matrix $W$. Table 20.12 summarizes the fast algorithm for calculation of the kinship matrices. Before using these kinship matrices, we need to normalize them by dividing the matrices by the averages of their diagonal elements.

### 20.1.6.4 Computational Time Complexity Analysis

Let $n$ be the sample size and $m$ be the number of markers. We ignore the cost of computing $Z$ and $W$ because they are the original data required for calculation of all kinship matrices. Computational time complexity is defined as the number of multiplications involved in the calculation. For $K_a = ZZ^T$ or $K_d = WW^T$, the number of multiplications is $mn^2$ and thus the time complexity is $O(mn^2)$. For $K_{aa} = 0.5 \times K_a * K_a - 0.5 \times (Z*Z)(Z*Z)^T$, the number of multiplications is $n^2 + nm + mn^2$. Therefore, the time complexity is still $O(mn^2)$. Similarly, the number of multiplications involved for calculating $K_{dd} =$

**Table 20.12** Summary of all five types of kinship matrices using the fast algorithm

| Kinship matrix | Expression |
|---|---|
| $K_a$ | $ZZ^T$ |
| $K_d$ | $WW^T$ |
| $K_{aa}$ | $\frac{1}{2}\left[K_a * K_a - (Z * Z)(Z * Z)^T\right]$ |
| $K_{(ad)}$ | $K_a * K_d - (Z * W)(Z * W)^T$ |
| $K_{dd}$ | $\frac{1}{2}\left[K_d * K_d - (W * W)(W * W)^T\right]$ |

**Table 20.13** Additive kinship matrix ($K_a$) for 10 plants calculated from three loci

| $K_a$ | Plant1 | Plant2 | Plant3 | Plant4 | Plant5 | Plant6 | Plant7 | Plant8 | Plant9 | Plant10 |
|---|---|---|---|---|---|---|---|---|---|---|
| Plant1 | 1 | −1 | 0 | −1 | 0 | 0 | −1 | 0 | 0 | 0 |
| Plant2 | −1 | 3 | 0 | 2 | 1 | 0 | 0 | 0 | 2 | −1 |
| Plant3 | 0 | 0 | 0 | 0 | 0 | 0 | 0 | 0 | 0 | 0 |
| Plant4 | −1 | 2 | 0 | 2 | 0 | −1 | 1 | −1 | 1 | 0 |
| Plant5 | 0 | 1 | 0 | 0 | 1 | 1 | −1 | 1 | 1 | −1 |
| Plant6 | 0 | 0 | 0 | −1 | 1 | 2 | −1 | 2 | 0 | −1 |
| Plant7 | −1 | 0 | 0 | 1 | −1 | −1 | 2 | −1 | −1 | 1 |
| Plant8 | 0 | 0 | 0 | −1 | 1 | 2 | −1 | 2 | 0 | −1 |
| Plant9 | 0 | 2 | 0 | 1 | 1 | 0 | −1 | 0 | 2 | −1 |
| Plant10 | 0 | −1 | 0 | 0 | −1 | −1 | 1 | −1 | −1 | 1 |

$0.5 \times K_d * K_d - 0.5 \times (W * W)(W * W)^T$ is $n^2 + nm + mn^2$ and thus the time complexity is $O(mn^2)$. Finally, the number of multiplications for $K_{(ad)} = K_a * K_d - (Z * W)(Z * W)^T$ is $n^2 + nm + mn^2$ again with the same time complexity $O(mn^2)$. However, using the direct method, the number of multiplications involved for calculating $K_{aa} = (Z \# Z)(Z \# Z)^T$ is $nm(m-1)/2 + n^2 m(m-1)/2 = nm(n+1)(m-1)/2$ and the time complexity is $O(n^2 m^2)$. The total number of multiplications to calculate the dominance by dominance kinship matrix $K_{dd} = (W \# W)(W \# W)^T$ is also $nm(m-1)/2 + n^2 m(m-1)/2$ with the same time complexity $O(m^2 n^2)$. The time complexity for calculating each of $K_{ad} = (Z \# W)(Z \# W)^T$ and $K_{da} = (W \# Z)(W \# Z)^T$ is again $O(m^2 n^2)$.

In summary, the time complexity to calculate $K_a$ or $K_d$ is $O(mn^2)$. Using the fast method, the time complexity for each of the three epistatic kinship matrices remains $O(mn^2)$. However, using the direct method, the time complexity to calculate each epistatic kinship matrix is $O(n^2 m^2)$.

### 20.1.6.5 Prediction of New Plants

In genomic selection, our main purpose is to predict the performance of future plants who have not been field evaluated and thus there are no phenotypic values available. These plants have genotypes and we can calculate the relationship matrix between plant of the training sample and the future plants. We now use the toy sample with ten plants as an example to demonstrate the method. We assume that the first three plants form the training sample and the last seven plants are the future plants whose phenotypes have not been available yet. From the genotypes of all plants, we calculate all kinship matrices. The prediction of future plants only requires the kinship matrix of the training sample (highlighted in gray of Table 20.13) and the kinship matrix between the training sample and the future plants (highlighted in yellow in Table 20.13). The kinship matrix of the future plants (highlighted in light blue in Table 20.13) is not needed at this stage. The number of future individuals is supposed to be extremely large in hybrid prediction. Therefore, we do not need to calculate the full kinship matrix; instead, we only need to calculate the $3 \times 3$ kinship matrix for the training sample and the $7 \times 3$ kinship matrix between the future plants and the plants in the training sample.

Let $Z_{n \times m}$ be the additive code and $W_{n \times m}$ be the dominance code, where $n$ is the total sample size and $m$ is the number of markers. Let $n_1$ and $n_2$ be the sizes of the training sample and the test sample, respectively, and $n_1 + n_2 = n$. We partition each of the two matrices into two blocks,

**Table 20.14** Kinship matrix for the training sample

| Kinship | Expression |
| --- | --- |
| $K_a^{11}$ | $Z_1 Z_1^T$ |
| $K_d^{11}$ | $W_1 W_1^T$ |
| $K_{aa}^{11}$ | $\frac{1}{2}\left[(Z_1 Z_1^T) * (Z_1 Z_1^T) - (Z_1 * Z_1)(Z_1 * Z_1)^T\right]$ |
| $K_{ad}^{11}$ | $(Z_1 Z_1^T) * (W_1 W_1^T) - (Z_1 * W_1)(Z_1 * W_1)^T$ |
| $K_{dd}^{11}$ | $\frac{1}{2}\left[(W_1 W_1^T) * (W_1 W_1^T) - (W_1 * W_1)(W_1 * W_1)^T\right]$ |

**Table 20.15** Kinship matrix between future plants and plants in the training sample

| Kinship | Expression |
| --- | --- |
| $K_a^{21}$ | $Z_2 Z_1^T$ |
| $K_d^{21}$ | $W_2 W_1^T$ |
| $K_{aa}^{21}$ | $\frac{1}{2}\left[(Z_2 Z_1^T) * (Z_2 Z_1^T) - (Z_2 * Z_2)(Z_1 * Z_1)^T\right]$ |
| $K_{ad}^{21}$ | $(Z_2 Z_1^T) * (W_2 W_1^T) - (Z_2 * W_2)(Z_1 * W_1)^T$ |
| $K_{dd}^{21}$ | $\frac{1}{2}\left[(W_2 W_1^T) * (W_2 W_1^T) - (W_2 * W_2)(W_1 * W_1)^T\right]$ |

$$Z = \begin{bmatrix} Z_1 \\ Z_2 \end{bmatrix} \text{ and } W = \begin{bmatrix} W_1 \\ W_2 \end{bmatrix}$$

where $Z_1$ is the additive code matrix for the training sample with dimension $n_1 \times m$, $Z_2$ is the additive code matrix for the test sample with dimension $n_2 \times m$, $W_1$ is the dominance code matrix for the training sample with dimension $n_1 \times m$, and $W_2$ is the dominance code matrix for the test sample with dimension $n_2 \times m$. Formulas for kinship matrices in the training sample are given in Table 20.14. The corresponding kinship matrices between future plants and the current plants are summarized in Table 20.15.

We now present the BLUP of genomic values of individuals in the future sample. Let $G_{11}$ be the genetic variance matrix of the training sample,

$$G_{11} = K_a^{11} \sigma_a^2 + K_d^{11} \sigma_d^2 + K_{aa}^{11} \sigma_{aa}^2 + K_{ad}^{11} \sigma_{(ad)}^2 + K_{dd}^{11} \sigma_{dd}^2$$

Let $G_{21}$ be the genetic covariance matrix between future plants and plants in the training sample.

$$G_{21} = K_a^{21} \sigma_a^2 + K_d^{21} \sigma_d^2 + K_{aa}^{21} \sigma_{aa}^2 + K_{ad}^{21} \sigma_{(ad)}^2 + K_{dd}^{21} \sigma_{dd}^2$$

The BLUP of all future individuals are

$$\widehat{\xi}_2 = \widehat{G}_{21}\left(\widehat{G}_{11} + I\widehat{\sigma}^2\right)^{-1}\left(y_1 - X_1\widehat{\beta}\right)$$

where $y_1$ and $X_1$ are the responses and the incidence matrix for individuals in the training sample.

## 20.2  Reproducing Kernel Hilbert Spaces (RKHS) Regression

### 20.2.1  RKHS Prediction

Reproducing kernel Hilbert spaces (Wahba 1990) was first introduced to the genomic prediction area by Gianola and his coworkers (Gianola et al. 2006; Gianola and van Kaam 2008). The presentation here follows closely the logic in de los Campos et al. (2009, 2010, 2013; de los Campos and Perez 2013).

Define the model for the array of phenotypic values of $n$ individuals by

$$y = X\beta + \xi(Z) + \varepsilon$$

where $X$ are an incidence matrix for fixed effects $\beta$, $Z$ is an $n \times m$ matrix of genotypic values for $m$ markers measured from $n$ individuals, $\xi(Z)$ is an $n \times 1$ vector of unknown function of $Z$, and $\varepsilon$ is an $n \times 1$ vector of residuals with an assumed $N(0, R\sigma^2)$ distribution. The model differs from the mixed model (8) by the unknown functions $\xi(Z)$, where it was linear $Z\gamma$ in the mixed model. In RKHS, the unknown functions are represented by functions linear on an $n \times 1$ vector of $\alpha$ but nonlinear on $Z$, i.e.,

$$\xi(Z) = K\alpha$$

where $K$ is an $n \times n$ kernel matrix built from $Z$ with some tuning parameters. For example, the single smoothing parameter Gaussian kernel is defined as $K = \{K_{ij}\}$. It means that the kernel matrix has $n$ rows and $n$ columns and the $i$th row and the $j$th column is

$$K_{ij} = \exp\left[-\frac{1}{h}\left(Z_i - Z_j\right)^T\left(Z_i - Z_j\right)\right]$$

where $Z_i = Z[i, ]$ and $Z_j = Z[j, ]$ are the $i$th row and the $j$th row of matrix $Z$, respectively, and $h$ is the bandwidth (a tuning parameter). The kernel matrix is symmetrical so that $K^T = K$. Intuitively, $K_{ij}$ is a kind of molecular similarity measurement between individuals $i$ and $j$. The bandwidth $h$ determines the decay of the similarity between $i$ and $j$ when $Z_i$ is different from $Z_j$. For example, if $h$ is large, $K_{ij}$ is not sensitive to the difference between $Z_i$ and $Z_j$. However, if $h$ is very small, the decay can be extremely fast. Imagine that if $h \to 0$, then $K_{ij} = 1$ when $Z_i = Z_j$ and $K_{ij} = 0$ otherwise, and the $K$ matrix would be an identity matrix. Since $\alpha$ is an $n \times 1$ vector of unknown coefficients, the solution will not be unique whether a least squares method or a maximum likelihood method is used. Therefore, a penalty is required to find a unique solution. The RKHS estimates of $\alpha$ are obtained by minimizing the following penalized sum of squares,

$$Q(\alpha) = (y - X\beta - K\alpha)^T(y - X\beta - K\alpha) + \lambda^{-1}\alpha^T K\alpha$$

where the second term, $\lambda^{-1}\alpha^T K\alpha$, is a penalty (de los Campos et al. 2010) and $\lambda^{-1}$ is another tuning parameter. The partial derivative of $Q(\alpha)$ with respect to $\beta$ and $\alpha$ are.

$$\frac{\partial Q(\beta)}{\partial \beta} = -2X^T(y - X\beta - K\alpha)$$

$$\frac{\partial Q(\alpha)}{\partial \alpha} = -2K^T(y - X\beta - K\alpha) + 2\lambda^{-1}K\alpha$$

Setting the above derivatives to zero leads to

$$\begin{bmatrix} X^TX & X^TK \\ K^TX & K^TK + \lambda^{-1}K \end{bmatrix}\begin{bmatrix} \beta \\ \alpha \end{bmatrix} = \begin{bmatrix} X^Ty \\ K^Ty \end{bmatrix} \tag{20.11}$$

The solutions are

$$\begin{bmatrix} \widehat{\beta} \\ \widehat{\alpha} \end{bmatrix} = \begin{bmatrix} X^TX & X^TK \\ K^TX & K^TK + \lambda^{-1}K \end{bmatrix}^{-1}\begin{bmatrix} X^Ty \\ K^Ty \end{bmatrix} \tag{20.12}$$

The predicted genomic values are

$$\widehat{\xi}(Z) = K\widehat{\alpha}$$

From a Bayesian perspective, estimates of (12) can be viewed as the posterior modes in the following model,

$$y = X\beta + K\alpha + \varepsilon \tag{20.13}$$

where $p(\varepsilon) = N(\varepsilon|\ 0,\ I\sigma^2)$, $p(\alpha) = N(\alpha|0, K^{-1}\sigma_\alpha^2)$ is the prior distribution of $\alpha$ and $\lambda = \sigma_\alpha^2/\sigma^2$. Therefore, RKHS is also called kernel regression or regression on the kernel.

RKHS is identical to GBLUP if the kernel is replaced by a marker-inferred kinship matrix used in GBLUP. The gain of RKHS is only due to the flexibility of kernel selection via changing the tuning parameter $h$. The expectation of model (13) is E $(y) = X\beta$ and the variance is

$$\text{var}(y) = \text{var}(\xi) + \text{var}(\varepsilon) = K\text{var}(\alpha)K^T + I\sigma^2 = KK^{-1}K^T\sigma_\alpha^2 + I\sigma^2 = K\sigma_\alpha^2 + I\sigma^2$$

So, the expectation and variance of RKHS are the same as the mixed model in GBLUP. The BLUP equations for RKHS (11) can be split into

$$X^TX\beta + X^TK\alpha = X^Ty \tag{20.14}$$

and

$$K^TX\beta + (K^TK + \lambda^{-1}K)\alpha = K^Ty \tag{20.15}$$

Multiplying both sides of Eq. (20.15) by $K^{-1}$ and recognizing the fact that $K = K^T$, we get

$$X\beta + (I + \lambda^{-1}K^{-1})K\alpha = y \tag{20.16}$$

Knowing the fact that $K\alpha = \xi$ and combing eqs. (14) and (16) together, we have

$$\begin{bmatrix} X^TX & X^T \\ X & I + \lambda^{-1}K^{-1} \end{bmatrix} \begin{bmatrix} \beta \\ \xi \end{bmatrix} = \begin{bmatrix} X^Ty \\ y \end{bmatrix}$$

which is identical to the BLUP equation presented in (20.4). The predicted phenotypic values for GBLUP and RKHS are the same also but with different expressions

$$\widehat{y} = \begin{cases} X\widehat{\beta} + Z\widehat{\gamma} & \text{for GBLUP} \\ X\widehat{\beta} + K\widehat{\alpha} & \text{for RKHS} \end{cases}$$

## 20.2.2 Estimation of Variance Components

The restricted maximum likelihood can be used to estimate the variance components. The expectation of $y$ is E$(y) = X\beta$ and the variance matrix is

$$\text{var}(y) = V = K\text{var}(\alpha)K + I\sigma^2 = KK^{-1}K\sigma_\alpha^2 + I\sigma^2 = K\sigma_\alpha^2 + I\sigma^2$$

The restricted likelihood function is

$$L_R(\theta) = -\frac{1}{2}\ln|V| - \frac{1}{2}\ln|X^T V^{-1} X| - \frac{1}{2}\left(y - X\widehat{\beta}\right)^T V^{-1}\left(y - X\widehat{\beta}\right)$$

where $\theta = \{\sigma_\alpha^2, \sigma^2\}$ are the parameters and

$$\widehat{\beta} = \left(X^T H^{-1} X\right)^{-1} X^T H^{-1} y$$

where $H = K\lambda + I$. Given $\lambda$, the residual variance can be absorbed using

$$\widehat{\sigma}^2 = \frac{1}{n-q}\left(y - X\widehat{\beta}\right)^T H^{-1}\left(y - X\widehat{\beta}\right)$$

Substituting this residual variance back to the restricted likelihood function yields the following profiled likelihood function,

$$L_R(\lambda) = -\frac{1}{2}\ln|H| - \frac{1}{2}\ln|X^T H^{-1} X| - \frac{1}{2\widehat{\sigma}^2}\left(y - X\widehat{\beta}\right)^T H^{-1}\left(y - X\widehat{\beta}\right) - \frac{n-q}{2}\ln\left(\widehat{\sigma}^2\right)$$

$$= -\frac{1}{2}\ln|H| - \frac{1}{2}\ln|X^T H^{-1} X| - \frac{n-q}{2}\ln\left(\widehat{\sigma}^2\right) - \frac{n-q}{2}$$

which contains only one parameter, $\lambda$, and a simple Newton iteration algorithm can be used to find the REML estimate of $\lambda$.

### 20.2.3 Prediction of Future Individuals

The linear models for the training and test samples are

$$\begin{bmatrix} y_1 \\ y_2 \end{bmatrix} = \begin{bmatrix} X_1\beta \\ X_2\beta \end{bmatrix} + \begin{bmatrix} \xi_1 \\ \xi_2 \end{bmatrix} + \begin{bmatrix} \varepsilon_1 \\ \varepsilon_2 \end{bmatrix}$$

where

$$\begin{bmatrix} \xi_1 \\ \xi_2 \end{bmatrix} = \begin{bmatrix} K_{11} & K_{12} \\ K_{21} & K_{22} \end{bmatrix}\begin{bmatrix} \alpha_1 \\ \alpha_2 \end{bmatrix}$$

The expectation of the model is

$$E\begin{bmatrix} y_1 \\ y_2 \end{bmatrix} = \begin{bmatrix} X_1\beta \\ X_2\beta \end{bmatrix}$$

and the variance matrix is

$$\text{var}\begin{bmatrix} y_1 \\ y_2 \end{bmatrix} = \begin{bmatrix} K_{11} & K_{12} \\ K_{21} & K_{22} \end{bmatrix}\sigma_\alpha^2 + \begin{bmatrix} I & 0 \\ 0 & I \end{bmatrix}\sigma^2$$

where $K_{11}$ is the kernel matrix corresponding to the training sample, $K_{21} = K_{12}^T$ is the kernel matrix corresponding to the similarity between the test sample and the training sample, and $K_{22}$ is the kernel matrix corresponding to the test sample. The variance of the kernel regression is

$$\text{var}\begin{bmatrix}\xi_1 \\ \xi_2\end{bmatrix} = \begin{bmatrix}K_{11} & K_{12} \\ K_{21} & K_{22}\end{bmatrix}\begin{bmatrix}K_{11} & K_{12} \\ K_{21} & K_{22}\end{bmatrix}^{-1}\begin{bmatrix}K_{11} & K_{12} \\ K_{21} & K_{22}\end{bmatrix}\sigma_\alpha^2 = \begin{bmatrix}K_{11} & K_{12} \\ K_{21} & K_{22}\end{bmatrix}\sigma_\alpha^2$$

From the multivariate normal theorem (Giri 1996), we predict the genomic value of individuals in the test sample by

$$\widehat{\xi}_2 = \widehat{\sigma}_\alpha^2 K_{21}\left(K_{11}\widehat{\sigma}_\alpha^2 + I\widehat{\sigma}^2\right)^{-1}\left(y_1 - X_1\widehat{\beta}\right)$$

where the parameters are assumed to be estimated from the training sample. It is important to remember that individuals in the test sample are ranked based on $\widehat{\xi}_2$, the predicted genomic values, not based on the predicted phenotypic values, which are

$$\widehat{y}_2 = X_2\widehat{\beta} + \widehat{\sigma}_\alpha^2 K_{21}\left(K_{11}\widehat{\sigma}_\alpha^2 + I\widehat{\sigma}^2\right)^{-1}\left(y_1 - X_1\widehat{\beta}\right)$$

because $\widehat{y}_2$ contain the fixed environmental effects. The predicted genomic values are subject to prediction errors, which are defined as

$$\text{var}\left(\widehat{\xi}_2\right) = K_{22}\widehat{\sigma}_\alpha^2 - \widehat{\sigma}_\alpha^2 K_{21}\left(K_{11}\widehat{\sigma}_\alpha^2 + I\widehat{\sigma}^2\right)^{-1}K_{12}\widehat{\sigma}_\alpha^2 \tag{20.17}$$

Square roots of the diagonal elements of $\text{var}\left(\widehat{\xi}_2\right)$ are the standard errors of the prediction, i.e., the prediction errors. The kinship matrix between individuals in the test sample is only used here in Eq. (20.17).

### 20.2.4 Bayesian RKHS Regression

The Bayesian approach is implemented via the Markov chain Monte Carlo sampling algorithm. All parameters are assumed to be random variables, and they are simulated from their conditional posterior distributions. Missing values in the response variable (missing phenotypes) can be easily sampled from their posterior distributions. Therefore, the phenotype of a future individual can be simulated and the posterior mode or posterior mean of the phenotype of the future individual is the predicted genomic value of that individual. Therefore, the Bayesian RKHS regression simply includes both the training sample and the test sample in one input dataset but setting the phenotypic values of the test sample to missing values.

The Bayesian method requires a model, a distribution of the data given parameters, a prior distribution of each parameter. From these pieces of information, a posterior distribution of a parameter is derived, from which a realized value of the parameter is sampled. The posterior mode or posterior mean is the Bayesian estimate of that parameter. The model is

$$y = X\beta + K\alpha + \varepsilon$$

The data are represented by $y$, the parameters include $\theta = \{\alpha, \beta, \sigma^2\}$. Given the parameters, the distribution of the data is multivariate normal,

$$p(y|\theta) = N\left(y|X\beta + K\alpha, I\sigma^2\right)$$

The prior for $\beta$ is flat represented by

$$p(\beta) = N\left(\beta|0, 10^{10}\right)$$

The variance is extremely large and the distribution is virtually flat. The distribution of $\alpha$ is multivariate normal,

$$p(\alpha) = N(\alpha|0, K^{-1}\sigma_\alpha^2)$$

The prior distribution of $\sigma^2$ is a scaled inverse Chi-square,

$$p(\sigma^2) = \frac{1}{\chi^2}(\sigma^2|df_0, S_0)$$

where $df_0 = 5$ is the degree of freedom and $S_0 = 1$ is the scale of the prior distribution. de los Campos et al. (de los Campos et al. 2010) suggested to use $S_0 = S_y^2(1 - R^2)(df_0 + 2)$ as the scale parameter, where $S_y^2$ is the observed variance of the phenotypic values and $R^2$ is some suggested proportion of phenotypic variance explained by the model effects. By default, $R^2 = 0.5$. Other values of $df_0$ and $S_0$ can also be chosen, for example, $df_0 = S_0 = 0$.

The prior distribution for $\alpha$ involves an unknown variance $\sigma_\alpha^2$, which also needs a prior distribution so that it can be estimated from the data. Such a system including prior of prior is called a hierarchical prior system. The prior for $\sigma_\alpha^2$ is a scaled inverse Chi-square also,

$$p(\sigma_\alpha^2) = \frac{1}{\chi^2}(\sigma_\alpha^2|df_\alpha, S_\alpha)$$

where $df_\alpha = 5$ and $S_\alpha = 1$. Of course, other values can also be assigned to the hyper parameters. For example, $S_\alpha = S_y^2 R^2(df_\alpha + 2)n/\text{tr}(K)$ (de los Campos et al. 2013).

Given the distribution of the data and the prior distributions of the parameters, posterior distributions of parameters are derived. In fact, we only need the fully conditional posterior distribution, i.e., the distribution of one parameter conditional on the data and all other parameters. From the fully conditional posterior distribution, the parameter is sampled. This process is called Gibbs sampler (Geman and Geman 1984), a special case of Markov chain Monte Carlo algorithm (Gilks and Wild 1992; Gilks et al. 1995, 1996).

The posterior distribution of the fixed effects ($\beta$) is multivariate normal, denoted by $p(\beta| \cdots) = N(\beta| \mu_\beta, V_\beta)$, where

$$\mu_\beta = (X^TX)^{-1}X^T(y - K\alpha)$$

and

$$\Sigma_\beta = (X^TX)^{-1}\sigma^2$$

The posterior distribution of the regression coefficients ($\alpha$) is also multivariate norm, denoted by $p(\alpha| \cdots) = N(\alpha| \mu_\alpha, \Sigma_\alpha)$, where

$$\mu_\alpha = (K^TK + \lambda K)^{-1}K^T(y - X\beta) = U(D + \lambda I)^{-1}U^T(y - X\beta)$$

and

$$\Sigma_\alpha = (K^TK + \lambda K)^{-1}\sigma^2 = UD^{-1}(D + I\lambda)^{-1}U^T\sigma^2$$

Application of the eigenvalue decomposition, $K = UDU^T$, is to avoid repeated calculation of $(K^TK + \lambda K)^{-1}$. The posterior distribution of $\sigma^2$ remains a scaled inverse Chi-square,

$$p(\sigma^2|\cdots) = \frac{1}{\chi^2}(n + df_0, SS_\varepsilon + df_0 S_0)$$

where

$$SS_\varepsilon = (y - X\beta - K\alpha)^T (y - X\beta - K\alpha)$$

is the residual sum of squares. The posterior distribution of $\sigma_\alpha^2$ is

$$p(\sigma_\alpha^2 | \cdots) = \frac{1}{\chi^2}(n + df_\alpha, SS_\alpha + df_\alpha S_\alpha)$$

where

$$SS_\alpha = \alpha^T [\text{var}(\alpha)]^{-1} \alpha + df_\alpha S_\alpha = \alpha^T K\alpha + df_\alpha S_\alpha$$

Missing phenotypic values are sampled from their fully conditional posterior distribution (just the likelihood). For example, if $y_j$ is missing, it will be sampled from

$$p(y_j | \cdots) = N(y_j | X_j\beta + K_j\alpha, \sigma^2)$$

where $X_j$ is the $j$th row of matrix $X$ and $K_j$ is the $j$th row of matrix $K$. The Gibbs sampler is summarized below,

Step 0: Set $t = 0$ and initialize all parameters with a random draw from the prior distribution, denoted by $\theta^{(t)} = \{\beta^{(t)}, \alpha^{(t)}, \sigma_\alpha^{2(t)}, \sigma^{2(t)}\}$.

Step 1: Sample missing phenotypic values from their distributions (likelihoods).

Step 2: Sample $\beta$ from its fully conditional posterior distribution, denoted by $\beta^{(t+1)}$.

Step 3: Sample $\alpha$ from its fully conditional posterior distribution, denoted by $\alpha^{(t+1)}$.

Step 4: Sample $\sigma_\alpha^2$ from its fully conditional posterior distribution, denoted by $\sigma_\alpha^{2(t+1)}$.

Step 5: Sample $\sigma^2$ from its fully conditional posterior distribution, denoted by $\sigma^{2(t+1)}$.

Step 6: Repeat Step 1 to Step 5 until a desirable number of iterations is reached.

The posterior sample contains all sampled parameters, including missing phenotypic values. The first 1000 draws (burn-ins) should be deleted. Thereafter, keep one in every 10 draws (thinning rate) until a desirable number of draws is reached in the posterior samples. The posterior mode or posterior mean of a parameter is considered the Bayesian estimate of that parameter. The posterior mean of a missing phenotypic value is the predicted trait value. The predicted genomic value $\widehat{\xi}_j = K_j\widehat{\alpha}$ for an individual with missing phenotype is the actual predicted genomic value.

## 20.2.5 Kernel Selection

The Gaussian kernel is defined as

$$K_{ij} = \exp\left(-\frac{1}{h}D_{ij}^2\right)$$

where

$$D_{ij}^2 = (Z_i - Z_j)^T (Z_i - Z_j)$$

is the squared Euclidean distance between individuals $i$ and $j$ measured from all markers. An alternative expression of the Euclidean distance is

$$D_{ij}^2 = \sum_{k=1}^{m} \left( Z_{ik} - Z_{jk} \right)^2$$

The optimal value of $h$ depends on many things, e.g., population size, number of markers, the genetic architecture of the trait under study, numeric codes of genotypes, and so on. It is better to normalize the squared Euclidean distance using

$$D_{ij}^2 = \frac{1}{\max \left( D_{ij}^2 \right)} D_{ij}^2$$

which is expressed in a pseudo code notation. This normalized distance takes a value between 0 and 1. The kernel matrix is defined as

$$K = \left\{ \exp \left( -\frac{1}{h} D_{ij}^2 \right) \right\}$$

Although a prior can be assigned to $h$ so that it can be sampled in the Bayesian method, this treatment is not realistic because every time when $h$ changes, the kernel has to be recalculated. The usual approach is to evaluate a grid of $h$ from a very small value to a very large value and select the one with the highest predictability using a ten-fold cross validation (CV) or the generalized cross validation (GCV) scheme. Under the normalized distance, we can conveniently choose a global set of $h$ for all data. For example, the grid for $h$ may start from 0.1 and end with 10 incremented by 0.5. CV and GCV will be introduced in the next section. de los Campos et al. (2013) proposed a multiple kernel approach and each kernel has a different tuning parameter. For example, we can take three kernels

$$y = X\beta + K_A \alpha_A + K_B \alpha_B + K_C \alpha_C + \varepsilon$$

where

$$K_A = \left\{ \exp \left( -\frac{1}{h_A} D_{ij}^2 \right) \right\} = \left\{ \exp \left( -\frac{1}{0.1} D_{ij}^2 \right) \right\}$$

$$K_B = \left\{ \exp \left( -\frac{1}{h_B} D_{ij}^2 \right) \right\} = \left\{ \exp \left( -\frac{1}{1.25} D_{ij}^2 \right) \right\}$$

$$K_C = \left\{ \exp \left( -\frac{1}{h_C} D_{ij}^2 \right) \right\} = \left\{ \exp \left( -\frac{1}{10} D_{ij}^2 \right) \right\}$$

For each set of kernel regression coefficients, a variance is assigned, $\alpha_A \sim N\left(0, K_A^{-1}\sigma_A^2\right)$, $\alpha_B \sim N\left(0, K_B^{-1}\sigma_B^2\right)$ and $\alpha_C \sim N\left(0, K_C^{-1}\sigma_C^2\right)$. The variance matrix for $y$ is

$$\text{var}(y) = K_A \sigma_A^2 + K_B \sigma_B^2 + K_C \sigma_C^2 + I\sigma^2$$

The multiple kernel (MK) approach is also called kernel average (KA) and each kernel is weighted by the kernel-specific variance. The epistatic GBLUP model introduced early can also be implemented using the kernel average approach. We simply replace each kernel by a kinship matrix. The result will be identical to the GBLUP of multiple kinship matrices, subject to sampling errors.

The BGLR (Bayesian generalized linear regression) software package in R (de los Campos and Perez 2013) is available to perform all kinds of Bayesian regression analysis for genomic prediction, including Bayes ridge regression, Bayes A, Bayes B, Bayes C, Bayesian LASSO, RKHS, and so on. The package is available in CRAN at the R-forge website https://r-forge.r-project.org/projects/bglr/.

Before we exit this section, let us introduce another general purpose kernel, the Laplace kernel defined as

$$K = \left\{ \exp\left(-\frac{1}{h}D_{ij}\right) \right\}$$

The only difference between the Laplace kernel and the Gaussian kernel is in the use of $D_{ij}$ or $D_{ij}^2$. The Laplace kernel uses the Euclidean distance while the Gaussian kernel uses the squared distance. Euclidean distance is just one kind of distance measurement. Other types of distance can also be used to build the kernel.

## 20.3  Predictability

### 20.3.1  Data Centering

Predictability is often defined as the squared correlation between the predicted and the observed trait values. Both the predicted and observed trait values must be adjusted by the fixed effects in genomic selection because the genomic values are treated as random effects. To measure the predictability of a model or a trait, we often use a cross validation approach, where the fixed effects should be removed from the phenotypic values prior to data analysis. If there are no fixed effects other than the grand mean or intercept, the phenotypic value of each individual should be subtract by the population mean. The response variable is the fixed-effect-adjusted phenotypic value. With multiple fixed effects, we can fit a linear fixed model without the polygenic effect

$$y = X\beta + e$$

and take the residual as the response variable. The residual is

$$r = y - X\widehat{\beta} = y - X(X^T X)^{-1} X^T y = \left[ I - X(X^T X)^{-1} X^T \right] y = Ly$$

where

$$L = I - X(X^T X)^{-1} X^T$$

Since we always use $y$ to represent the response variable, the pseudo code notation for a centered $y$ is

$$y = Ly$$

Thereafter, the response variable $y$ is always adjusted by the fixed effects so that the expectation of $y$ is always zero, i.e., $E(y) = 0$. One of the basic assumptions in the genomic selection model discussed so far is the i.i.d. assumption of the residual error, i.e., $\varepsilon \sim N(0, I\sigma^2)$. However, after data are centered, this assumption has been violated. We can look at the linear mixed model again,

$$y = X\beta + Z\gamma + \varepsilon \tag{20.18}$$

The centered model is

$$Ly = LX\beta + L(Z\gamma + \varepsilon) = L(Z\gamma + \varepsilon) \tag{20.19}$$

The variance of the centered data is

$$\text{var}(Ly) = L\text{var}(Z\gamma + \varepsilon)L^T = LKL^T\phi^2 + LL^T\sigma^2 \tag{20.20}$$

The covariance structure for the residual error is $R = LL^T$, not an identity matrix. However, we simply ignore this fact and assume $R = I$ in all subsequent discussions.

## 20.3.2 Maximum Likelihood Estimate of Parameters in Random Models

Estimation of variance components in a random model is much easier than that in a mixed model. Let $K = UDU^T$ be the eigenvalue decomposition for the kinship matrix ($K$). Define $y^* = U^Ty$ as linear transformations of the response variable and $q = r(X)$ as the column rank of $X$. The likelihood function for $\lambda = \phi^2/\sigma^2$ is.

$$L(\lambda) = -\frac{1}{2}\ln \mid D\lambda + I \mid -\frac{n-q}{2}\ln\left(\sigma^2\right) - \frac{1}{2\sigma^2}y^{*T}(D\lambda + I)^{-1}y^*$$

Let

$$\widehat{\sigma}^2 = \frac{1}{n-q}y^{*T}(D\lambda + I)^{-1}y^*$$

so that the likelihood function is rewritten as

$$L(\lambda) = -\frac{1}{2}\ln \mid D\lambda + I \mid -\frac{n-q}{2}\ln\left(\widehat{\sigma}^2\right) - \frac{n-q}{2} \tag{20.21}$$

The last term should be ignored because it is irrelevant to the maximum likelihood estimate of the variance ratio. Only one parameter ($\lambda$) is involved in (20.21) and the solution is very easy to obtain using the Newton method.

## 20.3.3 Predicted Residual Error Sum of Squares

We introduced two methods for genomic prediction, BLUP and RKHS. There are more than 20 other prediction methods available. How do we compare the performance of these methods? We often use the R-square to measure the model goodness of fit,

$$R^2_{GOOD} = 1 - \frac{RESS}{SS}$$

where

$$RESS = (y - Z\widehat{\gamma})^T(y - Z\widehat{\gamma}) = \left(y - \widehat{\xi}\right)^T\left(y - \widehat{\xi}\right)$$

is the residual error sum of squares (RESS) and

$$SS = y^Ty$$

is the total sum of squares (SS). In genomic selection, it is very common to observe $RESS = 0$ and thus $R^2_{GOOD} = 1$ when the number of markers is relatively large. Therefore, the model goodness of fit is useless. The predicted residual error sum of squares (PRESS) (Allen 1971; Allen 1974) is often used to measure a model predictability. Let us define the fitted residual error for individual $j$ by

$$\widehat{\varepsilon}_j = y_j - Z_j\widehat{\gamma}$$

The predicted residual error for individual $j$ is defined as

$$\widehat{e}_j = y_j - Z_j\widehat{\gamma}_{-j}$$

where $\widehat{\gamma}_{-j}$ is the BLUP of $\gamma$ estimated from a sample excluding $y_j$. In other words, if we want to predict the genomic value for individual $j$, the phenotypic value from individual $j$ should not contribute to parameter estimation. If it does, there is a conflict of interest. The predicted residual errors (not the fitted residual errors) are used to calculate the PRESS,

$$PRESS = \sum_{j=1}^{n}\widehat{e}_j^2 = \sum_{j=1}^{n}\left(y_j - Z_j\widehat{\gamma}_{-j}\right)^2$$

The predictability is defined as

$$R_{PRED}^2 = 1 - \frac{PRESS}{SS}$$

Another measurement of predictability is the squared correlation between the "observed" trait value ($y_j$) and the predicted trait value ($\widehat{y}_j$),

$$\widehat{y}_j = Z_j\widehat{\gamma}_{-j} \tag{20.22}$$

The squared correlation between the "observed" and the predicted genomic values is defined as

$$R_{CORR}^2 = \frac{\mathrm{cov}^2(y,\widehat{y})}{\mathrm{var}(y)\mathrm{var}(\widehat{y})}$$

Theoretically, the two measurements of predictability are the same. Note that the model for $y_j$ is.

$$y_j = Z_j\widehat{\gamma}_{-j} + e_j = \widehat{y}_j + e_j \tag{20.23}$$

Therefore,

$$\mathrm{cov}\left(y_j,\widehat{y}_j\right) = \mathrm{cov}\left(\widehat{y}_j + e_j,\widehat{y}_j\right) = \mathrm{cov}\left(\widehat{y}_j,\widehat{y}_j\right) + \mathrm{cov}\left(\widehat{y}_j,e_j\right) = \mathrm{var}\left(\widehat{y}_j\right)$$

The variance of the trait value is

$$\mathrm{var}\left(y_j\right) = \mathrm{var}\left(\widehat{y}_j\right) + \mathrm{var}\left(e_j\right)$$

Therefore,

$$R_{CORR}^2 = \frac{\mathrm{var}^2(\widehat{y})}{\mathrm{var}(y)\mathrm{var}(\widehat{y})} = \frac{\mathrm{var}(\widehat{y})}{\mathrm{var}(y)} = \frac{\mathrm{var}(y) - \mathrm{var}(e)}{\mathrm{var}(y)} = 1 - \frac{\mathrm{var}(e)}{\mathrm{var}(y)} \tag{20.24}$$

The predicted residual error variance is

$$\mathrm{var}(e) = \frac{1}{n}\sum_{j=1}^{n}\left(y_j - Z_j\widehat{\gamma}_{-j}\right)^2 = \frac{1}{n}PRESS$$

The denominator of Eq. (20.24) is

$$\text{var}(y) = \frac{1}{n} \sum_{j=1}^{n} y_j^2 = \frac{1}{n} SS$$

Therefore,

$$R_{CORR}^2 = 1 - \frac{\text{var}(e)}{\text{var}(y)} = 1 - \frac{PRESS/n}{SS/n} = 1 - \frac{PRESS}{SS} = R_{PRED}^2$$

### 20.3.4 Cross Validation

The PRESS discussed so far is obtained by the so-called leave-one-out-cross validation (LOOCV) (Craven and Wahba 1979). The parameters estimated from the sample leaving the $j$th observation out are denoted by $\widehat{\gamma}_{-j}$. To calculate the PRESS, we need $\widehat{\gamma}_{-j}$ for all $j = 1, \cdots, n$. This means that we have to perform data analysis $n$ times. If the sample size is large, the computational burden can be huge. Therefore, a K-fold cross validation is often practiced where K often takes 10 and the cross validation is then called a ten-fold cross validation (Picard and Cook 1984). The LOOCV is then called the n-fold cross validation. In a ten-fold cross validation, the sample is partitioned into 10 parts (roughly equal size per part) and 9 parts are used to estimate the parameters for prediction of the remaining part. Eventually, all parts will be predicted. Let $k = 1, \cdots, 10$, and $n_k$ be the size of the $k$th part so that $\sum n_k = n$. The PRESS and the total sum of squares in a ten-fold cross validation are

$$PRESS = \sum_{k=1}^{n} \left(y_k - Z_k \widehat{\gamma}_{-k}\right)^T \left(y_k - Z_k \widehat{\gamma}_{-k}\right)$$

where $y_k$ is an $n_k \times 1$ vector of the phenotypic values for individuals in the $k$th part and $\widehat{\gamma}_{-k}$ is the parameter estimated from the sample that excludes the $k$th part. The total sum of squares is

$$SS = \sum_{k=1}^{n} y_k^T y_k$$

A ten-fold cross validation scheme is illustrated in Fig. 20.2. We say a ten-fold cross validation because this is just one of many different ways of partitioning the sample. Different ways of partitioning the sample will generate slightly different predictabilities. We often repeat the ten-fold cross validation several times, say 20 times or more, depending on how much time we want to spend on the CV analysis. Our past experience indicates that replicating 20 times is sufficient. The average PRESS or the average squared correlation over the 20 replicated CV analyses is the reported predictability. The predictability drawn from a K-fold cross validation is denoted by $R_{CV}^2$ and the predictability in the special case of n-fold cross validation is denoted by $R_{LOOCV}^2$.

### 20.3.5 The HAT Method

Cook (Cook 1977; Cook 1979) developed an explicit method to calculate PRESS by correcting the deflated residual error of an observation using the leverage value of the observation without repeated analyses of the partitioned samples. This method applies to the least square regression under the fixed model framework, where the fitted $y$ is a linear function of the observed $y$ as shown below,

$$\widehat{y} = X\widehat{\beta} = X\left(X^T X\right)^{-1} X^T y = Hy$$

where

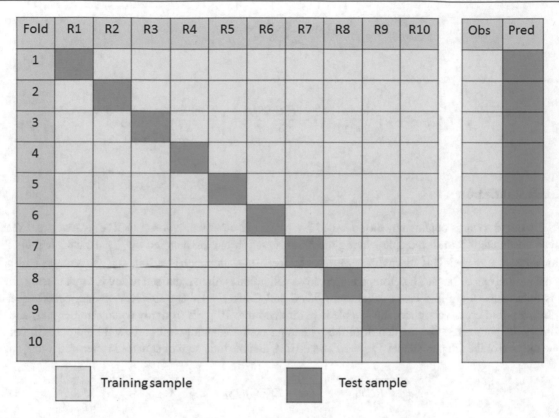

| Fold | R1 | R2 | R3 | R4 | R5 | R6 | R7 | R8 | R9 | R10 | | Obs | Pred |
|------|----|----|----|----|----|----|----|----|----|-----|--|-----|------|
| 1 | ▨ | | | | | | | | | | | | |
| 2 | | ▨ | | | | | | | | | | | |
| 3 | | | ▨ | | | | | | | | | | |
| 4 | | | | ▨ | | | | | | | | | |
| 5 | | | | | ▨ | | | | | | | | |
| 6 | | | | | | ▨ | | | | | | | |
| 7 | | | | | | | ▨ | | | | | | |
| 8 | | | | | | | | ▨ | | | | | |
| 9 | | | | | | | | | ▨ | | | | |
| 10 | | | | | | | | | | ▨ | | | |

☐ Training sample    ▨ Test sample

**Fig. 20.2** A ten-fold cross validation scheme. Each column represents an analysis with a training sample containing 9 parts (yellow) and a test sample with 1 part (green). After each part is predicted in turn with a sample excluding the part predicted, we place the predicted values together called "Pred" along with the observed values called "Obs". The PRESS or the squared correlation between the observed (yellow) and the predicted (green) values are calculated

$$H = X\left(X^T X\right)^{-1} X^T$$

is called the hat matrix, explaining why the method is called the HAT method. Let $\widehat{\varepsilon}_j = y_j - X_j \widehat{\beta}$ be the fitted residual error. The predicted residual error is $\widehat{e}_j = y_j - X_j \widehat{\beta}_{-j}$ but can be expressed as (Cook 1977; Cook 1979)

$$\widehat{e}_j = \widehat{\varepsilon}_j / \left(1 - h_{jj}\right) \tag{20.25}$$

where $h_{jj}$ is the leverage value of the $j$th observation (the $j$th diagonal element of the hat matrix). It is the contribution of the prediction for an individual from itself and may be called the *conflict of interest factor*. The predicted residual error is the fitted residual error after correction for the deflation. The sum of squares of the predicted residual errors over all individuals is the PRESS. The explicit relationship in (20.25) means that we do not need $\beta_{-j}$ to calculate the predicted residual error; rather, we calculate $\varepsilon_j$ from the whole data and then correct $\varepsilon_j$ for the deflation. The PRESS is

$$PRESS = \sum_{j=1}^{n} \widehat{e}_j^2 = \sum_{j=1}^{n} \widehat{\varepsilon}_j^2 / \left(1 - h_{jj}\right)^2 = \sum_{j=1}^{n} \left(y_j - X_j \widehat{\beta}\right)^2 / \left(1 - h_{jj}\right)^2$$

To find an explicit expression of PRESS for a random model, we need to identify a random effect version of the hat matrix and use the leverage value of the $j$th observation to correct for the deflated residual error. The BLUP of the polygenic effect in genomic selection (fitted genomic value) is

$$\widehat{y} = \widehat{\phi}^2 K V^{-1} y$$

where

$$V = K\phi^2 + I\sigma^2 = (K\lambda + I)\sigma^2$$

Therefore,

$$\widehat{y} = \phi^2 K V^{-1} y = \lambda K (K\lambda + I)^{-1} y = H^R y$$

where $H^R = \lambda K(K\lambda + I)^{-1}$ is a linear projection of the observed to the fitted trait value. The fitted and observed trait value are connected by a random version of the hat matrix. Let

$$\widehat{\varepsilon}_j = y_j - Z_j \widehat{\gamma}$$

be the fitted residual error. The predicted residual error may be expressed as

$$\widehat{e}_j = \widehat{\varepsilon}_j / \left(1 - h_{jj}^R\right)$$

where $h_{jj}^R$ is the $j$th diagonal element of the hat matrix $H^R$. The PRESS in the random model is

$$PRESS = \sum_{j=1}^{n} \widehat{e}_j^2 = \sum_{j=1}^{n} \widehat{\varepsilon}_j^2 / \left(1 - h_{jj}^R\right)^2$$

The predictability from the HAT method is

$$R_{HAT}^2 = 1 - PRESS/SS$$

The difference between $H^R$ in the random model and $H = X(X^TX)^{-1}X^T$ in the fixed model is that $H^R$ contains a parameter $\lambda$ while $H$ does not contain any parameters. If $\lambda$ is replaced by an estimated $\lambda$ from the whole sample, the PRESS of the random model is an approximation of the actual PRESS. As a result, $R_{HAT}^2$ will be biased upward compared to $R_{LOOCV}^2$ (the predictability drawn from LOOCV.

Although the HAT method does not need to refit the model for each part predicted, it still needs to partition the sample into $K$ parts if comparison with the traditional K-fold CV is of interest. Let $\varepsilon_k = y_k - Z_k \widehat{\gamma}$ be the estimated residual errors for all individuals in the $k$th part and $H_{kk}^R$ be the diagonal block of matrix $H^R$ corresponding to all individuals in the $k$th part. The predicted residual errors for the $k$th part are $\widehat{e}_k = \left(I - H_{kk}^R\right)^{-1}\widehat{\varepsilon}_k$. The PRESS under this random model becomes

$$\text{PRESS} = \sum_{k=1}^{K} \widehat{e}_k^T \widehat{e}_k = \sum_{k=1}^{K} \widehat{\varepsilon}_k^T \left(I - H_{kk}^R\right)^{-2} \widehat{\varepsilon}_k$$

The predictability is measured by

$$R_{\text{HAT}}^2 = 1 - \text{PRESS}/\text{SS}$$

The n-fold HAT approach is a special case where the $k$th part to be predicted contains only one individual, i.e., $H_{kk}^R = h_{jj}^R$ for $k = j$.

### 20.3.6 Generalized Cross Validation (GCV)

The generalized cross validation (Golab et al. 1979) is an alternative method to correct for the deflated residual error variance. The GCV calculated residual error sum of squares is called generalized residual error sum of squares (GRESS), which is defined by

$$
\text{GRESS} = \frac{(y - Z\hat{\gamma})^T (y - Z\hat{\gamma})}{[n^{-1}\text{tr}(I - H^R)]^2} = \frac{\hat{\varepsilon}^T \hat{\varepsilon}}{[n^{-1}\text{tr}(I - H^R)]^2}
$$

It is equivalent to dividing each estimated residual error by the average $\left(1 - h_{jj}^R\right)$ across all observations. Therefore, an intuitive expression of the above equation is

$$
\text{GRESS} = \sum_{j=1}^{n} \hat{\varepsilon}_j^2 / \left(1 - \overline{h}\right)^2
$$

where $\overline{h} = \sum_{j=1}^{n} h_{jj}/n$ is the average leverage value across all observations and

$$
\hat{\varepsilon}_j = y_j - Z_j \hat{\gamma}
$$

The predictability is defined as

$$
R_{\text{GCV}}^2 = 1 - \text{GRESS}/\text{SS}
$$

Golab et al. (1979) stated that GRESS is a rotation-invariant PRESS. It is not intuitive to interpret GRESS and therefore we prefer to report PRESS and thus $R_{\text{HAT}}^2$.

---

## 20.4   An Example of Hybrid Prediction

### 20.4.1 The Hybrid Data

The data were obtained from a hybrid population of rice (*Oryza sativa*) derived from the cross between Zhenshan 97 and Minghui 63 (Hua et al. 2002; Xing et al. 2002; Hua et al. 2003). This hybrid (Shanyou 63) is the most widely grown hybrid in China. A total of 210 RILs were derived by single-seed-descent from this hybrid. The original authors created 278 crosses by randomly pairing the 210 RILs. This population is called an immortalized $F_2$ (IMF2) because it mimics an $F_2$ population with a 1:2:1 ratio of the three genotypes (Hua et al. 2003). The 1000 grain weight (KGW) trait was analyzed as an example. The trait was measured in two consecutive years (1998 and 1999). Each year, eight plants from each cross were measured and the average KGW of the eight plants was treated as the original data point. The experiment was conducted under a randomized complete block design with replicates (years and locations) as the blocks. The genomic data are represented by 1619 bins inferred from ~270,000 SNPs of the rice genome (Yu et al. 2011). All SNPs within a bin have exactly the same segregation pattern (perfect LD). The bin genotypes of the 210 RILs were coded as 1 for the Zhenshan 97 genotype and 0 for the Minghui 63 genotype. Genotypes of the hybrids were deduced from genotypes of the two parents. The sample data were taken from the IMF2 population of hybrid rice with 278 hybrids genotyped for 1619 bins. The sample data are stored in three files: "IMF2-Genotypes.csv" for the genotype data, "IMF2-Phenotypes.csv" for the phenotype. The kinship matrix is stored in a file called "Kinship.csv". The Gaussian kernel matrix with bandwidth $h = 2$ is stored in a file called "Kernel.csv".

## 20.4.2  Proc Mixed

```
/*read data*/
proc import datafile="IMF2-Phenotypes.csv" out=phe dbms=csv replace;
proc import datafile="Kinship.csv" out=kinship dbms=csv replace;
proc import datafile="KernelSAS.csv" out=kernel dbms=csv replace;
run;



/*estimate parameters*/
proc mixed data=phe method=reml;
    class imf2;
    model kgw=/solution outp=pred;
    random imf2/type=lin(1) ldata=kinsip;
run;
```

Replacing the kinship in ldata = kinsip by ldata = kernel will generate the RKHS result.

```
proc means data=phe mean;
    var kgw;
    output out=new mean=mean;
run;

data phe1;
    set phe;
    if _n_=1 then set new;
    kgw = kgw - mean;
run;

/* define macro for cross validation*/
%macro CV;
    proc datasets;
        delete myOut;
    run;
    %do k=1 %to 10;
        data one;
            set phe1;
            y=kgw;
            if rep1=&k then y=.;
            drop rep2-rep10;
        run;
        proc mixed data=one method=reml;
            class imf2;
            model y=/noint solution outp=pred;
            random imf2/type=lin(1) ldata=kinship;
        run;
        data two;
            set pred;
            if rep1=&k;
            yhat=pred;
            yobs=kgw;
```

```
      run;
      proc append base=myOut data=two;
   %end;
%mend cv;


/*run cross validation macro*/
%cv


/*calculate correlation*/
proc corr data=myout;
   var yhat yobs;
run;


/*export output*/
proc export data=myout outfile="out\SAS\outBLUP.csv" dbms=csv replace;
run;
```

Using the above SAS programs, we estimated parameters for KGW of the hybrid rice. The estimated parameters for BLUP and RKHS are shown in Table 20.16, where the predictability was drawn from a ten-fold cross validation. The two different measurements of the predictability (squared correlation and PRESS) are virtually the same. Clearly, RKHS has a much higher predictability than BLUP.

We now demonstrate the variation of predictability in cross validation due to different ways of fold partitioning of the sample. We increased the number of folds from 2 to 50, and replicated the cross validation 10 times. Figure 20.3 shows the predictabilities of 10 replicated cross validation for BLUP and RKHS.

The RKHS method is clearly more efficient than the BLUP method with about $(0.70 - 0.58)/0.58 \approx 20\%$ improvement in predictability. For BLUP, increasing the number of folds progressively increases the predictability. The increase starts to slow down when the number of folds reaches about 10. Figure 20.4 shows the predictabilities of 10 replicated HAT predictions for BLUP and RKHS.

Figure 20.5 compares the average predictabilities over 10 replicates computed from the HAT method and the CV method. For BLUP, the HAT predictability is always greater than the CV predictability. In the end (at 50 folds), the HAT predictability is 0.6232 and the CV predictability is 0.5766 with a bias of $(0.6232 - 0.5766)/0.5766 = 8.1\%$. For RKHS, the HAT and CV predictabilities are 0.7176 and 0.7047, respectively, with a bias of $(0.7176 - 0.7047)/0.7047 = 1.83\%$.

The bandwidth of the kernel for the RKHS method was set at $h = 2$ and the corresponding predictability from the HAT method is 0.7176. As $h$ decreases, the predictability also decreases. When $h = 1000$, the predictability is 0.7274. Further increasing the $h$ value causes the program to crash.

The low predictability of BLUP in this example is due to a different way of kinship matrix calculation. We treated the genotype of a marker as a class variable with three levels (A, H, and B). We first created three dummy variables for each SNP. The three dummy variables are collectively stored in an $n \times 3$ matrix denoted by $Z_k$. For example, if individual 1 is of type A, individual 2 is of type B, and the last individual is of type H, the $Z_k$ matrix should looks like

**Table 20.16** Estimated parameters for KGW of the hybrid rice from BLUP and RKHS

| Estimated parameter | BLUP | RKHS |
|---|---|---|
| $\beta$ | 25.2392 | 24.8044 |
| $\phi^2$ | 3.9035 | 11.1781 |
| $\sigma^2$ | 0.3876 | 0.2901 |
| $R^2_{COR}$ (R-square) | 0.6278 | 0.7128 |
| $R^2_{CV}$ (PRESS) | 0.6351 | 0.7189 |

**Fig. 20.3** Plots of predictabilities of 10 replicated cross validations for BLUP (lower portion) and RKHS (upper portion) against the number of folds. The average predictability curves of the 10 replicates are highlighted in bold

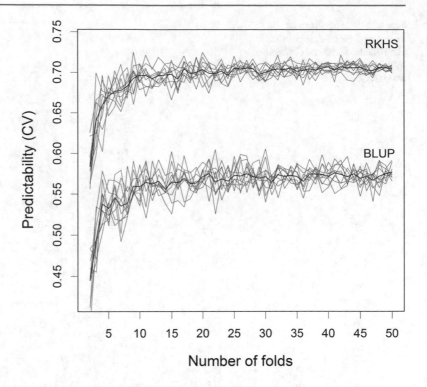

**Fig. 20.4** Plots of predictabilities of 10 replicated HAT prediction for BLUP (lower portion) and RKHS (upper portion) against the number of folds. The average predictability curves of the 10 replicates are highlighted in bold

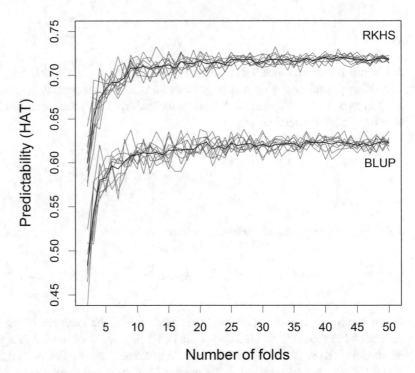

$$Z_k = \begin{bmatrix} 1 & 0 & 0 \\ 0 & 0 & 1 \\ \vdots & \vdots & \vdots \\ 0 & 1 & 0 \end{bmatrix}$$

The kinship matrix is calculated using

**Fig. 20.5** Comparison of average predictabilities of the HAT method with the CV method over 10 replicates for BLUP and RKHS

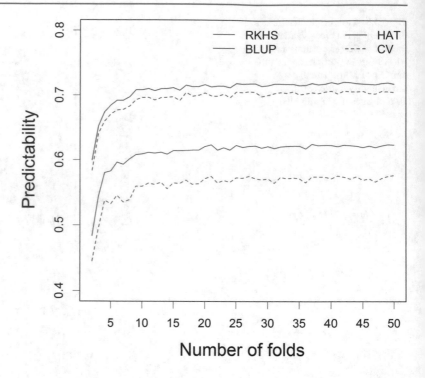

$$K_G = \frac{1}{m} \sum_{k=1}^{m} Z_k Z_k^T$$

A kinship matrix calculated from $Z_k$ coded as dummy variables will have all diagonal elements of 1 after normalization. Surprisingly, such a kinship matrix generated a very low predictability for the BLUP method. Therefore, this way of coding the genotype is not recommended. Coding the additive effect and dominance effect with $Z_{jk}$ and $W_{jk}$ is still the most efficient way for genomic selection, where

$$Z_{jk} = \begin{cases} +1 \text{ for } A \\ 0 \text{ for } H \\ -1 \text{ for } B \end{cases} \text{ and } W_{jk} = \begin{cases} -1 \text{ for } A \\ 1 \text{ for } H \\ -1 \text{ for } B \end{cases}$$

The additive and dominance kinship matrices are

$$K_A = \frac{1}{c_A} \sum_{k=1}^{m} Z_k Z_k^T \text{ and } K_D = \frac{1}{c_D} \sum_{k=1}^{m} W_k W_k^T$$

where $c_A$ and $c_D$ are normalization factors (often take the averages of the diagonal elements of the unnormalized kinship matrices). For the same trait (KGW) of the IMF2 rice, we fitted the following models, (1) RKHS with a single kernel of $h = 2$ denoted by RKHS-SK; (2) RKHS with multiple kernels of $h_1 = 0.1$, $h_2 = 0.5$, and $h_3 = 2.5$, which is denoted by RKHS-MK; (3) BLUP with a single kinship matrix created from the genotype class variable denoted by BLUP-$K_G$, (4) BLUP with a single additive kinship matrix denoted by BLUP-$K_A$, (5) BLUP with both kinship matrices designated as BLUP-$K_A$-$K_D$. The results of ten-fold cross validation replicated 10 times are illustrated in Fig. 20.6. Clearly, the two RKHS-based methods outperform the BLUP methods, especially the multiple kernel method (RKHS-MK). The dummy variable coding (BLUP-$K_G$) has a much lower predictability than BLUP-$K_A$ and BLUP-$K_A$-$K_D$. The difference between the last two BLUP methods is barely noticeable.

**Fig. 20.6** Comparison of five prediction methods for the KGW trait of rice, where RKHS-SK is the single kernel RKHS method, RKHS-MK is the multiple kernel method, BLUP-KG is the BLUP method with the kinship matrix generated from the dummy variable coding system of the genotypes, BLUP-KA is the BLUP method with the additive kinship matrix and BLUP-KA-KD is the BLUP method incorporating both the additive and dominance kinship matrices

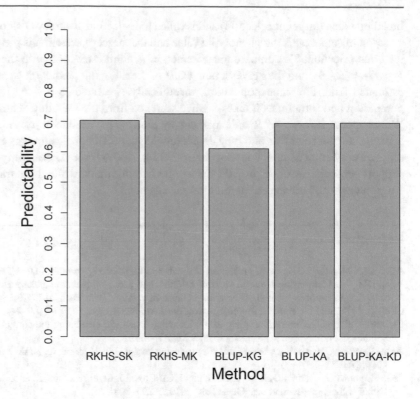

We define predictability as a function of the PRESS or the squared correlation between the observed and predicted trait values. Most people in the genomic selection community use the correlation itself to measure the prediction ability. One of the reasons to use the correlation rather than the squared correlation is that the former looks better than the latter in the reports. In the animal breeding community, prediction accuracy is often used to measure the relative efficiency of a genomic selection procedure in comparison to a conventional selection program. The prediction accuracy is defined as

$$\psi = \sqrt{R_{CV}^2/h^2} = R_{CV}/h$$

where $R_{CV}^2$ is the "predictability" defined in this chapter and $h^2$ is the heritability of the trait under study. The square root of the heritability is denoted by $h$ (this is not the bandwidth of the kernel matrix). The reported prediction accuracy ($\psi$) is often very high and sometime can be greater than 1. For example, if the predictability is $R_{CV}^2 = 0.25$ (looks low), the square root is $R_{CV} = 0.5$ (already looks relatively high). If the heritability of the trait is $h^2 = 0.35$, the prediction accuracy will be

$$\psi = \sqrt{R_{CV}^2/h^2} = \sqrt{0.25/0.35} = 0.8451543$$

looking very high. If $h^2 \leq 0.25$, the prediction accuracy will be $\psi \geq 1$, a perfect prediction accuracy after truncation at 1. The prediction accuracy is inversely proportional to the square root of heritability, implying that a low heritability trait gives high prediction accuracy. This explains why some people purposely choose low heritability traits for genomic selection. For example, people may use the high prediction accuracy argument to justify genomic selection for stress-related traits in plants and fertility-related traits in animals. The prediction accuracy is really a misleading term. In order for genomic selection to work, the trait must be heritable and the greater the heritability the better the prediction.

Recall that the breeders' equation says that the response to a mass selection ($R$) is

$$R = h^2 S = i\sigma_P \frac{\sigma_A^2}{\sigma_P^2} h^2 = i\sigma_A \frac{\sigma_A}{\sigma_P} = i\sigma_A h \tag{20.26}$$

where $S$ is the selection differential, $i = S/\sigma_P$ is the selection intensity, $\sigma_P^2$ is the phenotypic variance, and $\sigma_A^2$ is the additive variance. For mass selection, $h = r_{AP}$, which is the correlation between the phenotypic value (criterion of selection) and the breeding value (target of selection) and is called the selection accuracy (Falconer and Mackay 1996). In genomic selection, the criterion of selection is the phenotypic value and the target of selection is the genomic value. Therefore, $R_{CV}$ is the accuracy of genomic selection. To compare the accuracy of genomic selection with the accuracy of mass selection, one takes the ratio, $R_{CV}/r_{AP} = R_{CV}/h$, and thus gives a new name of the ratio, the prediction accuracy. If all genetic parameters (including $h^2$) are estimated from the same population, theoretically, we have $R_{CV} \leq h$. However, $h^2$ is often estimated from a different population and sometimes it takes a value searched from the literature. There is no guarantee that $R_{CV}$ is always less than $h$. What happens if $R_{CV} > h$? People may simply truncate the ratio to unity. Administrators and government officials of funding agencies are not scientists and they do not know that prediction accuracy is a relative term compared to mass selection. When they see prediction accuracy in the range of 90% to 100% for low heritability traits like fertility and stress resistance, they will support genomic selection projects for low heritability traits, while these traits are better improved through improvement of environments and advanced agricultural management.

# References

Allen DM. Mean square error of prediction as a criterion for selecting variables. Dent Tech. 1971;13:469–75.

Allen DM. The relationship between variable selection and data agumentation and a method for prediction. Dent Tech. 1974;16:125–7.

Cook D. Detection of influential observation in linear regression. Dent Tech. 1977;19:15–8.

Cook D. Influential observation in linear regression. J Am Stat Assoc. 1979;74:169–74.

Craven P, Wahba G. Smoothing noisy data with spline functions: estimating the correct degree of smoothing by the method fgeneralized cross-validation. Numerische Mathematika. 1979;31:377–403.

de los Campos G, Gianola D, Rosa GJM. Reproducing kernel Hilbert spaces regression: a general framework for genetic evaluation1. J Anim Sci. 2009;87:1883–7.

de los Campos G, Gianola D, Rosa GJM, Weigel KA, Crossa J. Semi-parametric genomic-enabled prediction of genetic values using reproducing kernel Hilbert spaces methods. Genet Res. 2010;92:295–308.

de los Campos G, Hickey JM, Pong-Wong R, Daetwyler HD, Calus MPL. Whole-genome regression and prediction methods applied to plant and animal breeding. Genetics. 2013;193:327–45.

de los Campos G, Perez P. BGLR: Bayesian generalized regression R package, version 1.0., p. R package, 2013.

Falconer DS, Mackay TFC. Introduction to quantitative genetics. Harlow: Addison Wesley Longman; 1996.

Geman S, Geman D. Stochastic relaxation, Gibbs distributions and the Bayesian restoration of images. IEEE Trans Pattern Anal Mach Intell. 1984;6: 721–41.

Gianola D. Priors in whole-genome regression: the bayesian alphabet returns. Genetics. 2013;194:573–96.

Gianola D, Fernando RL, Stella A. Genomic-assisted prediction of genetic value with semiparametric procedures. Genetics. 2006;173:1761–76.

Gianola D, van Kaam JBCHM. Reproducing kernel hilbert spaces regression methods for genomic assisted prediction of quantitative traits. Genetics. 2008;178:2289–303.

Gilks WR, Best NG, Tan KK. Adaptive rejection Metropolis sampling within Gibbs sampling. J R Stat Soc Ser C Appl Stat. 1995;44:455–72.

Gilks WR, Richardson S, Spiegelhalter DJ. Markov chain Monte Carlo in practice. London: Chapman & Hall; 1996.

Gilks WR, Wild P. Adaptive rejection sampling for Gibbs sampling. J R Stat Soc Ser C Appl Stat. 1992;41:337–48.

Giri NC. Multivariate statistical analysis. New York: Marcel Dekker, Inc.; 1996.

Golab GH, Heath M, Wahba G. Generalized cross-validation as a method for choosig a good ridge parameter. Dent Tech. 1979;21:215–23.

Golub GH, Van Loan CF. Matrix computations. Baltimore: Johns Hopkins University Press; 1996.

Hua J, Xing Y, Wu W, Xu C, Sun X, Yu S, Zhang Q. Single-locus heterotic effects and dominance by dominance interactions can adequately explain the genetic basis of heterosis in an elite rice hybrid. Proc Natl Acad Sci U S A. 2003;100:2574–9.

Hua JP, Xing YZ, Xu CG, Sun XL, Yu SB, Zhang Q. Genetic dissection of an elite rice hybrid revealed that heterozygotes are not always advantageous for performance. Genetics. 2002;162:1885–95.

Ithnin M, Xu Y, Marjuni M, Mohamed Serdari N, Din A, Low L, Tan Y-C, Yap S-J, Ooi L, Rajanaidu N, et al. Multiple locus genome-wide association studies for important economic traits of oil palm. Tree Genet Genomes. 2017;13:103.

Lande R, Thompson R. Efficiency of marker-assisted selection in the improvement of quantitative traits. Genetics. 1990;124:743–56.

Meuwissen THE, Hayes BJ, Goddard ME. Prediction of total genetic value using genome-wide dense marker maps. Genetics. 2001;157:1819.

Ober U, Ayroles JF, Stone EA, Richards S, Zhu D, Gibbs RA, Stricker C, Gianola D, Schlather M, Mackay TFC, et al. Using whole-genome sequence data to predict quantitative trait phenotypes in *Drosophila melanogaster*. PLoS Genet. 2012;8:e1002685.

Picard RR, Cook D. Cross-validation of regression models. J Am Stat Assoc. 1984;79:575–83.

Tibshirani R. Regression shrinkage and selection via the Lasso. J R Stat Soc Ser B. 1996;58:267–88.

VanRaden PM. Efficient methods to compute genomic predictions. J Dairy Sci. 2008;91:4414–23.

Wahba G. Spline models for observational data. Philadelphia: SIAM; 1990.

Xing Y, Tan F, Hua J, Sun L, Xu G, Zhang Q. Characterization of the main effects, epistatic effects and their environmental interactions of QTLs on the genetic basis of yield traits in rice. Theor Appl Genet. 2002;105:248–57.

Xu S. Mapping quantitative trait loci by controlling polygenic background effects. Genetics. 2013;195:1209–22.

Yu H, Xie W, Wang J, Xing Y, Xu C, Li X, Xiao J, Zhang Q. Gains in QTL detection using an ultra-high density SNP map based on population sequencing relative to traditional RFLP/SSR markers. PLoS One. 2011;6:e17595. https://doi.org/10.11371/journal.pone.0017595.

# Index

**A**

Additive×additive×additive variance, 79
Additive by additive (AA) effects, 69
Additive by dominance (AD) effects, 69
Additive effect, 65, 69, 73
Additive kinship matrix, 380, 383, 385
    additive, 383
Additive model, 313
Additive relationship matrix, 185, 191, 192
Additive variance, 78
AdditivityTests, 103
Aggregate breeding value, 287
Allelic interaction effect, *see* Dominance effect
Allelic values, 178
Allelic variance, 177
    genetic model, 177
Analysis of covariance (ANCOVA), 221, 222
Analysis of variance (ANOVA), 6, 86, 88, 221
    heritability, 168–171
*Aptenodytes forsteri*, 4
Artificial selection, 255
    breeding value, 255, 258
    directional selection, 234–236, 238–240
    evalution, 257
    family selection, 256
    index selection, 257
    individual selection method, 255
    methods, 255
    objective, 255
    pedigree selection, 257
    progeny testing, 256
    sib selection, 256
    sources, 256
    stabilizing selection, 240, 242
    within-family selection, 257
Average effect
    gene/allele, 67
    gene substitution, 67, 68

**B**

Bayesian generalized linear regression (BGLR), 393
Bayesian method, 128
Bayesian RKHS regression, 390–392
Beavis effect
    bias corrected heritability, 343, 344
    biased QTL heritability, 341, 342
    biased QTL variance, 341
    comparative genomics, 339
    justification, 339
    Mendelian segregation, 341

non-central Chi-square distribution, 339
    non-centrality parameter, 341, 343
    QTL heritability, 341, 342
    QTL mapping, 339
    residual variance, 341, 342
    theoretical bias, 342
    truncated central Chi-square distribution, 343
    truncated distributions, 339
    truncated non-central Chi-square distribution, 340
    truncated normal distribution, 339
    unbiased QTL heritability, 343
    Wald test, 340, 341, 343
    winner's curse, 339
Best linear unbiased estimation (BLUE), 6, 8, 322–324, 386, 388, 402
Best linear unbiased prediction (BLUP), 6, 8, 369, 370
    random effects, 370, 371
    and selection index, 271, 272
    theory of, 272, 273
Biallelic system, 67
Bio-Medical Data Package (BMDP), 9
Biometrical genetics, 6
Blending theory, 13, 16
Bonferroni correction, 332, 333
Breeding value, 3, 4, 68, 69, 178
Broad sense heritability, 77
Bulmer effect, 297

**C**

*Cavia porcellus*, 5
CentiMorgan (cM), 309
Chi-square distribution, 102, 315, 324, 325, 330, 332, 348, 352, 355
Chi-square test, 14, 15, 17, 18, 28
Cholesky decomposition, 210, 212
Classical quantitative genetics, 7, 8, 10, 11
Coancestry, 179, 181–186, 188
    calculation, 182
    tabular method, 181
Coancestry matrix, 189, 191
Coefficients of linear contrasts, 71
Combined selection, 263, 264
Complicated behavior traits, 4, 5
Composite interval mapping (CIM)
    candidate QTL, 319, 320
    chromosomes, 319
    co-factor/covariate, 319, 320
    control of polygenic background, 321–323
    ECM algorithm, 320
    genotypes, candidate QTL, 320
    interval mapping, 319
    LOD score profile, 330

Composite interval mapping (CIM) (*cont.*)
  maximum likelihood method, 320
  model selection approach, 319
  polygenic-based methods, 329
  posterior probabilities, 320
  procedure, 325, 329
  putative QTL, 320
  QTL effects, 330
  software packages, 321
Computational time complexity analysis, 384, 385
Computing formulas, 88
Conditional expectation, 45
Conditional likelihood function, 125
Conditional probability table, 20
Conditional variance, 45
Conditional variance-covariance matrix, 315
Confidence interval (CI), 201
Conventional mixed model, 272
Correlation analyses, 7
Correlation coefficient, 34, 35, 48, 218
Covariance, 103, 178
  between-generation change, 251, 252
  between two linear combinations, 60
  conditional expectation, 45
  conditional variance, 46
  definition, 43
  estimating covariance from a sample, 44
  properties, 43
  sample estimates, 47
  within-generation change, 250, 251
Cross covariance
  between relatives, 217
Cross validation, 397, 398, 403
Cunningham's multistage index selection, 304
Cunningham's multistage selection indices, 299


**D**
Decontaminated efficient mixed model association (DEMMA), 323, 325
  BLUP value, 353
  kinship matrix, 353, 354
  linear mixed model, 354
  polygenic effect, 354
  revised mixed model, 354
Definition formulas, 88
Degrees of freedoms, 91, 103
Delta method, 90
Deshrunk effect, 325
Deshrunken ridge regression (DRR) method, 325, 328, 330, 332
Desired gain selection index, 295–297
Determinant of matrix, 57
Diagonal element, 187
Diagonal matrix, 51
Disruptive selection, 242, 243
Dominance by additive (DA) effects, 69
Dominance by dominance (DD) effects, 69
Dominance deviation, 68
Dominance effect, 64, 65, 69, 71, 177
Dominance kinship matrix, 383
  additive, 384
  dominance, 384
Dominance variance, 78
Dwarfing gene, 64


**E**
Efficient mixed model association (EMMA), 323, 325
  covariance structure, 350

  eigenvalue decomposition, 352
  GEMMA, 350
  kinship matrix, 350
  quadratic form, 353
  ratio, 351
  REML, 351
Efficient mixed model association studies expedited (EMMAX) method, 352
  additive and dominance model, 364
  least squares method, 363
  polygenic effect, 365
  polygenic variance, 363
  residual variance, 364
  variance components, 364
Eigenvalue decomposition, 391
Eigenvalue decomposition algorithm, 374, 375
Eigenvalues, 58, 361
Eigenvectors, 58
Emperor penguins, 4
Environmentability, 95
Environmental covariance, 143, 215
Environmental effects
  variance, 86
Environmental errors
  ANOVA, 90, 91, 93
  definition, 85
  G×E interaction
    ANOVA table for testing, 99
    definition, 95, 97
    genotype plot by year interactions for rice yield trait, 99
    joint distribution (counts and frequencies), genotypes and two environments, 97
    overall ANOVA table, model and residual error terms, 99
    partitioning of phenotypic variance, 100, 102
    Tukey's one degree of freedom, 102, 103, 105, 106
    type I ANOVA table, 99
  general (permanent), 85
  intra-class correlation coefficient, 89
  rearrangement of SAT test score data, 93
  repeatability
    application, 94, 95
    estimation, 86–88, 90, 91, 93
    hypothetic SAT test scores, 91
    intra-class correlation coefficient, 89
    variance ratio, 86
  special (temporary), 85
Epistasis, test, 118, 119
Epistatic effects, 69
  two loci, 69, 70
    coefficients of linear contrasts, 71
    *H* matrix, 71
    inverse matrix of $H$-1, 71
    KGW genotypic values, 72
    population mean of multiple loci, 75, 76
    2D plot of interaction, 73, 74
    3D plot of interaction, 74
Epistatic GBLUP model, 393
Epistatic kinship matrices, 382, 384, 385
Epistatic variance
  between two loci, 79, 81
  definition, 79
Estimating equations (EE) algorithm, 318, 319, 326, 327, 329, 331
Estimating variance parameters, 372, 374
Euclidean distance, 392
Expectation
  definition, 39
  mean estimation, 40
  properties, 39

Expectation-conditional-maximization (ECM) algorithm, 320
Expectation and maximization (EM) algorithm, 7, 128, 198
Expectation and variance of a linear combination, 60

**F**

Family model, 206
Family segregation variance, 143–146
Family selection
   derivation, 259
   full-sib family, 258
   mean phenotypic value, 259
   objective, 258
Fast algorithm, 384
Female penguins, 4, 5
First genetic model, 64–66
Fisher scoring algorithm (FISHER), 307, 331
Fixation index, 33
Flanking marker genotypes, 313
FORTRAN programs, 9
Fraternity, 178
F statistics, 34–36
F-test, 8, 107–110, 314, 317, 334, 335
   for multiple samples, 110, 112
Full-sibs
   analysis, 158–160
   genetic covariance between, 140
      long way of derivation, 141
      short way of derivation, 140, 141

**G**

Gamma distribution, 324
Gaussian kernel, 387, 392
Gene frequency, 29, 31
Gene substitution
   average effect, 67, 81, 82
Generalized cross validation (GCV), 393, 400
Generalized inverse, 56
Generalized linear mixed models (GLMM), 6, 10
Generalized linear models (GLM), 6
Generalized residual error sum of squares (GRESS), 400
Genetic correlation
   causes, 216, 217
   definition, 215, 216
   linkage-effect-caused genetic correlation, 216
   nested mating design, 227–232
   parent-offspring correlation, 217, 219, 220
   from sib data, 221–224, 226, 227
Genetic covariance
   between full-siblings, 140
      long way of derivation, 141, 142
      short way of derivation, 140, 141
   between half-sibs, 139, 140
   between monozygotic twins, 142
   between offspring and mid-parent
      resemblance between relatives, 139
   long derivation, 137–139
   short derivation, 136, 137
Genetic drift, 25, 30, 31, 33
Genetic gain, 1
Genetic models, 178
   coancestry, 178
   fraternity, 178
   individuals, 178
Genetic variance, 77, 81

Genotype by environment (G×E) interaction
   ANOVA table for testing, 99
   average yield of rice, genotype and year combinations, 97
   definition, 95
   partitioning of phenotypic variance, 100, 101
   Tukey's one degree of freedom, 102, 106
   type I ANOVA table, 99
Genome scanning, 332, 334
Genome-wide association studies (GWAS), 347
   linear mixed model, 360
   Manhattan plot, 362
   population structure, 347
   QTL mapping population, 347
   random populations, 347
Genome-wide efficient mixed model association (GEMMA), 350
Genomic best linear unbiased prediction (GBLUP)
   BLUP, random effects, 370, 371
   eigenvalue decomposition, fast computing, 374, 375
   estimating variance parameters, 372, 374
   genomic prediction, 367
   kinship matrix (see Kinship matrix)
   maximum likelihood method, 367
   predicting genomic values, future individuals, 371, 372
   ridge regression, 367–370
Genomic data, 400
Genomic selection, 10
   Bayesian Alphabet series, 367
   GBLUP (see Genomic best linear unbiased prediction (GBLUP))
   heritability traits, 406
   hybrid prediction
      hybrid data, 400
      PROC MIXED, 401–402
   mass selection accuracy, 406
   phenotypic value vs. breeding value, 406
   predictability (see Predictability, genomic selection)
   RKHS regression (see Reproducing kernel Hilbert spaces (RKHS) regression)
   traditional marker-assisted selection, 367
Genotype frequency, 376
Genotype indicators, 307, 308
Genotypes, 307
Genotypic values, 64–66, 69–73
GLIMMIX procedure in SAS, 10
Graphical user interface (GUI), 9

**H**

Hadamard matrix multiplication, 383, 384
Haldane mapping function, 309
Half-sibs
   analysis, 160, 161
   genetic covariance between, 139, 140
Hardy-Weinberg equilibrium, 68, 80, 137, 378
   applications
      frequency of carriers calculation, 28
      frequency of the recessive allele, 27
      iteration process, 28
   different frequencies for male and female populations, 27
   from gene to genotype frequencies, 26
   from genotype to gene frequencies, 26
   square law, 25
   test
      Chi-square test for goodness of fit, 28
      MN blood group, 29
      2×2 contingency table association test, 29, 30
      Yates' continuity correction for 2×2 contingency table test, 30

Harville's restricted likelihood function, 200
The HAT method, 397–399
Henderson's mixed model equation (MME), 273, 370
Henderson's Type 1 method, 88
Heritability, 1, 94, 95, 147
    analysis of variance, 168–171
    cross of two inbred parents, $F_2$ derived from, 148
    multiple inbred lines/multiple hybrids, 148
        with replications, 149, 150
        without replications, 150–152
    nested mating design, 171–175
    parent-offspring regression, 152
        estimation, 155–157
        middle parent *vs.* mean offspring, 155
        middle parent *vs.* single offspring, 153, 154
        single parent *vs.* mean offspring, 154
        single parent *vs.* single offspring, 152, 153
    regression analysis, 167, 168
    sib analysis, 158–164
    standard error, 164
        analysis of variances, 166, 167
        regression method, 164, 165
    unbalanced data, 160
Human height prediction, 1
Hybrid data, 400


I
Identification number (ID), 185
Identity matrix, 52
Identity-by-descent (IBD), 6, 33, 179, 181
Identity-by-state (IBS), 179
Immortalized $F_2$ (IMF2), 400
Inbreeding coefficient, 33, 179, 181
Independent culling level selection, 283, 284
Index selection, 284
    breeders' equation, 288
    breeding values, 287
    genetic variance, 292
    Lagrange multiplier, 294
    objective, 287
    phenotypic values, 287
    variance-covariance matrices, 285, 291
    weights, 291
Individual selection method, 255
Interval mapping (IM), 323
    BC design, 307
    binary traits, 330
    composite interval mapping, 313
    conditional probability, 310–312
    datasets, 332
    EE algorithm, 331
    genome positions, 328
    genotype, 310
    individual marker analysis, 310
    IRWLS, 315, 317, 327, 331
    LOD test statistic profiles, 328, 329
    LOD test statistics, markers, 312
    LS method, 313–315, 330, 331
    ML method, 307, 317–319, 327, 331
    multipoint method, 332
    procedures, 325, 327
    putative QTL position, 310, 331
    QTL genotype, flanking markers, 332
    revolutionized genetic mapping, 307
    statistical methods, 312

transition probability matrix, 310
Intra-class correlation analysis, 86
Intra-class correlation coefficient, 89
IRWLS method, 327
Iteration process, 316
Iteratively reweighted least squares (IRWLS) method, 307, 315, 317,
    325–328, 331


J
Joint distribution, two gametes, 68


K
Kernel average (KA), 393
Kernel Hilbert spaces regression, 367
Kernel matrix, 389, 393
Kernel regression, 388
Kernel selection, 392, 394
Kinship matrix, 190–193, 323
    additive, 379, 383
    additive genotype indicators, 375
    computational time complexity analysis, 384, 385
    definition, 375
    direct method, 380, 382, 383
    dominance, 379, 383
    epistasis, 379
    fast algorithm, epistatic kinship matrix calculation, 382, 384
    frequency of allele, 376
    genotype frequency, 376
    Hardy-Weinberg equilibrium, 378
    Henderson's mixed model equations, 377
    marker-specific kinship matrix, 376
    normalization factor, 376
    numerical codes
        epistatic indicators, 380
        genotypes, 380, 381
    numerical coding system, 379
    polygenic effects, 379
    polygenic variance, 375, 376
    prediction, new plants, 385, 386
    primary, 375, 376
    regression coefficient, 377
    summary, 378
    training sample, 386
    VanRaden fitted, 377
    variance-covariance matrix, 375


L
Lagrange multipliers, 294
Law of Independent Assortment, 18
Law of Segregation, 16
Least squares (LS) method, 307, 313–315, 321, 325, 327, 330, 331
Leave-one-out-cross validation (LOOCV), 397, 399
Likelihood function, 338
Likelihood ratio test, 314, 315, 317, 319
    Chi-square, 202
    CI, 201
    p-value, 202
Linear combination, 59
Linear models, 8, 116, 371
    correlation coefficients, 48, 49
    regression coefficient, 48–50
Linkage disequilibrium (LD), 307–310, 355
Log likelihood function, 314

Long derivation, 137–139
Long way of derivation, 141

**M**
Major gene detection
    epistatic effects, 118
        epistatic variance components and significance test, 119–122
        test of epistasis, 118, 119
Manhattan plot, 355, 365
Marginal genotypic values, 69
Marker-inferred kinship matrix, 370
Markov chain Monte Carlo (MCMC), 128, 391
Mass selection, 255
Mating design, 21
MATLAB, 9
Matrix addition, 52
Matrix algebra
    definition, 51
    determinant of a matrix, 57
    diagonal matrix, 51
    eigenvalues and eigenvectors, 58, 59
    generalized inverse, 56
    identity matrix, 52
    matrix addition, 52
    matrix inverse, 55, 56
    matrix multiplication, 52, 54, 55
    matrix subtraction, 52
    matrix transpose, 55
    orthogonal matrix, 57
    square matrix, 51
    symmetric matrix, 51
    trace of a matrix, 57
Matrix inverse, 55, 56
Matrix multiplication, 52–55
Matrix notation, 195
Matrix subtraction, 52
Maximum likelihood (ML) method, 88, 307, 317–320, 325, 373
Maximum likelihood estimates (MLE), 198
    ANOVA, 199
    function, 198
    MIXED procedure, 199
    parameters, 198, 199
Mendel, Gregor, 14
Mendelian genetics
    departure from Mendelian ratio, 19
    environmental errors to normal distribution, 22, 23
    experiments, 13
    multiple loci with or without environmental effects, 21
    quantitative traits, 24
    traits of common pea, 13
    trait value distributions, 23
    two loci without environmental effects, 20
    variance of the trait value, 22
Mendelian inheritance, 20
Mendelian segregation ratio, 317
Mendel's $F_2$ cross experiment, 16
Mendel's first law, 16, 67
Mendel's laws of inheritance
    $F_2$ cross experiment, 16
    Law of Independent Assortment, 18, 19
    Law of Segregation, 16
        $F_2$ mating design for two traits, 18
        seed shape and seed color, 18
    Punnett square, 16
    seven cross experiments in common pea, 16

Mendel's second law, 18
Microsoft Excel, 9
Minimum variance quadratic unbiased estimation (MIVQUE0), 88
Mixed linear model (MLM)
    GWAS procedure, 349, 350
    kinship matrix, 350
Mixed model, 322
Mixed model analysis, 195, 277
    asymptotic covariance matrix, 207
    asymptotic variance matrix, 203
    coancestry matrix, 196
    diagonal elements, 196
    linear model, 195
    null model, 204
    pedigrees, 195, 197
    phenotypic data, 206
    phenotypic values, 196, 197, 202
    probability density, 196
    REML analysis, 205
    SAS code, 206
    scalar and matrix, 197
    variance components, 203
    variance method, 195
Mixed model methodology, 349
    MIXED procedure, 349
    polygenic effects, 349
MIXED procedure, 7, 89, 224, 231
Mixture distribution, 128, 130, 131
ML method, 327
Modern quantitative genetics, 10, 11
Molecular quantitative genetics, 10
Moment method, 88
Monozygotic twins
    genetic covariance between, 142
Monte Carlo algorithm, 181
Monte Carlo simulation, 207, 210
    coancestry, 208
    genetic value, 208
    mean and variance, 209
    pedigree, 208
    simulation, 210
Morgan (M), 309
Multiple inbred lines/multiple hybrids, 148
    with replications, 149, 150
    without replications, 150–152
Multiple kernel (MK) approach, 393
Multiple trait selection
    independent culling level selection, 283
    index selection, 284
    indirect selection, 284
    secondary trait, 285
    selection index, 283
    tandem selection, 283
Multistage index selection, 297
    Cunningham's weights, 298, 301
    objective, 297
    phenotypic variance matrix, 297, 300
    Xu and Muir' weights, 299
Multivariate normal distribution, 197

**N**
Narrow sense heritability, 77
Natural selection
    directional selection, 243
    disruptive selection, 245

Natural selection (*cont.*)
   stabilizing selection, 244
Nested mating design
   heritability, 171–175
Nested/hierarchical mating design, 161–164
Newton iteration, 374
Newton iteration algorithm, 389
Newton-Raphson algorithm, 198
Newton-Raphson iteration equation, 198
Non-allelic interaction effects, *see* Epistatic effects
Non-central Chi-square distribution, 340
Nonparametric method, 367
Normal distribution, 41
Null model, 129

**O**
Offspring and mid-parent
   genetic covariance between, 139
Ordinary ridge regression (ORR), 325, 329–331
   degrees of freedom, 331
   estimated effects, 331, 332
   Wald tests, 332
Orthogonal matrices, 57

**P**
Parent-offspring regression, 152
   estimation, 155–157
   middle parent *vs.* mean offspring, 155
   middle parent *vs.* single offspring, 153, 154
   single parent *vs.* mean offspring, 154
   single parent *vs.* single offspring, 152, 153
Parent-progeny regression analysis, 8
Partitioning of phenotypic variance, 100, 102
Paternal allele, 177
Path analysis, 182–184
   coancestry, 182
   double-headed curve, 185
   inbreeding coefficient, 183
   pedigree, 182
   tabular method, 184
Pearson correlation, 7
Pearson's product-moment correlation coefficient, 7
Pearson's Chi-square test, 30
Pedigree, 180, 181, 184, 186, 188, 209
Pedigree selection
   variance, 262
Penetrance, of genotype, 20
Permanent environmental effect, 87
Permutation test, 333, 334
Phenotypic covariance, 215
Phenotypic evolution, 305
Phenotypic resemblance, 143
Phenotypic value, 63, 271
Phenotypic variance, 336
Phenotypic variance matrix, 296
Physical linkage, 307
*Pisum sativum*, 16
Polygenic-based methods, 330
Polygenic effect, 321
Polygenic variance, 337, 375
Population genetics
   gene frequencies, 26
   genotype frequencies, 25, 26
   HW equilibrium (*see* Hardy-Weinberg equilibrium)

   primary founders, 25
Population mean, 66
Population structure, 358, 360
   consensus conclusion, 358
   design matrix, 358
   diagonal matrix, 359
   effects, 358
   eigenvector, 360
   GWAS model, 358
   linear mixed model, 359
   PCA, 359
   score matrix, 361
Predictability, genomic selection
   animal breeding community, 405
   BLUP, 402
   centered model, 394
   cross validation, 394, 397, 398, 402, 403
   CV *vs.* HAT method, 402, 404
   definition, 394
   GCV, 400
   HAT method, 397–399, 403
   heritability, 405
   KGW trait, rice, 405
   kinship matrix, 403, 404
   maximum likelihood estimate of parameters, random models, 395
   prediction accuracy, 405
   PRESS, 395, 396, 405
   random effects, 394
   response variable, 394
   RKHS, 403
Predicted genomic values, 372
Predicted random effects, 274
Predicted residual error sum of squares (PRESS), 395–400, 405
Prediction errors, 372
Primary kinship matrix, 376
Principal component, 361
Principal component analysis (PCA), 359, 360
Prior probabilities, 326
PROC ALLELE, 37
PROC INBREED statement, 276
PROC MIXED, 10
Progeny testing, 256
Pseudo marker, 325
Punnett square, 16

**Q**
Q matrix, 359
Q-Q plot, 355–357
QTL heritability, 334, 337
Qualitative traits, 5, 122–127
Quantitative genetics, 8
Quantitative trait locus (QTL) mapping, 6
   Beavis effect (*see* Beavis effect)
   Bonferroni correction, multiple tests, 332, 333
   CIM, 319–323, 330
   EM algorithm, 326, 329
   $F_2$ mice, 325
   genome positions, polygenic procedures, 329
   genotype probabilities, 327
   IM (*see* Interval mapping (IM))
   LD, 307–310
   markers, 325
   method, 307
   mixture density, 327
   mouse genome, 325

permutation test, 333, 334
polygenic background control, 328
posterior probabilities, 326, 327
pseudo marker, 325
QTL Cartographer software package, 325
ridge regression, 323–325
size, 334–338
technical details, 326, 327
Quantitative traits, 5, 6, 128–131

**R**
Random effect model, 195
Random effects, 272
Random environmental error, 86
Random model, 336
Recombinant inbred line (RIL), 108
Recombination fractions, 307–309, 325
Regression analysis, 8
    heritability, 167, 168
    three genotypes, 116–118
    two genotypes, 113–115
Regression coefficient, 48, 81, 82, 103
Regression line, 83
Regression method, 164, 165
Relatives, resemblance between, see Resemblance between relatives
Repeatability
    ANOVA analysis, 88, 90, 93
    ANOVA table for the SAT test score, 91
    application, 94, 95
    computing formulas, 88
    data structure for repeated measurement, 87
    estimation, 86
    hypothetic SAT test scores, 91
    intra-class correlation coefficient, 89
    rearrangement of SAT test score data, 93
    variance ratio, 86
Reproducing kernel Hilbert spaces (RKHS) regression
    Bayesian approach, 390–392
    future individuals prediction, 389, 390
    Kernel selection, 392, 394
    prediction, 386, 388
    variance components estimation, 388, 389
Resemblance between relatives, 135, 136
    environmental covariance, 143
    family segregation variance, derivation of within, 143–146
    genetic covariance
        between half-sibs, 139, 140
        between monozygotic twins, 142
        full-siblings, 140–142
        long derivation, 137–139
        short derivation, 136, 137
    genetic covariance between offspring and mid-parent, 139
    phenotypic resemblance, 143
Residual error, 48
Response to selection, 236
Restricted log likelihood function, 374
Restricted maximum likelihood (REML), 89, 200, 322, 337, 349, 351, 373, 374, 388
    expectation and variance, 200
    partial derivative, 201
Restricted selection index, 295
R Foundation for Statistical Computing, 9
Ridge regression, 323–325, 367–370
R package, 189, 190
    path values, 189

pedigree, 190
    tabular version, 190
R program, 15

**S**
SAS procedure, 191
    variables, 192
SAS program, 98
    British bull pedigree and simulated phenotypic values, 280
    estimated parameters, 278
    MIXED procedure, 277, 278
    numerator relationship matrix, 276
    predicted phenotypic effects, 281
    predicted phenotypic values, 279
    predicted random effects, 279, 280
    PROC MIXED, 281
    tabular pedigree and phenotypic values, 275
SAT test scores, 91
Segregation analysis
    qualitative traits, 122–127
    quantitative traits, 128–131
Selection
    artificial selection
        directional selection, 234–236, 238–240
        disruptive selection, 242, 243
        stabilizing selection, 240, 242
    definition, 233
    gene and genotype frequencies, 233, 234
    multiple traits, 246, 247
    natural selection
        directional selection, 243
        disruptive selection, 245
        stabilizing selection, 244
    single trait, 245, 246
    values, 248–250
Selection criterion, 255, 257
Selection differential, 1
Selection gradients, 305
Selection index, 289, 372
    candidate phenotype with family mean, 269, 270
    definition, 265
    derivation of index weights, 265, 266
    evaluation of index selection, 267
    simple combined selection, 267–269
Selection intensity, 236
Sherman-Morrison-Woodbury formulas, 369
Short derivation, 136, 137
Short way of derivation, 140, 141
Sib analysis, 8
    full-sib analysis, 158–160
    half-sib analysis, 160, 161
    mixed model, 164
    nested/hierarchical mating design, 161–164
Sib selection, 261
    covariance, 262
    variance, 261
Simple regression analysis, 347
    design matrix, 348
    GWAS, 349
    residual error variance, 348
    simple linear model, 347
Smith-Hazel index, 292, 294
Smith-Hazel index selection, 295
Smith-Hazel selection index, 283, 289, 290
Square law, 25, 26

Square matrix, 51
Standard error
    heritability, 164
        analysis of variances, 166, 167
        regression method, 164, 165
Statistical Analysis System (SAS), 9
Statistical model, 313
Statistical Package for the Social Sciences (SPSS), 9
Statistical software packages, 9
Sums of cross products (SSCP), 224
Symmetric matrix, 51


**T**
Tabular method, 185
Tallis's restriction selection index, 295
Tandem selection, 283
Three genotypes
    regression analysis, 116–118
Total genetic variance, 77
Trace of a matrix, 57
Traditional marker-assisted selection, 367, 368
Transpose, of matrix, 55
t-test, 8, 107–110
Tukey.test, 105
Tukey's constraint interaction model, 102
Tukey's one degree of freedom, 102–106
Two genotypes
    regression analysis, 113–115
2×2 contingency table association test, 29


**U**
Unbalance data, 90
Unit of additive distance, 309

**V**
VARCOMP procedure, 92
Variance
    application of the variance property, 42
    definition, 40
    estimating variance from a sample, 41
    normal distribution, 41
    properties, 41
Variance and covariance matrices
    definition, 287
    elements, 286
    environmental, 286
    environmental error effects, 285
    genetic, 285
    phenotypic, 285, 286
Variance-covariance matrix, 60, 101, 226, 314, 319


**W**
Wald test, 322, 324, 327, 329, 340, 343, 352, 362, 364
Weighted least squares method, 317
Within-family selection
    covariance, 261
    objective, 260
    phenotypic value, 260
    quantitative genetics, 260
    variance, 260
Woodbury matrix identity, 369
Wright-Fisher model, 30
Wright's F statistics, 33


**Y**
Yao Ming, 1–3

Printed in the United States
by Baker & Taylor Publisher Services